FISH, FISHING AND FISHERIES

SEXUAL PLASTICITY AND GAMETOGENESIS IN FISHES

FISH, FISHING AND FISHERIES

Additional books in this series can be found on Nova's website under the Series tab.

Additional e-books in this series can be found on Nova's website under the e-book tab.

MARINE BIOLOGY

Additional books in this series can be found on Nova's website under the Series tab.

Additional e-books in this series can be found on Nova's website under the e-book tab.

FISH, FISHING AND FISHERIES

SEXUAL PLASTICITY AND GAMETOGENESIS IN FISHES

BALASUBRAMANIAN SENTHILKUMARAN
EDITOR

New York

Library of Congress Cataloging-in-Publication Data

ISBN: 978-1-62618-848-8

Published by Nova Science Publishers, Inc. † New York

Dedicated to my beloved father late Shri. P. N. Balasubramaniam

&

my father-in-law late Shri. M. C. S. Nambiar

Contents

Preface

Fishes are extremely diverse which mainly includes osteichthyes, the bony fishes, and chondrichthyes, the cartilaginous fishes. Osteichthyes (mainly includes teleosts), represent the largest and most diverse class of vertebrates in existence today, with about 32,400 species, which is more than the combined total of mammals, birds, amphibians and reptiles. Teleosts are sexually labile which can display an astonishing variety of reproductive modes such as gonochorism, protandry, protogyny, true hermaphroditism, gynogenesis and androgenesis. Successful reproduction of a species mainly depends on organized reproductive strategies including gonadal differentiation/development and gametogenesis. Gonadal development/differentiation is governed by a delicate balance of genetic, hormonal and environmental factors. On the other hand, efficient and regulated gametogenesis is critical for the quality seed production, which is a key area of aquaculture research and development.

This book represents collective version of review chapters contributed by eminent scientists all over the world including India. The demand for the fishes is in rise every day, with the improved aquaculture research on controlling gametogenesis and spawning has provided possible options to increase the production, yet, the equilibrium never met. The application of modern molecular biology techniques and tools facilitate scientists to delineate several fundamental questions of endocrinology and reproductive biology, and thus assist in sustainable aquaculture management. Amazing expansion in the number of quality publication and scientific congregation concerning fish reproduction is a witness for the growing importance for the aquaculture research. Our knowledge of sexual plasticity and gamete maturation has been expanded rapidly since the recent advances in sexual development research using model organisms ranging from fishes to mammals. Sexual development research is on the top priority because of its great potential in finding sex determining genes. The recent surge of interest in identifying the mechanisms pertaining to endocrine disruption makes fisheries research to progress in sexual development and gamete maturation studies, which are the two emerging multidisciplinary research fields with wide range of future prospects.

This book is a tribute to those investigators who work on sex determination, differentiation, gamete maturation and fish reproductive physiology, who inspired a whole generation of researchers in these fields. The subject matter of the book is on the research areas of endocrinology, reproductive biology, aquaculture and sexual development, which form a basic study material for young and aspiring fish biologists.

Several renowned scientists from India and abroad have contributed their research findings. I am grateful to all the contributors for their tremendous help in preparing the articles for this book and would like to thank their funding agencies for supporting their research.

I am indebted and grateful to my mentors and teachers, Prof. H. Kagawa, Prof. Y. Nagahama, Prof. K. P. Joy, Prof. K. Ikenaka, Dr. S. Thangavelu, Dr. V. Veeraragavan and Dr. Gulam Mohideen. I also thank my collaborators and other scientists in India and abroad from whom I received moral support and help during my research career.

I sincerely thank to my first batch of Ph.D students, Dr. I. Swapna, Dr. G. Sreenivasulu, Dr. M. K. Rasheeda, Dr. K. Raghuveer and Dr. P. Sridevi for their research contribution and the present batch of Ph.D students, Dr. Mamta Sajwan, Mr. A. Rajakumar, Ms. Y. Prathibha, Mr. R. Murugananthkumar and Ms. C. Laldinsangi for their research contribution and careful assistance to bring out this book.

I wish and pray to Almighty God for prosperity and good health for all the contributors to continue their academic excellence in the years to come. I am profoundly thankful to my wife, Sudhakumari and Son, Vigneshvar for their constant support and encouragement to bring out this book. Last but not least, I wish to thank the Indian national funding agencies, DBT, DST, UGC and CSIR for their grant support to perform the research work described in review articles from my laboratory.

In: Sexual Plasticity and Gametogenesis in Fishes
Editor: Balasubramanian Senthilkumaran
ISBN: 978-1-62618-848-8

Chapter I

Involvement of Neuropeptides in Gonadotropin Secretion in Teleost Fish

Masafumi Amano[1]
School of Marine Biosciences, Kitasato University,
Sagamihara, Kanagawa, Japan

Abstract

The gonadotropin-releasing hormone (GnRH) is a key neuropeptide hormone that regulates gonadotropin (GTH) secretion in fish. Phylogenetic analyses have classified the GnRH precursors of vertebrates into 4 groups. Among them, the GnRH1 group is localized to the hypothalamus and acts on the pituitary to stimulate GTH secretion. In addition to GnRH, neuropeptide Y (NPY), which was originally found to be one of the most potent orexigenic neuropeptides in mammals, has been reported to be involved in GTH secretion. In some teleost fish, NPY not only stimulates food intake but also enhances luteinizing hormone (LH) secretion both *in vivo* and *in vitro*. In these fish, NPY-immunoreactive (ir) fibers innervate LH cells in the pituitary. NPY also stimulates GnRH release *in vitro*, and NPY-ir fibers are located in many brain regions, including the preoptic-anterior hypothalamus where GnRH-ir cell bodies are abundant. Neuronal interaction of GnRH and NPY in the brain has been shown in some teleost fish. Therefore, it is considered that NPY stimulates LH secretion by direct action on the gonadotroph and also by enhancing GnRH release; reproduction of teleost fish is considered to be regulated by several neuropeptides in the brain, such as GnRH and NPY.

Introduction

Neuropeptides are peptide hormones synthesized in the brain. Thyrotropin-releasing hormone (TRH) is a neuropeptide that was first identified from pig hypothalamus (Boler et

[1] E-mail: amanoma@kitasato-u.ac.jp.

al., 1969) and ovine hypothalamus (Burgus et al., 1969). Thereafter, various types of neuropeptides have been found in the brain of teleost fish.

Reproduction is one of the most important physiological events in animals. Understanding the reproductive mechanism of teleost fish is necessary for the establishment and development of fish aquaculture. Gonadal maturation of teleost fish is mainly regulated by the gonadotropin-releasing hormone (GnRH). GnRH is a neuropeptide comprising 10 amino acids and was originally isolated from pig and sheep hypothalami as a physiological regulator of luteinizing hormone (LH) release from the pituitary (Matsuo et al., 1971; Burgus et al., 1972). At present, it is generally accepted that GnRH regulates the synthesis and release of pituitary gonadotropins (GTHs), i.e., follicle-stimulating hormone (FSH) and LH (see Okubo & Nagahama, 2008). GTHs then stimulate the secretion of steroid hormones from the gonads. Steroid hormones feed back to the brain and pituitary. In mammalian, avian, reptilian, and amphibian species, GnRH is conveyed to the pituitary via the hypothalamo-hypophyseal portal vessels. In mammals, pulsatile release of mammalian GnRH (mGnRH) by hypothalamic neurons stimulates GTH secretion from the pituitary. However, teleost fish lack the median eminence. Instead, GnRH neurons directly innervate the pituitary. It is therefore interesting to examine GnRH systems in teleost fish in view of comparative endocrinology.

Recently, the Kiss1/ G-coupled protein receptor 54(GPR54) system was discovered in mammals (Ohtaki et al., 2001; de Roux et al., 2003; Seminara et al., 2003). It is now considered that kiss1/GPR54 signaling plays a key role in the regulation of GnRH and, consequently, GTH secretion in mammals (see Zohar et al., 2010). In teleost fish, the identification and characterization of a Kiss1 gene has been reported in zebrafish *Danio rerio* (van Aerle et al., 2008), medaka *Orizias latipes* (Kanda et al., 2008), and goldfish *Carassius auratus* (Yang et al., 2010). In addition to *kiss1*, a paralogue of *kiss1* called *kiss2* has been found in teleost fish such as zebrafish and medaka (Kitahashi et al., 2009), goldfish (Li et al., 2009), and chub mackerel *Scomber japonicus* (Selvaraj et al., 2010). In some fish such as grass puffer *Takifugu niphobles* (Shahjahan et al., 2010) and Senegalese sole *Solea senegalensis* (Mechaly et al., 2011), only *kiss2* has been found. Species differences have been reported in the distribution of Kiss1 and Kiss2 neurons in the brain of medaka and zebrafish (Kanda et al., 2008; Kitahashi et al., 2009; Mitani et al., 2010; Servili et al., 2011). Furthermore, species differences have been reported in the effects of kisspeptin administration on GTH secretion: Kiss2 was significantly more potent than Kiss1 in stimulating FSH and LH secretion in sea bass *Dicentrarchus labrax* (Felip et al., 2009) and Kiss2 but not Kiss1 administration significantly increased FSHβ and LHβ mRNA levels in the pituitary of sexually mature female zebrafish (Kitahashi et al., 2009). Thus, molecular species, central distribution, and the physiological significance of kisspeptins in teleost fish are not fully understood. Clarification of the relationship between the GnRH system and the Kiss1/GPR54 system in teleost fish is necessary to understand the reproductive biology of these species.

Neuropeptide Y (NPY) comprises 36 amino acids in both teleost fish and mammals. NPY was originally found to be one of the most potent orexigenic neuropeptides in mammals (Halford et al., 2004; Kalra et al., 1999). NPY mRNA or peptide sequences have been determined in a large number of fish species such as sea bass (Cerdá-Reverter et al., 2000a, b, c), rainbow trout *Oncorhynchus mykiss* (Doyon et al., 2003), and channel catfish *Ictalurus punctatus* (Leonard et al., 2001). Several studies have demonstrated that NPY also stimulates food intake in teleost fish (see Volkoff et al., 2005).

In this chapter, the involvement of neuropeptides in reproduction and/or GTH secretion in teleost fish is discussed by mainly focusing on GnRH and NPY. It has also been reported that NPY is pivotal for GTH secretion in some teleosts.

GnRH and GTH secretion

Thus far, 15 forms of GnRH molecules have been identified in vertebrates on the basis of data obtained from the comparison of their primary structures or complementary DNA (cDNA) sequences (Kavanaugh et al., 2008; see Okubo & Nagahama, 2008). It has been shown that 2 or 3 molecular forms of GnRH are present in a single species (see Amano et al., 1997; see Okubo & Nagahama, 2008). GnRH forms are traditionally named after the species from which they were first identified. For example, masu salmon *Oncorhynchus masou* and goldfish have salmon GnRH (sGnRH) and chicken GnRH-II (cGnRH-II), and red seabream *Pagrus major* and barfin flounder *Verasper moseri* have sGnRH, cGnRH-II, and seabream GnRH (sbGnRH). It is considered that only one form of GnRH is involved in GTH secretion, as described below.

In teleost fish, 2 forms of GTHs, namely GTH-I and GTH-II, were first isolated from chum salmon *Oncorhynchus keta* pituitary (Suzuki et al., 1988a, b, c; Kawauchi et al., 1989). These GTHs have also been isolated from several other fish species such as coho salmon *Oncorhynchus kisutch* (Swanson et al., 1989), bonito *Katsuwonus plelamis* (Koide et al., 1993), tuna *Thunnus obesus* (Okada et al., 1994), and yellowtail *Seriola dumerilii* (Garcia-Hernandez et al., 1997), and their cDNAs have been cloned from several teleost fish species such as masu salmon (Gen et al., 1993; Kato et al., 1993), goldfish (Kobayashi et al., 1997; Yoshiura et al., 1997), Siberian sturgeon *Acipenser baeri* (Quérat et al., 2000), Australian lungfish *Neoceratodus forsteri* (Quérat et al., 2004), and the elasmobranch *Scyliorhinus canicula* (Quérat et al., 2001). GTH-I and GTH-II are now commonly referred to as FSH and LH, respectively, as in mammals (see Kawauchi and Sower, 2006).

To clarify the function of each GnRH-immunoreactive (ir) neuronal group, it is necessary to examine its projection area in the brain and pituitary. Immunohistochemistry has been used for this purpose. We first demonstrated the distribution of sGnRH-ir and cGnRH-II-ir cell bodies and fibers in the brain and pituitary of the teleost fish masu salmon, using specific antisera raised against sGnRH and cGnRH-II (Amano et al., 1991). sGnRH-ir cell bodies were found to be distributed from the olfactory nerve through the preoptic area. sGnRH-ir fibers were distributed in various brain regions and directly innervated the pituitary. cGnRH-II-ir cell bodies were detected only in the midbrain tegmentum, and the distribution of cGnRH-II-ir fibers was similar to that of sGnRH-ir fibers, except for the absence of cGnRH-II-ir fibers in the pituitary. These results suggest that sGnRH not only regulates GTH secretion in the pituitary but also functions as a neuromodulator in the brain, whereas cGnRH-II functions only as a neuromodulator in masu salmon (Amano et al., 1991). Thereafter, a considerable number of immunohistochemical studies on GnRH have been eagerly conducted in various fish species. Distribution of GnRH-ir cell bodies and fibers in the brain and pituitary of teleost fish has been recently reviewed by Okubo and Nagahama (2008).

At present, phylogenetic analyses have classified the GnRH precursors of vertebrates into 4 groups (see Okubo & Nagahama, 2008). The GnRH1 group is localized to the hypothalamus and acts on the pituitary gland to stimulate GTH secretion. The GnRH2 group, found in the midbrain, includes cGnRH-II, which is thought to function as a neuromodulator in the brain. The GnRH3 group has only been found in the terminal nerves of teleost fish, includes sGnRH, and is thought to influence sex-associated behaviors (Fernald & White, 1999). Most cyprinid and salmonid species, including such as masu salmon, rainbow trout, goldfish and zebrafish, perhaps do not possess GnRH1. In these species, GnRH3 is likely to have taken on the role of GnRH1. The GnRH4 group comprises only the lamprey GnRHs (lGnRH-I and III), which are hypothalamic neuropeptides of diencephalic/ventricular origin (Silver et al., 2004). Again, it is considered that GnRH not only regulates GTH secretion but also functions as a neuromodulator in the brain of teleost fish. GnRH has also been implicated in reproductive behavior in many species, including teleost fish such as dwarf gourami *Colisa lalia* (Yamamoto et al., 1997) and goldfish (Volkoff & Peter, 1999).

The central involvement of GnRH in *in vivo* stimulation of LH release from the pituitary gland has been functionally established in all orders of teleost fish (see Zohar et al., 2010). However, *in vivo* stimulatory effects of GnRH on FSH release have been reported only limited fish species such as rainbow trout (Breton et al., 1998; Mananos et al., 1999; Weil et al., 1999), coho salmon (Dickey & Swanson, 1998), white sturgeon *Acipenser transmontanus* (Moberg et al., 1995), and Nile tilapia *Oreochromis niloticus* (Levavi-Sivan et al., 2006; Aizen et al., 2007a, b). Immunoassays for FSH of other teleost fish should be established in order to confirm the involvement of GnRH in *in vivo* FSH release in teleost fish.

It has been clarified that pituitary responsiveness to *in vivo* GnRH treatment, in terms of GTH subunit synthesis, changes during gonadal maturation. For example, when GnRH was administered *in vivo* to juvenile common carp *Cyprinus carpio*, no changes were observed in GTH subunit genes, while both FSHβ and LHβ subunit mRNAs increased in first-year maturing males (Kandel-Kfir et al., 2002). In contrast, long-term administration of GnRH analog (GnRHa) in male striped bass *Morone saxatilis* resulted in a significant increase in all 3 GTH subunit mRNAs in juveniles, but had no effect in maturing fish (Hassin et al., 2000). These results are supported by many *in vitro* studies using cultured pituitary cells or fragments: the effects of GnRH on GTH synthetic activity vary according to the particular reproductive physiology of fish. The involvement of GnRH in GTH secretion has been recently reviewed by Zohar et al. (2010).

NPY and GTH secretion

NPY is one of the most potent orexigenic neuropeptides expressed in mammals (Kalra et al., 1999; Halford et al., 2004). It has been reported that NPY also stimulates food intake in teleost fish, as in mammals (see Volkoff et al., 2005). For example, central injections of NPY cause a dose-dependent increase in food intake in goldfish (Lopez-Patino et al., 1999;De Pedro et al., 2000;Narnaware et al., 2000), coho salmon and channel catfish (Silverstain & Plisetskaya, 2000).

To understand the function of NPY in the brain of teleost fish, the distribution of NPY-ir cell bodies and fibers should be clarified, as in the case of GnRH. The distribution of NPY-ir cell bodies and fibers has been examined by immunohistochemistry in various fish species, including elasmobranchs, the cloudy dogfish *Scyliorhinus torazame* (Chiba, 2000), dipnoans, the African lungfish *Protopterus annectens* (Trabucchi et al., 2000), and actinopterygian fish such as sea bass (Cerdá-Reverter et al., 2000b), rainbow trout (Danger et al., 1991), brown trout *Salmo trutta fario* (Castro et al., 1999), catfish (Gaikwad et al., 2003, 2005), goldfish (Kah et al., 1989), common carp (Marchetti et al., 2000; Pirone et al., 2004), pejerrey *Odontesthes bonariensis* (Traverso et al., 2003), spotted gar *Lepisosteus oculatus* (Chiba, 2005), ayu *Plecoglossus altivelis* (Chiba et al., 1996), gilthead seabream *Sparus aurata* L. (Pirone et al., 2008), masu salmon (Amano et al., 2009), and the Siberian sturgeon *Acipenser baeri* (Amiya et al., 2011). In these fish species, NPY-ir cell bodies are generally located not only in the preoptic-anterior hypothalamic region but also in other brain regions such as the nervus terminalis and dorsolateral midbrain tegmentum. NPY-ir fibers are widely distributed in the central nervous system including the pituitary.

Distribution of neurons expressing NPY mRNA has also been examined by *in situ* hybridization in some teleost fish such as goldfish (Peng et al., 1994), chinook salmon *Oncorhynchus tshawytscha* and coho salmon (Silverstein et al., 1998). For example, in goldfish, neurons expressing NPY mRNA were detected mainly in the forebrain regions (particularly in the nucleus entopeduncularis of the ventral telencephalon, the preoptic area, the olfactory bulbs, and various thalamic regions) and also in the optic-tectum and locus coeruleus (Peng et al., 1994). These results are consistent with those obtained by immunohistochemistry (Kah et al., 1989).

Although the pivotal function of NPY in the fish brain is considered to be stimulation of food intake, it has been reported that both *in vivo* and *in vitro* treatment with NPY stimulates LH release in teleost fish. In rainbow trout and common carp, *in vivo* administration of NPY alone induced a 2-fold increase in LH release. In addition, stimulation with LH-releasing hormone (LHRH) analogue induced greater LH release from fish that were first treated with NPY (Breton et al., 1991). In rainbow trout, *in vitro* treatment with NPY induced dose-dependent stimulation of LH release and treatment with a GnRH antagonist blocked NPY-induced LH release, as revealed by a perifusion experiment, suggesting that NPY acts presynaptically on GnRH fibers to modulate LH release (Danger et al., 1991). In goldfish, intraperitoneal injection of NPY induced time- and dose-dependent increases in plasma LH and growth hormone (GH) levels (Peng et al., 1993a). Moreover, *in vitro* treatment with NPY induced a dose-dependent stimulation of LH and GH release, as revealed by a perifusion experiment, suggesting that NPY acts at the pituitary level to stimulate LH and GH release (Kah et al., 1989; Peng et al., 1990; Peng et al., 1993a, b). Furthermore, it has been reported that NPY stimulates LH synthesis in the pituitary in tilapia; exposure to NPY was followed by increased mRNA levels of glycoprotein α and LHβ, but NPY had no effect on FSHβ mRNA levels, as revealed by *in vitro* culture of pituitary cells (Gur et al., 2002).

To confirm that NPY physiologically stimulates LH secretion, connection of NPY-ir fibers with LH cells in the pituitary should be clarified by dual-label immunohistochemistry. In catfish, it has been reported that NPY-ir fibers directly innervate LH cells (Gaikwad et al., 2003). In contrast to mammals, FSH and LH are synthesized in 2 different cell-types in teleost fish; in salmon pituitary, FSH cells are located in the periphery of the glandular cord of the proximal pars distalis (PPD) and LH cells are located in the central parts of the glandular cord of the PPD (see Kawauchi and Sower, 2006). Antisera raised against synthetic fragment peptides of GTH subunits have been obtained from the mummichog *Fundulus heteroclitus* (Shimizu and Yamashita, 2002), by using cDNA sequencing data (Lin et al. 1992). Using these antisera, researchers have successfully detected FSH cells and LH cells in various fish species, suggesting the wide applicability of these antibodies in different orders of fish (Shimizu et al. 2003a, b, c). Therefore, it is necessary to examine whether NPY-ir fibers innervate FSH cells and/or LH cells in the pituitary by dual-label immunohistochemistry in teleost fish.

Involvement of steroid hormones has been reported in the NPY mRNA profiles of the brain. In the goldfish, pretreatment with testosterone or estradiol-17β (E_2) induced a 2- to 3-fold increase in NPY mRNA levels in the telencephalon/preoptic area, as revealed by a ribonuclease protection assay. Moreover, a study using *in situ* hybridization has shown that pretreatment with testosterone activates NPY synthetic activity in the preoptic region (Peng et al., 1994). However, whether increased NPY mRNA with steroids stimulates food intake or not has not been demonstrated.

Interaction between GnRH and NPY

NPY also stimulates GnRH release from preoptic-anterior hypothalamic slices and pituitary fragments with similar potency (Peng et al., 1993b). In goldfish implanted with testosterone and E_2 *in vivo*, NPY significantly stimulated the release of GnRH from hypothalamic slices and thus was considered to be one of the mechanisms mediating the feedback of these steroids (Peng et al., 1993c). Therefore, it is considered that NPY stimulates LH release by a direct action on the gonadotroph and also by enhancing GnRH release in goldfish (see Trudeau, 1997). Furthermore, given that NPY stimulates LH and GH release and that central injections of NPY cause a dose-dependent increase in food intake in goldfish (Lopez-Patino et al., 1999;De Pedro et al., 2000; Narnaware et al., 2000), NPY should be regarded as an element of the multifactorial systems regulating reproduction and somatic growth.

In vitro release of GnRH by NPY has also been reported in red seabream: NPY stimulates *in vitro* release of sbGnRH from slices of the preoptic-anterior hypothalamus during immaturity, recrudescence and spawning (Senthilkumaran et al., 2001). This action of NPY is suggested to be direct, because the stimulatory influence of NPY on release of sbGnRH was not altered by the addition of antagonists of serotonin (5-HT), gamma-aminobutyric acid (GABA), and noradrenaline.

Considering that GTH secretion is mainly controlled by GnRH, that NPY stimulates GnRH release *in vitro*, and that NPY-ir fibers are located in wide brain regions including the preoptic-anterior hypothalamus where GnRH-ir cell bodies are abundant, it is interesting to examine the neuronal interaction between NPY-ir neurons and GnRH-ir neurons in the brain by dual-label immunohistochemistry. Close apposition of NPY-ir fibers to GnRH-ir cell bodies in the preoptic area was revealed in ayu. In addition, NPY labeling in the nucleus tuberis lateralis became intense concomitantly with an increase in the labeled varicosities in the middle region of the neurohypophysis during gonadal maturation. These results suggest that there is a correlative involvement of NPY and GnRH in gonadal maturation possibly through the control of GTH secretion (Chiba et al., 1996). In the spotted gar, dual-label immunohistochemistry suggests that NPY directly regulates the function of GnRH in the hypothalamus, although no data is available for pituitary GTH (Chiba, 2005). Moreover, it has been reported that NPY is involved in the up-regulation of the sGnRH system in the forebrain of catfish *Clarias batrachus*; intracranial injection of NPY increased sGnRH-like peptide content in the telencephalon plus preoptic area, pituitary, olfactory organ and olfactory bulb, as revealed by high performance liquid chromatography – electrospray ionization – mass spectrometric analysis (Gaikwad et al., 2005). The Siberian sturgeon has mGnRH and cGnRH-II in the brain. It has been reported that mGnRH-ir cell bodies are located in the olfactory nerves and bulbs, telencephalon, preoptic area, and mediobasal hypothalamus, that mGnRH-ir fibers are detected in the pituitary, and that cGnRH-II-ir cell bodies are located in the midbrain tegmentum (Leprêtre et al., 1993). Recently, we found that, in the brain of the Siberian sturgeon, some NPY-ir fibers were in close contact with GnRH (possibly mGnRH)-ir cell bodies in the preoptic area, and some GnRH-ir fibers were in close contact with NPY-ir cell bodies in the ventral part of the ventral telencephalon. This suggests that reciprocal connections exist between the GnRH and NPY neurons in the brain of the Siberian sturgeon. Considering that NPY-ir fibers were not detected in the pituitary in the Siberian sturgeon, it is possible that NPY indirectly regulates GTH secretion via GnRH neurons, although innervations of GnRH to GTH cells in the pituitary has yet to be clarified in this species (Amiya et al., 2011). FSH and LH are considered to exist also in the Siberian sturgeon, since two forms of GTH were identified in the white sturgeon and *in vivo* treatment of GnRH analog stimulated the release of both GTHs in mature males and preovulatory females (Moberg et al., 1995).

Conclusion

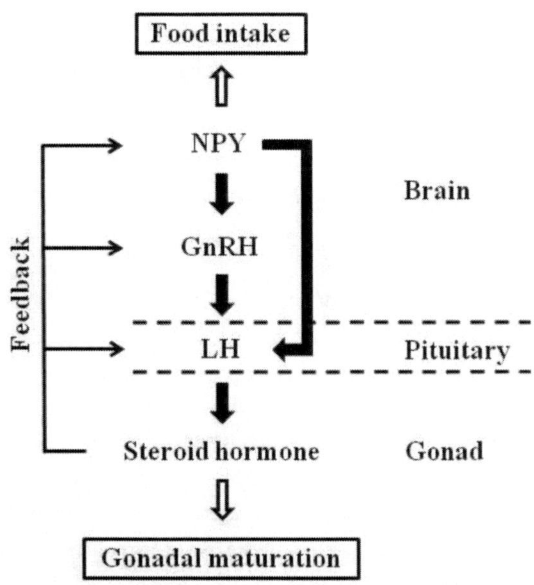

Figure 1. A schematic drawing of the relation between GnRH and NPY pathway in the brain of teleost fish on GTH secretion, including food intake regulation. Steroid hormones are considered to feedback to brain (hypothalamus) and pituitary.

It has been demonstrated that GnRH and NPY play an important role in GTH secretion in teleost fish. Neuronal interaction between GnRH and NPY has been clarified in some teleost fish. In vivo and in vitro experiments have shown that NPY stimulates LH secretion by direct action on the gonadotroph and also by enhancing GnRH release. A schematic drawing of the relation between GnRH and NPY pathway in the brain of teleost fish on GTH secretion, including food intake regulation, is shown in Fig. 1. Other neuropeptides, for example, pituitary adenylate cyclase-activating polypeptide (PACAP), which was originally isolated from ovine hypothalami that could activate adenylate cyclase in cultured rat pituitary cells (Miyata et al., 1989, 1990), has also been reported to be involved in GTH secretion in teleost fish such as goldfish (Chang et al., 2010). Clarification of the relationship between GnRH and such neuropeptides is necessary to completely understand the reproductive biology of fish.

References

Aizen, J., Kasuto, H., Golan, M., Zakay, H. & Levavi-Sivan, B. (2007a). Tilapia follicle-stimulating hormone (FSH): immunochemistry, stimulation by gonadotropin-releasing hormone, and effect of biologically active recombinant FSH on steroid secretion. *Biol Reprod*, *76*, 692-700.

Aizen, J., Kasuto, H. & Levavi-Sivan, B. (2007b). Development of specific enzyme-linked immunosorbent assay for determining LH and FSH levels in tilapia, using recombinant gonadotropins. *Gen Comp Endocrinol*, *153*, 323-332.

Amano, M., Oka, Y., Aida, K., Okumoto, N., Kawashima, S. & Hasegawa, Y. (1991). Immunocytochemical demonstration of salmon GnRH and chicken GnRH-II in the brain of masu salmon, *Oncorhynchus masou*. *J Comp Neurol*, *314*, 587-597.

Amano, M., Urano, A. & Aida, K. (1997). Distribution and function of gonadotropin-releasing hormone (GnRH) in the teleost brain. *Zool Sci*, *14*, 1-11.

Amano, M., Amiya, N., Hiramatsu, M., Tomioka, T. & Oka, Y. (2009). Interaction between neuropeptide Y immunoreactive neurons and galanin immunoreactive neurons in the brain of the masu salmon, *Oncorhynchus masou*. *Neurosci Lett*, *462*, 33-38.

Amiya, N., Amano, M., Tabuchi, A. & Oka, Y. (2011). Anatomical relations between neuropeptide Y, galanin, and gonadotropin-releasing hormone in the brain of chondrostean, the Siberian sturgeon *Acipenser baeri*. *Neurosci Lett*, *503*, 87-92.

Boler, J., Enzmann, F., Folkers, K., Bowers, C.Y. & Schally, A. V. (1969). The identity of chemical and hormonal properties of the thyrotropin releasing hormone and pyroglutamyl-histidyl-proline amide. *Biochem Biophys Res Commun*, *37*, 705-710.

Breton, B., Mikolajczyk, T., Popek, W., Bieniarz, Z. & Epler, P. (1991). Neuropeptide Y stimulates *in vivo* gonadotropin secretion in teleost fish. *Gen Comp Endocrinol*, *84*, 277-283.

Breton, B., Govoroun, M. & Mikolajczyk, T. (1998). GTH I and GTH II secretion profiles during the reproductive cycle in female rainbow trout: relationship with pituitary responsiveness to GnRH-A stimulation. *Gen Comp Endocrinol*, *111*, 38-50.

Burgus, R., Dunn, T., Desiderio, D. & Guillemin, R. (1969). Molecular structure of the hypothalamic hypophysiotropic TRF factor of ovine origin: mass spectrometry demonstration of the PCA-His-Pro-NH_2 sequence. *C R Acad Sci Hebd Seances Acad Sci D*, *269*, 1870-1873. (in French).

Burgus, R., Butcher, M., Amoss, M., Ling, N., Monahan, M., River, J., Fellows, R., Blackwell, R., Vale, W. & Guillemin, R. (1972). Primary structure of the ovine hypothalamic luteinizing hormone-releasing factor (LRF). *Proc Natl Acad Sci USA*, *69*, 278-282.

Castro, A., Becerra, M., Manso, M. J. & Anadón, R. (1999). Development of immunoreactivity to neuropeptide Y in the brain of brown trout (*Salmo trutta fario*). *J Comp Neurol*, *414*, 13-32.

Cerdá-Reverter, J. M., Anglade, I., Martinez-Rodriguez, G., Mazurais, D., Munoz-Cueto, J. A., Carrillo, M., Kah, O. & Zanuy, S. (2000a). Characterization of neuropeptide Y expression in the brain of a perciform fish, the sea bass (*Dicentrarchus labrax*). *J Chem Neuroanat*, *19*, 197-210.

Cerdá-Reverter, J. M., Martinez-Rodriguez, G., Anglade, I., Kah, O. & Zanuy, S. (2000b). Peptide YY (PYY) and fish pancreatic peptide Y (PY) expression in the brain of the sea bass (*Dicentrarchus labrax*) as revealed by in situ hybridization. *J Comp Neurol*, *426*, 197-208.

Cerdá-Reverter, J. M., Martinez-Rodriguez, G., Zanuy, S., Carrillo, M. & Larhammar, D. (2000c). Molecular evolution of the neuropeptide Y (NPY) family of peptides: cloning of three NPY-related peptides from the sea bass (*Dicentrarchus labrax*). *Regul Pept*, *95*, 25-34.

Chang, J. P., Sawisky, G. R., Mitchell, G., Uretsky, A. D., Kwong, P., Grey, C. L., Meints, A. N. & Booth, M. (2010). PACAP stimulation of maturational gonadotropin secretion in goldfish involves extracellular signal-regulated kinase, but not nitric oxide or guanylate cyclase, signaling. *Gen Comp Endocrinol*, *165*, 127-135.

Chiba, A., Sohn, Y. C. & Honma, Y. (1996). Distribution of neuropeptide Y and gonadotropin-releasing hormone immunoreactivities in the brain and hypophysis of the ayu, *Plecoglossus altivelis* (Teleostei). *Arch Histol Cytol*, *59*, 137-148.

Chiba, A. (2000). Immunohistochemical cell types in the terminal nerve ganglion of the cloudy dogfish, *Scyliorhinus torazame*, with special regard to neuropeptide Y/FMRFamide-immunoreactive cells. *Neurosci Lett*, *286*, 195-198.

Chiba, A. (2005). Neuropeptide Y-immunoreactive (NPY-ir) structures in the brain of the gar *Lepisosteus oculatus* (Lepisosteiformes, Osteichthyes) with special reference to their anatomical relations to gonadotropin-releasing hormone (GnRH)-ir structures in the hypothalamus and the terminal nerve. *Gen Comp Endocrinol*, *142*, 336-346.

Danger, J. M., Breton, B., Vallarino, M., Fournier, A., Pelletier, G. & Vaudry, H. (1991). Neuropeptide-Y in the trout brain and pituitary: localization, characterization, and action on gonadotropin release. *Endocrinology*, *128*, 2360-2368.

de Pedro, N., Lopez-Patino, M. A., Guijarro, A. I., Pinillos, M. L., Delgado, M. J. & Alonso-Bedate, M. (2000). NPY receptors and opioidergic system are involved in NPY-induced feeding in goldfish. *Peptides*, *21*, 1495-1502.

de Roux, N., Genin, E., Carel, J. C., Matsuda, F., Chaussain, J. L. & Milgrom, E. (2003). Hypogonadotropic hypogonadism due to loss of function of the KiSS1-derived peptide receptor GPR54. *Proc Natl Acad Sci USA*. *100*, 10972–10976.

Dickey, J. T. & Swanson, P. (1998). Effects of sex steroids on gonadotropin (FSH and LH) regulation in coho salmon (*Oncorhynchus kisutch*). *J Mol Endocrinol*, *21*, 291-306.

Doyon, C., Gilmour, K. M., Trudeau, V. L. & Moon, T. W. (2003). Corticotropin-releasing factor and neuropeptide Y mRNA levels are elevated in the preoptic area of socially subordinate rainbow trout. *Gen Comp Endocrinol*, *133*, 260-271.

Felip, A., Zanuy, S., Pineda, R., Pinilla, L., Carrillo, M., Tena-Sempere, M., Gómez, A. (2009). Evidence for two distinct *KiSS* genes in non-placental vertebrates that encode kisspeptins with different gonadotropin-releasing activities in fish and mammals. *Mol Cell Endocrinol*, *312*, 61−71.

Fernald, R. D. & White, R. B. (1999). Gonadotropin-releasing hormone genes: phylogeny, structure, and functions. *Front Neuroendocrinol*, *20*, 224-240.

Gaikwad, A., Biju, K. C. & Subhedar, N. (2003). GnRH-LH secreting cells axis in the pituitary of the teleost *Clarias batrachus* responds to neuropeptide Y treatment: an immunocytochemical study. *Gen Comp Endocrinol*, *131*, 126-133.

Gaikwad, A., Biju, K. C., Muthal, P. L., Saha, S. & Subhedar, N. (2005). Role of neuropeptide Y in the regulation of gonadotropin releasing hormone system in the forebrain of *Clarias batrachus* (Linn.): immunocytochemistry and high performance liquid chromatography-electrospray ionization-mass spectrometric analysis. *Neuroscience*, *133*, 267-279.

Garcia-Hernandez, M. P., Koide, Y., Diaz, M. V. & Kawauchi, H. (1997). Isolation and characterization of two distinct gonadotropins from the pituitary gland of Mediterranean Yellowtail, *Seriora dumerilli* (Risso, 1810). *Gen Comp Endocrinol*, *106*, 389-399.

Gen, K., Maruyama, O., Kato, T., Tomozawa, K., Wakabayashi, K. & Kato, Y. (1993). Molecular cloning of cDNAs encoding two types of gonadotropin alpha subunit from the masu salmon, *Oncorhynchus masou*; construction of specific oligonucleotides for the alpha 1 and alpha 2 subunits. *J Mol Endocrinol*, *11*, 265−273.

Gur, G., Bonfil, D., Safarian, H., Naor, Z. & Yaron, Z. (2002). Pituitary adenylate cyclase activating polypeptide and neuropeptide Y regulation of gonadotropin subunit gene expression in tilapia: role of PKC, PKA and ERK. *Neuroendocrinology*, *75*, 164−174.

Halford, J. C., Cooper, G. D. & Dovey, T. M. (2004). The pharmacology of human appetite expression. *Curr Drug Targets*, *5*, 221-240.

Hassin, S., Holland, M. C., Zohar, Y. (2000). Early maturity in the male striped bass, *Morone saxatilis*: follicle-stimulating hormone and luteinizing hormone gene expression and their regulation by gonadotropin-releasing hormone analogue and testosterone. *Biol Reprod*, *63*, 1691-21697.

Kah, O., Danger, J. M., Dubourg, P., Pelletier, G., Vaudry, H. & Calas, A. (1989). Characterization, cerebral distribution and gonadotrophin-release activity of neuropeptide Y (NPY) in the goldfish. *Fish Physiol Biochem*, *7*, 69-76.

Kalra, S. P., Dube, M. G., Pu, S., Xu, B., Horvath, T. L. & Kalra, P. S. (1999). Interacting appetite-regulating pathways in the hypothalamic regulation of body weight. *Endocr Rev*, *20*, 68-100.

Kanda, S., Akazome, Y., Matsunaga, T., Yamamoto, N., Yamada, S., Tsukamura, H., Maeda, K. & Oka, Y. (2008). Identification of KiSS-1 product kisspeptin and steroid-sensitive sexually dimorphic kisspeptin neurons in medaka (*Oryzias latipes*). *Endocrinology*, *149*, 2467−2476.

Kandel-Kfir, M., Gur, G., Melamed, P., Zilberstein, Y., Cohen, Y., Zmora, N., Kobayashi, M., Elizur, A. & Yaron, Z. (2002). Gonadotropin response to GnRH during sexual ontogeny in the common carp, *Cyprinus carpio*. *Comp Biochem Physiol B Biochem Mol Biol*, *132*, 17–26.

Kato Y, Gen K, Maruyama O, Tomizawa K, Kato T. Molecular cloning of cDNAs encoding two gonadotrophin β subunits (GTH-Iβ and GTH-IIβ) from the masu salmon, *Oncorhynchus masou*: Rapid divergence of the GTH-Iβ gene. *J Mol Endocrinol*, 1993; 11: 275–282.

Kavanaugh, S. I,. Nozaki, M. & Sower, S. A. (2008). Origins of gonadotropin-releasing hormone (GnRH) in vertebrates: identification of a novel GnRH in a basal vertebrate, the sea lamprey. *Endocrinology*, *149*, 3860-3869.

Kawauchi, H., Suzuki, K., Itoh, H., Swanson, P., Naito, N., Nagahama, Y., Nozaki, M., Nakai, Y. & Itoh, S. (1989). The duality of teleost gonadotropins. *Fish Physiol Biochem*, *7*, 29–38.

Kawauchi, H. & Sower, S. A. (2006). The dawn and evolution of hormones in the adenohypophysis. *Gen Comp Endocrinol*, *148*, 3–14.

Kitahashi,T., Ogawa, S. & Parhar, I. S. (2009). Cloning and expression of kiss2 in the zebrafish and medaka. *Endocrinology*, *150*, 821-831.

Kobayashi, M., Kato, Y., Yoshiura, Y. & Aida, K. (1997). Molecular cloning of cDNA encoding two types of pituitary gonadotropin α subunit from the goldfish, *Carassius auratus*. *Gen Comp Endocrinol*, *105*, 372-378.

Koide, Y., Itoh, H. & Kawauchi, H. (1993). Isolation and characterization of two distinct gonadotropins, GTH I and GTH II, from bonito (*Katsuwonus plelamis*) pituitary glands. *Int J Pept Prot Res*, *41*, 52-65.

Leonard, J. B., Waldbieser, G. C. & Silverstein, J. T. (2001). Neuropeptide Y sequence and messenger RNA distribution in channel catfish (*Ictalurus punctatus*). *Mar Biotechnol*, *3* 111-118.

Leprêtre, E., Anglade, I., Williot, P., Vandesande, F., Tramu, G. & Kah, O. (1993). Comparative distribution of mammalian GnRH (gonadotropin releasing hormone) and chicken GnRH-II in the brain of the immature Siberian sturgeon (*Acipenser baeri*). *J Comp Neurol*, *337*, 568-583.

Levavi-Sivan, B., Biran, J. & Fireman, E. (2006). Sex steroids are involved in the regulation of gonadotropin-releasing hormone and dopamine D2 receptors in female tilapia pituitary. *Biol Reprod*, *75*, 642-650.

Li, S., Zhang, Y., Liu, Y., Huang, X., Huang, W., Liu, D., Zhu, P., Shi, Y., Cheng, C. H. K., Liu, X. & Lin, H. (2009). Structural and functional multiplicity of the kisspeptin/GPR54 system in goldfish (*Carassius auratus*). *J Endocrinol*, *201*, 407–418.

Lin, Y-WP., Rupnow, B. A., Price, D. A., Greenberg, R. M. & Wallace, R. A. (1992). *Fundulus heteroclitus* gonadotropins. 3. Cloning and sequencing of gonadotropic hormone (GTH) I and II β-subunits using the polymerase chain reaction. *Mol Cell Endocrinol*, *85*, 127-139.

Lopez-Patino, M. A., Guijarro, A. I., Isorna, E., Delgado, M. J., Alonso-Bedate, M. & de Pedro, N. (1999). Neuropeptide Y has a stimulatory action on feeding behavior in goldfish (*Carassius auratus*). *Eur J Pharmacol*, *377*, 147-153.

Mananos, E. L., Anglade, I., Chyb, J., Saligaut, C., Breton, B. & Kah, O. (1999). Involvement of gamma-aminobutyric acid in the control of GTH-1 and GTH-2 secretion in male and female rainbow trout. *Neuroendocrinology*, *69*, 269-280.

Marchetti, G., Cozzi, B., Tavanti, M., Russo, V., Pellegrini, S. & Fabiani, O. (2000). The distribution of neuropeptide Y-immunoreactive neurons and nerve fibers in the forebrain of the carp *Cyprinus carpio* L. *J Chem Neuroanat*, *20*, 129-139.

Matsuo, H., Baba, Y., Nair, R. M. G., Arimura, A. & Schaly, A.V. (1971). Structure of the porcine LH- and FSH-releasing hormone. I. The proposed amino acid sequence. *Biochem Biophys Res Commun*, *43*, 1334-1339.

Mechaly, A. S., Viñas, J. & Piferrer, F. (2011). Gene structure analysis of kisspeptin-2 (Kiss2) in the Senegalse sole (*Solea senegalensis*): Characterization of two splice variants of Kiss2 and novel evidence for metabolic regulation of kisspeptin signaling in non-mammalian species. *Mol Cell Endocrinol*, *339*, 14−24.

Mitani, Y., Kanda, S., Akazome, Y., Zempo, B. & Oka, Y. (2010). Hypothalamic Kiss1 but not Kiss2 neurons are involved in estrogen feedback in medaka (*Oryzias latipes*). *Endocrinology*, *151*, 1751−1759.

Miyata, A., Arimura, A. Dahl, D. H,. Minamino, N., Uehara, A., Jiang, L., Culler, M. D. & Coy, D. H. (1989). Isolation of a novel 38 residue hypothalamic peptide which stimulates adenylate cyclase in pituitary cells. *Biochem Biophys Res Commun*, *164*, 567-574.

Miyata, A., Dahl, D. H,. Jiang, L., Kitada, C., Kubo, K., Fujino, M., Minamino, N. & Arimura, A. (1990). Isolation of a neuropeptide corresponding to the N-terminal 27 residues of the pituitary adenylate cyclase activating polypeptide with 38 residues (PACAP38). *Biochem Biophys Res Commun*, *170*, 643-648.

Moberg, G. P., Watson, J. G., Doroshov, S., Papkoff, H. & Pavlick Jr, R. J. (1995). Physiological evidence for two sturgeon gonadotropins in *Acipenser transmontanus*. *Aquaculture*, *135*, 27-39.

Narnaware, Y. K., Peyon, P. P., Lin, X. & Peter, R. E. (2000). Regulation of food intake by neuropeptide Y in goldfish. *Am J Physiol Regul Integr Comp Physiol*, *279*, R1025-R1034.

Ohtaki, T., Shintani, Y., Honda, S., Matsumoto, H., Hori, A., Kanehashi, K., Terao, Y., Kumano, S., Takatsu, Y., Masuda, Y., Ishibashi, Y., Watanabe, T., Asada, M., Yamada, T., Suenaga, M., Kitada, C., Usuki, S., Kurokawa, T., Onda, H., Nishimura, O. & Fujino, M. (2001). Metastasis suppressor gene KiSS-1 encodes peptide ligand of a G-protein-coupled receptor. *Nature*, *411*, 613−617.

Okada, T., Kawazoe, I., Kimura, S., Sasamoto, Y., Aida, K. & Kawauchi, H. (1994). Purification and characterization of gonadotropin I and II from pituitary glands of tuna (*Thunnus obesus*). *Int J Pept Prot Res*, 43, 69−90.

Okubo, K. & Nagahama, Y. (2008). Structural and functional evolution of gonadotropin-releasing hormone in vertebrates. *Acta Physiol*, *193*, 3-15

Peng, C., Huang, Y. P. & Peter, R. E. (1990). Neuropeptide Y stimulates growth hormone and gonadotropin release from the goldfish pituitary in vitro. *Neuroendocrinology*, *52*, 28-34.

Peng, C., Humphires, S., Peter, R. E., Rivier, J. E., Blomqvist, A. G. & Larhammar, D. (1993a). Actions of goldfish neuropeptide Y on the secretion of growth hormone and gonadotropin-II in female goldfish. *Gen Comp Endocrinol*, *90*, 306-317.

Peng, C., Chang, J. P., Yu, K. L., Wong, A. O. L., van Goor, F., Peter, R. E. & Rivier, J. E. (1993b). Neuropeptide Y stimulates growth hormone and gonadotropin-II secretion in the goldfish pituitary: involvement of presynaptic and pituitary cell actions. *Endocrinology*, *132*, 1820-1829.

Peng, C., Trudeau, V. L. & Peter, R. E. (1993c). Seasonal variation of neuropeptide Y actions on growth hormone and gonadotropin-II secretion in the goldfish: effects of sex steroids. *J Neuroendocrinol*, *5*, 273-280.

Peng, C., Gallin, W., Peter, R. E., Blomqvist, A. G. & Larhammar, D. (1994). Neuropeptide-Y gene expression in the goldfish brain: distribution and regulation by ovarian steroids. *Endocrinology*, *134*, 1095-1103.

Pirone, A., Lenzi, C., Betti, L., Giannaccini, G., Lucacchini, A., Marroni, P. & Fabiani, O. (2004). Immunohistochemical distribution of neuropeptide Y in the mesencephalon and rhombencephalon of carp, *Cyprinus carpio* L. (Cyprinidae: Teleostei). *Comp Biochem Physiol A*, 138, 175-185.

Pirone, A., Lenzi, C., Marroni, P., Betti, L., Mascia, G., Giannaccini, G., Lucacchini, A. & Fabiani, O. (2008). Neuropeptide Y in the brain and retina of the adult teleost gilthead seabream (*Sparus aurata* L.). *Anat Histol Embryol*, *37*, 231-240.

Quérat, B., Arai, Y., Henry, A., Akama, Y., Longhurst, T. J. & Joss, J. M. P. (2004). Primary glycoprotein hormone β subunits in the Australian lungfish and estimation of the relative evolution rate of these subunits within vertebrates. *Biol Reprod*, *70*, 356-363.

Quérat, B., Sellouk, A. & Salmon, C. (2000). Phylogenetic analysis of the vertebrate glycoprotein hormone family including new sequences of sturgeon (*Acipenser baeri*) β subunits of the two gonadotropins and the thyroid-stimulating hormone. *Biol Reprod*, *63*, 222-228.

Quérat, B., Tonnerre-Donicarli, C., Genies, F. & Salmon, C. (2001). Duality of gonadotropins in gnathostomes. Gen *Comp Endocrinol*, 124, 308-314.

Selvaraj, S., Kitano, H., Fujinaga, Y., Ohga, H., Yoneda, M., Yamaguchi, A., Shimizu, M. & Matsuyama, M. (2010). Molecular characterization, tissue distribution, and mRNA expression profiles of two *Kiss* genes in the adult male and female chub mackerel (*Scomber japonicus*) during different gonadal stages. *Gen Comp Endocrinol*, *169*, 28–38.

Seminara, S. B., Messager, S., Chatzidaki, E. E., Thresher, R. R., Acierno, J. S. Jr., Shagoury, J. K., Bo-Abbas, Y., Kuohung, W., Schwinof, K. M., Hendrick, A. G., Zhan, D., Dixon, J., Kaiser, U. B., Slaugenhaupt, S. A., Gusella, J. F., O'Rahilly, S., Carlton, M. B., Crowley, W. F. Jr., Aparicio, S.A. & Colledge, W. H. (2003). The GPR54 gene as a regulator of puberty. *N Engl J Med*, *349*, 1614–1627.

Senthilkumaran, B., Okuzawa, K., Gen, K. & Kagawa, H. (2001). Effects of serotonin, GABA and neuropeptide Y on seabream gonadotropin releasing hormone release *in vitro* from preoptic-anterior hypothalamus and pituitary of red sea bream, *Pagrus major*. *J Neuroendocrinol*, *13*, 395-400.

Servili. A., Page, Y. L., Leprince, J., Caraty, A., Escobar, S., Parhar, I. S., Seong, J. Y., Vaudry, H. & Kah, O. (2011). Organization of two independent kisspeptin systems derived from evolutionary-ancient kiss genes in the brain of zebrafish. *Endocrinology*, *152*, 1527–1540.

Shahjahan, Md., Motohashi, E., Doi, H. & Ando, H. (2010). Elevation of Kiss2 and its receptor gene expression in the brain and pituitary of grass puffer during the spawning season. *Gen Comp Endocrinol*, *169*, 48–57.

Shimizu, A. & Yamashita, M. (2002). Purification of mummichog (*Fundulus heteroclitus*) gonadotropins and their subunits, using an immunochemical assay with antisera raised against synthetic peptides. *Gen Comp Endocrinol*, *125*, 79-91.

Shimizu, A., Kagawa, H. & Tanaka, H. (2003a). Immunocytochemical identification of gonadotrophs (FSH cells and LH cells) in various perciform fishes using antisera raised against synthetic peptides. *Fish Physiol Biochem*, *28*, 109-110.

Shimizu A, Sakai T, Nashida K, Honda H (2003b). Universal antisera for immunocytochemical identification of two different gonadotrophs in acanthopterygian fishes. *Fish Physiol Biochem*, *29*, 275-287.

Shimizu, A., Tanaka, H. & Kagawa, H. (2003c). Immunocytochemical applications of specific antisera raised against synthetic fragment peptides of mummichog GTH subunits: examining seasonal variations of gonadotrophs (FSH cells and LH cells) in the mummichog and applications to other acanthopterygian fishes. *Gen Comp Endocrinol*, *132*, 35-45.

Silver, M. R., Kawauchi, H., Nozaki, M. & Sower, S. A. (2004). Cloning and analysis of the lamprey GnRH-III cDNA from eight species of lamprey representing the three families of Petromyzoniformes. *Gen Comp Endocrinol*, *139*, 85-94.

Silverstein, J. T., Breininger, J., Baskin, D. G. & Plisetskaya, E. M. (1998) Neuropeptide Y-like gene expression in the salmon brain increases with fasting. *Gen Comp Endocrinol*, *110*, 157-165.

Silverstain, J. T. & Plisetskaya, E. M. (2000). The effects of NPY and insulin on food intake regulation in fish. *Am Zool*, *40*, 296-308.

Suzuki, K., Kawauchi, H. & Nagahama, Y. (1988a). Isolation and characterization of two distinct gonadotropins from chum salmon pituitary glands. *Gen Comp Endocrinol*, *71*, 292−301.

Suzuki, K., Kawauchi, H. & Nagahama, Y. (1988b). Isolation and characterization of subunits from two distinct salmon gonadotropins. *Gen Comp Endocrinol*, *71*, 302−306.

Suzuki, K., Kanamori, A., Nagahama,Y. & Kawauchi, H. (1988c). Development of salmon GTHI and GTHII radioimmunoassays. *Gen Comp Endocrinol*, *71*, 459−467.

Swanson, P., Bernard, M., Nozaki, M., Suzuki, K., Kawauchi, H. & Dickhoff, W. W. (1989). Gonadotropins I and II in juvenile coho salmon. *Fish Physiol Biochem*, *7*, 169−176.

Trabucchi, M., Chartrel, N., Pelletier, G., Vallarino, M. & Vaudry, H. (2000). Distribution of GAD-immunoreactive neurons in the diencephalon of the African lungfish *Protopterus annectens*: colocalization of GAD and NPY in the preoptic area. *J Comp Neurol*, *419*, 223-232.

Traverso, J. M., Ravaglia, M. A., Vissio, P. G., Maggese, M. C. & Paz, D. A. (2003). Localization of neuropeptide Y-like immunoreactive structures in the brain of the pejerrey, *Odontesthes bonariensis* (Teleostei, Atheriniformes). *Anat Histol Embryol*, *32*, 29-35.

Trudeau, V. L. (1997). Neuroendocrine regulation of gonadotrophin II release and gonadal growth in the goldfish, *Carassius auratus*. *Rev Reprod*, *2*, 55-68.

van Aerle, R., Kille, P., Lange, A. & Tyler, C. R. (2008). Evidence for the existence of a functional Kiss1/Kiss1 receptor pathway in fish. *Peptides*, *29*, 57−64.

Volkoff, H., Canosa, L. F., Unniappan, S., Cerdá-Reverter, J. M., Bernier, N. J., Kelly, S. P. & Peter, R. E. (2005). Neuropetides and the control of food intake in fish. *Gen Comp Endocrinol*, *142*, 3-19.

Volkoff, H. & Peter, R. E. (1999). Actions of two forms of gonadotropin releasing hormone and a GnRH antagonist on the spawning behavior of the goldfish *Carassius auratus*. *Gen Comp Endocrinol*, *116*, 347–355.

Weil, C., Carre, F., Blaise, O., Breton, B. & Le Bail, P. Y. (1999). Differential effect of insulin-like growth factor I on in vitro gonadotropin (I and II) and growth hormone secretions in rainbow trout (*Oncorhynchus mykiss*) at different stages of reproductive cycle. *Endocrinology*, *140*, 2054-2062.

Yamamoto, N., Oka, Y. & Kawashima, S. (1997). Lesions of gonadotropin-releasing hormone-immunoreactive terminal nerve cells: effects on the reproductive behavior of male dwarf gouramis. *Neuroendocrinol*, *65*, 403-412.

Yang, B., Jiang, Q., Chan, T., Ko, W. K. W. & Wong, A. O. L. (2010). Goldfish kisspeptin: molecular cloning, tissue distribution of transcript expression, and stimulatory effects on prolactin, growth hormone and luteinizing hormone secretion and gene expression via direct actions at the pituitary level. *Gen Comp Endocrinol*, *165*, 60–71.

Yoshiura, Y., Kobayashi, M., Kato, Y. & Aida, K. (1997). Molecular cloning of the cDNAs encoding two gonadotropin beta subunits (GTH-Iβ and -IIβ) from the goldfish, *Carassius auratus*. *Gen Comp Endocrinol*, *105*, 379–389.

Zohar, Y., Muñoz-Cueto, J. A., Elizur, A. & Kah, O. (2010). Neuroendocrinology of reproduction in teleost fish. *Gen Comp Endocrinol*, *165*, 438-455.

In: Sexual Plasticity and Gametogenesis in Fishes
Editor: Balasubramanian Senthilkumaran
ISBN: 978-1-62618-848-8

Chapter II

Neuroendocrine Control of Lunar-Synchronized Spawning Rhythm in Grass Puffer

Hironori Ando[1,2,*], Taro Ikegami[2,†], Yusuke Maruyama[3], Md. Shahjahan[2,‡] and Atsuhiko Hattori[3]

[1] Sado Marine Biological Station, Faculty of Science, Niigata University, Sado, Niigata, Japan

[2] Laboratory of Advanced Animal and Marine Bioresources, Graduate School of Bioresource and Bioenvironmental Sciences, Kyushu University, Fukuoka, Japan

[3] Department of Biology, College of Liberal Arts and Sciences, Tokyo Medical and Dental University, Ichikawa, Chiba, Japan

Abstract

Grass puffer (*Takifugu niphobles*) exhibits a unique spawning rhythm: it spawns on the beach in semilunar cycles during spring tide in early summer. To explore the molecular and neuroendocrine mechanisms underlying the regulation of the semilunar-synchronized spawning rhythm, cyclic variations in melatonin secretion from the pineal gland and in the expression of four subtypes of melatonin receptor genes in the diencephalon were examined. The hypothalamic expression of genes for neuropeptides related to reproduction and their receptors were also examined during the reproductive cycle. The secretion of melatonin from the pineal gland showed diurnal and circadian variations with a significant peak in the nighttime. The nocturnal secretion of melatonin

[*] Correspondence to: Hironori Ando, Sado Marine Biological Station, Faculty of Science, Niigata University, 87 Tassha, Sado, Niigata 952-2135, Japan. Tel: (+81)-259-75-2012, Fax: (+81)-259-75-2870, E-mail: hando311@cc.niigata-u.ac.jp

[†] Present address: Department of Chemistry, Biology, and Marine Science, Faculty of Science, University of the Ryukyus, Okinawa 903-0213, Japan.

[‡] Present address: Department of Fisheries Management, Bangladesh Agricultural University, Mymensingh 2202, Bangladesh.

and the synchronous rise in the expression of four subtypes of melatonin receptor genes during the nighttime indicate that the action of melatonin on the hypothalamus is highly dependent on time and light. In the hypothalamus, the expression of genes for gonadotropin-releasing hormone, kisspeptin, LPXRFamide peptide and PQRFamide peptide and their receptors showed seasonal variations with increased expression in the spawning period. In particular, the expression of LPXRFa and LPXRFa-R genes showed diurnal and circadian variations in association with the expression of the four subtypes of melatonin receptor genes, indicating that the action of LPXRFa is most probably cyclic due to the regulation by melatonin. Although the link between the lunar cycle and the melatonin signal is missing at present, melatonin may play an important role in transmitting the photoperiodic information of moonlight to the reproductive neuroendocrine center in the hypothalamus of grass puffer.

Introduction

Synchronous reproduction is crucial to reproductive success in most vertebrate species. This enables the partners of different sexes in full maturation to meet at a specific phase that is set by periodic environmental and internal conditions such as time, season and age. Thus, the reproductive function of most animals is tightly dependent on environmental condition. In teleosts, photoperiod and water temperature are the important environmental factors that regulate reproductive function. These factors have an influence on the brain–pituitary–gonad (BPG) axis, the central regulatory system in vertebrates, although the molecular and neuroendocrine mechanisms underlying the regulation of the BPG axis by these factors are largely unknown. Our recent studies on the molecular neuroendocrine regulation of grass puffer (*Takifugu niphobles*), a semilunar spawner, provide valuable information on the photoperiodic regulation of the BPG axis. We propose a possible role of melatonin, the pineal neurohormone, as a signal molecule that links between the BPG axis and lunar cycle. The aim of this chapter is an overview of molecular aspects of lunar-synchronized reproduction in grass puffer with special reference to cyclic expression of genes for neuropeptides related to reproduction and melatonin receptors in the diencephalon.

GnRH-GTH Axis and Hypothalamic Neuropeptides Related to Reproduction

Among various kinds of hormones involved in the BPG axis, gonadotropin-releasing hormone (GnRH), a hypothalamic decapeptide, and gonadotrpropin (GTH), a pituitary glycoprotein hormone, form a principal pair of chemical mediators of the BPG axis. GnRH acts on the pituitary to regulate the synthesis and release of two types of gonadotropins (GTHs), follicle-stimulating hormone (FSH) and luteinizing hormone (LH) (Ando et al., 2001; Yaron et al., 2003; Ando & Urano, 2005; Kah et al., 2007; Zohar et al., 2009). FSH and LH are heterodimeric glycoproteins that share a common α subunit (GPα) and possess a hormone–specific β subunit (FSHβ & LHβ). Although their roles during the reproductive cycle in fish have not yet been sufficiently established, there may be two different trends in expression pattern in association to their functions. First, single spawners, like salmonids, have shown a predominance of FSH during early gonadal recrudescence and gametogenesis and a clear prevalence of LH expression at the final stage of gonadal maturation (Prat et al.,

1996; Gomez et al., 1999; Swanson et al., 2003; Ando & Urano, 2005). Secondly, multiple spawners seem to express a progressive and simultaneous increase in both FSH and LH expression during the gonadal maturation (Elizur et al., 1996; Sohn et al., 1999; Kajimura et al., 2001; Weltzien et al., 2003; Meiri et al., 2004).

There are two or three GnRH forms in a single vertebrate species. One form of GnRH is hypophysiotropic GnRH, GnRH1, which regulates GTH secretion from the pituitary. The second form of GnRH is GnRH2 (formerly known as chicken GnRH-II), which is highly conserved from fish through mammals. GnRH2 neurons are localized in the midbrain tegmentum and project their axons widely throughout the brain. GnRH2 regulates sexual and feeding behavior (Millar, 2003; Matsuda et al., 2008). In some groups of teleosts such as perciformes, pleuronoctiformes and tetraodontiformes, GnRH3 (formerly known as salmon GnRH) is produced as the third form in the neuronal groups that are localized in the ventral forebrain along the terminal nerve. These neurons project their axons throughout the various brain loci, and GnRH3 plays a neuromodulatory role in relation to sexual behavior (Oka, 2009). The external and internal signals involved in the regulation of reproductive function are integrated in the central nervous system and primarily alter the activity of GnRH neurons.

Recent studies on the RFamide peptides containing a C terminal -Arg-Phe-NH_2 sequence have revealed that there are five groups of RFamide peptide family and several members of the family play an important role in the control of GnRH secretion in vertebrates (Tsutsui, 2009). Kisspeptin and its receptor, G-protein coupled receptor 54 (GPR54), activate GnRH neurons to stimulate GnRH release in mammals (Oakley et al., 2009). The ortholog genes for kisspetin and GPR54 have been identified in teleosts, and studies on their neuroanatomical structure, expression profiles and gene regulation showed that the kisspeptin/GPR54 system is also important in the control of reproduction in teleosts, possibly through the control of GnRH secretion (Elizur, 2009; Zohar et al., 2009; Tena-Sempere et al., 2012). The second group of the RFamide peptide family involved in the control of reproduction is LPXRFamide peptide (LPXRFa) group (Tsutsui, 2009), including gonadotropin-inhibitory hormone (GnIH) and RFamide-related peptide (RFRP). LPXRFa regulates GTH secretion directly and indirectly via GnRH neurons. However, the effect of LPXRFa on GTH secretion varies distinctly among animal classes. LPXRFa inhibits LH secretion in goldfish (Zhang et al., 2010), quail (Tsutsui et al., 2000) and rats (Hinuma et al., 2000; Johnson et al., 2007), whereas it stimulates the release of LH and FSH in sockeye salmon (Amano et al., 2006) and the expression of FSHβ and LHβ genes in grass puffer (Shahjahan et al., 2011). The third group is PQRFa peptide group, including neuropeptide FF (NPFF), neuropeptide AF (NPAF) and neuropeptide SF (NPSF). PQRFa has recently been shown to be involved in the regulation of reproduction in lower vertebrates (Saito et al. 2010; Osugi et al. 2011). These hypothalamic neuropeptides are considered to be involved in the photoperiodic control of GnRH neurons.

Lunar-Synchronized Spawning and Timing Mechanisms

Lunar- or semilunar-synchronized reproduction has been seen in a wide variety of organisms, particularly those living in shallow waters and reef areas. These include corals (Harrison et al., 1984), marine insects (Kaiser et al., 2011) and many marine teleost species, such as the California grunion (*Leuresthes tenuis*) (Clark, 1925), damselfish (*Pomacentrus nagasakiensis*) (Moyer, 1975), mummichog (*Fundulus heteroclitus*) (Tayler et al., 1979), groupers (Collin et al. 1987), rabbitfishes (Rahman et al., 2000; Takemura et al., 2004b), mudskipper (*Bolephthalamus perctinirostris*) (Wang et al., 2008) and grass puffer (Motohashi et al., 2010). In these organisms, changes in moonlight and tide are main environmental cues (often called a Zeitgeber or environmental synchronizer) that entrain a biological rhythm for the synchronization of reproductive functions (Leatherland et al., 1992; Takemura et al., 2004b).

Several studies demonstrated an endogenous biological rhythm for the lunar-synchronized spawning experimentally under constant environmental conditions (Hsiao & Meier, 1986, 1989; Rahman et al., 2000). There is a paucity of information on the molecular mechanism of the lunar-related biological rhythm, although that of circadian rhythm has been the focus of considerable attention with a particular emphasis on the molecular regulation of circadian clock genes. The lunar-related biological rhythm may be driven by circa(semi)lunar and/or circatidal clocks. Several studies have tried to identify signal molecules that are connected to these clocks. In corals, which spawn on full moon nights, moonlight regulates this spawning and a blue light-sensing photoreceptor, cryptochrome, exhibits lunar-dependent expression with higher expression levels on the full moon nights compared to new moon nights (Levy et al., 2007). Similar involvement of the moonlight-cryptochrome interaction was suggested in the lunar-synchronized spawning of the golden rabbitfish (*Siganus canaliculatus*) that occurs around the first quarter moon (Fukushiro et al., 2011). Cryptochrome has also a role in the oscillation and resetting of circadian clock by light in insects and mammals (Liu et al., 2007). Based upon these results, it is proposed that cryptochrome plays an important role in lunar phase-recognition machinery in a wide range of organisms.

On the other hand, in the golden rabbitfish, the changes in the intensity of moonlight are important for the lunar-synchronized spawning, possibly through lunar phase-dependent variations in the plasma melatonin concentrations (Takemura et al., 2004a). The plasma melatonin levels at midnight of new moon were higher than those of full moon. Melatonin is the pineal neurohormone and its secretion shows an apparent diurnal variation, being high in the nighttime and low during the daytime (Reiter, 1993). Therefore, melatonin, "the nocturnal hormone" transmits the photoperiodic information to the central and peripheral organs (Falcón et al., 2007). Importantly, melatonin has been shown to regulate GnIH and RERP expression and thus participates in the neuroendocrine control of seasonal reproduction in birds and mammals (Ubuka et al., 2005; Revel et al., 2008). It is therefore conceivable that, in the golden rabbitfish, the information of moonlight intensity may be transmitted to the BPG axis by the lunar-cycle dependent plasma melatonin concentrations and then melatonin controls the reproductive activity via the hypothalamic reproductive neuropeptides.

Grass Puffer as a Model Animal for Studying the Neuroendocrine Control of Lunar-Synchronized Spawning

The grass puffer is a common intertidal puffer species in Japan, and spawning occurs at certain seashore locations over several days of spring tide in early summer. Before spawning, these fish aggregate at the spawning ground 2–3.5 hours before high tide, and several tens of males actively pursue one female and finally spawning takes place in groups of 10–60 individuals, of which one is female. Spawning usually continues for 1–2 hours during the rising tidal phase (Uno, 1955; Nozaki et al., 1976; Honma & Kitami, 1980; Honma et al., 1980; Yamahira, 1994; Motohashi et al., 2010). Since the spawning occurs at the same timing and location every year, every two weeks and every day of spring tide during the spawning season (Nozaki et al., 1976; Motohashi et al., 2010), we are aware of time and place of the spawning and thus can obtain spawning fish as well as non-spawning fish that usually subsist in the coastal area.

Also noted is that grass puffer is closely related to tiger puffer (*Takifugu rublipes*). The whole genome sequence of tiger puffer is available on the websites. Takifugu species share very high similarity in their genome sequences (Yamanoue et al., 2009). Indeed, we have shown that the nucleotide sequence similarity between grass and tiger puffers for neuropeptide genes is higher than 90% even in their non-coding regions (Motohashi et al., 2008; Shahjahan et al., 2010a,b; Shahjahan et al., 2011). Taken together, these facts indicate that grass puffer provides an excellent animal model for studying the molecular and neuroendocrine mechanisms of the semilunar-synchronized spawning rhythm.

Melatonin Secretion and its Receptor Gene Expression

To examine a possible role of melatonin in the control of the semilunar spawning rhythm in grass puffer, melatonin secretion from the pineal gland was examined in vitro using a primary culture system. Under natural light/dark (LD) condition, the levels of melatonin secretion were high in the nighttime and almost no secretion was observed during the daytime (Fig. 1). Under constant darkness (DD), the secretion pattern showed an apparent circadian variation with one peak at the end of subjective nighttime. In addition, the expression levels of gene encoding arylalkylamine-*N*-acetyltransferase (AANAT), a rate-limiting enzyme in melatonin synthesis, showed similar diurnal and circadian variations in the pineal gland. Therefore, nocturnal secretion of melatonin and its circadian regulation is apparent in grass puffer like other vertebrate species.

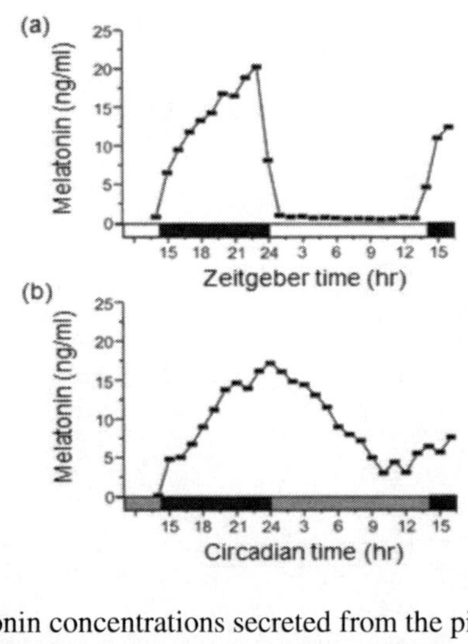

Figure 1. Changes in the melatonin concentrations secreted from the pineal gland in vitro under LD (a) and DD (b) conditions.

The action of melatonin is mediated via melatonin receptors that belong to the G protein-coupled receptor superfamily (Reppert et al., 1996). In vertebrates, there are three melatonin receptor types, Mel1a (also known as MT1), Mel1b (also known as MT2) and Mel1c. Furthermore, two different subtypes of Mel1a, Mel1a 1.4 and Mel1a 1.7, have been identified in fish (Reppert et al., 1995; Mazurais et al., 1999; Ikegami et al., 2009a, b). Accordingly, there are four subtypes of melatonin receptors in teleost fish including grass puffer. In the diencephalon of grass puffer, the four subtype genes are synchronously expressed with diurnal and circadian variations under LD and DD conditions, respectively (Ikegami et al., 2009b). Under DD conditions, Mel genes showed one peak at circadian time (CT) 18 during the subjective nighttime (Fig. 2a). Taking together with the nocturnal rise in the plasma melatonin concentrations, the action of melatonin on the diencephalon is certainly dependent on time and light, being high in the nighttime and dark condition.

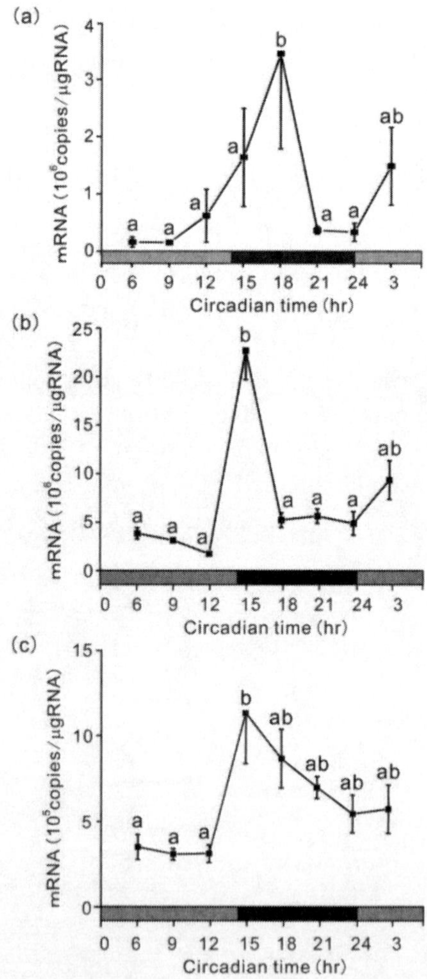

Figure 2. Circadian variations in the expression levels of genes for a melatonin receptor subtype (Mel1a1.4) (a), LPXRFa (b) and LPXRFa receptor (c) in the diencephalon of grass puffer. Values accompanied by different letters are statistically significantly different among circadian time ($p < 0.05$).

Cyclic Expression of Genes for Reproductive Neuropeptides an Their Receptors

To investigate the neuroendocrine mechanism of the semilunar spawning rhythm, we examined cyclic changes in the expression levels of genes for the above mentioned neuropeptides related to reproduction and their receptors. We first determined magnitudes of changes in the brain amounts of mRNAs for GnRH1, GnRH2, GnRH3, kisspeptin, GPR54, LPXRFa, LPXRFa receptor (LPXRFa-R) over several months during the reproductive cycle. The expression levels of these genes significantly varied depending on reproductive stage and gender. In both sexes, kisspeptin gene (*kiss2*) was predominantly expressed during the course of sexual maturation with a significant elevation from the pre-spawning to post-spawning stages (Shahjahan et al., 2010b). Kisspeptin most probably plays a central role in the control of gonadal maturation from the early stage of gametogenesis to the post-spawning stage. It should also be noted that kisspeptin and GPR54 genes showed similar expression patterns during the gonadal maturation, suggesting a common regulatory mechanism of gene transcription for the ligand and receptor. On the other hand, the expression pattern of the three GnRH genes was quite different from each other. The levels of GnRH1 gene expression were extensively increased from the pre-spawning to spawning stages and decreased to almost no expression in the post-spawning stage in both sexes, in association with the substantial increases in FSHβ and LHβ subunit gene expression and the plasma sex steroids in the spawning stage (Shahjahan et al., 2010a). GnRH1 plays a crucial role in the regulation of final sexual maturation through activation of GTH secretion. In contrast, GnRH2 showed no noticeable changes during the gonadal maturation, and GnRH3 showed an apparent sexually dimorphic expression: in the males the expression of GnRH3 gene was significantly increased in the spawning stage like GnRH1, whereas in the females it increased along with the gonadal maturation and reached a maximum in the post-spawning stage (Shahjahan et al., 2010a). LPXRFa and LPXRFa-R genes were synchronously expressed with a gradual increase during the gonadal maturation and a decrease in the post-spawning stage in both sexes, suggesting their roles in gonadal maturation (Shahjahan et al., 2011). Actually, we showed that LPXRFa stimulated the expression of FSHβ and LHβ subunit genes in the pituitary of grass puffer.

The augmented expression of these neuropeptide and its receptor genes in the spawning stage is indicative of their essentiality in the semilunar spawning rhythm. It is of considerable importance to examine whether they show cyclic expression in relation to diurnal and lunar changes. LPXRFa and LPXRFa-R genes showed diurnal and circadian variations under LD and DD conditions, respectively (Shahjahan et al., 2011). Under DD conditions, both LPXRFa and LPXRFa-R genes showed one peak at CT15 during the subjective nighttime (Figs. 2b and c). The synchronous and diurnal variations of LPXRFa, LPXRFa-R and melatonin receptor genes suggest that the expression of LPXRFa and LPXRFa-R genes is regulated by melatonin. The photoperiodic regulation of LPXRFa gene through melatonin has been reported in quail (Ubuka et al., 2005) and Syrian hamster (Revel et al., 2008). Considering that melatonin could be a signal molecule that transmits the information of moonlight intensity, melatonin may have a pivotal role in the control of the semilunar-synchronized spawning in grass puffer. The lunar-phase dependent variations of the plasma melatonin levels are currently under investigation.

Perspectives

We currently hypothesize that melatonin most probably transmits photoperiodic information of moonlight to the hypothalamus and regulates the activity of the hypothalamic neuroendocrine center for reproduction (Fig. 3). In the hypothalamus, LPXRFa neurons may have complex interaction with neurons that secrete GnRH, kisspeptin and PQRFa. In particular, GnRH, kisspeptin and PQRFa neurons are potentially target of melatonin through direct action or indirect action via LPXRFa neurons, and may exert cyclic oscillation of their activities. As for the melatonin secretion, the involvement of moonlight needs to be investigated. It should be of considerable interest and importance to determine the functional and molecular interplay between melatonin and possible circa(semi)lunar and circatidal clocks in the pineal gland. We have shown that the semilunar reproductive rhythm is endogenously maintained with surprising precision during the spawning period in grass puffer by careful observation of their spawning behavior in aquarium under laboratory conditions without tidal stimuli (Motohashi et al., 2010). The endogenous semilunar rhythm and moonlight are therefore most probably important for making the semilunar variations of the melatonin signal in the brain of grass puffer.

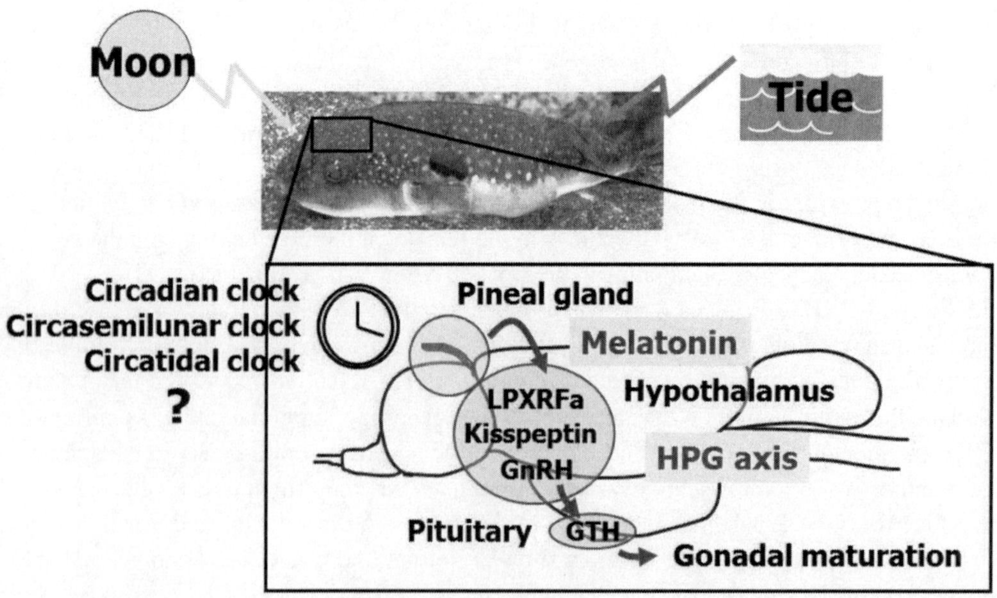

Figure 3. Schematic representation of the proposed mechanisms underlying the neuroendocrine control of semilunar-synchronized spawning rhythm in grass puffer. See the text for explanation.

Conclusion

In grass puffer, a semilunar spawner, melatonin secretion from the pineal gland showed diurnal and circadian variations with a significant peak in the nighttime. The nocturnal secretion in melatonin and the synchronous rise in the expression levels of four subtypes of melatonin receptor genes in the diencephalon during the nighttime indicate that the action of melatonin on the hypothalamus is highly dependent on time and light, being high in the nighttime and dark condition. In the hypothalamus, the expression of neuropeptides related to reproduction such as GnRH, kisspeptin and LPXRFa and of their receptors showed seasonal variations during the reproductive cycle with increased expression in the spawning period. In particular, the expression of LPXRFa and LPXRFa-R genes showed diurnal and circadian variations in association with the expression of the four melatonin receptor subtype genes, indicating that the action of LPXRFa is most probably cyclic due to the regulation by melatonin. Although the link between the lunar cycle and the melatonin signal is missing at present, these coincident molecular events may be important at least for the regulation of the lunar-related spawning in grass puffer.

Acknowledgments

We are grateful to Mr. H. Doi (Shimonoseki Academy of Marine Science), and Mr. S. Maruo for their help in sampling fish. We also thank Prof. K. Tsutsui (Waseda University), Dr. T. Osugi (Waseda University) and Dr. K. Ukena (Hiroshima University) for providing LPXRFa peptide. Our thanks are extended to all of former and present laboratory members for their help and discussion. Our research cited in this review was supported by Grants-in-Aid from the Ministry of Education, Culture, Sports, Science and Technology, Japan.

References

Amano, M., Moriyama, S., Iigo, M., Kitamura, S., Amiya, N., Yamamori, K., Ukena, K. & Tsutsui, K. (2006). Novel fish hypothalamic neuropeptides stimulate the release of gonadotrophins and growth hormone from the pituitary of sockeye salmon. *J Endocrinol, 188*, 417–423.

Ando, H. & Urano, A. (2005). Molecular regulation of gonadotropin secretion by gonadotropin-releasing hormone in salmonid fishes. *Zool Sci, 22*, 379-389.

Ando, H., Hew, C. L. & Urano, A. (2001). Signal transduction pathways and transcription factors involved in the gonadotropin-releasing hormone-stimulated gonadotropin subunit gene expression. *Comp Biochem Physiol Part B, 129*, 525-532.

Clark, F. N. (1925). The life history of *Leuresthes tenuis*, an atherine fish with tide controlled spawning habits. *Fish Bull, 10*, 1–51.

Collin, P. L., Shapiro, D. Y. & Weiler, D. (1987). Aspects of the reproduction of two groupers, *Epinephelus guttatus* and *E. striatus* in the West Indies. *Bull Mar Sci, 40*, 220–230.

Elizur, A. (2009). The KiSS1/GPR54 system in fish. *Peptides, 30*, 164-70.

Elizur, A., Zmora, N., Rosenfeld, H., Meiri, I., Hassin, S., Gordin, H. & Zohar, Y. (1996). Gonadotropins β-GtHI and β-GtHII from the gilthead seabream, *Sparus aurata*. *Gen Comp Endocrinol, 102*, 39–46.

Falcón, J., Besseau, L., Sandrine, S. & Boeuf, G. (2007). Melatonin effects on the hypothalamo-pituitary axis in fish. *Trends Endocrinol Metab, 18*, 81-88.

Fukushiro, M., Takeuchi, T., Takeuchi, Y., Hur, S.-P., Sugama, N., Takemura, A., Kubo, Y., Okano, K. & Okano, T. (2011). Lunar phase-dependent expression of cryptochrome and a photoperiodic mechanism for lunar phase-recognition in a reef fish, goldlined spinefoot. *PLoS ONE*, *6*,e28643.

Gomez, J. M., Weil, C., Ollitrault, M., Le Bail, P.-Y., Breton, B. & Le Gac, F. L. (1999). Growth hormone (GH) and gonadotropin subunit gene expression and pituitary and plasma changes during spermatogenesis and oogenesis in rainbow trout (*Oncorhynchus mykiss*). *Gen Comp Endocrinol*, *113*, 413-428.

Harrison, P. L., Babcock, R. C., Bull, G. D., Oliver, J. K., Wallace, C. C. & Willis, B. L. (1984). Mass spawning in tropical reef corals. *Science, 223*, 1186–1189.

Hinuma, S., Shintani, Y., Fukusumi, S., Iijima, N., Matsumoto, Y., Hosoya, M., Fujii, R., Watanabe, T., Kikuchi, K., Terao, Y., Yano, T., Yamamoto, T., Kawamata, Y., Habata, Y., Asada, M., Kitada, C., Kurokawa, T., Onda, H., Nishimura, O., Tanaka, M., Ibata, Y. & Fujino, M. (2000). New neuropeptides containing carboxy-terminal RFamide and their receptor in mammals. *Nat Cell Biol*, *2*, 703–708.

Honma, Y. & Kitami, T. (1980). Spawning behavior of the grass puffer, *Fugu niphobles* (Jordan et Snyder), on the coast of Sado Island in the Japan Sea. *Aquabiology*, *8*,168–173.

Honma, Y., Ozawa, T. & Chiba, A. (1980). Maturation and spawning behavior of the puffer, *Fugu niphobles*, occurring on the coast of Sado Island in the Sea of Japan (a preliminary report). *Japan J Ichthyol*, *27*, 129–138.

Hsiao, S. M. & Meier, A. H. (1986). Spawning cycles of the Gulf killifish, *Fundulus grandis*, in closed circulation systems. *J Exp Zool*, *240*, 105–112.

Hsiao, S. M. & Meier, A. H. (1989). Comparison of semilunar cycles of spawning activity in *Fundulus grandis* and *F. heteroclitus* held under constant laboratory conditions. *J Exp Zool*, *252*, 213–218.

Ikegami, T., Azuma,K., Nakamura, M., Suzuki,N., Hattori, A. & Ando, H. (2009a). Diurnal expressions of four subtypes of melatonin receptor genes in the optic tectum and retina of goldfish. *Comp Biochem Physiol Part A*, *152*, 219-224.

Ikegami, T., Motohashi, E., Doi, H., Hattori, A. & Ando, H. (2009b). Synchronized diurnal and circadian expressions of four subtypes of melatonin receptor genes in the diencephlon of a puffer fish with lunar-related spawning cycles. *Neurosci Lett*, *462*, 58-63.

Johnson, M. A., Tsutsui, K. & Fraley, G. S. (2007). Rat RFamide-related peptide-3 stimulates GH secretion, inhibits LH secretion, and has variable effects on sex behavior in the adult male rat. *Horm Behav*, *51*, 171–180.

Kah, O., Lethimonier, C., Somoza, G., Guilgur, L. G., Vaillant, C. & Lareyre, J.J. (2007). GnRH and GnRH receptors in metazoa: a historical, comparative, and evolutive perspective. *Gen Comp Endocrinol*, *153*, 346-364.

Kaiser, T. S., Neumann, D. & Heckel, D. G. (2011). Timing the tides: genetic control of diurnal and lunar emergence times is correlated in the marine midge *Clunio marinus*. *BMC Genet*, *12*, 49.

Kajimura, S., Yoshiura, Y., Suzuki, M. & Aida, K. (2001). cDNA cloning of two gonadotropin β subunits (GTH-Iβ and -IIβ) and their expression profiles during gametogenesis in the Japanese flounder (*Paralichthys olivaceus*). *Gen Comp Endocrinol*, *122*, 117–129.

Leatherland, J. F., Farbridge, K. J. & Boujard, T. (1992). Lunar and semilunar rhythms in fishes. In M. A. Ali (Eds.), *Rhythms in Fishes* (pp. 83–107). New York: Plenum Press.

Levy, O., Appelbaum, L., Leggat, W., Gothlif, Y., Hayward, D. C., Miller, D. J. & Hoegh-Guldberg, O. (2007). Light-responsive cryptochromes from a simple multicellular animal, the coral *Acropora millepora*. *Science*, *318*, 467–470.

Liu, A. C., Lewis, W. G. & Kay, S. A. (2007). Mammalian circadian signaling networks and therapeutic targets. *Nat Chem Biol*, *3*, 630–639.

Matsuda, K., Nakamura, K., Shimakura, S., Miura, T., Kageyama, H., Uchiyama, M., Shioda, S. & Ando, H. (2008). Inhibitory effect of chicken gonadotropin-releasing hormone II on food intake in the goldfish, *Carassius auratus*. *Horm Behav*, *54*, 83-89.

Mazurais, D., Brierley, I., Anglade, I., Drew, J., Randall, C., Bromage, N., Michel, D., Kah, O. & Williams, L. M. (1999). Central melatonin receptors in the rainbow trout: comparative distribution of ligand binding and gene expression. *J Comp Neurol*, *409*, 313-324.

Meiri, I., Knibb, W. R., Zohar, Y. & Elizur, A. (2004). Temporal profile of β follicle-stimulating hormone, β luteinizing hormone, and growth hormone gene expression in the protandrous hermaphrodite, gilthead seabream, *Sparus aurata*. *Gen Comp Endocrinol*, *137*, 288–299.

Millar, R. P. (2003). GnRH II and type II GnRH receptors. *Trends Endocrinol Metab*, *14*, 35-43.

Motohashi, E., Hamabata, T. & Ando, H. (2008). Structure of neurohypophysial hormone genes and changes in the levels of expression during spawning season in grass puffer (*Takifugu niphobles*). *Gen Comp Endocrinol*, *155*, 456-463.

Motohashi, E., Yoshihara, T., Doi, H. & Ando, H. (2010). Aggregating behavior of grass puffer, *Takifugu niphobles*, observed in aquarium during the spawning period. *Zool Sci*, *27*, 559-564.

Moyer, J. T. (1975). Reproductive behavior of the damselfish, *Pomacentrus nagasakiensis* at Miyake-jima, Japan. *Japan J Ichthyol*, *22*, 151–163.

Nozaki, M., Tsutsumi, T., Kobayashi, H., Takei, Y., Ichikawa,T., Tsuneki, K., Miyagawa, K., Uemura, H. & Tatsumi, Y. (1976). Spawning habit of the puffer, *Fugu niphobles* (Jordan et Snyder) I. *Zool Mag (Tokyo)*, *85*, 156–168

Oakley, A. E., Clifton, D. K., Steiner, R. A. (2009). Kisspeptin signaling in the brain. *Endocrinol Rev*, *30*, 713-743.

Oka, Y. (2009). Three types of gonadotropin-releasing hormone neurons and steroid-sensitive sexually dimorphic kisspeptin neurons in teleosts. *J Neuroendocrinol*, *21*, 334-338.

Osugi, T., Uchida, K., Nozaki, M. & Tsutsui, K. (2011). Characterization of novel RFamide peptides in the central nervous system of the brown hagfish: isolation, localization, and functional analysis. *Endocrinology*, *152*, 4252-4264.

Prat, F., Sumpter, J. P. & Tyler, C. R. (1996). Validation of radioimmunoassays for two salmon gonadotropins (GTH I and GTH II) and their plasma concentrations throughout the reproductive cycle in male and female rainbow trout (*Oncorhynchus mykiss*). *Biol Reprod*, *44*, 29-38.

Rahman, M. S., Takemura, A. & Takano, K. (2000). Correlation between plasma steroid hormones and vitellogenin profiles and lunar periodicity in the female golden rabbitfish, *Siganus guttatus* (Bloch). *Comp Biochem Physiol B Biochem Mol Biol*, *127*, 113-122.

Reiter, R. J. (1993). The melatonin rhythm: both a clock and a calendar. *Experientia*, *49*, 654–664.

Reppert, S. M., Weaver, D. R. & Godson, C. (1996). Melatonin receptors step into the light: cloning and classification of subtypes. *Trends Pharmacol Sci*, *17*, 100-102.

Reppert, S. M., Weaver, D. R., Cassone, V. M., Godson, C. & Kolakowski Jr, L. F. (1995). Melatonin receptors are for the birds: molecular analysis of two receptor subtypes differentially expressed in chick brain. *Neuron*, *15*, 1003-1015.

Revel, F. G., Saboureau, M., Pevet, P., Simonneaux, V. & Mikkelsen, J. D. (2008). RFamide-related peptide gene is a melatonin-driven photoperiodic gene. *Endocrinology*, *149*, 902–912.

Saito, T. H., Nakane, R., Akazome, Y., Abe, H. & Oka, Y. (2010). Electrophysiological analysis of the inhibitory effects of FMRFamide-like peptides on the pacemaker activity of gonadotropin-releasing hormone neurons. *J Neurophysiol*, *104*, 3518-3529.

Shahjahan, Md., Hamabata, T., Motohashi, E., Doi, H. & Ando, H. (2010a). Differential expression of three types of gonadotropin-releasing hormone genes during the spawning season in grass puffer, *Takifugu niphobles*. *Gen Comp Endocrinol*, *167*, 153-163.

Shahjahan, Md., Motohashi, E., Doi, H. & Ando, H. (2010b). Elevation of Kiss2 and its receptor gene expression in the brain and pituitary of grass puffer during the spawning season. *Gen Comp Endocrinol*, *169*, 48-57.

Shahjahan, Md., Ikegami, T., Osugi, T., Ukena, K., Doi, H., Hattori, A., Tsutsui, K. & Ando, H. (2011). Synchronised expressions of LPXRFamide peptide and its receptor genes: seasonal, diurnal and circadian changes during spawning period in grass puffer. *J Neuroendocrinol*, *23*, 39-51.

Sohn, Y. C., Yoshiura, Y., Kobayashi, M. & Aida, K. (1999). Seasonal changes in mRNA levels of gonadotropin and thyrotropin subunits in the goldfish, *Carassius auratus*. *Gen Comp Endocrinol*, *113*, 436–444.

Swanson, P., Dickey, J. T. & Campbell, B. (2003). Biochemistry and physiology of fish gonadotropins. *Fish Physiol Biochem*, *28*, 53–59.

Takemura, A., Susilo, E. S., Rahman, M. S. & Morita, M. (2004a). Perception and possible utilization of moonlight intensity for reproductive activities in a lunar-synchronized spawner, the golden rabbitfish. *J Exp Zool*, *301*, 844–851.

Takemura, A., Rahman, M. S., Nakamura, S., Park, Y. J. & Takano, K. (2004b). Lunar cycles and reproductive activity in reef fishes with particular attention to rabbitfishes. *Fish Fisheries*, *5*, 317–328.

Tayler, M. H., Leach, G. J., DiMichele, L., Levitan, W. M. & Jacob, W. F. (1979). Lunar spawning cycle in the mummichog, *Fundulus heteroclitus* (Pisces: Cyprinodontidae). *Copea*, *1979*, 291–297.

Tena-Sempere, M., Felip, A., Gómez, A., Zanuy, S. & Carrillo, M. (2012). Comparative insights of the kisspeptin/kisspeptin receptor system: lessons from non-mammalian vertebrates. *Gen. Comp Endocrinol*, *175*, 234-243.

Tsutsui, K. (2009). A new key neurohormone controlling reproduction, gonadotropin-inhibitory hormone (GnIH): Biosynthesis, mode of action and functional significance. *Prog Neurobiol*, *88*, 76–88.

Tsutsui, K., Saigoh, E., Ukena, K., Teranishi, H., Fujisawa, Y., Kikuchi, M., Ishii, S. & Sharp, P. J. (2000). A novel avian hypothalamic peptide inhibiting gonadotropin release. *Biochem. Biophys Res Commun*, *275*, 661–667.

Ubuka, T., Bentley, G. E., Ukena, K., Wingfield, J. C. & Tsutsui, K. (2005). Melatonin induces the expression of gonadotropin-inhibitory hormone in the avian brain. *Proc Natl Acad Sci USA*, *102*, 3052-3057.

Uno, Y. (1955). Spawning habit and early development of a puffer, *Fugu* (*Torafugu*) *niphobles* (Jordan et Snyder). *J Tokyo Univ Fish*, *42*, 169–183.

Wang, Q., Hong, W., Chen, S. & Zhang, Q. (2008). Variation with semilunar periodicity of plasma steroid hormone production in the mudskipper *Boleophthalmus pectinirostris*. *Gen Comp Endocrinol*, *155*, 821–826.

Weltzien, F. A., Norberg, B. & Swanson, P. (2003). Isolation and characterization of FSH and LH from pituitary glands of Atlantic halibut (*Hippoglossus hippoglossus* L.). *Gen Comp Endocrinol*, *131*, 97–105.

Yamahira, K. (1994). Combined effects of tidal and diurnal cycles on spawning of the puffer, *Takifugu niphobles* (Tetraodontidae). *Environ Biol Fish*, *40*, 255–261.

Yamanoue, Y., Miya, M., Matsuura, K., Miyazawa, S., Tsukamoto, N., Doi, H., Takahashi,H., Mabuchi, K., Nishida, M. & Sakai, H. (2009). Explosive speciation of *Takifugu*: Another use of fugu as a model system for evolutionary biology. *Mol Biol Evol*, *26*, 623-629.

Yaron, Z., Gur, G., Melamed, P., Rosenfeld, H., Elizur, A. & Levavi-Sivan, B. (2003). Regulation of fish gonadotropins. *Int Rev Cytol*, *225*, 131-185.

Zhang, Y., Li, S., Liu, Y., Lu, D., Chen, H., Huang, X., Liu, X., Meng, Z., Lin, H. & Cheng, C. H. K. (2010). Structural diversity of the gnih/gnih receptor system in teleost: Its involvement in early development and the negative control of LH release. *Peptides*, *31*, 1034–1043.

Zohar, Y., Muñoz-Cueto, J. A., Elizur, A. & Kah, O. (2009). Neuroendocrinology of reproduction in teleost fish. *Gen Comp Endocrinol*, *165*, 438-455.

In: Sexual Plasticity and Gametogenesis in Fishes
Editor: Balasubramanian Senthilkumaran
ISBN: 978-1-62618-848-8

Chapter III

Cytokines as Intraovarian Mediators of Luteinizing Hormone-Induced Ovulation in Fish

Diego Crespo, Kousik Pramanick and Josep V. Planas*

Departament de Fisiologia i Immunologia, Facultat de Biologia,
Institut de Biomedicina de la Universitat de Barcelona (IBUB),
Universitat de Barcelona, Barcelona, Spain

Abstract

Ovulation is the process by which the mature oocyte is expelled from the surrounding layers in the ovarian follicle. This process involves the proteolytic degradation of a specific site in the follicle wall and its contraction to allow the release of the oocyte through the rupture site. The hormone that orchestrates all the events leading to ovulation is luteinizing hormone (LH). In fish, the specific actions of LH in its induction of ovulation are not well understood, despite evidence for the involvement of prostaglandins in ovulation. In mammals, ovulation is considered an inflammatory-like response due to the involvement of prostaglandins and many other immune mediators (e.g. cytokines and chemokines) in this process. The main objective of this chapter is to review available data on the processes that take place in the follicle wall during ovulation in fish and to present recent data showing that the stimulatory effects of LH on ovulation are accomplished through the stimulation of ovarian cytokines and prostaglandins.

Introduction

In fish, as in the rest of vertebrates, ovulation is the culmination of oogenesis and involves the release of the mature oocyte from its surrounding follicular layers. For ovulation

* E-mail: jplanas@ub.edu.

to occur successfully, several events must take place in the fish ovarian follicle. First, the follicular layers must undergo specific and localized degradation through proteolysis and possibly through cell death without damaging the enclosed oocyte while ensuring that the rupture of the follicle occurs only in one particular site of the follicular wall through which the oocyte will be expelled. Second, the ruptured follicular wall must retract to assist in the expulsion of the oocyte. Retraction of the follicular wall is also referred to as follicle contraction and is accomplished due to the presence of contractile smooth muscle cells as well as collagen fibres in the theca layer. Finally, although it has not clearly been established if this is a general mechanism among fish species, oocyte hydration could also contribute to the release of the oocyte through the ruptured follicular envelope by allowing the oocyte to increase its pressure onto the surrounding follicular layers (Milla et al., 2006).

It is believed that the hormone that orchestrates all the events leading from oocyte maturation to the expulsion of the oocyte in the fish ovary is the pituitary gonadotropin luteinizing hormone (LH). To date, native or recombinant LH has been shown to regulate ovarian steroid production (Swanson et al., 1991; Okada et al., 1994;Govoroun et al., 1997; Zmora et al., 2007; Kazeto et al., 2008), the expression of steroidogenic enzymes (Kobayashi et al., 2006; Kazeto et al., 2008) and gonadotropin receptor activity (Kazeto et al., 2008; Molés et al., 2011). However, reports on the effects of LH on ovulation in fish are rather limited. For example, an induction of ovulation in bitterling (*Rhodeusocellatus ocellatus*) females using recombinant goldfish (*Carassius auratus*) LH has been reported, albeit with lower potency than with goldfish pituitary extracts (Kobayashi et al., 2006). Furthermore, despite fragmented information on the involvement of specific ovarian factors (e.g. prostaglandins, cytokines; see below) on follicular contraction and ovulation, there is a remarkable lack of knowledge on the precise actions and targets of LH in its induction of ovulation in fish. In this review, we will attempt to expose available data on the factors that are produced in the ovary and that regulate ovulation in fish and we will also provide unpublished information from our laboratory on the specific effects of LH on the preparatory events leading to ovulation in salmonid fish with emphasis on the involvement of cytokines and prostaglandins.

Ovulation as an Inflammatory-Like Reaction in Fish

For almost three decades, ovulation in mammals has been considered an inflammatory-like response because ovulatory follicles have numerous features characteristic of inflammation, including production of high levels of prostaglandins, synthesis of a hyaluronan (HA)-rich matrix and increased vascularization (Espey, 1980; Richards et al., 2008). This hypothesis is supported by extensive literature describing the expression of innate immune cell-related surveillance proteins (Toll-like receptors, TLRs (Takeda et al., 2003;Richards et al., 2008) and cytokines (such as interleukin 1β, IL-1β (Brännström et al., 1993a; Peterson et al., 1993), interleukin 6, IL-6 (Machelon et al., 1994; Van der Hoek et al., 1998), and tumor necrosis factor α, TNFα (Brännström et al., 1995; Murdoch et al., 1997)) during ovulation. Granulosa and cumulus cells express receptors of the TLR family, known to be important for the stimulatory action of lipopolysaccharide (LPS) on cytokine production from immune cells (Jiang et al., 2006). There is evidence indicating that TLR2 and TLR4 are functional in granulosa and cumulus cells because LPS can induce the expression of known TLR2 and TLR4 target genes such as IL-6 and TNFα (Shimada et al., 2006). Specifically, IL-1β has been suggested to be an important regulator of ovulation since this cytokine potentiates the LH-induced ovulatory rate in the perfused rat ovary preparation (Brännstrom et al., 1993a). In addition, the natural IL-1β receptor antagonist (IRAP) inhibits ovulation (Peterson et al., 1993).

In fish, there is also evidence, mostly derived from gene expression analyses, for the appearance of inflammatory-like features during ovulation. Specifically, several classes of proteases with an important proteolytic action during ovulation (including serine proteases, MMPs, members of the ADAMTS family and plasminogen activators/plasmin system) have been found to be up-regulated in the fish preovulatory ovary (Bobe et al., 2006; Crespo et al., 2010a). In addition to proteases, pro-inflammatory immune factors, such as cytokines and chemokines are also known to be expressed in the fish ovary during the periovulatory period. One of the first cytokines to be identified in the fish ovary was TNFα (Bobe and Goetz, 2001a), a pro-inflammatory cytokine that in mammals is known to be involved in the control of ovulation (Murdoch et al., 1997). Recent data demonstrated that TNFα at the mRNA level is expressed in the theca layer but not in the granulosa layer of trout preovulatory follicles and that the TNFα protein is produced by preovulatory ovarian follicles (Crespo et al., 2012). Also, members of the TNF receptor families were found to be expressed in the fish periovulatory ovary (Bobe and Goetz, 2000; Bobe and Goetz, 2001a; Crespo et al., 2010a). Other pro-inflammatory cytokines that are expressed in the fish ovary include IL-6 (Iliev et al., 2007; Pramanick and Planas, unpublished data). In addition, chemokines, coagulation factors and vasoconstrictor agents have also been found to be expressed during the preovulatory period in salmonids (Bobe et al., 2006; Crespo et al., 2010a). This set of genes includes CC (CCL4 and cc chemokine SCYA110-2) (Crespo et al., 2010a) and CXC chemokines (CXCL14) (Bobe et al., 2006), known to have an important chemotactic activity in mammals (Yoshimura et al., 1987; Ren et al., 2010). Furthermore, the expression of coagulation factor V (F5) and coagulation factor X (F10) precursor has been shown to be enhanced in the rainbow trout (Bobe et al., 2006) and brown trout (Crespo et al., 2010a) preovulatory ovaries, respectively. In zebrafish (*Danio rerio*), coagulation factor V has been previously characterized (Hanumanthaiah et al., 2002), suggesting that modern coagulation pathways found in mammals could also be functional in fish and that their secretion prior to ovulation could prevent bleeding from ruptured ovarian follicles (Bobe et al., 2006). Furthermore, a strong up-regulation of angiotensin I-converting enzyme 2 (ACE2) was observed in the rainbow trout ovary throughout the preovulatory period (Bobe et al., 2006). In fact, salmon angiotensin I and human angiotensin II are both able to increase the level of *in vitro* spontaneous ovulation in brook trout (Hsu and Goetz, 1992). It can therefore be hypothesized that the observed increase of ACE2 gene expression in the trout preovulatory ovary could be involved in the changes in vascular dynamics that could occur during the ovulatory process. Despite the above-mentioned changes in the expression of genes involved in inflammation, these data only indirectly suggest the involvement of immune-like mediators in the control of ovulation in fish. Among the various inflammatory mediators, prostaglandins have been the best studied to date in fish.

The Role of Prostaglandins in Ovulation

Arachidonic acid and its metabolites, including prostaglandins, have been demonstrated to be involved in ovulation in different fish species (Cetta and Goetz, 1982; Goetz et al., 1982; Goetz and Cetta, 1983; Goetz et al., 1989; Bradley and Goetz, 1994; Lister and Van Der Kraak, 2008). Initial *in vitro* studies have shown that in the yellow perch (*Perca flavescens*), 17,20β-P-induced ovulation is inhibited by indomethacin (INDO), a prostaglandin synthesis inhibitor (Goetz and Garczynski, 1997). In addition, prostaglandin levels increase in yellow perch ovarian follicles that have been stimulated to ovulate with 17,20β-P and ovulation can be restored in INDO-blocked incubates by addition of primary prostaglandins (Goetz and Garczynski, 1997). Thus, it appears that in the yellow perch, 17,20β-P stimulates ovulation through the production of an eicosanoid that is most likely a primary prostaglandin. In other fish species in which 17,20β-P does not stimulate ovulation *in vitro*, such as the brook trout (Goetz and Bergman, 1978), prostaglandins also play a key role in ovulation. In a number of teleost fish species examined to date, there is a significant increase in the production of prostaglandin F2α (PGF2α) during ovulation (Goetz and Garczynski, 1997; Goetz and Cetta, 1983) and several studies have shown that PGF2α is the most effective inducer of *in vitro* ovulation (Goetz et al., 1982; Goetz et al., 1989; Kagawa et al., 2003; Patiño et al., 2003a). However, the large increase in the plasma levels of F-series prostaglandins (PGFs) during ovulation in fish may not be exclusively responsible for oocyte expulsion due to fluctuations in the levels of other important prostaglandins during ovulation. For example, prostaglandin E2 (PGE2) is effective in inducing *in vitro* ovulation in the Japanese eel (Kagawa et al., 2003) and the inhibitory effect of INDO on *in vitro* ovulation is completely blocked in the presence of PGE2 in medaka (Fujimori et al., 2011). However, in other species, such as brook trout, PGE2 levels decrease significantly just prior to and during ovulation as compared with gravid control females (Cetta and Goetz, 1982). The decrease in ovarian PGE2 levels may enhance ovulation since E prostaglandins inhibit *in vitro* ovulation of brook trout oocytes (Goetz et al., 1982). Interestingly, mature follicles can produce prostaglandins E and F; however, it is apparent that several other tissues within the fish ovary also produce primary prostaglandins, including the stroma and connective tissue (Goetz, 1991). Furthermore, it has been reported that PGE2 and PGF2α production in brook trout follicles is stimulated by orthovanadate, phorbol-12 myristate 13-acetate ester and the calcium ionophore A23187, suggesting that signal transduction pathways, including G-proteins, inositol phosphate turnover, PKC and transmembrane calcium movement, are involved in mediating follicular prostaglandin production in fish ovarian follicles (Hsu and Goetz, 1991; Kellner *and* Van Der Kraak, 1992). Regarding the specific expression of enzymes involved in prostaglandin synthesis (cyclooxygenases), recent studies revealed the enhanced expression of *cox-2* (also named *ptgs2*) at the time of ovulation in zebrafish (Lister and Van Der Kraak, 2009) and medaka (Fujimori et al., 2011). Evidence for the inducible nature of *cox-2* has been reported in fish (Lister and Van Der Kraak, 2009), like in mammals (Espey, 2006). However, the higher expression of *cox-2* compared with *cox-1* in the ovary of zebrafish and medaka is rather distinct from the findings in other teleosts, including brook trout (Roberts et al., 2000). In zebrafish, *in vivo* treatment with human chorionic gonadotropin (hCG) resulted in increased ovarian expression of *ptgs2* and increased ovarian levels of PGF2α and 17,20β-P (Lister and Van Der Kraak, 2009). Although indirectly, these results suggest that prostaglandin production by the zebrafish ovary could be under the control of pituitary gonadotropins.

Surprisingly little is known regarding the biological effects of prostaglandins in the fish ovary. Like in mammals, there is solid evidence indicating that prostaglandins regulate follicle contraction in fish. In the brook trout model, PGF2α can directly stimulate follicular contraction in ovarian follicles *in vitro* (Hsu and Goetz, 1992). On the other hand, PGE2 has inhibitory effects on follicular contraction in brook trout ovarian follicles (Hsu and Goetz, 1992). Therefore, a nice correlation between the effects of prostaglandins on follicle contraction and ovulation can be observed, indicating that the role of prostaglandins regulating ovulation in fish can be explained, at least in part, through their effects on follicle contraction. Although the precise target of prostaglandins is not known, it is suspected that prostaglandins could affect the contractibility of smooth muscle cell fibres present in the follicle wall. Taken together, the evidence to date suggests that prostaglandins are probably one of the final hormonal inducers of ovulation in fish and that there appears to be a differential balance in the production and biosynthesis of prostaglandins and their actions on this complex process depending on the fish species studied.

Besides regulating follicle contraction and oocyte expulsion, prostaglandins may have other functions at the time of ovulation, including the control of steroidogenesis (Van Der Kraak and Chang, 1990) and spawning behaviour (Sorensen et al., 1988). In fact, it is known that female sexual behaviour can be induced by prostaglandin injection in several fish species (Munakata and Kobayashi, 2010) and that PGF2α is released into the water as a "postovulatory prostaglandin pheromone" (Sorensen and Goetz, 1993).

Proteolytic Events Taking Place during Ovulation

As stated above, the degradation of the follicle wall, which involves primarily the degradation of the extracellular cell matrix (ECM), plays a crucial role in the ovulatory process, facilitating the rupture of the follicle and the subsequent expulsion of the mature oocyte through the rupture site. Prior to ovulation, the microvillous connections between the oocyte and granulosa cells are broken, proteolytic enzymes digest the follicular layers and cause follicle rupture, allowing the oocyte to be released. The mechanism of follicle wall rupture has been the subject of intensive investigation in vertebrates (Curry and Osteen, 2003; Curry and Smith, 2006; Murdoch and McDonnel, 2002; Robker et al., 2000). It is well known since the late 80's that there is an increase in proteolytic activity in the follicular wall during ovulation in fish (Berndtson and Goetz, 1988; Berndtson and Goetz, 1990). A number of proteases have been characterized from the vertebrate ovarian follicle and these may be involved in the separation between the oocyte and the follicular wall and in the subsequent rupture of the ovarian follicle. Several studies have shown that a wide variety of proteases and anti-proteases are expressed in the fish ovary around the time of ovulation (Garczynski and Goetz, 1997; Hajnik et al., 1998; Bobe and Goetz, 2001b; Ogiwara et al., 2005; Bobe et al., 2006; Crespo et al., 2010a). Among the identified proteases, serine proteases seem to play an important role in the ovulatory process. In particular, a serine protease with high homology to kallikrein identified in the brook trout ovary (named KT-14) has been hypothesized to participate in the contraction of the ovarian follicle (Hajnik et al., 1998). Furthermore, plasmin has been recently shown in the medaka ovary to degrade laminin and fibronectin, suggesting that the plasminogen activator/plasmin system may also be involved in the degradation of the follicular wall in the teleost ovary (Ogiwara et al., 2012). In rainbow trout, profiling studies have revealed up-regulation of serine protease 23 (SP23) during the natural preovulatory period (Bobe et al., 2006) and in brown trout preovulatory follicles treated *in vitro* with salmon LH (Crespo and Planas, unpublished data). In addition, other members of this family of proteases, such as serine protease-like proteins 1 and 3, have been found to be differentially expressed in the brown trout ovarian follicle wall just prior to ovulation (Crespo et al., 2010a). Among the anti-proteases, the best characterized are the trout ovulatory proteins (TOPs) that have been shown to be produced by brook trout granulosa cells during and after ovulation (Garczynsky and Goetz, 1997; Coffman et al., 2000). Their anti-protease and anti-bacterial activities may serve to protect ovulated eggs in the coelomic fluid from the action of proteases released during ovulation (Coffman and Goetz, 1998; Coffman et al., 2000).

In medaka, matrix metalloproteinase 2 (MMP-2; also called gelatinase A) and membrane-type matrix metalloproteinases 1 (MT1-MMP) and 2 (MT2-MMP) have been identified as the hydrolytic enzymes responsible for the rupture of the ovarian follicle during ovulation (Ogiwara et al., 2005). Matrix metalloproteinases (e.g. gelatinases, collagenases and stromelysins), which digest collagen, gelatin (denatured collagen) and other components of the ECM (such as fibronectin, laminin and elastin), are well-known to be critical in the control of reproductive function in mammals (Curry and Osteen, 2003). Assuming that collagen types I and IV are the major ECM proteins in the follicle wall layers, and particularly in the basement membrane separating the granulosa and theca layers, a mechanistic model of ovulation has been proposed in medaka, in which MMP-2 and MT2-MMP differentially and critically contribute to the degradation of the follicle wall (Ogiwara et al., 2005). This model of degradation of the follicle wall in the medaka ovary is based on the findings that MT2-MMP preferentially degrades collagen type I, while MMP-2 hydrolyzes collagen type IV.

Gonadotropin Regulation of the Ovulation Process in Fish

In fish, like in other vertebrates, hormones secreted from the anterior pituitary gland are involved in the control of different reproductive events, including ovulation (Nagahama and Yamashita, 2008; Levavi-Sivan et al., 2010). The main reproductive hormones are FSH and LH and they directly control many aspects of gonadal development, oocyte maturation and ovulation and are structurally and functionally homologous to their counterparts in other vertebrates (Planas et al., 2000; Planas and Swanson, 2007). Like tetrapod gonadotropins (GTHs), fish FSH and LH are heterodimeric glycoproteins formed by the non-covalent association of a common α subunit (CGα) with distinct β subunits (FSHβ and LHβ) that confer hormone specificity (Boime and Ben-Menahem, 1999). Initial studies on fish GTHs suggested the existence of a single LH-like gonadotropin regulating gametogenesis (Burzawa-Gerard, 1982). The duality of the gonadotropic hormones was later demonstrated from chum salmon (*Oncorhynchusketa*) pituitary extracts (Kawauchi et al., 1989). The two bioactive

GTHs, initially termed GTH I (FSH) and GTH II (LH), were subsequently purified in different fish species including coho salmon (*Oncorhynchus kisutch*; (Swanson et al., 1991)), tuna (*Thunnus obesus*; (Okada et al., 1994)), trout (*Oncorhynchus mykiss*; (Govoroun et al., 1997)) and seabass (*Dicentrarchus labrax*; (Mañanos et al., 1997; Molés et al., 2008)). In the case of Japanese eel (*Anguilla japonica*; (Kamei et al., 2005)), only FSH has been purified. Biochemical purification of the native hormones from pituitary glands is a highly demanding process with regard to time and cost, requiring large amounts of biological material and in many cases it has not been successful (Kamei et al., 2003; Vischer et al., 2003). During the last decade, the isolation and characterization of cDNAs coding for fish GTH subunits has increased considerably, providing the possibility to produce recombinant fish GTHs (Kobayashi et al., 2003; Zmora et al., 2007; Kazeto et al., 2008; Kobayashi et al., 2010; Molés et al., 2011).

The role of pituitary GTHs regulating gonadal function and gamete production has been best described in salmonid fish, due mostly to the availability of purified FSH and LH for *in vitro* and *in vivo* studies (Planas and Swanson, 2007). In female salmonid fish, it is generally well established that LH is primarily responsible for the induction of oocyte maturation and ovulation of mature oocytes (Nagahama and Yamashita, 2008). It is believed that these two processes are not temporally separated but occurring almost simultaneously (Patiño et al., 2003b). As indicated earlier, although the actions of LH inducing oocyte maturation are relatively well understood, much less is known regarding the specific effects of LH in inducing ovulation. In fact, the effects of LH on ovulation *in vitro* in fish have only been recently reported in the literature for two species: medaka (Ogiwara et al., 2013) and brown trout (Crespo et al., 2013). Furthermore, the possible regulation by LH of events responsible for the rupture of the follicular wall and the resulting release of the mature oocyte from the follicle enclosure in teleost fish has not been documented. Our laboratory has recently investigated the direct effects of purified salmon LH (sLH) on trout preovulatory follicles. Results obtained to date strongly support the notion that LH is directly involved in inducing several events that lead to the preparation of the trout ovarian follicle for ovulation (Crespo and Planas, unpublished data). First, exposure of brown trout preovulatory follicles to sLH *in vitro* significantly results in an induction of follicle contraction. Second, sLH stimulates proteolytic (gelatinase/collagenase) activity in brown trout ovarian tissue *in vitro*. The proteolytic action of sLH on the brown trout follicle is likely the result of the stimulatory effects of sLH on the expression of genes known to be involved in the regulation of proteolysis in the teleost ovary, such as MMP-2, KT-14, TOP-2, as measured by qPCR. In line with these findings, gene expression profiling by microarray of brown trout ovarian follicles exposed *in vitro* to physiological concentrations of sLH (25 ng/ml) evidenced important changes in the expression of additional genes involved in proteolysis and remodelling of the ECM. In addition to confirming the up-regulation of MMP-2 by sLH, our microarray data evidenced the up-regulation by sLH of additional genes involved in proteolysis, such as cathepsin L1, plasminogen activator inhibitor-1, MMP-19, serine protease 23 and ADAM metallopeptidase with thrombospondin type 1 (ADAMTS-1). Interestingly, cathepsin L1 and ADAMTS-1 are induced after LH stimulation in mice preovulatory follicles and their abnormal expression in anovulatory progesterone receptor knockout mice suggests a critical role for these factors in follicular rupture (Robker et al., 2000). Therefore, sLH acts on trout preovulatory follicles by regulating the expression of the different proteolytic systems known to be involved in the controlled degradation of the follicle wall in the mammalian ovary: MMPs, plasminogen activators/plasmin, cathepsins and ADAMTs (Ohnishi et al., 2005). These observations suggest that the mechanisms involved in the controlled degradation of the ovarian follicle during ovulation may have been established early during vertebrate evolution.

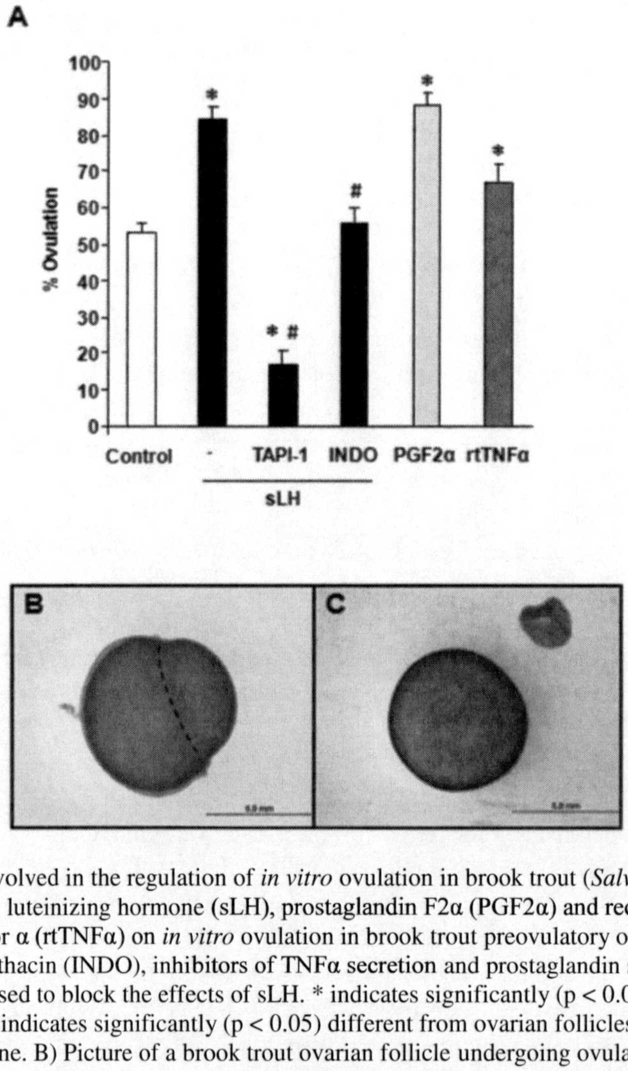

Figure 1. Factors involved in the regulation of *in vitro* ovulation in brook trout (*Salvelinus fontinalis*). A)Effects of salmon luteinizing hormone (sLH), prostaglandin F2α (PGF2α) and recombinant trout tumor necrosis factor α (rtTNFα) on *in vitro* ovulation in brook trout preovulatory ovarian follicles. TAPI-1 and indomethacin (INDO), inhibitors of TNFα secretion and prostaglandin synthesis, respectively, were used to block the effects of sLH. * indicates significantly ($p < 0.05$) different from the control group. # indicates significantly ($p < 0.05$) different from ovarian follicles incubated in the presence of sLH alone. B) Picture of a brook trout ovarian follicle undergoing ovulation. C) Picture of an ovulated brook trout oocyte already expelled from the surrounding follicle wall, which is shown contracted in the upper right hand side of the image.

Concomitant with the increased mRNA levels of proteolytic enzymes and of the proteolytic activity in preovulatory brown trout ovarian follicles exposed to sLH, microarray analyses evidenced the up-regulation by sLH of a number of genes encoding ECM components, including several types of collagens, fibronectin and laminin. Specifically, sLH increased the mRNA levels of collagen type IV, the major constituent of basement membranes and the substrate of MMP-2, as well as collagens type I, VI, VII, VIII, XI, XII and XVIII. In the medaka ovary, type IV collagen was localized in the basement membrane and was shown to be synthesized by theca cells (Kato et al., 2010). The comprehensive effects of LH regulating the expression of proteases and ECM substrates in trout preovulatory follicles strongly suggest that LH dramatically impacts the biomechanics of follicle weakening just prior to ovulation in fish.

An important demonstration of the key role of LH in regulating the ovulatory process in fish comes from data from our laboratory showing that sLH is able to stimulate *in vitro* ovulation of preovulatory brook trout follicles (Crespo et al., 2013) (Figure 1). Therefore, these and previous data confirm the importance of LH as an essential hormone for the completion of the ovulatory process in fish.

Cytokine Mediators of the Effects of LH on the Preparatory Events Leading to Ovulation in Trout

As indicated above, there is clear evidence for the involvement of cytokines, such as TNFα and members of the interleukin family, in ovulation in mammals. One of the most studied cytokines in the mammalian ovulatory process is TNFα. This cytokine is a pleiotropic pro-inflammatory cytokine produced by numerous immune cells during acute inflammation (Wang et al., 2003) but is also known to affect the growth, differentiation, survival and physiological function of a variety of different cell types. In fact, TNFα induces ovulation in the perfused rat ovary (Brännström et al., 1995) and intrafollicular injection of TNFα antiserum blocks ovulation in the sheep ovary (Gottsch et al., 2000). Furthermore, TNFα has been shown to be secreted by preovulatory follicles in vertebrates (Zolti et al., 1990; Brännström et al., 1994). Interestingly, this pro-inflammatory cytokine stimulates the biosynthesis of prostaglandins in rat preovulatory follicles (Brännström et al., 1993b) and the activity of MMP-2 in follicular explants (Gottsch et al., 2000). TNFα has also been suggested as a critical mediator in the organization and structure of the preovulatory follicle and ovarian degradation and subsequent rupture due to its action inducing proteolysis and cellular apoptosis in mammals (Murdoch and McDonnel, 2002).

In fish, there is little published data on the effects of TNFα prior to or during ovulation. One of the first indications on the possible involvement of TNFα in the ovulatory process in fish arose from experiments in which preovulatory brook trout females were treated *in vivo* with LPS, used to mimic a bacterial infection in fish. LPS administration *in vivo* results in the induction of apoptosis in ovarian follicle cells and in the potentiation of LH-stimulated testosterone production by brook trout preovulatory follicles (MacKenzie et al., 2006). Furthermore, brook trout preovulatory trout follicles incubated with conditioned medium from LPS-stimulated trout macrophages show an increase in follicle contraction, suggesting that factors produced by trout macrophages in response to LPS may have stimulated follicle contraction (MacKenzie et al., 2006). It was later demonstrated that trout macrophage LPS-conditioned medium is enriched in biologically active TNFα (Roher et al., 2011; Vraskou et al., 2011). In addition, LPS administration *in vivo* stimulates TNFα mRNA levels in the trout ovary and, importantly, causes an advancement in the time of ovulation and, consequently, a decrease in egg quality as shown by a lower survival at the eyed and yolk sac resorption stages (Crespo et al., 2010a; Crespo et al., 2010b). Taken together, these observations suggest that LPS administration during the preovulatory period in trout could result, among in other factors, in the production of TNFα that, in turn, could induce the ovarian follicle to ovulate.

In view of the evidence in trout linking LPS-induced production of TNFα and advanced ovulation and also in view of the known physiological role of TNFα in ovulation in mammals (Murdoch et al., 1997), we hypothesized that TNFα may also play a normal, physiological role in the trout ovary during the periovulatory period. An important finding supporting the notion that TNFα may constitute a physiological regulator of ovarian function is the demonstration of the stimulation of TNFα expression by sLH at the mRNA and protein levels in brown trout follicles prior to and after oocyte maturation (Crespo et al., 2012; Crespo and Planas, unpublished data). Analyses of TNFα mRNA expression in isolated brown trout theca and granulosa layers show that TNFα is expressed in theca layers and not in granulosa layers, suggesting that the site of TNFα production in the brown trout follicles is the theca layer. Furthermore, incubation of brown trout preovulatory follicles with sLH in the presence of TAPI-1, an inhibitor of TNFα converting enzyme (TACE) also known as a disintegrin and metalloproteinase (ADAM17), results in the abrogation of the stimulatory effects of sLH on follicular contraction, proteolysis of the follicular wall, expression of genes involved in the regulation of proteolysis (e.g. KT-14, TOP-2 and MMP-2) and, importantly, on *in vitro* ovulation (Crespo and Planas, unpublished data; Figure 1). Although it is well recognized that TAPI-1 can shed other transmembrane proteins (e.g. EGF-like growth factors), TAPI-1 is able to completely inhibit sLH-induced secretion of TNFα by brown trout preovulatory follicles (Crespo et al., 2012). Furthermore, sLH stimulates the mRNA expression of TACE/ADAM17 in brown trout follicles before and after oocyte maturation (Crespo et al., 2012; Crespo and Planas, unpublished data). Therefore, we can conclude that LH may control ovulation in trout through the production of TNFα.

Direct evidence for the actions of TNFα in the trout ovary is derived from studies evaluating the *in vitro* effects of recombinant trout TNFα (rtTNFα) on trout ovarian follicles. As in mammals (Murdoch et al., 1997), one of the actions of rtTNFα is the stimulation of apoptosis in trout granulosa cells and the modulation of the expression of pro- and anti-apoptotic genes in trout preovulatory follicles, suggesting a tight control of apoptosis by TNFα in the trout ovary during the periovulatory period (Crespo et al., 2010a). We hypothesize that TNFα-induced apoptosis of granulosa cells could contribute to the debilitation of the follicular wall and its subsequent rupture, allowing the release of the mature oocyte. The pro-ovulatory action of TNFα is supported by our observations that rtTNFα stimulates follicle contraction and proteolysis of the follicle wall (Crespo et al., 2010a; Crespo and Planas, unpublished data). Furthermore, rtTNFα induces the expression of genes involved in the control of proteolysis in trout ovarian follicles *in vitro*, including several genes whose expression is also up-regulated by LH such as MMP-2, TOP-2, KT-14. Microarray analyses of sLH- and rtTNFα-stimulated brown trout preovulatory follicles show a remarkable similarity in the transcriptional events stimulated by both factors. Most importantly, rtTNFα is able to stimulate *in vitro* ovulation in preovulatory brook trout follicles (Figure 1). With all these data combined, we can conclude that the ovulatory effects of LH on the trout ovary may be mediated by the intraovarian production of TNFα.

Given the well-known stimulation of ovulation by prostaglandins in fish (see above), one important question that needs to be addressed is whether the ovulatory effects of LH, and of TNFα, are mediated by prostaglandins in the fish ovary. Recent data from our laboratory clearly shows that the stimulatory effects of sLH on follicle contraction and ovulation of trout preovulatory follicles are completely abolished by INDO (Crespo and Planas, unpublished data; Figure 1). Furthermore, the stimulatory effects of rtTNFα on follicle contraction are also completely blocked by INDO (Crespo and Planas, unpublished data). These observations, coupled with the stimulation by sLH and rtTNFα of the *in vitro* production of PGF2α, as well as the mRNA levels of COX-1 and COX-2, and by the stimulation of follicle contraction and *in vitro* ovulation by PGF2α (Crespo and Planas, unpublished data; Figure 1), suggest that LH and TNFα may stimulate ovulation in trout through the production of PGF2α. Therefore, this suggests that most, but not all, actions of LH and TNFα on ovulation are mediated by PGF2α.

In mammals, cytokines other than TNFα have also been reported to play a role in the regulation of ovulation (see above). In particular, IL-6 has received considerable attention as a factor important for ovarian cumulus cell function and the expansion of cumulus cell-oocyte complex and, consequently, for ovulation in mammals (Liu et al., 2009). In fish, the relatively high expression of IL-6 in the trout ovary (Iliev et al., 2007) represented the first evidence for a possible role of IL-6 in the regulation of ovarian function in fish. In order to investigate the possible role of IL-6 in ovulation in trout, our laboratory has evaluated the *in vitro* effects of recombinant trout IL-6 (rtIL-6) on follicle contraction, steroid production, PGF2α production, proteolytic activity and expression of genes involved in proteolysis in trout preovulatory follicles. To date, our results indicate that rtIL-6 can stimulate follicle contraction *in vitro* and that its effects are blocked by prostaglandin synthesis inhibitors, suggesting the involvement of prostaglandins in its stimulation of follicle contraction (Pramanick and Planas, unpublished data).

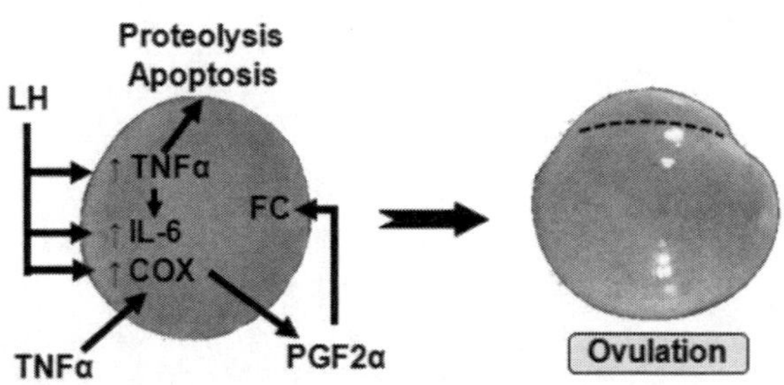

Figure 2. Working model highlighting the involvement of cytokines in luteinizing hormone (LH)-induced ovulation in the trout ovary. According the proposed model, LH may stimulate proteolytic activity and apoptosis in the follicle wall through the intraovarian production of TNFα. In addition, LH may stimulate follicle contraction (FC) again by engaging TNFα, which may stimulate the production of interleukin-6 (IL-6), which, in turn, stimulates the expression of enzymes responsible for the production of PGF2α, the likely final effector on FC.

The mediatory role of prostaglandins in IL-6 action is supported by our demonstration of the stimulation of PGF2α production by rtIL-6 in trout preovulatory follicles. Interestingly, rtIL-6 does not stimulate proteolytic activity nor the expression of various genes involved in proteolysis (e.g. MMP2, TIMP-2, KT-14, NOP and TOP-2). In this regard, rtIL-6 differs in relation to TNFα in its inability to affect proteolysis of the follicle wall in trout preovulatory follicles but has a similar stimulatory effect on follicle contraction. Furthermore, the mRNA expression levels of IL-6 in trout preovulatory follicles are increased in response to sLH and rtTNFα, suggesting that IL-6 could represent an intraovarian mediator of the effects of TNFα on PGF2α production. It is interesting that rtIL-6 is unable to stimulate proteolytic activity. Clearly, further studies are needed in order to understand the apparently complex interplay between intraovarian cytokines and prostaglandin production and action in the fish ovary.

Conclusion

The impact of immune-relevant mediators in the regulation of the ovulatory process in fish is currently being unraveled. Recent studies have shown that TNFα can mediate the stimulatory effects of LH on various processes involved in the preparation of the trout follicle for ovulation, including the contraction and proteolysis of the follicle wall, granulosa cell apoptosis and, importantly, *in vitro* ovulation (Figure 2). In addition, TNFα can exert its effects on follicle contraction and ovulation through the stimulation of the local production of PGF2α. In addition to TNFα, IL-6, another pro-inflammatory cytokine produced by the fish ovary, appears also to be involved in the regulation of events leading to ovulation. Specifically, IL-6 stimulates follicle contraction and PGF2α production but not proteolytic activity in the trout ovarian follicle wall. Furthermore, IL-6 expression is induced by LH and TNFα, leading to the suggestion that the effects of LH on follicle contraction could be mediated by TNFα that, in turn, would stimulate IL-6 that, in turn, would stimulate the production of PGF2α, the final effector (Figure 2). Interestingly, the effects of LH inducing proteolysis of the ovarian follicle wall are mediated by TNFα but apparently not by IL-6. These observations suggest that LH triggers ovulation through the stimulation of the production of various cytokines (i.e. TNFα and IL-6, possibly among others) with specific actions within the ovary.

Obviously, a number of important questions regarding the mechanisms by which LH triggers ovulation in the fish ovary still remain unanswered. For example, the effects of LH, IL-6 and PGF2α on apoptosis in the ovarian follicle wall have not been investigated, nor the importance of apoptotic cell death in the weakening of the follicular wall allowing its rupture in a single, localized spot. It will be interesting to determine in future studies whether if apoptosis of follicular cells is localized to the point of follicle rupture during ovulation. By the same token, it will be worth investigating the proteolytic mechanisms operating in the follicular wall that result in the focal degradation of the follicle in one particular spot. It will be important to study the spatial and temporal distribution of the mediators of LH on follicle rupture during the ovulatory process by *in situ* hybridization and immunocytochemistry.

Acknowledgments

We would like to thank Dr. Penny Swanson (Northwest Fisheries Science Center, NOAA Fisheries, Seattle, WA, USA) for her kind gift of salmon LH and her support of our work over the years, Dr. Rick Goetz (NOAA Northwest Fisheries Science Center, Manchester Research Station, Port Orchard, WA, USA) for his always stimulating collaboration and Dr. Aleksei Krasnov for his generous assistance with microarray analysis. The work from our laboratory described in this chapter was supported by grants from the Ministerio de Ciencia e Innovación (MICINN), Spain, to J.V.P. (CSD2007-0002 and AGL2009-07006). K.P. was supported by a BOYSCAST fellowship from the Ministry of Science and Technology, Government of India. Current address for K.P. is the Department of Zoology, Presidency University, Kolkata 700073, India.

References

Berndtson, A. K. and Goetz, F. W. (1988). Protease activity in brook trout (*Salvelinus fontinalis*) follicle walls demonstrated by substrate-polyacrylamide gel electrophoresis. *Biol. Reprod.,* 38, 511-516.

Berndtson, A. K. and Goetz, F. W. (1990). Metallo-protease activity increases prior to ovulation in brook trout (*Salvelinus fontinalis*) and yellow perch (*Perca flavescens*) follicle walls. *Biol. Reprod.,* 42, 391-398.

Bobe, J. and Goetz, F. W. (2000). A tumor necrosis factor decoy receptor homologue is up-regulated in the brook trout (*Salvelinus fontinalis*) ovary at the completion of ovulation. *Biol. Reprod.,* 62, 420-426.

Bobe, J. and Goetz, F. W. (2001a). Molecular cloning and expression of a TNF receptor and two TNF ligands in the fish ovary. *Comp. Biochem. Physiol. B Biochem. Mol. Biol.,*129, 475-481.

Bobe, J. and Goetz, F. W., (2001b). A novel osteopontin-like protein is expressed in the trout ovary during ovulation. *FEBS Lett.,* 489, 119-124.

Bobe, J., Montfort, J., Nguyen, T. and Fostier, A. (2006). Identification of new participants in the rainbow trout (*Oncorhynchus mykiss*) oocyte maturation and ovulation processes using cDNA microarrays. *Reprod. Biol. Endocrinol.,* 4, 39.

Boime I. and Ben-Menahem, D. (1999). Glycoprotein hormone structure-function and analog design. *Recent Prog. Horm. Res.,* 54, 217–288.

Bradley, J. A. and Goetz, F. W. (1994). The inhibitory effects of indomethacin, nordihydroguaiaretic acid, and pyrrolidinedithiocarbamate on ovulation and prostaglandin synthesis in yellow perch (*Perca flavescens*) follicle incubates. *Prostaglandins,* 48, 11-20.

Brännström, M., Bonello, N., Wang, L. J. and Norman, R. J. (1995). Effects of tumour necrosis factor alpha (TNF alpha) on ovulation in the rat ovary. *Reprod. Fertil. Dev.,* 7, 67-73.

Brännström, M., Norman, R. J., Seamark, R. F. and Robertson, S. A. (1994). Rat ovary produces cytokines during ovulation. *Biol. Reprod.,* 50, 88-94.

Brännström, M., Wang, L. and Norman, R. J. (1993a). Ovulatory effect of interleukin-1 beta on the perfused rat ovary. *Endocrinology,* 132, 399-404.

Brännström, M., Wang, L. and Norman, R. J. (1993b). Effects of cytokines on prostaglandin production and steroidogenesis of incubated preovulatory follicles of the rat. *Biol. Reprod.,*48, 165-171.

Burzawa-Gerard, E. (1982). Chemical data on pituitary gonadotropins and their implication to evolution. *Can. J. Fish. Aquat. Sci.,* 39, 80-91.

Cetta, F. and Goetz, F. W. (1982). Ovarian and plasma prostaglandin E and F levels in brook trout (*Salvelinus fontinalis*) during pituitary-induced ovulation. *Biol. Reprod.,* 27, 1216-1221.

Coffman, M. A. and Goetz, F. W. (1998). Trout ovulatory proteins are partially responsible for the anti-proteolytic activity found in trout coelomic fluid. *Biol. Reprod.,* 59, 497-502.

Coffman, M. A., Pinter, J. H. and Goetz, F. W. (2000). Trout ovulatory proteins: site of synthesis, regulation, and possible biological function. *Biol. Reprod.* 62, 928-938.

Crespo, D., Bonnet, E., Roher, N., MacKenzie, S. A., Krasnov, A., Goetz, F. W., Bobe, J. and Planas, J. V. (2010a). Cellular and molecular evidence for a role of tumor necrosis factor alpha in the ovulatory mechanism of trout. *Reprod. Biol. Endocrinol.,* 8, 34.

Crespo, D., Planas, J. V. and Bobe, J. (2010b). Lipopolysaccharide administration in preovulatory rainbow trout (*Oncorhynchus mykiss*) reduces egg quality. *Aquaculture,* 300, 240-242.

Crespo, D., Mañanos, E. L., Roher, N., MacKenzie, S. A. and Planas, J. V. (2012). Tumor necrosis factor alpha may act as an intraovarian mediator of luteinizing hormone-induced oocyte maturation in trout. *Biol. Reprod.,* 86, 1-12.

Crespo, D., Pramanick, K., Goetz, F. W. and Planas, J. V. (2013). Luteinizing hormone stimulation of in vitro ovulation in brook trout (*Salvelinus fontinalis*) involves follicle contraction and activation of proteolytic genes. *Gen. Comp. Endocrinol.*, doi: 10.1016/j.ygcen.2013.02.031.

Curry, T. E., Jr. and Osteen, K. G. (2003). The matrix metalloproteinase system: changes, regulation, and impact throughout the ovarian and uterine reproductive cycle. *Endocr. Rev.,* 24, 428-465.

Curry, T. E., Jr. and Smith, M. F., (2006). Impact of extracellular matrix remodeling on ovulation and the folliculo-luteal transition. *Semin. Reprod. Med.,* 24, 228-241.

Espey, L. L. (1980). Ovulation as an inflammatory reaction - a hypothesis. *Biol. Reprod.,* 22, 73-106.

Espey, L. L. (2006). Comprehensive analysis of ovarian gene expression during ovulation using differential display. *Methods Mol. Biol.,* 317, 219-241.

Fujimori, C., Ogiwara, K., Hagiwara, A., Rajapakse, S., Kimura, A. and Takahashi, T. (2011). Expression of cyclooxygenase-2 and prostaglandin receptor EP4b mRNA in the ovary of the medaka fish, *Oryzias latipes*: possible involvement in ovulation. *Mol. Cell Endocrinol.,* 332, 67-77.

Garczynski, M. A. and Goetz, F. W. (1997). Molecular characterization of a ribonucleic acid transcript that is highly up-regulated at the time of ovulation in the brook trout (*Salvelinus fontinalis*) ovary. *Biol. Reprod.,* 57, 856-864.

Goetz, F. W. (1991). Compartmentalization of prostaglandin synthesis within the fish ovary. *Am. J. Physiol.,* 260, 862-865.

Goetz, F. W. and Bergman, H. L. (1978). The effects of steroids on final maturation and ovulation of oocytes from brook trout (*Salvelinus fontinalis*) and yellow perch (*Perca flavescens*). *Biol. Reprod.,* 18, 293-298.

Goetz, F. W. and Cetta, F. (1983). Ovarian and plasma PGE and PGF levels in naturally ovulating brook trout (*Salvelinus fontinalis*) and the effects of indomethacin on prostaglandin levels. *Prostaglandins,* 26, 387-395.

Goetz, F. W., Duman, P., Berndtson, A., Janowsky, E. G. (1989). The role of prostaglandinds in the control of ovulation in yellow perch, *Perca flavescens*. *Fish Physiol Biochem,* 7, 163-168.

Goetz, F. W. and Garczynski, M. (1997). The ovarian regulation of ovulation in teleost fish. *Fish. Physiol. Biochem.,* 17, 33-38.

Goetz, F. W., Smith, D. C. and Krickl, S. P., (1982). The effects of prostaglandins, phosphodiesterase inhibitors, and cyclic AMP on ovulation of brook trout (*Salvelinus fontinalis*) oocytes. *Gen. Comp. Endocrinol.,* 48, 154-60.

Gottsch, M. L., Van Kirk, E. A. and Murdoch, W. J. (2000). Tumour necrosis factor alpha up-regulates matrix metalloproteinase-2 activity in periovulatory ovine follicles: metamorphic and endocrine implications. *Reprod. Fertil. Dev.,* 12, 75-80.

Govoroun, M. S., Huet, J., C., Pernollet, J., C. and Breton, B. (1997). Use of immobilized metal ion affinity chromatography and dye-ligand chromatography for the separation and purification of rainbow trout putuitary gonadotropins, GTH I and GTH II. *J. Chromatogr. B. Biomed. Sci. Appl.,* 698, 35-46.

Hajnik, C. A., Goetz, F. W., Hsu, S. Y. and Sokal, N., (1998). Characterization of a ribonucleic acid transcript from the brook trout (*Salvelinus fontinalis*) ovary with structural similarities to mammalian adipsin/complement factor D and tissue kallikrein, and the effects of kallikrein-like serine proteases on follicle contraction. *Biol. Reprod.,* 58, 887-897.

Hanumanthaiah, R., Day, K. and Jagadeeswaran, P. (2002). Comprehensive analysis of blood coagulation pathways in teleostei: evolution of coagulation factor genes and identification of zebrafish factor VIIi. *Blood Cells Mol. Dis.,* 29, 57-68.

Hsu, S. Y. and Goetz, F. W. (1991). Synergistic induction of ovulation and prostaglandin synthesis in goldfish (*Carassius auratus*) follicles by sodium orthovanadate and hydrogen peroxide. *Biol. Reprod.,* 45, 912-917.

Hsu S. Y. and Goetz, F. W. (1992). Angiotensins stimulate in vitro ovulation and contraction of brook trout (*Salvelinus fontinalis*) follicles. *Fish. Physiol. Biochem.,* 10, 277-282.

Iliev, D. B., Castellana, B., Mackenzie, S., Planas, J. V. and Goetz, F. W. (2007). Cloning and expression analysis of an IL-6 homolog in rainbow trout (*Oncorhynchus mykiss*). *Mol. Immunol.,* 44, 1803-1807.

Jiang, D., Liang, J., Li, Y. and Noble, P. W. (2006). The role of Toll-like receptors in non-infectious lung injury. *Cell Res.,* 16, 693-701.

Kagawa, H., Tanaka, H., Unuma, T., Ohta, H., Gen, K. and Okuzawa, K. (2003). Role of prostaglandin in the control of ovulation in the Japanese eel *Anguilla japonica. Fish. Sci.,* 69, 234-241.

Kamei, H., Kawazoe, I., Kaneko, T. and Aida, K. (2005). Purification of follicle-stimulating hormone from immature Japanese eel, *Anguilla japonica*, and its biochemical properties and steroidogenic activities. *Gen. Comp. Endocrinol.,* 143, 257-266.

Kamei, H., Ohira, T., Yoshiura, Y., Uchida, N., Nagasawa, H. and Aida, K. (2003). Expression of a biologically active recombinant follicle stimulating hormone of Japanese Eel *Anguilla japonica* using methylotropic yeast, *Pichia pastoris. Gen. Comp. Endocrinol.,* 134, 244-254.

Kato, Y., Ogiwara, K., Fujimori, C., Kimura, A. and Takahashi, T. (2010). Expression and localization of collagen type IV alpha1 chain in medaka ovary. *Cell Tissue Res.,* 340, 595-605.

Kawauchi, H., Suzuki, K., Itoh, H., Swanson, P., Naito, N., Nagahama, Y., Nozaki, M., Nakai, Y., and Itoh, S. (1989). The duality of teleost gonadotropins. *J. Fish. Physiol.,* 7, 29–38.

Kazeto, Y., Kohara, M., Miura, T., Miura, C., Yamaguchi, S., Trant, J. M., Adachi, S. and Yamauchi, K. (2008). Japanese eel follicle-stimulating hormone (Fsh) and luteinizing hormone (Lh): production of biologically active recombinant Fsh and Lh by *Drosophila* S2 cells and their differential actions on the reproductive biology. *Biol. Reprod,* 79, 938-946.

Kellner, R. G. and Van Der Kraak, G. (1992). Multifactorial regulation of prostaglandin synthesis in preovulatory goldfish ovarian follicles. *Biol. Reprod.,* 46, 630-635.

Kobayashi, M., Hayakawa, Y., Park, W., Banba, A., Yoshizaki, G., Kumamaru, K., Kagawa, H., Kaki, H., Nagaya, H. and Sohn, Y. C. (2010). Production of recombinant Japanese eel gonadotropins by baculovirus in silkworm larvae. *Gen. Comp. Endocrinol.,* 167, 379-386.

Kobayashi, M., Morita, T., K Ikeguchi, G Yoshizaki, and T Suzuki, S. W., (2006). In vivo biological activity of recombinant goldfish gonadotropins produced by baculovirus in silkworm larvae. *Aquaculture,* 256, 433-442.

Kobayashi, M., Morita, T., Ikeguchi, K., Yoshizaki, G., Suzuki, T. and Watabe, S., (2003). Production of recombinant goldfish gonadotropins by baculovirus in silkworm larvae. *Fish Physiol. Biochem.,* 28, 469-471.

Levavi-Sivan, B., Bogerd, J., Mañanos, E. L., Gómez, A. and Lareyre, J. J. (2010). Perspectives on fish gonadotropins and their receptors. *Gen. Comp. Endocrinol.,* 165, 412-437.

Lister, A. L. and Van Der Kraak, G. (2008). An investigation into the role of prostaglandins in zebrafish oocyte maturation and ovulation. *Gen. Comp. Endocrinol.,* 159, 46-57.

Lister, A. L. and Van Der Kraak, G. J. (2009). Regulation of prostaglandin synthesis in ovaries of sexually-mature zebrafish (*Danio rerio*). *Mol. Reprod. Dev.,* 76, 1064-1075.

Liu, Z., de Matos, D. G., Fan, H. Y., Shimada, M., Palmer, S. and Richards, J. S. (2009). Interleukin-6: an autocrine regulator of the mouse cumulus cell-oocyte complex expansion process. *Endocrinology,* 150, 3360-3368.

MacKenzie, S., Montserrat, N., Mas, M., Acerete, L., Tort, L., Krasnov, A., Goetz, F. W. and Planas, J. V. (2006). Bacterial lipopolysaccharide induces apoptosis in the trout ovary. *Reprod. Biol. Endocrinol.,* 4, 46.

Machelon, V., Emilie, D., Lefevre, A., Nome, F., Durand-Gasselin, I. and Testart, J. (1994). Interleukin-6 biosynthesis in human preovulatory follicles: some of its potential roles at ovulation. *J. Clin. Endocrinol. Metab.,* 79, 633-642.

Mañanos, E. L., Swanson, P., Stubblefield, J. and Zohar, Y. (1997). Purification of gonadotropin II from a teleost fish, the hybrid striped bass, and development of a specific enzyme-linked immunosorbent assay. *Gen. Comp. Endocrinol.,* 108, 209-222.

Milla, S., Jalabert, B., Rime, H., Prunet, P. and Bobe, J. (2006). Hydration of rainbow trout oocyte during meiotic maturation and in vitro regulation by 17,20beta-dihydroxy-4-pregnen-3-one and cortisol. *J. Exp. Biol.,* 209, 1147-1156.

Molés, G., Gómez, A., Rocha, A., Carrillo, M. and Zanuy, S. (2008). Purification and characterization of follicle-stimulating hormone from pituitary glands of sea bass (*Dicentrarchus labrax*). *Gen. Comp. Endocrinol.,* 158, 68-76.

Molés, G., Zanuy, S., Muñoz, I., Crespo, B., Martínez, I., Mañanos, E. and Gómez, A. (2011). Receptor specificity and functional comparison of recombinant sea bass (*Dicentrarchus labrax*) gonadotropins (FSH and LH) produced in different host systems. *Biol. Reprod.,* 84, 1171-1181.

Munakata, A. and Kobayashi, M. (2010). Endocrine control of sexual behavior in teleost fish. *Gen. Comp. Endocrinol.,* 165, 456-468.

Murdoch, W. J., Colgin, D. C. and Ellis, J. A. (1997). Role of tumor necrosis factor-alpha in the ovulatory mechanism of ewes. *J. Anim. Sci.,* 75, 1601-1605.

Murdoch, W. J. and McDonnel, A. C. (2002). Roles of the ovarian surface epithelium in ovulation and carcinogenesis. *Reproduction,* 123, 743-750.

Nagahama, Y. and Yamashita, M. (2008). Regulation of oocyte maturation in fish. *Dev. Growth Differ. 50 Suppl 1*, 195-219.

Ogiwara, K., Takano, N., Shinohara, M., Murakami, M. and Takahashi, T. (2005). Gelatinase A and membrane-type matrix metalloproteinases 1 and 2 are responsible for follicle rupture during ovulation in the medaka. *Proc. Natl. Acad. Sci. USA,* 102, 8442-8447.

Ogiwara, K., Minagawa, K., Takano, N., Kageyama, T. and Takahashi, T. (2012). Apparent involvement of plasmin in early-stage follicle rupture during ovulation in medaka. *Biol. Reprod,* 86, 1-10.

Ogiwara, K., Fujimori, C., Rajapakse, S. and Takahashi, T. (2013). Characterization of luteinizing hormone and luteinizing hormone receptor and their indispensable role in the ovulatory process of the medaka. *PloS ONE,* 8, E54482.

Ohnishi, J., Ohnishi, E., Shibuya, H. and Takahashi, T. (2005). Functions for proteinases in the ovulatory process. *Biochim. Biophys. Acta,* 1751, 95-109.

Okada, T., Kawazoe, I., Kimura, S., Sasamoto, Y., Aida, K. and Kawauchi, H. (1994). Purification and characterization of gonadotropin I and II from pituitary glands of tuna (*Thunnus obesus*). *Int. J. Pept. Protein Res.,* 43, 69-80.

Patiño, R., Yoshizaki, G., Bolamba, D. and Thomas, P. (2003a). Role of arachidonic acid and protein kinase C during maturation-inducing hormone-dependent meiotic resumption and ovulation in ovarian follicles of Atlantic croaker. *Biol. Reprod.,* 68, 516-523.

Patiño R., Thomas, P., Yoshizaki G, (2003b). Ovarian follicle maturation and ovulation: an integrated perspective. *Fish Physiol. Biochem.,* 28, 305-308.

Peterson, C. M., Hales, H. A., Hatasaka, H. H., Mitchell, M. D., Rittenhouse, L. and Jones, K. P. (1993). Interleukin-1 β (IL-1 β) modulates prostaglandin production and the natural IL-1 receptor antagonist inhibits ovulation in the optimally stimulated rat ovarian perfusion model. *Endocrinology,* 133, 2301-2306.

Planas, J. V., Athos, J., Goetz, F. W. and Swanson, P., (2000). Regulation of ovarian steroidogenesis *in vitro* by follicle-stimulating hormone and luteinizing hormone during sexual maturation in salmonid fish. *Biol. Reprod.,* 62, 1262-1269.

Planas, J. V. and Swanson, P. (2007). Physiological function of gonadotropins in fish. In: Rocha, M. J., Arukwe, A. and Kapoor, B. G. (Eds.), *Fish Reproduction.*, Science Publishers. 37-66.

Ren, M., Guo, Q., Guo, L., Lenz, M., Qian, F., Koenen, R. R., Xu, H., Schilling, A. B., Weber, C., Ye, R. D., Dinner, A. R. and Tang, W. J. (2010). Polymerization of MIP-1 chemokine (CCL3 and CCL4) and clearance of MIP-1 by insulin-degrading enzyme. *EMBO J,* 29, 3952-3966.

Richards, J. S., Liu, Z. and Shimada, M. (2008). Immune-like mechanisms in ovulation. *Trends Endocrinol. Metab.,* 19, 191-6.

Roberts, S. B., Langenau, D. M. and Goetz, F. W. (2000). Cloning and characterization of prostaglandin endoperoxide synthase-1 and -2 from the brook trout ovary. *Mol. Cell Endocrinol.,* 160, 89-97.

Robker, R. L., Russell, D. L., Espey, L. L., Lydon, J. P., O'Malley, B. W. and Richards, J. S. (2000). Progesterone-regulated genes in the ovulation process: ADAMTS-1 and cathepsin L proteases. *Proc. Natl. Acad. Sci. USA,* 97, 4689-4694.

Roher, N., Callol, A., Planas, J. V., Goetz, F. W. and MacKenzie, S. A. (2011). Endotoxin recognition in fish results in inflammatory cytokine secretion not gene expression. *Innate Immun.*,17, 16-28.

Shimada, M., Hernandez-Gonzalez, I., Gonzalez-Robanya, I. and Richards, J. S. (2006). Induced expression of pattern recognition receptors in cumulus oocyte complexes: novel evidence for innate immune-like functions during ovulation. *Mol. Endocrinol.,* 20, 3228-3239.

Sorensen, P. W. and Goetz, F. W. (1993). Pheromonal and reproductive function of F prostaglandins and their metabolites in teleost fish. *J. Lipid Mediat.,* 6, 385-393.

Sorensen, P. W., Hara, T. J., Stacey, N. E. and Goetz, F. W. (1988). F prostaglandins function as potent olfactory stimulants that comprise the postovulatory female sex pheromone in goldfish. *Biol. Reprod.,* 39, 1039-1050.

Swanson, P., Suzuki, K., Kawauchi, H. and Dickhoff, W. W. (1991). Isolation and characterization of two coho salmon gonadotropins, GTH I and GTH II. *Biol. Reprod.,* 44, 29-38.

Takeda, K., Kaisho, T. and Akira, S. (2003). Toll-like receptors. *Annu. Rev. Immunol.,* 21, 335-376.

Van der Hoek, K. H., Woodhouse, C. M., Brännström, M. and Norman, R. J. (1998). Effects of interleukin (IL)-6 on luteinizing hormone- and IL-1β-induced ovulation and steroidogenesis in the rat ovary. *Biol. Reprod.,* 58, 1266-1271.

Van Der Kraak, G. and Chang, J. P. (1990). Arachidonic acid stimulates steroidogenesis in goldfish preovulatory ovarian follicles. *Gen Comp Endocrinol,* 77, 221-228.

Vischer, H. F., Granneman, J. C., Linskens, M. H., Schulz, R. W. and Bogerd, J. (2003). Both recombinant African catfish LH and FSH are able to activate the African catfish FSH receptor. *J. Mol. Endocrinol.,* 31, 133-140.

Vraskou, Y., Roher, N., Diaz, M., Antonescu, C. N., MacKenzie, S. A. and Planas, J. V. (2011). Direct involvement of tumor necrosis factor alpha; in the regulation of glucose uptake in rainbow trout muscle cells. *Am. J. Physiol. Regul. Integr. Comp. Physiol.,* 300, 716-723.

Wang, H., Czura, C.J., Tracey, K.J. (2003). Tumor necrosis factor. In: Thomson, A.W., Lotze, M.T. (Eds.), *The Cytokine Handbook.*, fourth ed. Academic Press. 837-860.

Yoshimura, T., Matsushima, K., Tanaka, S., Robinson, E. A., Appella, E., Oppenheim, J. J. and Leonard, E. J. (1987). Purification of a human monocyte-derived neutrophil chemotactic factor that has peptide sequence similarity to other host defense cytokines. *Proc. Natl. Acad. Sci. USA,* 84, 9233-9237.

Zmora, N., Kazeto, Y., Kumar, R. S., Schulz, R. W. and Trant, J. M. (2007). Production of recombinant channel catfish (*Ictalurus punctatus*) FSH and LH in S2 *Drosophila* cell line and an indication of their different actions. *J. Endocrinol.,* 194, 407-416.

Zolti, M., Meirom, R., Shemesh, M., Wollach, D., Mashiach, S., Shore, L. and Rafael, Z. B., (1990). Granulosa cells as a source and target organ for tumor necrosis factor-alpha. *FEBS Lett.,* 261, 253-255.

In: Sexual Plasticity and Gametogenesis in Fishes
Editor: Balasubramanian Senthilkumaran
ISBN: 978-1-62618-848-8

Chapter IV

Teleosts as Models for Environmental Risk Assessment

Aparna Dutta-Gupta[*]
Department of Animal Sciences, School of Life Sciences-Centre for Advanced Studies, University of Hyderabad, Central University, Hyderabad, Andhra Pradesh, India

Abstract

Over the past four decades aquatic organisms have emerged as promising models for the evaluation of possible toxic effects of a variety of compounds including pesticides, herbicides, xenobiotics, hormones and their analogs, agonists, antagonists etc. In this chapter, we discuss and provide an understanding of the biological system of bony fishes (teleosts) inhabiting aquatic environment and their interaction with environmental contaminants with possibilities of using them as potential models as well as tools for toxicity testing and environmental risk assessment. Further viewed from an overall perspective, the threat posed from environmental contaminants with hormonal activity either agonists or antagonists is of serious concern with the intensity of effect depending on the concentration, duration and timing of exposure.

Introduction

It has been well documented that anthropogenic compounds including pesticides, herbicides, their metabolites and hormones including their analogs, agonists and antagonists from agricultural, industrial, pharmaceutical and house hold sources find their way into the environment including water bodies, and pose potential threats to aquatic organisms as well as humans. Further, a large number of these compounds exhibit endocrine disrupting behavior. Endocrine disrupting chemicals (EDCs) interfere with the endocrine/physiological system and disrupt either hormonal balance or action of hormones. EDCs are diverse in their

[*] Email: aparnaduttagupta@gmail.com.

classification based on their chemical structure as well as mode of action (Ankley et al., 2009). These chemicals either have the ability to mimic or block the action of natural hormones or may directly or indirectly cause disruption in living systems. Their toxicological effects are exerted on various tissues/organs including brain, endocrine glands, liver, gonads etc. of non-target species as well. EDCs also have deleterious effects on reproductive behavior of mammals including humans. Chinese hamster ovary cell *in vitro* assays with 200 different pesticides including few of their isomers and metabolites demonstrated (Kojima et al., 2004) that a large number of these possesses estrogenic activity, although their potency differs from one another. Prospective ecological risk assessment for various xenobiotics and environmental contaminants is routinely carried out using various methods including analysis of biomolecular signals, biological kinetics, transportation and toxicity at both cell and organism level (Ankley et al., 2009). Development of gonad intersex and alteration in secondary sex characteristics have been observed in bony fishes (teleosts) after exposure to estrogenic and anti-androgenic compounds including pesticides and herbicides at fairly low and ecologically relevant doses, clearly suggesting that bony fishes are extremely sensitive to physiological and endocrine disruption (Kiparissis et al., 2003; Balch et al., 2004, 2006; Raghuveer and Senthilkumaran, 2009; Tetreault et al., 2011; Wu et al., 2012; Kruger et al., 2013; Zhang et al., 2013). In this context, consumer products and pesticides pose greater threats.

Bisphenol A (BPA) is a widely used constituent in a variety of consumer products including polycarbonate plastics, epoxy resins, flame retardants, sealants etc. that is present in surface waters as well as in waste water effluents (Furhacker et al., 2000; Staples et al., 2000; Koplin et al., 2002). BPA exposure interferes with reproductive function of male fishes. The adverse effect ranges from severe alteration in testicular morphology at reported environmental concentration (1-10 μg BPA/L) to reduction in number of spermatogenic cysts and gonado-somatic index (GSI) as well as lowered serum testosterone levels (Sohoni et al., 2001; Kinnberg and Toft, 2003, Mandich et al., 2007). In addition, environmental anti-androgens have now been recognized as factors that largely contribute in chemical-induced feminization of wild fish producing phenotypic effects similar to environmental estrogens primarily by blocking androgen action (Filby et al., 2006). The reproductive toxicity of various EDCs in the fathead minnow was shown to be associated with their effect on hypothalamic-pituitary-gonadal axis (Ankley et al., 2008) and exert their effects indirectly on the gonads via gonadotropins (Da Cuna et al., 2012). In female teleosts, the ovary undergoes a seasonal reproductive cycle divided into various stages during which immature oocytes develop into functional follicles (Kime, 1995). The exposure of fish to EDCs at any stage of the cycle inhibits gonadal growth and function. BPA treatment inhibited egg production and reduced egg hatchability in F_1 generation in fathead minnow (Sohini et al., 2001). In the blue gill fish, females exposed to endosulfan showed empty ovarian follicles, thinning of the ovarian wall and elevated atresia. The extent of damage was proportional to the concentration of the pesticide (Dutta and Dulal, 2008). Similar disruptive effects were seen in female zebrafish with endosulfan (Han et al., 2011). Interestingly, pesticides such as endosulfan affect both sexes. Studies from our laboratory revealed precocious ovarian development and impaired testicular development in catfish after exposure to endosulfan (Chakrabarty et al., 2012; Rajakumar et al., 2011).

The mode of action of EDCs is often routed through endocrine or hormonal systems. Extensive studies using teleosts clearly suggest that the relative concentration and the ratio of sex steroids provide a male versus female hormonal milieu, while androgens facilitate the development of males, estrogens favor female development. The EDCs present in aquatic environments have great potential in disrupting this hormonal balance and in turn affect sexual differentiation (Filby et al., 2007). Sexual plasticity seen during the gonadal development and differentiation is known to be controlled by various sex determining genes such as *cyp19a1, dmrt1, foxl 2* and *sox9* (Nagahama et al., 2004; Guo et al., 2005, Filby et al., 2007; Liu et al., 2007; Ijiri et al., 2008; Raghuveer and Senthilkumaran, 2009; Wang et al., 2010; Leet et al., 2011; Rasheeda et al., 2010; Raghuveer et al., 2011; Sridevi and Senthilkumaran, 2011). In addition to these transcription factor genes, several other genes that primarily regulate steroidogenesis were differentially expressed during sex differentiation such as *cyp17α*, *cyp11α*, *3β-hsd*, and *star* (Baron and Guiguen, 2003; Baron et al., 2005; Nakamura et al., 2008; Raghuveer et al., 2012). The analysis and profiling of the genes associated with gonadal differentiation revealed the role of insulin-like growth factor (Schlueter et al., 2007) in embryonic germ line development in zebrafish, and follistatin, *sox23*, *sox24*, *bmp7*, *bcl21b*, *fancl*, *gcl* and *inha* in ovarian differentiation of rainbow trout (Baron et al., 2005; Vizziano et al., 2007). Although these genes belong to various families, their precise temporal dimorphic expression governs the sex differentiation. Estrogenic EDCs were shown to promote feminization of Japanese flounder larvae reared at masculinizing temperature (Nagahama et al., 2004). This is most likely manifested by the involvement of genes belonging to several families including transcription factors, germ cell proteins, steroidogenic enzymes, hormones, growth factors and their receptors (Leet et al., 2011).

Among various fish species used presently, zebrafish (*Danio rerio*), with its complete genome sequence known, has emerged as an ideal choice for a large number of investigators. Further, it has a great potential to facilitate research into endocrine disruption at the level of gene expression and this approach would provide an open platform for understanding the mode as well as a mechanism of action of different EDCs. Earlier studies using zebrafish showed that each EDC group with similar mode of action exhibits specific gene expression profile that could be used as a "signature/finger print" for environmental risk evaluation (Watanabe et al., 2003; Moggs et al., 2004). Further, the zebrafish model is already being used in various toxicological programs worldwide in assessing human risk and preclinical drug discovery and screening (Zhang et al., 2003; Sipes et al., 2011; Molina et al., 2013). In addition, a zerbrafish embryo is also emerging as a low cost high-throughput model with an impressive range of possible applications in environmental risk assessment beyond simple acute toxicity testing (Scholz et al., 2008; Ali et al., 2011). A recent trend in biological, biomedical, pharmacological and agricultural areas is the utilization of nanoparticle-based formulations for targeted delivery. Few ecotoxicological studies reveal that under environmental conditions, engineered nanoparticles (ENP) of pesticides have toxic effects on various organisms (Nowack and Bucheli, 2007) including plants (Ma et al., 2010). In addition to their organismal effects, this could also be propagated through population (Thomas et al., 2011), which would subsequently alter communities comprising of many organisms (e.g., bacteria, plant, fish etc.) and have a serious deleterious effect on the food web and ecosystem. To a certain extent, the physicochemical properties of these nano-sized complexes influencing the biological functions have been elucidated. However, many questions regarding the *in vivo* behavior of nanoparticle-based formulation and their interactions with each other in the environment as well as during exposure still need to be addressed (Holden et al., 2012). Further, the horizontal as well as generational transfer of ENPs in the environment makes it necessary to evaluate ENP based formulations through extensive biophysical, biochemical and biological studies (Ke and Lamm, 2011). Bony fish models would not only facilitate systematic analysis of organism-level effects of nanoparticle based formulations on metabolism and reproduction, but would also enable researcher to assess the pharmokinetics and tissue distribution under diverse aquatic environments. This would help to further streamline the engineering of complex nanostructures that ensure higher efficacy in a cost effective manner, as well as predict the engineered nanomaterial characteristic to be avoided (Alpanese et al., 2012). Although a range of teleost species are used for toxicity assays as well as tested for detection of environmental contaminants it is impossible to identify a particular species as a standard model. However, one could select various fish models based on several criteria ranging from size to life cycle pattern, spawning season, breeding pattern, larval growth pattern etc.

Conclusion

Teleost fishes provide a good example to demonstrate consistent physiological, biochemical, and molecular changes upon exposure to various chemicals and formulations. Further these molecular interactions not only alter the physiology but also have a direct bearing on the development as well as reproduction of bony fish. Taken together, our research findings as well as reported work of others in this article suggest that fishes could be used as powerful model for toxicity assessment at organism level as well as a sensitive tool for analyzing the contamination of a variety of environmental toxicants including EDCs.

Acknowledgments

I (co-PI) would like to thank the Grant-in-Aid (BT/PR11263/AAQ/03/419/2008) from the Department of Biotechnology, India, which allowed me to use bony fishes as test models in collaboration with Prof. B. Senthilkumaran (PI). The support received from Ms. C. Laldinsangi in the preparation of this manuscript is greatly acknowledged.

References

Albanese, A., Tang, P. S. and Chan, W. C. W. (2012). The effect of nanoparticle size, shape and surface chemistry on biological systems. *Annu. Rev. Biomed. Eng., 14,* 1-16.

Ankley, G. T., Bencic, D. C., Breen, M. S., Collette, T. W., Conolly, R. B., Denslow, N. D, Edwards, S. W., Ekman, D. R, Garcia-Reyero, N., Jensen, K. M., Lazorchak, J. M., Marlinovic, D., Mitler, D. H., Perkins, E. J., Orlando, E. F., Villeneuve, D. L., Wang, R. L. and Watanabe, K. H. (2009). Endocrine disrupting chemicals in fish: Developing exposure indicators and predictive models of effects based on mechanism of action. *Aquat. Toxicol., 92,* 168-178.

Ankley, G. T., Miracle, A. L., Perkins, E. J. and Daston, G. P. (2008). Genomics in regulatory ecotoxicology: Applications and challenges, Pensacola, FL: SETAC Press.

Balch, G. and Metcalfe, C. (2006). Developmental effects in Japanese medaka (*Oryzias latipes*) exposed to nonylphenol ethoxylates and their degradation products. *Chemosphere, 62,* 1214-1223.

Balch, G. C., Mackenzie, C. A. and Metcalfe, C. D. (2004). Alterations to gonadal development and reproductive success in Japanese medaka (*Oryzias latipes*) exposed to 17-α ethinylestradiol. *Environ. Toxicol. Chem., 23,* 782-791.

Baron, D. and Guiguen, Y. (2003). Gene expression during gonadal sex differentiation in rainbow trout (*Onchorynchus mykiss*): from candidate genes studies to high throughput genomic approach. *Fish Physiol. Biochem., 28,* 119-123.

Chakrabarty, S., Rajakumar, A., Raghuveer, K., Sridevi, P., Mohanachary, A., Prathibha, Y., Bhasyam, L., Dutta-Gupta, A. and Senthikumaran, B. (2012) Endosulfan and flutamide, alone and in combination, target ovarian growth in juvenile catfish, *Clarias batrachus. Comp. Biochem. Physiol. C Toxicol. Pharmacol., 155,* 491-7.

Chakravorty, S., Lal, B. and Singh, T. P. (1992). Effect of endosulfan (thiodan) on vitellogenesis and its modulation by different hormones in the vitellogenic catfish *Clarias batrachus*. *Toxicology, 75,* 191-198.

Chang, X., Kobayashi, T., Senthilkumaran, B., Kabayashi-Kajura, H., Sudhakumari, C. C. and Nagahama, Y. (2005). Two types of aromatase with different encoding genes, tissue distribution and developmental expression in nile tilapia (*Oreochromis niloticus*). *Gen. Comp. Endocrinol., 141,* 101-105.

Dutta, H. M. and Dalal, R. (2008). The effect of endosulfan on the ovary of the bluegill sunfish: A histopathological study. *Int. J. Environ. Res., 2,* 215-224.

Da Cuna, R. H., Vasquez, G. R., Piol, M. N., Guerrero, N. V., Maggese, M. C. and Lo Nostro, F. L. (2011). Assessment of the acute toxicity of the organochlorine pesticide endosulfan in *Cichlosoma dimeris*. *Ecotoxicol. Environ. Saf., 74,* 1065-1073.

Furhacker, M., Scharf, S. and Weber, H. (2000). Bisphenol A: emissions from point sources. *Chemosphere, 41,* 751-56.

Filby, A. L., Thorpe, K. L., Maack, G. and Tyler, C. R. (2006). Gene expression profiles revealing the mechanism of anti-androgen and estrogen induced feminization in fish. *Aquat. Toxicol., 81,* 219-231.

Guo, Y., Cheng, H., Huang, X., Gao, S., Yu, H. and Zhou, R. (2005). Gene structure, multiple alternative splicing and expression in gonads of zebrafish Dmrt1. *Biochem. Biophys. Res. Commun., 330,* 950-957.

Holden, P. A., Nesbet, R. M., Lenihan, H. S., Miller, R. J., Cherr, G. N., Schimel, J. P. and Gardea-Torresday J. L. (2012). Ecological nanotoxicology: Integrating nanomaterial hazard considerations across the subcellular, population, community, and ecosystems levels. *Acc. Chem. Res.*

Han, X. B., Lei, E. N., Lam, M. H and Wu, R. S. (2011). A whole life cycle assessment on effects of waterborne PBDEs on gene expression profile along the brain-pituitary-gonadal axis and in the liver of zebrafish. *Mar. Pollut. Bull, 63,* 160-165.

Ijiri, S., Kaneko, H., Kobayashi, T., Wang, D. S., Sakai, F., Paul-Prasanth, B., Nakamura, M. and Nagahama, Y. (2008). Several dimorphic expression of genes in gonads during early differentiation of a teleost fish, the nile tilapia *Oreochromis niloticus*. *Biol. Reprod., 78,* 333-341.

Ke, P. C. and Lamm, M. H. (2011). A biophysical perspective of understanding nanoparticles at large. *Phys. Chem. Chem. Phys., 13,* 7273-7283.

Kinnberg, K. and Toft, G. (2003). Effects of estrogenic and antiandrogenic compounds on the testis structure of the adult guppy (*Poecilia reticulata*). *Ecotoxicol. Environ. Saf., 54,* 16-24.

Kruger, T., Barhoorn, I., Jansen Van Vuren, J. and Borman, R. (2013). The use of the urogenital papilla of male feral African sharptooth catfish (*Clarias gariepinus*) as an indicator of exposure to estrogenic chemicals in two polluted dams in an urban nature reserve, Gauteng, South Africa. *Ecotoxicol. Environ. Saf., 87,* 98-107.

Kolpin, D. W., Furlong, E. T., Meyer, M. T., Thurman, E. M., Zaugg, S. D., Barber, L. B. and Buxton, H. T. (2002). Pharmaceuticals, hormones and other organic wastewater contaminants in U.S. streams, 1999-2000: a national reconnaissance. *Environ. Sci. Technol., 36,* 1202-1211.

Kojima, H., Katura, E., Takeuchi, S., Nuyama, K and Kobayashi, K., (2001). Screening for estrogen and androgen receptor activities in 200 pesticides by in vitro reporter gene assays using Chinese hamster ovary cells. *Environ Health Perspect., 112,* 524-531.

Kiparissis, Y., Balch, G. C., Matcalfe, T. L. and Metcalfe, C. D. (2003). Effects of the isoflavones genestein and equol on the gonadal development of Japanese medaka *Oryzias latipes*. *Environ. Health Perspect., 111,* 1158-1163.

Kime, D. E. (1995). The effects of pollution on reproduction in fish. *Rev. Fish. Biol., 5,* 52-95.

Kosmehl, T., Otte, J. C., Yang, L., Legradi, J., Bluhm, K., Zinsmeister, C., Keiter, S. H., Reifferscheid, G., Manz, W., Braunbeck, T., Strahle, U. and Hollert, H. (2012). A combined DNA-microassay and mechanism specific toxicity approach with zebrafish embryos to investigate the pollution of river sediments. *Reprod. Toxicol., 33,* 245-253.

Leet, J. K., Gall, H. E. and Sepulveda, M. S. (2011). A review of studies on androgen and estrogen exposure in fish early life stages: effects on gene and hormonal control of sexual differentiation. *J. Appl. Toxicol., 31,* 379-398.

Liu, Z., Wu, F., Jiao, B., Zhang, X., Hu, C., Huang, B., Zhou, L., Hung, X., Wang, Z., Zhang, Y., Nagahama, Y., Cheng, C. H. K. and Wang, D. (2007). Molecular cloning of doublesex and mab-3-related transcription factor 1, forkhead transcription factor gene 2, and two types of cytochrome p450 aromatase in Southern catfish and their possible role in sex differentiation. *J. Endocrinol., 194,* 223-241.

Ma, X., Geiser-Lee, J., Deng, Y. and Kolmakov, A. (2010). Interactions between engineered nanoparticles (ENPs) and plants: phytotoxicity, uptake and accumulation. *Sci. Total. Environ.,* 408, 3053-3061.

Mandich A., Bottero S., Benfenati E., Cevasco A., Erratico C., Maggioni S., Massari A., Pedemonte F. and Vigano L. (2007). In vivo exposure of carp to graded concentrations of bisphenol A. *Gen Comp Endocrinol., 153,* 15-24.

Molina, A. M., Lora, A. J., Blanco, A., Monterde, J. G., Ayala, N. G. and Moyano, R. (2013). Endocrine-active compound evaluation: Qualitative and quantitative histomorphological assessment of zebrafish gonads after Bisphenol A exposure. *Ecotoxicol. Environ. Saf., 88,* 155-62.

Nowack, B. and Bucheli, T. D. (2007). Occurrence, behavior and effects of nanoparticles in the environment. *Environ. Pollut., 150,* 5-22.

Nagahama, Y., Nakamura, M., Kitano, T. and Tokumoto, T. (2004). Several plasticity in fish: a possible target of endocrine disruptor action. *Environ. Sci., 11,* 73-82.

Nakamura, S., Aoki, Y., Saito, D., Kiroki, Y., Fujiyama, A., Naruse, K. and Tanaka, M. (2008). Sox9b/Sox9aZ-EGFP transgenic medaka reveals the morphological reorganization of the gonads and a common precursor of both the female and male supporting cells. *Mol. Reprod., 75,* 772-76.

Raghuveer, K. and Senthilkumaran, B. (2009). Identification of multiple dmrt1s in catfish: localization, dimorphic expression pattern, changes during testicular cycle and after methyl testosterone treatment. *J. Mol. Endocrinol., 42,* 437-48.

Raghveer, K., Senthilkumaran, B., Sudhakumari, C. C., Sridevi, P., Rajakumar, A., Singh, R., Murugananthkumar, R., and Majumdar, K. C. (2011). Dimorphic expression of various transcription factor and steroidogenic enzyme genes during gonadal ontogeny in the air breathing catfish, *Clarias gariepinus. Gen. Comp. Endocrinol., 167,* 234-45.

Raghveer, K. and Senthilkumaran, B. (2012). Cloning and expression of 3β-hydroxysteroid dehydrogenase during gonadal recrudescence and after hCG induction in the air-breathing catfish, *Clarias gariepinus. Steroids, 77,* 1133-40.

Rajakumar, A., Singh, R., Chakrabarty, S., Murugananthkumar, R., Laldinsangi, C., Prathibha, Y., Sudhakumari, C. C., Dutta-Gupta, A. and Senthilkumaran, B. (2012). Endosulfan and flutamide impair testicular development in the juvenile Asian catfish, *Clarias batrachus. Aquat. Toxicol., 110,* 123-32.

Rasheeda, M. K., Sridevi, P. and Senthilkumaran, B. (2010). Cytochrome P450 aromatases: Impact on gonadal development, recrudescence and effect of hCG in the catfish, *Clarias gariepinus. Gen. Comp. Endocrinol., 167,* 234-245.

Staples, C. A., Dorn, P. B., Kleeka, G. M., O'Block, S. T., Branson, D. R. and Harris, L. R. (2000). Bisphenol A concentrations in receiving waters near US manufacturing and processing facilities. *Chemosphere, 40,* 521-25.

Sohoni, P., Tyler, C. R., Hurd, K., Caunter, J., Hetheridge, M., Williams, T., Woods, C., Evans, M., Toy, R., Gargas, M. and Sumpter, J. P. (2001). Reproductive effects of long-term exposure to Bisphenol A in the fathead minnow (*Pimephales promelas*). *Environ. Sci. Technol., 35,* 2917-2925.

Sridevi, P. and Senthilkumaran, B. (2011). Cloning and differential expression of Foxl2 during ovarian development and recrudescence of the catfish *Clarias gariepinus. Gen. Comp. Endocrinol., 174,* 259-268.

Schlueter, P. J., Sang, X., Duan, C. and Wood, A. W. (2007). Insulin-like growth factor receptor 1b is required for zebrafish primordial germ cell migration and survival. *Dev. Biol., 305,* 377-387.

Sipes, N. S., Padilla, S. and Knudsen, T. B. (2011). Zebrafish: as an integrative model for twenty-first century toxicity testing. *Birth Defects Res. C Embryo. Today, 93,* 256-267.

Thomas, C. R., George, S., Horst, A. M., Ji, Z., Miller, R. J., Peralta-Uidea, J. R., Xia, T., Pokhrel, S., Madler, L., Gardea-Torresdey, J. L., Holden, P. A., Keller, A. A., Lenichan, H. S. Nel, A. E. and Zink, J. I. (2011). Nanomaterials in the environment: From materials to high-throughput screening to organisms. *ACS Nano*, *5,* 13-20.

Tetreault, G. R., Bennet, C. J., Shires, K., Knight, B., Servos, M. R., and McMaster, M. E. (2011). Intersex and reproductive impairment of wild fish exposed to multiple municipal waste water discharges. *Aquat. Toxicol., 104,* 278-290.

Vizziano, D., Randuineau, G., Baron, D., Carly, C, and Guiguen, Y. (2007). Characterization of early molecular sex differentiation in rainbow trout, *Onchorynchus mykiss. Dev. Dyn., 236,* 2198-2206.

Wang, J., Lin, X., Wang, H., Wu, T., Hu, X., Qin, F. and Wang, Z. (2010). Expression of two cytochrome p450 aromatase gene is regulated by endocrine disrupting chemicals in rare minnow *Gobiocypris rarus* juveniles. *Comp. Biochem. Physiol. C Toxicol. Pharmacol., 152,* 313-320.

Wu, G. C., Chiu, P. C., Lin, C. J. Lyu, L. S., Lan, D. S. and Chang, F. C. (2012). Testicular dmrt1 is involved in the sexual fate of the ovotestis in the protandrous black porgy. *Biol. Reprod., 41*, 1-11.

Yokota, H., Miyashita, N. and Yuasa, A. (2002). High glucuronidation activity of environmental estrogens in the carp (*Cyprinus carpio*) intestine. *Life Sci., 71,* 887-898.

Zhang, X., Gao, L., Yang, K., Tian, H., Ulang, W. and Ru, S. (2013). Monocrotophos pesticide modulates the expression of several differentiation genes and causes phenotypic feminization in zebrafish (*Danio rerio). Comp. Biochem. Physiol. C Toxicol. Pharmacol., 157,* 33-40.

Zhang, C., Willett, C. and Fremgen, T. (2003). Zebrafish: An animal model for toxicological studies. *Curr. Protoc. Toxicol.,* 1-7, New York: Wiley.

In: Sexual Plasticity and Gametogenesis in Fishes
Editor: Balasubramanian Senthilkumaran
ISBN: 978-1-62618-848-8

Chapter V

Atherinopsid Fishes as Models for the Study of Temperature-Dependent Sex Determination: Physiology of Gonadal Sex Differentiation in Pejerrey *Odontesthes Bonariensis*

Juan I. Fernandino[1], Ricardo S. Hattori[2], Carlos A. Strüssmann[2] and Gustavo M. Somoza[1,*]

[1]Instituto de Investigaciones Biotecnológicas-Instituto Tecnológico de Chascomús. (CONICET-UNSAM), Chascomús, Argentina

[2]Graduate School of Marine Science and Technology, Tokyo University of Marine Science and Technology, Tokyo Japan

Abstract

The development of a functional gonad implies consequences not only for the phenotypic sex but also for the reproductive capacity during adulthood. In recent years, atherinopsid fish have been extensively studied because they present a variety of and sometimes contrasting mechanisms of sex determination/differentiation.

In the present work we review the sex determining mechanisms in different species belonging to this family. As an example, *Odontesthes bonariensis*, commonly known as pejerrey, was singled out in order to discuss morphological aspects of the gonadal differentiation process in males and females and its relation to water temperature, the expression profile of some selected genes involved in this process, and the involvement of cortisol in the masculinizing process induced by warm temperature.

* Contact Information and Reprint Request: Gustavo Manuel Somoza. Av. Intendente Marino Km. 8.2 (B7130IWA). Chascomús. Provincia de Buenos Aires. Argentina. E-mail: somoza@intech.gov.ar.

The well described process of sex determination/differentiation in *O. bonariensis* and other atherinopsid species makes this family an emerging model to study the interactions between the environment and genotypic factors that ultimately lead to determination of the gonadal fate.

Keywords: Atherinopsidae, GSD, TSD, apoptosis, steroids, stress, cortisol

1. Mechanisms of Sex Determination in Atherinopsid Fish

Atherinopsid fish are native from continental and marine waters of the Americas and they are usually referred to as "New World silversides" (Dyer, 2006; Bloom et al., 2012). In the last three decades several species belonging to this group have been intensively investigated with the focus on the sex determination and gonadal differentiation processes.

Sex in vertebrates is generally determined by two mechanisms: genetic (GSD) and environmental (ESD) sex determination. In species with GSD, sex is genetically programmed at fertilization and follows the actions of specific gene(s) usually located in the sex chromosomes, whereas in ESD sex is determined by environmental factors (Valenzuela et al., 2003; Ospina-Álvarez and Piferrer, 2008; Mank and Avise, 2009; Angelopoulou et al., 2012). The best known environmental factor involved in ESD is temperature, defined as temperature-dependent sex determination (TSD). In this mechanism the phenotypic sex is driven by temperature during a sensitive period early in development (Devlin and Nagahama, 2002; Grossen et al., 2011; Matsumoto and Crews, 2012).

TSD was first discovered in reptiles (see Bull and Vogt, 1979) and later found for the first time in fish by Conover and Kynard (1981) in the Atlantic Silverside *Menidia menidia* (L.), an atherinopsid fish. Since then it has been demonstrated that temperature can affect in different degrees the sex determination in other Atherinposid species such as *Menidia peninsulae* (Middaugh and Hemmer, 1987), *Odontesthes bonariensis* (Strüssmann et al., 1996a), *Odontesthes argentinensis* and *Odontesthes hatcheri* (Strüssmann et al., 1996b). TSD has also been demonstrated in many other fish species (Stüssmann and Patiño, 1999; Ospina-Álvarez and Piferrer, 2008; Baroiller et al., 2009a,b; Luckenbach et al., 2009; Siegfried, 2010) and it has been reported as the most important, or at least, the best studied environmental factor influencing sex differentiation in teleosts including tilapia (Devlin and Nagahama, 2002; Baroiller et al., 2009a,b). Although TSD has been described in different fish orders, no relationship has been found with the phylogenetic position, suggesting that such a mechanism has emerged more than once during teleost evolution (Devlin and Nagahama, 2002; Mank et al., 2006).

It is now also clear that there are some teleost species in which GSD can be influenced by environmental factors and thus the phenotypic sex is determined as a result of a combination of the two mechanisms (Goto-Kazeto et al., 2006; Baroiller et al., 2009a, b; 2009b; Luckenbach et al., 2009). This is also true for those species in which a sex determining gene was identified as in medaka *Oryzias latipes* (Matsuda et al., 2002) and rainbow trout *Onchorynchus mykiss* (Yano et al., 2012), where high temperature has been demonstrated to produce skewed sex ratios (Sato et al., 2005; Hattori et al., 2007; Magerhans et al., 2009; Magerhans and Hörstgen-Schwark, 2010).

However, atherinopsids provide an excellent material to study the interactions of GSD/TSD because members of this family span from a "pure" TSD to a classical GSD model. For example, the brackish water species pejerrey *O. bonariensis* presents a very strong TSD whereby all-female and all-male populations can be obtained when larvae are raised at low (17° C) and high (29° C) temperatures, respectively (Strüssmann et al., 1996a, 1997). The marine pejerrey *O. argentinensis*, also shows a clear correlation between sex ratios and temperature: more females are produced at lower (18 and 21° C) than at higher (25° C) temperatures (Strüssmann et al., 1996b). The situation in the Patagonian pejerrey *O. hatcheri*, in which a clear GSD was described (Hattori et al., 2010; 2012; Koshimizu et al., 2010) is different. In this species, the larvae reared at intermediate temperatures (17-23° C) show sex ratios around 1:1 but low (13-15° C) and high (25° C) temperatures produce female- and male-skewed sex ratios, respectively (Strüssmann et al., 1996b, 1997). This is important because Hattori et al. (2012) described the sex determining gene (*amhy*) in this species. This indicates, as already said, that the temperature during the period of sex determination can affect sex ratios, even in cases where a clear and defined genetic component exists (Hattori et al., 2010; 2012; Koshimizu et al., 2010). It is important to note that *amhy* has been found also in *O. bonariensis* (unpublished results). Although its role in this species is still unknown, it supports the notion that TSD and GSD are the extremes of a continuum (Strüssmann and Patiño, 1999).

The objective of the present chapter is to review the physiology of the sex differentiation process in the best studied of these species, the pejerrey *O. bonariensis*. The pejerrey is a neotropical silverside species which inhabits brackish waters in the southern region of the South American continent. The strong sexual thermolability makes of pejerrey a unique and ideal model to study how temperature can influence the feminization and masculinization processes.

2. Histological Features of Gonadal Sex Differentiation

In differentiated gonochoristic species, i.e., those species in which the gonads differentiate directly into either an ovary or a testis, sex is determined during a critical period of embryonic and/or larval stages that precedes the appearance of sex-specific histological or cellular differences. In pejerrey, this time frame was estimated to be between 1 and 3 weeks after hatching (wah) at 27° C, 2 and 4 wah at 24° C, and 3 and 5 wah at 17 °C (Strüssmann et al., 1997), which in a size scale ranges from 11 to 18 mm of body length. Thus, the temperature experienced during these early developmental stages determines whether the primordial gonad will follow the male or female pathway.

The first signals of morphological sex differentiation start to be detected shortly after this period and appear first in females and then in males (at intermediate temperatures). The primordial ovary is characterized by stromal cell outgrowths, which then will lead to the formation of the ovarian cavity, by the onset of meiosis in the primordial germ cells, and by the formation of blood vessels.

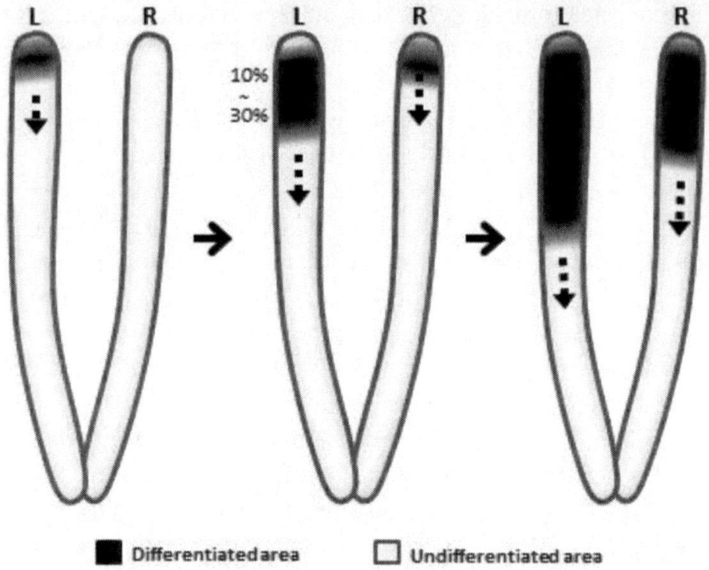

Figure 1. Schematic representation of the gradient of histological sex differentiation in pejerrey gonads. The dotted arrows indicate the wave of differentiation. L: left; R: right.

On the other hand, the testis differentiation is characterized by the formation of rudiments of the main sperm duct in the medullary region of the gonad and formation of characteristic germ cell cysts in its periphery. The germ cells in the male gonad enter meiosis a few weeks later (Ito et al., 2005).

An intriguing aspect of the gonadal differentiation in this species is the presence of a cephalo-caudal, left-to-right gradient of differentiation (Strüssmann and Ito, 2005). Regardless of sex, the left-anterior area is the region from where the onset of gonad differentiation begins toward the posterior part of the gonad. On the opposite side, the right gonad begins to differentiate only when 10-30% of the left side is already differentiated (Figure 1).

It is also important to stress that blood vessels are formed prior to sex differentiation (Ito et al., 2005), suggesting that extra-gonadal factors may be carried by the bloodstream into the left-anterior region of the gonad to trigger the sex differentiation process. This mechanism is not well understood but one explanation is based on the rarity of intersex gonads not only in wild animals but especially in laboratory-reared larvae that were subjected to shifts between low-feminizing and high-masculinizing temperatures, which would be expected given the high thermal plasticity of sex in this species. According to this idea, this gradient might be important to prevent the occurrence of gonadal sex ambiguity and therefore the appearance of intersex animals with ovarian and testicular tissue in the same lobe or with a testis in one lobe and an ovary in the other (Strüssmann and Ito, 2005).

Another feature of sex differentiation in pejerrey gonads is the occurrence of programmed cell death (apoptosis) linked to testicular differentiation. Thus, a preliminary analysis of apoptosis by the TUNEL (terminal deoxynucleotidyl transferase-mediated dUTP nick-end labeling) assay revealed that larvae reared at female-producing temperatures (FPT) had none to very few apoptotic cells whereas those from the male-producing temperatures (MPT) had widespread apoptosis in the anterior region of the right lobe between 4 or 5 and 10 wah. In larvae from mixed-sex-producing temperatures (MixPT), both patterns were observed and the proportion of animals with intense apoptosis correlated well with that of animals which differentiated as males (Strüssmann et al., 2008). Nevertheless the role of apoptosis during testis differentiation in pejerrey is still not clear. In the undifferentiated gonochorist zebrafish, *Danio rerio*, programmed cell death seems to have a clear role in the removal of oocyte-like cells in genotypic males and therefore in testis formation (Uchida et al., 2002; 2004). On the other hand, in medaka, increased apoptosis during differentiation has been shown to produce gonadal sex inversion from female to male (Kurokawa et al., 2007) and is associated with the regulation of the number of germ cells necessary for testicular development (Tanaka et al., 2008).The detection of apoptosis in the lobe with delayed differentiation (right side) and its occurrence mainly in somatic cells point out to a different mechanism or role in pejerrey testis differentiation. For example, as with the histological gradient of sex differentiation, the incidence of apoptosis in the anterior part of the right lobe could be also part of a mechanism to avoid gonadal ambiguity in larvae experiencing temperature changes during and immediately after the critical period of sex determination.

3. Genes Involved in Gonadal Differentiation

Since the ´90s and with the identification of the major sex determining gene SRY in mammals (Sinclair et al., 1990) the interest in the study of the molecular processes involved in gonadogenesis increased rapidly not only in mammals but also in other vertebrates. Interestingly, although the sex determining mechanisms or genes might differ in different species, the downstream molecular pathways involved in the gonadal fate seem to be highly conserved across vertebrates (Barske and Capel, 2008). These genes include for example *cyp19a1a* or gonadal aromatase, *dmrt1* and *amh*. This section presents an overview of the knowledge gathered so far on these genes in relation to the gonadal sex differentiation in pejerrey.

3.1. Gonadal Aromatase (*cyp19a1a*)

One of the most studied genes on the gonadal network is the cytochrome P450 aromatase enzyme, involved in the conversion of androgens into estrogens. This enzyme determines the balance between these two groups of steroids and it is essential for gonadal function in vertebrates (Carreau et al., 2002). Aromatase is also important at early stages of development; its expression and/or high activity have been associated with the process of ovarian differentiation in birds (Yoshida et al., 1996; Yang et al., 2008), reptiles (Gabriel et al., 2001; Pieau and Dorizzi, 2004) and also in teleost fish species (Guiguen et al., 2010). It is also important to highlight that there are two different genes that express aromatase enzymes in teleosts, the gonadal form or *cyp19a1a*, and the referred to as brain aromatase or *cyp19a1b* (Kishida and Callard, 2001; Jeng et al., 2012). The inhibition of the aromatase enzyme activity during the period of sex differentiation results in genotypic females developing as phenotypic males (Piferrer and Blázquez, 2005; Piferrer and Guiguen, 2008). These results emphasize the importance of the aromatase enzyme and estrogens as a pivotal role to the ovarian fate.

In pejerrey, the expression of *cyp19a1a* was studied in relation to the gonadal sex differentiation process as well as to the rearing temperature. Gonadal aromatase showed to be over-expressed at FPT compared to MPT-reared larvae (Karube et al., 2007), whereas at MixPT some fish showed a female- and other a male-like expression pattern. The pharmacological inhibition of aromatase activity induced a male skewed sex ratio in this species (Fernandino et al., 2008a), confirming the importance of this gene on ovarian development.

3.2. *Dmrt1*

One of the genes initially characterized in the process of sex determination and gonadal differentiation was the transcription factor DMRT1 (Doublesex Male abnormal-3 Related Transcription Factor-1). This transcription factor contains a DNA binding motif of zinc finger (DM domain) and it was first identified in nematodes and fruit flies (Shen and Hodgkin, 1988; Burtis and Baker, 1989). This gene is considered a key factor in male gonadal development from invertebrates to humans (Ferguson-Smith, 2007; Herpin and Schartl, 2011). In humans, for example, deletion of a copy of *DMRT1* is associated with male sex reversal in XY phenotypic females (Raymond et al., 1999a). Moreover, male *DMRT1* null mice exhibit severe defects in postnatal growth of the testis, while females are normal and fertile (Raymond et al., 2000). In chickens, *dmrt1* expression is higher in ZZ embryos (males) than in ZW (female) from the early stages of development of the genital ridge (Raymond et al., 1999b). A higher expression in testicles during gonadal development has been also described in reptiles (Sreenivasulu et al., 2002; Murdock and Wibbels, 2003; Valenzuela, 2010), amphibians (Shibata et al., 2002) and several teleost fish species (Marchand et al., 2000; He et al., 2003; Kobayashi et al., 2004; Guo et al., 2005; Raghuveer and Senthilkumaran, 2009; Cao et al., 2012; Masuyama et al., 2012).

In pejerrey, *dmrt1* showed a dimorphic expression pattern during early larval development displaying high expression at MPT, low at FPT and a bimodal expression at MixPT (Fernandino et al. 2008a). At this temperature regime, individuals with high *dmrt1* expression and conversely low *cyp19a1a* expression are considered to be those differentiating males whereas the opposite pattern is taken as an indication of female differentiation (Fernandino et al., 2008a).

3.3. *Amh*

Another gene that has generated great interest is the *AMH* (anti-Müllerian hormone, also known as Müllerian inhibiting Substance or Mis). AMH is a glycoprotein which belongs to the TGF-β family of growth and differentiation factors (Josso et al., 2001). In amniotes AMH is responsible of the regression of the Müllerian duct in male embryos, the control and evolution of the genital tract, and the regulation of follicular development (Josso et al., 2006).

Although teleost fish do not have Müllerian ducts (Wrobel, 2003), the expression of *amh* and its receptor (*amhrII*) have been detected in both sexes early in development during the gonadal differentiation period (Rodríguez-Marí et al., 2005; Nakamura et al., 2006; Klüver et al., 2007; Vizziano et al., 2007). The function of the amh/amhrII system was then established using a mutant for the receptor *amhrII* in medaka (*hotei* mutant), in which it was demonstrated that this system inhibits the proliferation of germ cells (Morinaga et al., 2007). Therefore, the male pathway regulates the germ cell proliferation in early stages of the developing gonad.

In pejerrey the *amh* expression profile, together with its regulation by gonadal steroids and glucocorticoids, was analyzed during the gonadal sex differentiation period. *Amh* is first expressed at low levels in the undifferentiated gonads of both sexes but becomes dimorphic, with higher expression in putative male gonads compared to females, during the process of gonadal differentiation (Fernandino et al., 2008b; Hattori et al., 2008; Hattori et al., 2009). Thus, *amh* shows the opposite expression pattern compared to *cyp19a1a*. These studies also demonstrated that the onset of *cyp19a1a* expression at the MixPT preceded that of *amh*. Moreover, larvae treated with estradiol during the critical time of sex differentiation consistently show low *amh* expression levels similar to those found at the FPT (Fernandino et al., 2008b). On the other hand, cortisol administration caused masculinization of pejerrey larvae (see subsequent section) and it was associated with *amh* and *cyp19a1a* profiles resembling those at the MPT (Hattori et al., 2009).

4. The role of Cortisol and Androgens in Gonadal Masculinization

In teleost fish the production of glucocorticoids (GCs) by the interrenal gland, located in the anterior kidney, is primarily regulated by the adrenocorticotropic hormone (ACTH), considered to be its main secretagogue (Mommsen et al., 1999). ACTH release in turn is regulated by the hypothalamic peptide corticotrophin releasing hormone (CRH).

The mechanism of action of GCs involves their passage through the plasma membrane and subsequent binding to cytoplasmic receptors (GRs). The hormone-receptor complex is transported into the nucleus where it binds to a specific DNA segment, the glucocorticoid response site (GRE), resulting in additional production of mRNA and subsequent synthesis of specific proteins in the cell (Bury and Sturm, 2007). The main GC in teleost fish is cortisol, which plays an important role in the modulation of the adaptive intermediary metabolism (Vijayan et al., 1994), ionic regulation (Sakamoto and McCormick, 2006), and immune function (Salas-Leiton et al., 2012).

Pejerrey reared at high temperatures eventually display signs of thermal stress. These include alterations in body shape, reduced swimming and feeding activity, changes in their pigmentation pattern, loss of weight and progressive germ cell degeneration (Ito et al., 2008). The thermal profile of the natural habitat of pejerrey (Gómez et al., 2007) also suggests that this species prefers cold to temperate waters and then high temperatures could act as a stress factor.

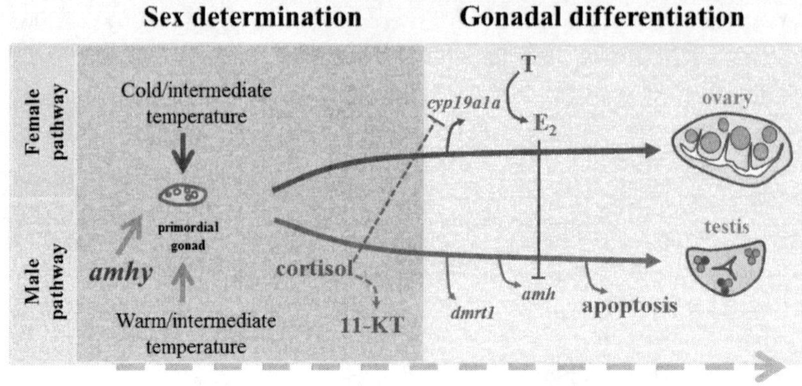

Figure 2. Proposed mechanism of the sex differentiation process in pejerrey *Odontesthes bonariensis*.

Within this context, studies were conducted to critically asses the association between high temperature and stress (cortisol levels) during gonadal differentiation and masculinization in pejerrey. These studies showed that body cortisol levels during the critical period of sex determination were positively correlated with rearing temperature (Hattori et al., 2009). Also, pejerrey larvae reared at MixPT and treated with cortisol and dexamethasone, a cortisol agonist, showed significantly male-biased sex-ratios in relation to the control. It is important to note that similar results were observed in the Japanese flounder *Paralichthys olivaceus* (Yamaguchi et al., 2010) and in medaka (Hayashi et al., 2010), evidencing that GCs also play key role(s) in gonadal fate in other species.

The exact mode of action of cortisol during pejerrey masculinization still needs to be scrutinized. It is well known that GCs are responsible for triggering the apoptotic cascade in mammals (Schmidt et al., 2004). In fact, treatment of pejerrey larvae with cortisol induced a significant increase in the incidence of apoptosis (Hattori et al., 2009), as observed in animals reared at the MPT (Strüssmann et al., 2008; Fernandino et al., 2011). Another possible pathway is the modulation of steroidogenesis. For instance, cortisol levels in pejerrey larvae were correlated with those of the main bioactive androgen in fish, 11-ketotestosterone (11-KT), and testosterone (T) and, cortisol administration caused inhibition of the expression of *cyp19a1a* (Hattori et al., 2009). Interestingly, relatively high levels of T and 11-KT were observed at MixPT even before the appearance of a sexually dimorphic (high/low) pattern of *cyp19a1a* expression. These results may have a bearing on the controversial issue of the involvement of 11-oxygenated androgens in fish gonadal sex determination/differentiation. Thus, it is currently believed that estrogens are essential for female sex differentiation whereas androgens are products or the consequence of testicular differentiation (Ijiri et al., 2008; Vizziano et al., 2008; Guiguen et al., 2010). However, high levels of 11-oxygenated androgens or the expression of enzymes involved in their synthesis have been detected during the critical period of sex determination/differentiation in some teleosts (Miura et al., 2008) and exposure to androgens during the early stages of fish development usually produces male-biased progeny (Devlin and Nagahama, 2002). The high 11-KT levels during the sex determination period of pejerrey, even before the first evidences of molecular and morphological evidences of gonadal differentiation (Ito et al., 2005; Fernandino et al., 2008a,b; Fernandino et al., 2011), suggests that testicular formation is the result of a proactive process and not merely the alternative outcome for ovarian differentiation (Figure 2). Taken together, these results are indications that GCs can act at earlier stages of gonadal differentiation and point to their strong effects on steroidogenesis at this critical period in life.

Conclusion

The fact that pejerrey sex determination shows a marked sensitiveness to temperature, such that all-female and all-male populations can be reliably produced at environmentally relevant temperatures, and the wealth of information on the morphological, endocrine, and molecular mechanisms involved in the determination/differentiation process make this species an excellent model to study the physiology of gonadal fate determination in fish. Looking ahead, the recent discovery of the *amhy* gene in the pejerrey genome (unpublished results) may provide even further avenues to explore the interactions between the environment and genotypic factors on the canalization of gonadal development. Further studies on those processes can undoubtedly contribute to the clarification of the complex gonadal fate framework in fish and possibly in other vertebrates.

References

Angelopoulou, R., Lavranos, G. and Manolakou, P. (2012). Sex determination strategies in 2012: towards a common regulatory model? *Repr. Biol. Endocrinol.*, 10, 13.

Baroiller, J. F., D'Cotta, H., Bezault, E., Wessels, S. and Hoerstgen-Schwark, G. (2009a). Tilapia sex determination: Where temperature and genetics meet. *Comp. Biochem. Physiol.*, 153A, 30-38.

Baroiller, J. F., D'Cotta, H. and Saillant, E. (2009b). Environmental effects on fish sex determination and differentiation. *Sex. Dev.*, 3, 118-135.

Barske, L. A. and Capel, B. (2008). Blurring the edges in vertebrate sex determination. *Curr. Opin. Genet. Dev.*, 18, 499-505.

Bloom, D. D., Unmack, P. J., Gosztonyi, A. E., Piller, K. R. and Lovejoy, N. R. (2012). It's a family matter: Molecular phylogenetics of Atheriniformes and the polyphyly of the surf silversides (Family: Notocheiridae). *Mol. Phylogenet. Evol.*, 62, 1025-1030.

Bull, J. J and Vogt, R. C. (1979). Temperature-dependent sex determination in turtles. *Science,* 206, 1186-1188.

Burtis, K. C. and Baker, B. S. (1989). Drosophila doublesex gene controls somatic sexual differentiation by producing alternatively spliced mRNAs encoding related sex-specific polypeptides. *Cell,* 56, 997-1010.

Bury, N. R. and Sturm, A. (2007). Evolution of the corticosteroid receptor signalling pathway in fish. *Gen. Comp. Endocrinol.*, 153, 47-56.

Devlin, R. H. and Nagahama, Y. (2002). Sex determination and sex differentiation in fish: an overview of genetic, physiological, and environmental influences. *Aquaculture,* 208, 191-364.

Dyer, B. S. (2006). Systematic revision of the South American silversides (Teleostei, Atheriniformes). *Biocell,* 30, 69-88.

Cao, M., Duan, J., Cheng, N., Zhong, X., Wang, Z., Hu, W. and Zhao, H. (2012). Sexually dimorphic and ontogenetic expression of *dmrt1*, *cyp19a1a* and *cyp19a1b* in *Gobiocypris rarus*. *Comp. Biochem. Physiol.,* 162A, 303-309.

Carreau, S., Bourguiba, S., Lambard, S., Galeraud-Denis, I., Genissel, C. and Levallet, J. (2002). Reproductive system: aromatase and estrogens. *Mol. Cell. Endocrinol.,* 193, 137-143.

Conover, D. O. and Kynard, B. E. (1981). Environmental sex determination: interaction of temperature and genotype in a fish. *Science,* 213, 577-579.

Ferguson-Smith, M. (2007). The evolution of sex chromosomes and sex determination in vertebrates and the key role of DMRT1. *Sex. Dev.,* 1, 2-11.

Fernandino, J. I., Hattori, R. S., Kimura, H., Strüssmann, C. A. and Somoza, G. M. (2008b). Expression profile and estrogenic regulation of anti-Müllerian hormone during gonadal development in pejerrey *Odontesthes bonariensis*, a teleost fish with strong temperature-dependent sex determination. *Dev. Dyn.,* 237, 3192–3199.

Fernandino, J. I., Hattori, R. S., Shinoda, T., Kimura, H., Strobl-Mazzulla, P. H., Strüssmann, C. A. and Somoza, G. M. (2008a). Dimorphic expression of *dmrt1* and *cyp19a1* (ovarian aromatase) during early gonadal development in pejerrey, *Odontesthes bonariensis*. *Sex. Dev.,* 2, 316-324.

Fernandino, J. I., Popesku, J. T., Paul-Prasanth, B., Xiong, H-L., Hattori, R. S., Strüssmann, C. A., Somoza, G. M., Matsuda, M., Nagahama, Y. and Trudeau, V. L. (2011). Analysis of sexually dimorphic expression of genes at early gonadogenesis of pejerrey *Odontesthes bonariensis* using a heterologous microarray. *Sex. Dev.,* 5, 89-101.

Gabriel, W. N., Blumberg, B., Sutton, S., Place, A. R. and Lance, V. A. (2001). Alligator aromatase cDNA sequence and its expression in embryos at male and female incubation temperatures. *J. Exp. Zool.,* 290, 439-448.

Gómez, S. E., Menni, R. C., Naya, J. G. and Ramirez, L. (2007). The physical-chemical habitat of the Buenos Aires pejerrey, *Odontesthes bonariensis* (Teleostei, Atherinopsidae), with a proposal of a water quality index. *Environ. Biol. Fish,* 78, 161-171.

Goto-Kazeto, R., Abe, Y., Masai, K., Yamaha, E., Adachi, S. and Yamauchi, K. (2006). Temperature-dependent sex differentiation in goldfish: Establishing the temperature-sensitive period and effect of constant and fluctuating water temperatures. *Aquaculture,* 254, 617-624.

Grossen, C., Neuenschwander, S. and Perrin, N. (2011). Temperature-dependent turnovers in sex-determination mechanisms: a quantitative model. *Evolution,* 65, 64-78.

Guiguen, Y., Fostier, A., Piferrer, F. and Chang, C.F. (2010). Ovarian aromatase and estrogens: A pivotal role for gonadal sex differentiation and sex change in fish. *Gen. Comp. Endocrinol.,* 165, 352-366.

Guo, Y., Cheng, H., Huang, X., Gao, S., Yu, H. and Zhou. R. (2005). Gene structure, multiple alternative splicing, and expression in gonads of zebrafish *Dmrt1*. *Biochem. Biophys. Res. Commun.,* 330, 950-957.

Hattori, R. S., Fernandino, J. I., Kishii, A., Kimura, H., Kinno, T., Oura, M., Somoza, G. M., Yokota, M., Strüssmann, C. A. and Watanabe, S. (2009). Cortisol-induced masculinization: does thermal stress affect gonadal fate in pejerrey, a teleost fish with temperature-dependent sex determination? *PLoS ONE,* 4, e6548.

Hattori, R. S., Fernandino, J. I., Strüssmann, C. A., Somoza, G. M., Yokota, M. and Watanabe, S. (2008). Characterization and expression profiles of DMRT1, AMH, SF1 and P450aro genes during gonadal sex differentiation in Patagonian pejerrey *Odontesthes hatcheri*. *Cybium,* 32, 95-96.

Hattori, R. S., Gould, R. J., Fujioka, T., Saito, T., Kurita, J., Strüssmann, C. A., Yokota, M. and Watanabe, S. (2007). Temperature-dependent sex determination in Hd-rR medaka *Oryzias latipes*: Gender sensitivity, thermal threshold, critical period, and DMRT1 expression profile. *Sex Dev.,* 1, 138-146.

Hattori, R. S., Murai, Y., Oura, M., Masuda, S., Majhi, S. K., Sakamoto, T., Fernandino, J. I., Somoza, G. M., Yokota, M. and Strüssmann, C. A. (2012). A Y-linked anti-Müllerian hormone duplication takes over a critical role in sex determination. *Proc. Natl. Acad. Sci. USA*, 109, 2955-2959.

Hattori, R. S., Oura, M., Sakamoto, T., Yokota, M., Watanabe, S. and Strüssmann, C. A. (2010). Establishment of a strain inheriting a sex-linked SNP marker in Patagonian pejerrey (*Odontesthes hatcheri*), a species with both genotypic and temperature-dependent sex determination. *Anim. Genet.,* 41, 81-84.

Hayashi, Y., Kobira, H., Yamaguchi, T., Shiraishi, E., Yazawa, T., Hirai, T., Kamei, Y. and Kitano, T. (2010). High temperature causes masculinization of genetically female medaka by elevation of cortisol. *Mol. Reprod. Dev.,* 77, 679-686.

He, C. L., Du, J. L., Wu, G. C., Lee, Y. H., Sun, L. T. and Chang, C. F. (2003). Differential Dmrt1 transcripts in gonads of the protandrous black porgy, *Acanthopagrus schlegeli. Cytogenet. Genome Res,* 101, 309-313.

Herpin, A. and Schartl, M. (2011). *Dmrt1* genes at the crossroads: A widespread and central class of sexual development factors in fish. *FEBS J.,* 278, 1010-1019.

Ijiri, S., Kaneko, H., Kobayashi, T., Wang, D. S., Sakai, F., Paul-Prasanth, B., Nakamura, M. and Nagahama, Y. (2008). Sexual dimorphic expression of genes in gonads during early differentiation of a teleost fish, the Nile tilapia *Oreochromis niloticus*. *Biol. Reprod.,* 78, 333-341.

Ito, L. S., Takahashi, C., Yamashita, M. and Strüssmann, C. A. (2008). Warm water induces apoptosis, gonadal degeneration, and germ cell loss in subadult pejerrey *Odontesthes bonariensis* (Pisces, Atheriniformes). *Physiol. Biochem. Zool.,* 81, 762-774.

Ito, L. S., Yamashita, M., Takashima F. and Strüssmann C. A. (2005). Dynamics and histological characteristics of gonadal sex differentiation in pejerrey (*Odontesthes bonariensis*) at feminizing and masculinizing temperatures. *J. Exp. Zool.,* 303A, 504-514.

Jeng, S-R., Yueh, W-S., Pen, Y-T., Gueguen, M. M., Pasquier, J., Dufour, S., Chang, C-F. and Kah, O. (2012). Expression of aromatase in radial glial cells in the brain of the Japanese eel provides insight into the evolution of the *cyp191a* gene in actinopterygians. *PLOS One,* 7, e44750.

Josso, N., Di Clemente, N. and Gouédard, L. (2001). Anti-Müllerian hormone and its receptors. *Mol. Cell. Endocrinol.,* 179, 25-32.

Josso, N., Picard, J. Y., Rey, R. and Di Clemente, N. (2006). Testicular anti-Müllerian hormone: History, genetics, regulation and clinical applications. *Pediatr. Endocr. Rev.,* 3, 347-358.

Karube, M., Fernandino, J. I., Strobl-Mazzulla, P. H., Strüssmann, C. A., Yoshizaki, G., Somoza, G. M. and Patiño, R. (2007). Characterization and expression profile of the ovarian cytochrome p-450 aromatase (*cyp19a1*) gene during the thermolabile sex determination period in pejerrey, *Odontesthes bonariensis*. *J. Exp. Zool.*, 307A, 625–636.

Kishida, M. and Callard, G. V. (2001). Distinct cytochrome P450 aromatase isoforms in zebrafish (*Danio rerio*) brain and ovary are differentially programmed and estrogen regulated during early development. *Endocrinology,* 142, 740-750.

Klüver, N., Pfennig, F., Pala, I., Storch, K., Schlieder, M., Froschauer, A., Gutzeit, H. O. and Schartl, M. (2007). Differential expression of anti-Müllerian hormone (*amh*) and anti-Müllerian hormone receptor type II (*amhrII*) in the teleost Medaka. *Dev. Dyn.,* 236, 271-281.

Kobayashi, T., Matsuda, M., Kajiura-Kobayashi, H., Suzuki, A., Saito, N., Nakamoto, M, Shibata, N. and Nagahama Y. (2004). Two DM domain genes, *DMY* and *DMRT1*, involved in testicular differentiation and development in the medaka, *Oryzias latipes*. *Dev. Dyn.,* 231, 518-526.

Koshimizu, K., Strüssmann, C. A., Okamoto, N., Fukuda, H. and Sakamoto, T. (2010). Construction of a genetic map and development of DNA markers linked to the sex-determining locus in the Patagonian pejerrey (*Odontesthes hatcheri*). *Mar. Biotechnol.,* 12, 8-13.

Kurokawa, H., Saito, D., Nakamura, S., Katoh-Fukui, Y., Ohta, K., Baba, T., Morohashi, K. I. and Tanaka, M. (2007). Germ cells are essential for sexual dimorphism in the medaka gonad. *Proc. Natl. Acad. Sci. USA*, 104, 16958-16963.

Luckenbach, J. A., Borski, R. J., Daniels, H. V. and Godwin, J. (2009). Sex determination in flatfishes: Mechanisms and environmental influences. *Semin. Cell Dev. Biol.,* 20, 256-263.

Magerhans, A. and Hörstgen-Schwark, G. (2010). Selection experiments to alter the sex ratio in rainbow trout (*Oncorhynchus mykiss*) by means of temperature treatment. *Aquaculture,* 306, 63-67.

Magerhans, A., Müller-Belecke, A. and Hörstgen-Schwark, G. (2009). Effect of rearing temperatures post hatching on sex ratios of rainbow trout (*Oncorhynchus mykiss*) populations. *Aquaculture,* 294, 25-29.

Mank, J. E. and Avise, J. C. (2009). Evolutionary diversity and turn-over of sex determination in teleost fishes. *Sex. Dev.,* 3, 60-67.

Mank, J. E., Promislow, D. E. L. and Avise, J. C. (2006). Evolution of alternative sex-determining mechanisms in teleost fishes. *Biol. J. Linn. Soc.,* 87, 83-93.

Marchand, O., Govoroun, M., D'Cotta, H., McMeel, O., Lareyre, J. J., Bernot, A., Laudet, V. and Guiguen, Y. (2000). DMRT1 expression during gonadal differentiation and spermatogenesis in the rainbow trout, *Oncorhynchus mykiss*. *BBA-Gene Struct. Expr.,* 1493, 180-187.

Masuyama, H., Yamada, M., Kamei, Y., Fujiwara-Ishikawa, T., Todo, T., Nagahama, Y. and Matsuda, M. (2012). *Dmrt1* mutation causes a male-to-female sex reversal after the sex determination by *Dmy* in the medaka. *Chromosome Res.,* 20, 163-176.

Matsuda, M., Nagahama, Y., Shinomiya, A., Sato, T., Matsuda, C., Kobayashi, T., Morrey, C. E., Shibata, N., Asakawa, S., Shimizu, N., Hori, H., Hamaguchi, S. and Sakaizumi, M. (2002). *DMY* is a Y-specific DM-domain gene required for male development in the medaka fish. *Nature,* 417, 559-563.

Matsumoto, Y. and Crews, D. (2012). Molecular mechanisms of temperature-dependent sex determination in the context of ecological developmental biology. *Mol. Cell. Endocrinol.*, 354, 103-110.

Middaugh, D. P. and Hemmer, M. J. (1987). Influence of environmental temperature on sex-ratios in the tidewater silverside *Menidia peninsulae* (Pisces: Atherinidae). Copeia, 1987, 958-964.

Miura, S., Horiguchi, R. and Nakamura, M. (2008). Immunohistochemical evidence for 11β-hydroxylase (P45011β) and androgen production in the gonad during sex differentiation and in adults in the protandrous anemonefish *Amphiprion clarkii*. *Zool. Sci.*, 25, 212-219.

Mommsen, T. P., Vijayan, M. M. and Moon, T. W. (1999). Cortisol in teleosts: dynamics, mechanisms of action, and metabolic regulation. *Fish Physiol. Biochem.*, 9, 211-268.

Morinaga, C., Saito, D., Nakamura, S., Sasaki, T., Asakawa, S., Shimizu, N., Mitani, H., Furutani-Seiki, M., Tanaka, T. and Kondoh, H. (2007). The *hotei* mutation of medaka in the anti-Müllerian hormone receptor causes the dysregulation of germ cell and sexual development. *Proc. Natl. Acad. Sci. USA*, 104, 9691–9696.

Murdock, C. and Wibbels, T. (2003). Expression of Dmrtl in a turtle with temperature-dependent sex determination. *Cytogenet. Genome Res.*, 101, 302-308.

Nakamura, S., Kobayashi, D., Aoki, Y., Yokoi, H., Ebe, Y., Wittbrodt, J. and Tanaka, M. (2006). Identification and lineage tracing of two population of somatic gonadal precursor in medaka embryos. *Dev. Biol.*, 295, 678-688.

Ospina-Álvarez, N. and Piferrer, F. (2009). Temperature-dependent sex determination in fish revisited: prevalence, a single sex ratio response pattern, and possible effects of climate change. *PLOS One*, 3, e2837.

Pieau, C. and Dorizzi, M. (2004). Oestrogens and temperature-dependent sex determination in reptiles: all is in the gonads. *J. Endocrinol.*, 181, 367-377.

Piferrer, F. and Blázquez, M. (2005). Aromatase distribution and regulation in fish. *Fish Physiol. Biochem.*, 31, 215-226.

Piferrer, F. and Guiguen, Y. (2008). Fish gonadogenesis. Part II: Molecular biology and genomics of sex differentiation. *Rev. Fisheries Sci.*, 16, 33-53.

Raghuveer, K. and Senthilkumaran, B. (2009). Identification of multiple dmrt1s in catfish: localization, dimorphic expression pattern, changes during testicular cycle and after methyltestosterone treatment. *J. Mol. Endocrinol.*, 42, 437-448.

Raymond, C. S., Murphy, M. W., O'Sullivan, M. G., Bardwell, V. J. and Zarkower, D. (2000). *Dmrt1*, a gene related to worm and fly sexual regulators, is required for mammalian testis differentiation. *Genes Dev.*, 14, 2587-2595.

Raymond, C. S., Kettlewell, J. R., Hirsch, B., Bardwell, V. J. and Zarkower, D. (1999b). Expression of Dmrtl in the genital ridge of mouse and chicken embryos suggests a role in vertebrate sexual development. *Dev. Biol.*, 215, 208-220.

Raymond, C. S., Parker, E. D., Kettlewell, J. R., Brown, L. G., Page, D. C., Kusz, K., Jaruzelska, J., Reinberg, Y., Flejter, W. L., Bardwell, V. J., Hirsch, B. and Zarkower, D. (1999a). A region of human chromosome 9p required for testis development contains two genes related to known sexual regulators. *Hum. Mol. Genet.*, 8, 989-996.

Rodríguez-Marí, A., Yan, Y. L., BreMiller, R. A., Wilson, C., Cañestro, C. and Postlethwait, J. H. (2005). Characterization and expression pattern of zebrafish anti-Müllerian hormone (*amh*) relative to *sox9a*, *sox9b*, and *cyp19a1a*, during gonad development. *Gene Expr. Patterns,* 5, 655-667.

Sakamoto, T. and McCormick, S. D. (2006). Prolactin and growth hormone in fish osmoregulation. *Gen. Comp. Endocrinol.,* 147, 24-30.

Salas-Leiton, E., Coste, O., Asensio, E., Infante, C., Cañavate, J. P. and Manchado, M. (2012). Dexamethasone modulates expression of genes involved in the innate immune system, growth and stress and increases susceptibility to bacterial disease in Senegalese sole (*Solea senegalensis* Kaup, 1858). *Fish Shellfish Immun.,* 32, 769-778.

Sato, T., Endo, T., Yamahira, K., Hamaguchi, S. and Sakaizumi, M. (2005). Induction of female-to-male sex reversal by high temperature treatment in medaka, *Oryzias latipes*. *Zool. Sci.,* 22, 985-988.

Schmidt, S., Rainer, J., Ploner, C., Presul, E., Riml, S. and Kofler, R. (2004). Glucocorticoid-induced apoptosis and glucocorticoid resistance: molecular mechanisms and clinical relevance. *Cell Death Differ.,* 11, S45-S55.

Shen, M. M. and Hodgkin, J. (1988). mab-3, a gene required for sex-specific yolk protein expression and a male-specific lineage in *C. elegans*. *Cell,* 54, 1019-1031.

Shibata, K., Takase, M. and Nakamura, M. (2002). The Dmrt1 expression in sex-reversed gonads of amphibians. *Gen. Comp. Endocrinol.,* 127, 232-241.

Siegfried, K. R. (2010). In search of determinants: gene expression during gonadal sex differentiation. *J. Fish Biol.,* 76, 1879-1902.

Sinclair, A. H., Berta, P., Palmer, M. S. Hawkins, J. R., Griffiths, B. L., Smith, M. J., Foster, J. W., Frischauf, A. M., Lovell-Badge, R. and Goodfellow, P. N. (1990). A gene from the human sex-determining region encodes a protein with homology to a conserved DNA-binding motif. *Nature,* 346, 240–244.

Sreenivasulu, K., Ganesh, S. and Raman, R. (2002). Evolutionarily conserved, *DMRT1*, encodes alternatively spliced transcripts and shows dimorphic expression during gonadal differentiation in the lizard, *Calotes versicolor*. *Gene Expr. Patterns,* 2, 51-60.

Strüssmann, C. A., Calsina Cota, J. C., Phonlor, G., Higuchi, H. and Takashima, F. (1996b). Temperature effects on sex differentiation of two South American atherinids, *Odontesthes argentinensis* and *Patagonina hatcheri*. *Environ. Biology Fish,* 47, 143-154.

Strüssmann, C. A. and Ito, L. S. (2005). Where does gonadal sex differentiation begin? Gradient of histological sex differentiation in the gonads of pejerrey, *Odontesthes bonariensis* (Pisces, Atherinidae). *J. Morphol.,* 265, 190-196.

Strüssmann, C. A., Kitahara, A. and Yamashita, M. (2008). Role of apoptosis in temperature-dependent sex determination of pejerrey *Odontesthes bonariensis*. *Cybium,* 32, 77-79.

Strüssmann, C. A., Moriyama, S., Hanke, E., Calsina Cota, J. C. and Takashima, F. (1996a). Evidence of thermolabile sex determination in pejerrey. *J. Fish Biol.,* 48, 643-651.

Strüssmann, C. A. and Patiño, R. (1999). Sex determination, environmental. In E. Knobil and J.D. Neill (Eds.), Encyclopedia of Reproduction (Vol 4, pp. 402-409). New York: Academic Press.

Strüssmann, C. A., Saito, T., Usui, M., Yamada, H. and Takashima, F. (1997). Thermal thresholds and critical period of thermolabile sex determination in two atherinid fishes, *Odontesthes bonariensis* and *Patagonina hatcheri*. *J. Exp. Zool.,* 278, 167-177.

Tanaka, M, Saito, D., Morinaga, C. and Kurokawa, H. (2008). Cross talk between germ cells and gonadal somatic cells is critical for sex differentiation of the gonads in the teleost fish, medaka (*Oryzias latipes*). *Dev. Growth Differ.,* 50, 273–278.

Uchida, D., Yamashita, M., Kitano, T. and Iguchi, T. (2002). Oocyte apoptosis during sex differentiation of juvenile zebrafish. *J. Exp. Biol.,* 205, 711-718.

Uchida, D., Yamashita, M., Kitano, T. and Iguchi, T. (2004). An aromatase inhibitor or high water temperature induce oocytes apoptosis and depletion of P450 aromatase activity in the gonads of genetic female zebrafish during sex-reversal. *Comp. Biochem. Physiol.,* 137A, 11-20.

Valenzuela, N. (2010). Multivariate expression analysis of the gene network underlying sexual development in turtle embryos with temperature-dependent and genotypic sex determination. *Sex. Dev.,* 4, 39-49.

Valenzuela, N., Adams, D. C. and Janzen, F. J. (2003). Pattern does not equal process: exactly when is sex environmentally determined? *Am. Nat.,* 161, 676–683.

Vijayan, M. M., Reddy, P. K., Leatherland, J. F. and Moon, T. W. (1994). The effects of cortisol on hepatocyte metabolism in rainbow trout: a study using the steroid analogue RU486. *Gen. Comp. Endocrinol.,* 96, 75-84.

Vizziano, D., Baron, D., Randuineau, G., Mahe, S., Cauty, C. and Guiguen, Y. (2008). Rainbow trout gonadal masculinization induced by inhibition of estrogen synthesis is more physiological than masculinization induced by androgen supplementation. *Biol. Reprod.,* 78, 939-946.

Vizziano, D., Randuineau, G., Baron, D., Cauty, C. and Guiguen, Y. (2007). Characterization of early molecular sex differentiation in rainbow trout, *Oncorhynchus mykiss*. *Dev. Dyn.,* 236, 2198-2206.

Wrobel, K. H. (2003). The genus *Acipenser* as a model for vertebrate urogenital development: The müllerian duct. *Anat. Embryol.,* 206, 255-271.

Yamaguchi, T., Yoshinaga, N., Yazawa, T., Gen, K. and Kitano, T. (2010). Cortisol is involved in temperature-dependent sex determination in the Japanese flounder. *Endocrinology,* 151, 3900-3908.

Yang, X., Zheng, J., Na, R., Li, J., Xu, G., Qu, L. and Yang, N. (2008). Degree of sex differentiation of genetic female chicken treated with different doses of an aromatase inhibitor. *Sex Dev.,* 2, 309-315.

Yano, A., Guyomard, R., Nicol, B., Jouanno, E., Quillet, E., Klopp, C., Cabau, C., Bouchez, O., Fostier, A. and Guiguen, Y. (2012). An immune-related gene evolved into the master sex-determining gene in rainbow trout, *Oncorhynchus mykiss*. *Curr. Biol.,* 22, 1423-1428.

Yoshida, K., Shimada, K. and Saito, N. (1996). Expression of P450(17α) hydroxylase and P450 aromatase genes in the chicken gonad before and after sexual differentiation. *Gen. Comp. Endocrinol.,* 102, 233-240.

In: Sexual Plasticity and Gametogenesis in Fishes
Editor: Balasubramanian Senthilkumaran

ISBN: 978-1-62618-848-8

Chapter VI

Developmental Aspects of Reproductive Hormones in Fish

Jon Hildahl
Norwegian School of Veterinary Science,
Department of Basic Sciences and Aquatic Medicine, Oslo, Norway

Abstract

Sexual maturation and reproduction are critical life history processes largely controlled by endocrine factors organized along the brain-pituitary-gonad axis. This is a highly conserved system across vertebrate phyla. Kisspeptin and gonadotropin releasing hormones are key neuroendocrine hormones produced by distinct neurons in the ventral hypothalamus, which stimulate pituitary gonadotrope cells in the anterior pituitary. These cells synthesize and secrete gonadotropic hormone, luteinizing hormone and follicle stimulating hormone, which are transported to the gonad in the circulation and stimulate steroidogenesis and gametogenesis responsible for the control of puberty and reproduction. This is a complex pathway regulated at each level by stimulatory and inhibitory factors. The main reproductive functions of this axis are well studied, but much less is known concerning the non-reproductive functions of these hormones. In particular, recent data demonstrates the involvement of Kiss, GnRH and LH in embryological and larval development. This chapter will discuss these finding and their implications for fish reproduction.

Introduction

Reproductive function in vertebrates is controlled by a pathway of neuroendocrine and endocrine factors organized along the brain-pituitary-gonad (b-p-g) axis. This is a complex endocrine system involving multiple ligands and receptors, positive and negative feedback loops by hormones and gonadal steroids as well as paracrine and autocrine signaling. Gonadotropin releasing hormone (GnRH) expressed in the brain is central to this axis, controlling multiple functions in reproduction, most importantly the stimulation of pituitary gonadotropins, luteinizing hormone (LH) and follicle stimulating hormone (FSH). Hypophysiotrophic GnRH neurons (usually GnRH1 neurons) are primarily localized to the ventral preoptic area and extend to either the portal blood system in tetrapods or into the pituitary of fish. Upon secretion, LH and FSH are transported to the gonad in the circulation where they stimulate steroidogenesis and gametogenesis. Since the discovery of GnRH in 1971 (Burgus et al., 1971; Matsuo et al., 1971), it has been widely established that this hormone, produced by cells in the basal forebrain, serves as a key molecular signal linking the brain to gonadotropin secretion involved in sexual development and reproduction. Early studies showed that hypothalamic factors stimulatepituitary gonadotropic hormones in carp (Breton et al., 1971) and subsequent identification of a specific GnRH form in fish, sGnRH/GnRH3 (Sherwood et al., 1983), laid the foundation for studies concerning the central role of GnRH in fish reproduction. Until the identification of a second gonadotropin in salmon in the late 1980s, the reproductive effects of pituitary gonadotropic hormones were considered to be from a single hormone, gonadotropin (GtH). This was later identified to be two distinct heterodimeric hormones, FSH/GtHI and LH/GtHII (Suzuki et al., 1988; Swanson et al., 1991). Today, with the power of molecular genetics and advanced physiological methods, we have witnessed an equally important shift in focus to Kisspeptin (Kiss) neurons, which represent an integrated target for inputs from sex steroids and environmental signals. This neuroendocrine system, at least in part, governs GnRH neuron activity. The critical role of Kiss in reproduction is emphasized by the fact that loss of function mutations in either *kiss1* (the gene encoding kisspeptin), or the gene encoding the Kiss receptor (*kiss1r*, originally named *gpr54*), results in failure of puberty and infertility in both humans and mice. By 1997, the Kiss1r and its ligand had been discovered with early studies focusing on its role in cancer. Subsequently in 2003, three papers described key reproductive function for Kiss1r (de Roux et al., 2003, Funes et al., 2003, Seminara et al., 2003). It has now become clear that Kiss provides an important, perhaps essential, signal to GnRH neurons in mammals.

There is a considerable variation in the control of GnRH neurons, providing a fragmented understanding of the neuroendocrine regulation of reproduction in fish. Multiple inhibitory and stimulatory factors control GnRH neurons, including dopamine (Dufour et al., 2010), gonadotropin-inhibiting hormone (Zhang et al., 2010), neuropeptide Y (Breton et al., 1991; Gaikwad et al., 2003; Senthilkumaran et al., 2001) and Kiss (Akazome et al., 2010), the latter of which has received the most attention. The Kiss system has been characterized in fish with the identification of *kiss1r* gene in multiple fish species in 2007 (Mohamed et al., 2005; Nocillado et al., 2007; Martinez-Chavez et al., 2008;). The *kiss1* gene was then identified in fish (Biran et al., 2008; Kanda et al., 2008; van et al., 2008) in 2008, and an additional ligand, *kiss2* (Felip et al., 2009; Kitahashi et al., 2009; Lee et al., 2009; Mitani et al., 2010), and a second receptor, *kiss2r* (Lee et al., 2009; Mechaly et al., 2010), genes were identified in fish a year later. Kiss has been implicated in the control of the onset of puberty and reproduction, similar to mammals, as both kiss ligands' gene expression (Biran et al., 2008; Kanda et al., 2008; Kitahashi et al., 2009; Mitani et al., 2010; Selvaraj et al., 2010; Shahjahan et al., 2010) and receptors' gene expression (Filby et al., 2008; Martinez-Chavez et al., 2008; Mechaly et al., 2009; Mechaly et al., 2010; Mohamed et al., 2007; Nocillado et al., 2007; Shahjahan et al., 2010) have been shown to be up-regulated during puberty and/or spawning. More recently studies in mammals have identified two key neuronal populations, closely associated with Kiss neurons in the basal hypothalamus, expressing neurokinin or dynorphin. All three of these neurons are co-localized in the hypothalamus and are implicated in the regulation

GnRH neuronal activity (Lehman et al., 2010). Multiple neurokinin genes have also recently been identified in fish, and they are suggested to be involved in the control of GnRH secretion and the onset of puberty based on their gene expression and neuroanatomy (Levavi-Sivan et al., 2011).

Functional Organization of The BPG Axis in Teleosts

Teleost fish are believed to have undergone an additional round of whole genome duplication (WGD), following the two vertebrate specific rounds of WGD (Meyer and Van de Peer, 2005; Kah et al., 2007). These WGD events are expected to be followed by substantial gene diversification and/or gene loss due to genetic redundancy. This means that non-teleost vertebrates have potentially four parologues of a gene, while fish have up to eight. Molecular gene evolution in fish is further complicated by partial or complete tetraploidization in some groups of fish (Bailey et al., 1978). Indeed fish are often found to have additional functional forms of endocrine genes relative to non-teleost vertebrates. This is the case for GnRH which has 1-2 forms in mammals and 2-3 forms in fish (Hildahl et al., 2011b). Similarly, fish have up to five GnRH-Rs (Ikemoto and Park, 2005; Moncaut et al., 2005) in contrast to two variants in non-teleost vertebrates. Additionally some fish possess an additional Kiss ligand and/or receptor gene (Akazome et al., 2010). In most fish species there is only one LH and FSH variant, although multiple isoforms of LH have been reported in zebrafish (So et al., 2005). The diversity of hormone ligand and receptors in fish complicates their functional understanding, such that there is often incomplete knowledge of the differential function of ligand and receptor variants.

Kiss

The function of Kiss is still poorly understood in fish. Species for which only one ligand and receptor have been identified include Atlantic cod (unpublished results), orange spotted grouper (Shi et al., 2010) and grass pufferfish (Shahjahan et al., 2010). All of these species have lost *kiss1*and *kiss-r2*, although the distinct function of the Kiss ligands and as an extension their functional conservation remains to be elucidated. Studies suggest that there is some degree of functional redundancy of the two independent Kiss systems. Kiss2 gene expression is more closely linked to spawning in multiple fish species (Kitahashi et al., 2009; Selvaraj et al., 2010; Shahjahan et al., 2010;Servili et al., 2011;), however in medaka, Kiss1 is the likely functional ligand based on ovx and estrogen treatment studies (Kanda et al., 2008; Mitani et al., 2010). The Kiss ligands appear to be differentially regulated in different fish species such that the *kiss1* but not *kiss2* gene is up-regulated in breeding and E_2 treated medaka (Kanda et al., 2008; Mitani et al., 2010). In zebrafish *kiss1*, *kiss2* and *kiss2r* mRNA expression increases in response to E_2 treatment (Servili et al., 2011), while Kiss2 but not Kiss1 treatment stimulated *lhb* and *fshb* gene expression in adult zebrafish (Kitahashi et al., 2009) and sea bass (Felip et al., 2009). However, Kiss1 and Kiss2 had a two-phase effect in hybrid bass such that Kiss2 was more effective in stimulating LH and influencing *gnrh1* and *kiss-r2* gene expression in pre-pubertal fish and Kiss1 was more effective in recrudescent fish (Zmora et al., 2012).

The Kiss ligand genes also show sex specific regulation between male and female chub mackerel (Selvaraj et al., 2010). Kiss likely has a direct action on the pituitary as Kiss receptors are expressed in the pituitary of multiple fish species (Martinez-Chavez et al., 2008; Shahjahan et al., 2010; Yang et al., 2010; Pasquier et al., 2011) and Kiss-10 treatment changes *in vitro* pituitary gene expression in primary cell culture, albeit differently between species (Yang et al., 2010; Pasquier et al., 2011). Thus, the diverse make-up and regulation of the Kiss system in fish reinforces the fact that much of the basic function of the Kiss system in fish remains to be revealed. For example, the means of action of the endogenous Kiss ligands during fish puberty and reproductive remains to be clarified, and virtually nothing is known about the upstream regulation of Kiss1 or Kiss 2 neurons in fish.

GnRH

The functions of the multiple GnRH ligands in fish have been extensively studied, although a comprehensive understanding of their diverse functions is limited.Most fish have the primarily hypophysiotrophic GnRH1 (Senthilkumaran et al., 1999) which show the greatest species diversity, the completely conserved midbrain GnRH2 and the fish-specific GnRH3. The later shows partially overlapping distribution with GnRH1 in the forebrain and it appears that these can functionally compensate for the loss of the other in species that possess only two of the hormones (Okuboand Nagahama, 2008; Hildahl et al., 2011b; Hildah et al., 2012b). The differential function of GnRH peptides is not yet fully understood. GnRH2 and GnRH3 are widely held to act as neuromodulatory factors due to the widespread distribution of neural projections and electrophysiological characteristics in the brain (Oka, 2010). In addition, GnRH2 has been implicated in the regulation of reproductive behavior (Volkoff and Peter, 1999; Temple et al., 2003; Kauffman and Rissman, 2004a; Barnett et al., 2006), food intake (Kauffman and Rissman, 2004b) and energy balance (Temple et al., 2003; Kauffman and Rissman, 2004a). GnRH3 is also believed to be involved in the control of reproductive behavior (Yamamoto et al., 1997; Ogawa et al., 2006). *In vitro* binding studies, show that all three GnRH bind to all the investigated GnRH-R. In general, GnRH2 binds with the highest affinity, but the relative affinities differ for different receptors (Illing et al., 1999; Okubo et al., 2001; Robison et al., 2001; Bogerd et al., 2002; Flanagan et al., 2007; Servili et al., 2010). This emphasizes the need to consider both ligand and receptor tissue distribution to distinguish physiologically relevant GnRH function. Further comparative studies in fish are therefore needed to elucidate the function of GnRH variants in different species. There is considerable diversity in the number and distribution of GnRH-R in fish, leaving their differential functionality largely unknown. Multiple variants are expressed in the brain and pituitary of diverse fish species (Illing et al., 1999; Bogerd et al., 2002; Ikemoto et al., 2004; Moncaut et al., 2005; Flanagan et al., 2007; Guilgur et al., 2009; Lin et al., 2010; Servili et al., 2010; Hildahl et al., 2011a), indicating their involvement in both neuroendocrine and hypophysiotriophic regulation. GnRH-R2a appears to be the most prominent variant involved in the regulation of sexual development and reproduction (Flanagan et al., 2007; Guilgur et al., 2009; Lin et al., 2010; Hildahl et al., 2011a; Hildah et al., 2012b), however, a detailed understanding of the physiology of these diverse receptor variants remains to be developed.

LH/FSH

The majority of vertebrates possess two pituitary gonadotropic hormones, which are secreted from specialized cells, gonadotropes, in the proximal pars distalis of the anterior pituitary. In fish, as opposed to mammals, LH and FSH are synthesized and secreted from distinct cells in the pituitary (Weltzien et al., 2003). LH and FSH are part of a larger family of cysteine knot-forming polypeptide glycoproteins which form non-covalently linked heterodimers between a common α, glycoprotein alpha (GPA), and a hormone-specific β subunit. Other members of the family include thyroid stimulating hormone (TSH) and the placenta specific chorionic gonadotropin.

LH and FSH have distinct receptors which are primarily localized to the gonad. In contrast to mammals, fish receptors show some cross-reactivity (Yan et al., 1992; Miwa et al., 1994; Oba et al., 1999a; Oba et al., 1999b) such that LH-R is more selective and the FSH-R is more promiscuous. The main site of action for FSH is in the granulosa cells of the ovary and Sertoli cells in the testis reflect its role in gametogenesis as well as steroidogenesis. LH on the other hand primarily targets ovarian

Theca cells and Leydig cells in the testis, reflecting its major role in fish steroidogenesis. Additionally, gonadotropin receptors are expressed in multiple tissues outside the b-p-g axis (Kumar et al., 2001; Vischer and Bogerd, 2003; Wong and Van Eenennaam, 2004; So et al., 2005; Rocha et al., 2007; Mittelholzer et al., 2009), suggesting LH and FSH could be involved in many physiological processes, although non-reproductive functions are largely unexplored.

Development of the B-P-G Axis

Proper development of the b-p-g axis is essential for the reproduction and propagation of a species. Developmental aspects of reproductive endocrinology, however, have received less attention compared to its control of puberty and spawning. Multiple hormone and receptor genes are expressed early during embryonic and larval development, suggesting a potential role for these factors during early life history. Interestingly, recent finding have identified developmental functions for multiple hormones in this axis. Neuroendocrine and endocrine hormones appear to be involved not only in establishing the b-p-g axis through autocrine and paracrine regulation, but are also implicated in embryonic and larval development.

Kiss

Both kiss ligands and receptors are expressed early in multiple fish species (Mohamed et al., 2007; Kitahashi et al., 2009; Hodne et al., 2011), providing the first indication that these systems could be important for early development. Antibodies have been established for both Kiss ligands in zebrafish (Servili et al., 2011), which were used to characterize the distribution of Kiss neurons in the adult fish brain, but corresponding analysis of the ontogeny of Kiss neuron and receptors during early development is currently lacking. Experiments in medaka reveal that both kiss ligand genes are expressed maternally and show expression peaks during key stages of early brain development (Hodne et al., 2011). This suggests that both Kiss ligands could be involved in embryonic development.

GnRH

The ontogeny of GnRH genes and neurons has been well studied using *in situ* hybridization (Parhar et al., 1998; Gonzalez-Martinez et al., 2002b; Gopinath et al., 2004;Wong et al., 2004; Kuo et al., 2005), antibodies targeted to both the GAP (Gonzalez-Martinez et al., 2002a) and decapeptide (Kuo et al., 2005; Pandolfi et al., 2002; Parhar et al., 1998; Swapna et al., 2008), as well as by lineage tracing studies linked to immunohistochemistry (Whitlock et al., 2003). This led to the initial identification of GnRH1 and GnRH3 in the olfactory placode of anterior neural plate origin and the latter with additional cranial neural crest origin (Whitlock et al., 2003). In adults, these two neuron populations have partially overlapping distribution in the terminal nerve region. GnRH2 neurons have also been found to be derived from the neural crest based on gene knockdown studies (Whitlock et al., 2005).

These cells are initially detected in the hindbrain and midbrain from which they extend projections throughout the brain (Van and Wong Eenennaam, 2004). More recently transgenic lines of zebrafish and medaka have been made linking GnRH to green fluorescent protein (GFP). Zebrafish and medaka provide intriguing comparative models as medaka possess distinct hypophysiotrophic GnRH1 neurons and terminal nerve GnRH3 neurons whereas zebrafish have only GnRH3. Tracing studies using these lines in medaka clearly identified four independent populations of GnRH1 neurons with different origins. One population originates in the olfactory region and then migrates along the terminal nerve tract to their final location in the ventralpreoptic area from which they extend processes to the anterior pituitary similar to previous reports.

In addition, three populations of GnRH1 neurons that do not extend projections into the pituitary were identified. These include one population originating in the dorsal telencephalon, which migrates to the dorsal preopticarea, another which migrates from the anterior telencephalon to the medial ventral telencephalon, and a thirdnon-migratory population is found in the ventral hypothalamus (Okubo et al., 2006).On the other hand, GnRH3 neurons migrate from the olfactory region to their final position at the terminal nerve extending to the mesencephalon, in addition to a trigeminal ganglion population (Okubo et al., 2006).

In contrast, zebrafish GnRH3 neurons originate in the olfactory placode and extend axon projections to the preoptic area and pituitary, establishing an extensive neuronal network throughout the brain including the telencephalon, hypothalamus, midbrain tegmentum hindbrain as well as into the eye (Palevitch et al., 2007;Abraham et al., 2008). The more widespread distribution of GnRH3 in zebrafish supports the notion that GnRH3 functional compensates for the loss of GnRH1. This is further supported by the finding that in Atlantic cod GnRH1 is a pseudogene and GnRH3 gene expression increases during spawning suggesting that it is the hypophysiotrophic variant in this species (Hildah et al., 2012b; Hildahl et al., 2011b).

Pituitary Development

The pituitary is a central endocrine gland producing and secreting six to eight hormones from specific cell types, driving a wide range of essential physiological processes including growth, metabolism and reproduction. The pituitary is composed of an anterior lobe, the adenohypophysis, and a posterior lobe called the neurohypophysis. These structures have different embryological origins. In tetrapods the anterior pituitaryis derived from an invagination of the oral ectoderm called Rathke's pouch. In fish, however, the adenohypophysis arises by anterior – posterior patterning of pre-placodal ectoderm, which maintains a linear sub-epithelial organization (Pogoda and Hammerschmidt, 2009). The adenohypophysis is composed mostly of different hormone producing cells, critical for the maintenance of homeostasis. The neurohypophysis, on the other hand, is derived from a ventral outgrowth of the brain originating from neuroectoderm and is comprised mainly of the nerve terminals of neuroendocrine cells.

Pituitary cell types are differentiated by the differential expression of transcription factors during early embryonic development (Pogoda and Hammerschmidt, 2009). Cell differentiation of many fish adenohypophysis cell types have been clearly defined in zebrafish by forward mutagenesis screens and gene knockdown experiments (Herzog et al., 2004; Nica et al., 2004; Nica et al., 2006; Pogoda et al., 2006). These authors identified a temporal and spatial gradient of regulatory factors, leading to the majority of anterior pituitary cell types.

Gonadotrope Development

Information concerning the early development of pituitary gonadotropes producing LH and FSH is limited relative to other pituitary cell typesdue to the lack of early indicators of *lhb* and *fshb* gene expression. Gonadotropes and thyrotropes are present in the adenohypophysis anlage of zebrafish by 32 hpf as determined by *gpa*gene expression (Nica et al., 2006), however, *fshb* is first detected later at 4 days post-fertilization (dpf) and *lhb* is not detected until around 25 dpf by *in situ* hybridization (Chen andGe, 2011).

The development of pituitary cell types has also been described in tilapia (Sakai et al., 2005), rainbow trout (Saga et al., 1993), American shad (Laiz-Carrion et al., 2003) and Ayu (Saga et al., 1999) by immunohistochemistry, which reveal the general trend toward later onset of pituitary gonadotrope expression at or following hatching. *lhb* transcripts, however, have been detected maternally and throughout embryonic development in medaka (Hildah et al., 2012a) and by 72 hpf in whole zebrafish larvae (Nica et al., 2006)using the more sensitive PCR method, so the ontogeny of *fshβ* and *lhβ* during fish larval development remains to be fully elucidated.

A transgenic line of medaka expressing GFP under the control of the *lhβ* promotor (*lhb*-Gfp) has recently been established (Hildah et al., 2012a). *In vivo* and *in situ* fluorescent imaging of GFP in these fish reveal novel regulation of the *lhβ* gene during early larval development, indicating a possible developmental function for LH.These data show that *lhβ* is first expressed in the pituitary of medaka by approximately 14 dpf, following first feeding and the complete absorption of the yolk. Developmental tracing in transgenic GnRH-Gfp medaka reveals that GnRH neurons extend ventrally into the pituitary between 10 - 20 dpf (Okubo et al., 2006), at approximately the same time as*lhβ*-Gfp is detected in the pituitary. LH gonadotropes could, therefore, require hypothalamic inputs for final gonadotrope activation similar to in sheep (Brooks et al., 1992; Szarek et al., 2008) and rat (Aubert et al., 1985) gonadotrope maturation. This is in line with the observation that gonadotropin expression starts later than other anterior pituitary hormones in multiple fish and mammal species (Asa et al., 1988; Saga et al., 1993; Japon et al., 1994; Saga et al., 1999; Laiz-Carrion et al., 2003;).

Zebrafish, however, provides partially conflicting results. GnRH3 neurons reach the pituitary in zebrafish already by 5 dpf, compared to the later migration of GnRH1 neurons in medaka. Interestingly, this corresponds to the start of *fshβ* pituitary expression but not *lhβ* pituitary expression in zebrafish (Chen and Ge, 2011). Thus the b-p-g axis is established earlier in zebrafish, although details of early gonadotrope development remain to be clarified in this species.

The expression of *gpα* in the absence of *tshβ* is an indication of early gonadotropes, but final maturation when synthesis and production of functional LH occurs may come later similar to mammals where *gpα* expression precedes *lhβ* and *fshβ* expression in mouse (Japon et al., 1994) and human gonadotropes (Pope et al., 2006). Downregulation of *lhb* in *eya1* mutant zebrafish suggests that *lhβ* is being produced in gonadotropes by 72 hpf, although since *lhβ* tissue specific expression could not be determined (Nica et al., 2006), this could represent extra-pituitary expression of LH. In addition, the presence of two *lhβ* genes in zebrafish (So et al., 2005) could lead to partially divergent function or tissue distribution, possibly explaining the proposed earlier onset of *lhβ* gene expression relative to medaka.

Developmental Function of the B-P-G Axis: Effect of Experimental Attenuation

The transparent embryos and advanced molecular techniques available in multiple fish models provide ideal systems to investigate early gene functions *in vivo*. Developmental tracing studies coupled with methods to attenuate hormone and receptor genes and neurons have begun to reveal novel functions of the b-p-g axis. These establish a basis for comparative studies into the consequence of perturbation of endocrine systems during development on later life history.

Embryogenesis

The putative developmental function of the Kiss system has been identified by a series of eloquent knockdown experiments in medaka targeting maternal and zygotic kiss ligand and receptor transcripts (Hodne et al., 2011). These studies found a two-step effect of *kiss1* knockdown, indicating early larval survival effects and later brain patterning function. Receptor expression and knockdown experiments further provided insight into different putative functions of *kiss-r1* and *kiss-r2* in fish.

Neurogenesis

Mechanisms controlling the development of GnRH neurons have been elucidated using transgenic and knockdown techniques in two powerful fish models, zebrafish and medaka. The importance of proper GnRH neuron development can be seen in humans with x-linked Kallmann syndrome which results in the absence of gonadal function due to the lack of gonadotropin secretion. This is due to a defect in the *KAL1* gene which encodes an extracellular matrix protein (Franco et al., 1991; Legouis et al., 1991). A model for this disease has been established in medaka and zebrafish, both of which, in contrast to rodents, possess two endogenous paralogues to the human gene, *kal1.1* and *kal1.2*. Knockdown experiments show that Kal 1.1 is involved in the migration of GnRH neurons such that attenuation of this protein inhibits their migration into the forebrain (Whitlock et al., 2005; Okubo et al., 2006). One factor, called nasal embryonic LHRH factor (NELF), involved in the migration of GnRH neurons in mice (Kramer and Wray, 2000) has been found to be mutated in two cases of hypogonadotropichypogonadism in humans (Miura et al., 2004; Pitteloud et al., 2007).

A paralogue of this gene has been identified in zebrafish and knockdown of this gene inhibits the migration of GnRH3 neurons similar to the phenotype of *kal1.1* knockdown (Palevitch et al., 2009). In addition, attenuation of GnRH during early larval development has established that it plays an autocrine function in neuron migration. Knockdown of the *gnrh3* gene in zebrafish disrupts normal GnRH3 neuron migration, showing defects in fiber pathfinding and localization (Abraham et al., 2008). Taken together these studies emphasize the value of fish for characterizing developmental processes and novel developmental endocrine function.

Late Phenotypic Effects

The failure of GnRH1 and GnRH3 neurons to migrate to their final position in the preoptic area and ventral hypothalamus leads to late phenotypic effects on reproductive development and spawning. This has been nicely shown in zebrafish using GnRH3 neuron ablation (Abraham et al., 2010).

Ablation of olfactory region derived GnRH3 neurons after 4 dpf lead to adult zebrafish lacking GnRH3 neurons up to 12 week after ablation. These fish did not spawn and had arrested oocyte development and reduced oocyte diameter, whereas fish with partial ablation had significantly reduced fecundity and normal oocyte development. This emphasizes the importance of hypothalamic GnRH for fish reproduction. Interestingly, the timing of GnRH neuron ablation is important such that the percent of larvae that regenerated GnRH neurons following ablation was more than 2 fold higher when ablation was performed after 2 days compared to 4 days.

This provides an interesting model for neuron plasticity and regenerative capacity. These studies are unable to distinguish GnRH hypophysiotrophic and behavioral effects since there is only a single hypothalamic ligand in zebrafish and *kal* knockdown attenuates both GnRH1 and GnRH3 neurons (Okubo et al., 2006). Comparative studies in medaka, selectively targeting the independent GnRH1 and GnRH3 neuron populations could, therefore, provide valuable insight into their distinct functions in the future.

Putative Peripheral Developmental Function of LH

There is limited data concerning the expression of *lhβ* outside the pituitary, especially during early development. In medaka, *lhβ*-Gfp is not co-localized with pituitary marker genes during early embryonic stages. Although, the reporter construct is not detected in the pituitary of medaka until after hatching, it is identified in the developing gut tube by 32 hpf (Hildah et al., 2012a). The digestive tract is derived from endoderm similar to the pituitary. Interestingly, an increase in *lhβ* transcript content and GFP initial detected corresponds to initiation of gut tube formation at stage 22. The developing gut tube starts by migration of endodermal cells to the midline. These cells then extends caudally, reaching the cloaca by stage 26 (54 – 58 hpf) (Kobayashi et al., 2006). *lhβ*-GFP development closely mirrors this early gut development, strongly suggesting that LH could be involved in gut development in this species. The significant decrease in *lhβ* expression at 120 hpf when gut tube formation is complete further suggests that LH could be important for early development of the gastrointestinal tract.

Conclusion

Sexual maturation and reproduction are controlled by complex endocrine regulation at the level of the brain, pituitary and gonad of fish. The increased diversity and species variation in terms of multiple ligands and receptors responsible for mediating these pathways likely provides increased adaptive capacity for fish. This, however, presents a significant challenge to the scientific community to elucidate the basic mechanisms of fish reproduction. The development of neuroendocrine and pituitary cells provide the foundation for later life history stages, although the role of reproductive hormones in fish development is only now starting to come to light. The dawn of molecular genetic techniques has provided powerful tools to detect early indicators of endocrine function. This is reflected in the early expression of GnRH, Kiss and pituitary gonadotropin genes. Gene expression and attenuation experiments have revealed interesting finding concerning the function of these systems in embryonic and larval development as well as neurogenesis. This should lay the foundation for future studies into the specific cellular functions of reproductive hormones in fish development and their significance for fish reproductive capacity. This is of particular significance to the challenges of environmental change on wild fish populations as well as the aquaculture industry.

References

Abraham, E., Palevitch, O., Gothilf, Y., and Zohar, Y. (2010). Targeted gonadotropin-releasing hormone-3 neuron ablation in zebrafish: effects on neurogenesis, neuronal migration, and reproduction. *Endocrinology,*151, 332-340.

Abraham, E., Palevitch, O., Ijiri, S., Du, S. J., Gothilf, Y. and Zohar, Y. (2008). Early development of forebrain gonadotrophin-releasing hormone (GnRH) neurones and the role of GnRH as an autocrine migration factor. *J. Neuroendocrinol.,* 20, 394-405.

Akazome, Y., Kanda, S., Okubo, K. and Oka, Y. (2010). Functional and evolutionary insights into vertebrate kisspeptin systems from studies of fish brain. *J. Fish Biol.,* 76, 161-182.

Asa, S. L., Kovacs, K., Horvath, E., Losinski, N. E., Laszlo, F. A., Domokos, I. and Halliday,W. C. (1988). Human fetal adenohypophysis. Electron microscopic and ultrastructural immunocytochemical analysis. *Neuroendocrinology, 48*, 423-431.

Aubert, M. L., Begeot, M., Winiger, B. P., Morel, G., Sizonenko, P. C. and Dubois, P. M. (1985). Ontogeny of hypothalamic luteinizing hormone-releasing hormone (GnRH) and pituitary GnRH receptors in fetal and neonatal rats. *Endocrinology, 116*, 1565-1576.

Bailey, G. S., Poulter, R. T.and Stockwell, P. A. (1978). Gene duplication in tetraploid fish: model for gene silencing at unlinked duplicated loci. *Proc. Natl. Acad. Sci. USA,* 75, 5575-5579.

Barnett, D. K., Bunnell, T. M., Millar, R. P. and Abbott, D. H. (2006). Gonadotropin-releasing hormone II stimulates female sexual behavior in marmoset monkeys. *Endocrinology,* 147, 615-623.

Biran, J., Ben-Dor, S. and Levavi-Sivan, B. (2008). Molecular identification and functional characterization of the kisspeptin/kisspeptin receptor system in lower vertebrates. *Biol. Reprod.,* 79, 776-786.

Bogerd, J, Diepenbroek, W. B., Hund, E., van, O. F., Teves, A. C., Leurs, R. and Blomenrohr,M. (2002). Two gonadotropin-releasing hormone receptors in the African catfish: no differences in ligand selectivity, but differences in tissue distribution. *Endocrinology,*143, 4673-4682.

Breton, B., Jalabert,B., Billard, R. and Weil,C. (1971). [In vitro stimulation of the release of pituitary gonadotropic hormone by a hypothalamic factor in the carp *Cyprinus carpio* L]. *C R Acad. Sci. Hebd. Seances Acad. Sci. D,* 273, 2591-2594.

Breton, B., Mikolajczyk,T., Popek,W., Bieniarz, K. and Epler,P. (1991). Neuropeptide Y stimulates in vivo gonadotropin secretion in teleost fish. *Gen. Comp. Endocrinol.,* 84, 277-283.

Brooks, A. N., Currie, I. S., Gibson, F. and Thomas, G. B. (1992). Neuroendocrine regulation of sheep fetuses. *J. Reprod. Fertil. Suppl.,* 45, 69-84.

Burgus, R., Butcher, M., Ling, N., Monahan, M., Rivier, J., Fellows, R., Amoss, M., Blackwell, R., Vale, W. and Guillemin, R. (1971). Molecular structure of the hypothalamic factor (LRF) of ovine origin monitoring the secretion of pituitary gonadotropic hormone of luteinization (LH). *C R Acad. Sci. Hebd. Seances. Acad. Sci. D,* 273, 1611-1613.

Chen, W. and Ge, W. (2012). Ontogenic Expression Profiles of Gonadotropins (fshβ and lhβ) and Growth Hormone (gh) During Sexual Differentiation and Puberty Onset in Female Zebrafish. *Biol. Reprod.,86, 73.*

Dufour, S., Sebert, M. E., Weltzien, F. A., Rousseau, K. and Pasqualini, C. (2010). Neuroendocrine control by dopamine of teleost reproduction. *J. Fish Biol.,* 76, 129-160.

Felip, A., Zanuy, S., Pineda, R., Pinilla, L., Carrillo, M., Tena-Sempere, M. and Gomez, A. (2009). Evidence for two distinct KiSS genes in non-placental vertebrates that encode kisspeptins with different gonadotropin-releasing activities in fish and mammals. *Mol. Cell Endocrinol.,* 312, 61-71.

Filby, A. L., van, A. R., Duitman, J. and Tyler, C. R. (2008). The kisspeptin/gonadotropin-releasing hormone pathway and molecular signaling of puberty in fish. *Biol. Reprod.,*78, 278-289.

Flanagan, C. A., Chen, C. C., Coetsee, M., Mamputha, S., Whitlock, K. E., Bredenkamp, N., Grosenick, L., Fernald, R. D. and Illing, N. (2007). Expression, structure, function, and evolution of gonadotropin-releasing hormone (GnRH) receptors GnRH-R1SHS and GnRH-R2PEY in the teleost, *Astatotilapia burtoni. Endocrinology,*148, 5060-5071.

Franco, B., Guioli, S., Pragliola, A., Incerti, B., Bardoni, B., Tonlorenzi, R., Carrozzo, R., Maestrini, E., Pieretti, M. and Taillon-Miller, P. (1991). A gene deleted in Kallmann's syndrome shares homology with neural cell adhesion and axonal path-finding molecules. *Nature,* 353, 529-536.

Gaikwad, A., Biju, K. C. and Subhedar, N. (2003). GnRH-LH secreting cells axis in the pituitary of the teleost *Clarias batrachus* responds to neuropeptide Y treatment: an immunocytochemical study. *Gen. Comp. Endocrinol.,* 131, 126-133.

Gonzalez-Martinez, D., Zmora, N., Mananos, E., Saligaut, D., Zanuy, S., Zohar, Y., Elizur, A., Kah, O. and Munoz-Cueto, J. A. (2002a). Immunohistochemical localization of three different prepro-GnRHs in the brain and pituitary of the European sea bass (*Dicentrarchus labrax*) using antibodies to the corresponding GnRH-associated peptides. *J. Comp. Neurol.,* 446, 95-113.

Gonzalez-Martinez, D., Zmora, N., Zanuy, S., Sarasquete, C., Elizur, A., Kah, O. and Munoz-Cueto, J. A. (2002b). Developmental expression of three different prepro-GnRH (gonadotrophin-releasing hormone) messengers in the brain of the European sea bass (*Dicentrarchus labrax*). *J. Chem. Neuroanat.,23*, 255-267.

Gopinath, A., Andrew, T. L. and Whitlock, K. E. (2004). Temporal and spatial expression of gonadotropin releasing hormone (GnRH) in the brain of developing zebrafish (*Danio rerio*). *Gene Expr. Patterns,4*, 65-70.

Guilgur, L. G., Strussmann, C. A. and Somoza, G. M. (2009). mRNA expression of GnRH variants and receptors in the brain, pituitary and ovaries of pejerrey (*Odontesthes bonariensis*) in relation to the reproductive status. *Fish Physiol. Biochem., 35*, 157-166.

Herzog, W., Sonntag, C., Walderich, B., Odenthal, J., Maischein, H. M. and Hammerschmidt, M. (2004). Genetic analysis of adenohypophysis formation in zebrafish. *Mol. Endocrinol., 18*, 1185-1195.

Hildah, J., Sandvik, G. K., Lifjeld, R., Hodne, K., Nagahama, Y., Haug,T. M., Okubo, K. and Weltzien, F. A. (2012a). Developmental tracing of luteinizing hormone beta-subunit gene expression using green fluorescent protein transgenic medaka (*Oryzias latipes*) reveals a putative novel developmental function. *Dev. Dyn.241*, 1665-1677.

Hildah, J., Taranger, G. L., Norberg, B., Haug, T. M. and Weltzien,F.A. (2012b). Differential regulation of GnRH ligand and receptor genes in the brain and pituitary of Atlantic cod exposed to different photoperiod. *Gen. Comp. Endocrinol., 180*, 7-14.

Hildahl, J., Sandvik, G. K., Edvardsen, R. B., Norberg, B., Haug, T. M. and Weltzien, F. A. (2011a). Four gonadotropin releasing hormone receptor genes in Atlantic cod are differentially expressed in the brain and pituitary during puberty. *Gen. Comp. Endocrinol.,*173, 333-345.

Hildahl, J., Sandvik, G. K., Edvardson, R. B., Fargernes, C., Norberg, B., Haug, T. M. and Weltzien, F. A. (2011b). Identification and gene expression analysis of three GnRH genes in Atlantic cod during puberty provides insight into transition from three to two GnRH variants in fish. *Gen. Comp. Endocrinol.,* 172, 458-467.

Hodne, K., Weltzien, F. A. and Okubo, K. (2011). A novel function of Kiss. *Indian. J. Sci. Technol.,* 4, 218.

Ikemoto, T., Enomoto, M. and Park, M. K. (2004). Identification and characterization of a reptilian GnRH receptor from the leopard gecko. *Mol. Cell Endocrinol.,* 214, 137-147.

Ikemoto, T. and Park, M. K. (2005). Identification and molecular characterization of three GnRH ligands and five GnRH receptors in the spotted green pufferfish. *Mol. Cell Endocrinol.,* 242, 67-79.

Illing, N., Troskie, B. E., Nahorniak, C. S., Hapgood, J. P., Peter, R. E. and Millar, R. P. (1999). Two gonadotropin-releasing hormone receptor subtypes with distinct ligand selectivity and differential distribution in brain and pituitary in the goldfish (*Carassius auratus*). *Proc. Natl. Acad. Sci. USA* , 96, 2526-2531.

Japon, M. A., Rubinstein, M. and Low, M. J. (1994). In situ hybridization analysis of anterior pituitary hormone gene expression during fetal mouse development. *J. Histochem. Cytochem., 42*, 1117-1125.

Kah, O., Lethimonier, C., Somoza, G., Guilgur, L. G., Vaillant, C. and Lareyre, J. J. (2007). GnRH and GnRH receptors in metazoa: a historical, comparative, and evolutive perspective. *Gen. Comp. Endocrinol.,*153, 346-364.

Kanda, S., Akazome, Y., Matsunaga, T., Yamamoto, N., Yamada, S., Tsukamura, H., Maeda, K. and Oka,Y. (2008). Identification of KiSS-1 product kisspeptin and steroid-sensitive sexually dimorphic kisspeptin neurons in medaka (*oryzias latipes*). *Endocrinology,149*, 2467-2476.

Kauffman, A. S. and Rissman, E. F. (2004a). A critical role for the evolutionarily conserved gonadotropin-releasing hormone II: mediation of energy status and female sexual behavior. *Endocrinology,*145, 3639-3646.

Kauffman, A. S. and Rissman, E. F. (2004b). The evolutionarily conserved gonadotropin-releasing hormone II modifies food intake. *Endocrinology,145*, 686-691.

Kitahashi, T., Ogawa, S. and Parhar, I. S. (2009). Cloning and expression of kiss2 in the zebrafish and medaka. *Endocrinology,150*, 821-831.

Kobayashi, D., Jindo,T., Naruse, K. and Takeda, H. (2006). Development of the endoderm and gut in medaka, *Oryzias latipes*. *Dev. Growth Differ., 48*, 283-295.

Kramer, P. R. and Wray, S. (2000). Novel gene expressed in nasal region influences outgrowth of olfactory axons and migration of luteinizing hormone-releasing hormone (LHRH) neurons. *Genes. Dev., 14*, 1824-1834.

Kumar, R. S., Ijiri, S. and Trant, J. M. (2001). Molecular biology of channel catfish gonadotropin receptors: 1. Cloning of a functional luteinizing hormone receptor and preovulatory induction of gene expression. *Biol. Reprod., 64*, 1010-1018.

Kuo, M. W., Lou, S. W., Postlethwait, J. and Chung, B. C. (2005). Chromosomal organization, evolutionary relationship, and expression of zebrafish GnRH family members. *J. Biomed. Sci., 12*, 629-639.

Laiz-Carrion, R., del, M. S.-N., del Pilar Martin del R. and Mancera, J. M. (2003). Ontogeny of adenohypophyseal cells in the pituitary of the American shad (*Alosa sapidissima*). *Gen. Comp. Endocrinol., 132*, 454-464.

Lee, Y. R., Tsunekawa, K., Moon, M. J., Um, H. N., Hwang, J. I., Osugi, T., Otaki, N., Sunakawa, Y., Kim, K., Vaudry, H. (2009). Molecular evolution of multiple forms of kisspeptins and GPR54 receptors in vertebrates. *Endocrinology,150*, 2837-2846.

Legouis, R., Hardelin, J. P., Levilliers, J., Claverie, J. M., Compain, S., Wunderle, V., Millasseau, P., Le, P. D., Cohen, D., Caterina, D. (1991). The candidate gene for the X-linked Kallmann syndrome encodes a protein related to adhesion molecules. *Cell, 67*, 423-435.

Lehman, M. N., Coolen, L. M. and Goodman, R. L. (2010). Minireview: kisspeptin /neurokinin B/dynorphin (KNDy) cells of the arcuate nucleus: a central node in the control of gonadotropin-releasing hormone secretion. *Endocrinology,*151, 3479-3489.

Levavi-Sivan, B., Biran, J., Ben-Dor, S. and Palevitch, O. (2011). Novel neuropeptides involved in the control of puberty and reproduction in fish. *Indian. J. Sci. Technol.,* 4, 12.

Lin, C. J., Wu, G. C., Lee, M. F., Lau, E. L., Dufour, S. and Chang, C. F. (2010). Regulation of two forms of gonadotropin-releasing hormone receptor gene expression in the protandrous black porgy fish, *Acanthopagrus schlegeli. Mol. Cell Endocrinol.,323*, 137-146.

Martinez-Chavez, C. C., Minghetti, M. and Migaud, H. (2008). GPR54 and rGnRH I gene expression during the onset of puberty in Nile tilapia. *Gen. Comp. Endocrinol.,156*, 224-233.

Matsuo, H., Baba, Y., Nair, R. M., Arimura, A. and Schally, A. V. (1971). Structure of the porcine LH- and FSH-releasing hormone. I. The proposed amino acid sequence. *Biochem. Biophys. Res. Commun.,43*, 1334-1339.

Mechaly, A. S., Vinas, J., Murphy, C., Reith, M. and Piferrer, F. (2010). Gene structure of the Kiss1 receptor-2 (Kiss1r-2) in the Atlantic halibut: insights into the evolution and regulation of Kiss1r genes. *Mol. Cell Endocrinol.,317*, 78-89.

Mechaly, A. S., Vinas, J. and Piferrer, F. (2009). Identification of two isoforms of the Kisspeptin-1 receptor (kiss1r) generated by alternative splicing in a modern teleost, the Senegalese sole (*Solea senegalensis*). *Biol. Reprod., 80*, 60-69.

Meyer, A. and Van de Peer, Y. (2005). From 2R to 3R: evidence for a fish-specific genome duplication (FSGD). *Bioessays,27*, 937-945.

Mitani, Y., Kanda, S., Akazome, Y., Zempo, B. and Oka, Y. (2010). Hypothalamic Kiss1 but not Kiss2 neurons are involved in estrogen feedback in medaka (*Oryzias latipes*). *Endocrinology,* 151, 1751-1759.

Mittelholzer, C., Andersson,E., Taranger,G.L., Consten,D., Hirai,T., Senthilkumaran, B., Nagahama,Y. and Norberg,B. (2009). Molecular characterization and quantification of the gonadotropin receptors FSH-R and LH-R from Atlantic cod (*Gadus morhua*). *Gen. Comp. Endocrinol.,160*, 47-58.

Miura, K., Acierno, J. S., Jr. and Seminara, S. B. (2004). Characterization of the human nasal embryonic LHRH factor gene, NELF, and a mutation screening among 65 patients with idiopathic hypogonadotropic hypogonadism (IHH). *J. Hum. Genet., 49*, 265-268.

Miwa, S., Yan, L. and Swanson, P. (1994). Localization of two gonadotropin receptors in the salmon gonad by in vitro ligand autoradiography. *Biol. Reprod.,50*, 629-642.

Mohamed, J. S., Benninghoff, A. D., Holt, G. J. and Khan, I. A. (2007). Developmental expression of the G protein-coupled receptor 54 and three GnRH mRNAs in the teleost fish cobia. *J. Mol. Endocrinol.,38*, 235-244.

Mohamed, J. S., Thomas, P. and Khan, I. A. (2005). Isolation, cloning, and expression of three prepro-GnRH mRNAs in Atlantic croaker brain and pituitary. *J. Comp. Neurol.,488*, 384-395.

Moncaut, N., Somoza, G., Power, D. M. and Canario, A. V. (2005). Five gonadotrophin-releasing hormone receptors in a teleost fish: isolation, tissue distribution and phylogenetic relationships. *J. Mol. Endocrinol.,34*, 767-779.

Nica, G., Herzog, W., Sonntag, C. and Hammerschmidt, M. (2004). Zebrafish pit1 mutants lack three pituitary cell types and develop severe dwarfism. *Mol Endocrinol,18*, 1196-1209.

Nica, G., Herzog, W., Sonntag, C., Nowak, M., Schwarz, H., Zapata, A. G., and Hammerschmidt, M. (2006). Eya1 is required for lineage-specific differentiation, but not for cell survival in the zebrafish adenohypophysis. *Dev. Biol., 292*, 189-204.

Nocillado, J. N., Levavi-Sivan, B., Carrick, F. and Elizur, A. (2007). Temporal expression of G-protein-coupled receptor 54 (GPR54), gonadotropin-releasing hormones (GnRH), and dopamine receptor D2 (drd2) in pubertal female grey mullet, *Mugil cephalus. Gen. Comp. Endocrinol.,150*, 278-287.

Oba, Y., Hirai, T., Yoshiura, Y., Yoshikuni, M., Kawauchi, H. and Nagahama, Y. (1999a). Cloning, functional characterization, and expression of a gonadotropin receptor cDNA in the ovary and testis of amago salmon (*Oncorhynchus rhodurus*). *Biochem. Biophys. Res. Commun.,* 263, 584-590.

Oba, Y., Hirai, T., Yoshiura, Y., Yoshikuni, M., Kawauchi, H. and Nagahama,Y. (1999b). The duality of fish gonadotropin receptors: cloning and functional characterization of a second gonadotropin receptor cDNA expressed in the ovary and testis of amago salmon (*Oncorhynchus rhodurus*). *Biochem. Biophys. Res. Commun., 265*, 366-371.

Ogawa, S., Akiyama, G., Kato, S., Soga, T., Sakuma, Y. and Parhar, I. S. (2006). Immunoneutralization of gonadotropin-releasing hormone type-III suppresses male reproductive behavior of cichlids. *Neurosci. Lett., 403*, 201-205.

Oka, Y. (2010). Electrophysiological characteristics of gonadotrophin-releasing hormone 1-3 neurones: insights from a study of fish brains. *J. Neuroendocrinol.,22*, 659-663.

Okubo, K. and Nagahama, Y. (2008). Structural and functional evolution of gonadotropin-releasing hormone in vertebrates. *Acta. Physiol. (Oxf),193*, 3-15.

Okubo, K., Nagata, S., Ko, R., Kataoka, H., Yoshiura, Y., Mitani, H., Kondo, M., Naruse, K., Shima, A. and Aida, K. (2001). Identification and characterization of two distinct GnRH receptor subtypes in a teleost, the medaka Oryzias latipes. *Endocrinology, 142*, 4729-4739.

Okubo, K., Sakai, F., Lau, E. L., Yoshizaki, G., Takeuchi, Y., Naruse, K., Aida, K. and Nagahama, Y. (2006). Forebrain gonadotropin-releasing hormone neuronal development: insights from transgenic medaka and the relevance to X-linked Kallmann syndrome. *Endocrinology,147*, 1076-1084.

Palevitch, O., Abraham, E., Borodovsky, N., Levkowitz, G., Zohar, Y. and Gothilf, Y. (2009). Nasal embryonic LHRH factor plays a role in the developmental migration and projection of gonadotropin-releasing hormone 3 neurons in zebrafish. *Dev. Dyn., 238*, 66-75.

Palevitch, O., Kight, K., Abraham, E., Wray, S., Zohar, Y. and Gothilf, Y. (2007). Ontogeny of the GnRH systems in zebrafish brain: in situ hybridization and promoter-reporter expression analyses in intact animals. *Cell Tissue Res., 327*, 313-322.

Pandolfi, M., Parhar, I. S., Ravaglia, M. A., Meijide, F. J., Maggese, M. C. and Paz, D. A. (2002). Ontogeny and distribution of gonadotropin-releasing hormone (GnRH) neuronal systems in the brain of the cichlid fish *Cichlasoma dimerus*. *Anat. Embryol. (Berl),205*, 271-281.

Parhar, I. S., Soga, T., Ishikawa, Y., Nagahama, Y. and Sakuma,Y. (1998). Neurons synthesizing gonadotropin-releasing hormone mRNA subtypes have multiple developmental origins in the medaka. *J. Comp. Neurol.,* 401, 217-226.

Pasquier, J., Lafont, A. G., Leprince, J., Vaudry, H., Rousseau, K., and Dufour, S. (2011). First evidence for a direct inhibitory effect of kisspeptins on LH expression in the eel, *Anguilla anguilla*. *Gen. Comp. Endocrinol.*173, 216-225.

Pitteloud, N., Quinton, R., Pearce, S., Raivio, T., Acierno, J., Dwyer, A., Plummer, L., Hughes, V., Seminara, S., Cheng, Y. Z. (2007). Digenic mutations account for variable phenotypes in idiopathic hypogonadotropic hypogonadism. *J. Clin. Invest., 117*, 457-463.

Pogoda, H. M. and Hammerschmidt, M. (2009). How to make a teleost adenohypophysis: molecular pathways of pituitary development in zebrafish. *Mol Cell Endocrinol,* 312, 2-13.

Pogoda, H. M., von der, H. S., Herzog, W., Kramer, C., Schwarz, H. and Hammerschmidt,M. (2006). The proneural gene ascl1a is required for endocrine differentiation and cell survival in the zebrafish adenohypophysis. *Development, 133*, 1079-1089.

Pope, C., McNeilly, J. R., Coutts, S., Millar, M., Anderson, R. A. and McNeilly,A.S. (2006). Gonadotrope and thyrotrope development in the human and mouse anterior pituitary gland. *Dev. Biol., 297*, 172-181.

Robison, R. R., White, R. B., Illing, N., Troskie, B. E., Morley, M., Millar, R. P. and Fernald,R.D. (2001). Gonadotropin-releasing hormone receptor in the teleost *Haplochromis burtoni*: structure, location, and function. *Endocrinology,142*, 1737-1743.

Rocha, A., Gomez, A., Zanuy, S., Cerda-Reverter, J. M., and Carrillo, M. (2007). Molecular characterization of two sea bass gonadotropin receptors: cDNA cloning, expression analysis, and functional activity. *Mol. Cell Endocrinol., 272*, 63-76.

Saga, T., Oota,Y., Nozaki, M. and Swanson,P. (1993). Salmonid pituitary gonadotrophs. III. Chronological appearance of GTH I and other adenohypophysial hormones in the pituitary of the developing rainbow trout (*Oncorhynchus mykiss irideus*). *Gen. Comp. Endocrinol., 92*, 233-241.

Saga, T., Yamaki, K., Doi, Y. and Yoshizuka, M. (1999). Chronological study of the appearance of adenohypophysial cells in the ayu (*Plecoglossus altivelis*). *Anat. Embryol. (Berl),200*, 469-475.

Sakai, F., Swapna, I., Sudhakumari, C. C., Ganesh, M. V., Kagawa, H., Kobayashi, T., Fan, H., Nagahama, Y. and Senthilkumaran, B. (2005). Immunocytochemical localization of gonadotropins during the development of XX and XY Nile tilapia. *Fish Physiol. Biochem.* 31, 177-181.

Selvaraj, S., Kitano, H., Fujinaga, Y., Ohga, H., Yoneda, M., Yamaguchi, A., Shimizu, A. and Matsuyama, M. (2010). Molecular characterization, tissue distribution, and mRNA expression profiles of two Kiss genes in the adult male and female chub mackerel (*Scomber japonicus*) during different gonadal stages. *Gen. Comp. Endocrinol.,* 169, 28-38.

Senthilkumaran, B., Okuzawa, K., Gen, K. and Kagawa, H. (2001). Effects of serotonin, GABA and neuropeptide Y on seabream gonadotropin releasing hormone release in vitro from preoptic-anterior hypothalamus and pituitary of red seabream, *Pagrus major. J. Neuroendocrinol.,13*, 395-400.

Senthilkumaran, B., Okuzawa, K., Gen, K., Ookura, T. and Kagawa, H. (1999). Distribution and seasonal variations in levels of three native GnRHs in the brain and pituitary of perciform fish. *J. Neuroendocrinol.,* 11, 181-186.

Servili, A., Le, P. Y., Leprince, J., Caraty, A., Escobar, S., Parhar, I. S., Seong, J. Y., Vaudry, H. and Kah, O. (2011). Organization of two independent kisspeptin systems derived from evolutionary-ancient kiss genes in the brain of zebrafish. *Endocrinology, 152*, 1527-1540.

Servili, A., Lethimonier, C., Lareyre, J. J., Lopez-Olmeda, J. F., Sanchez-Vazquez, F. J., Kah, O. and Munoz-Cueto, J. A. (2010). The Highly conserved gonadotropin-releasing hormone-2 form acts as a melatonin-releasing factor in the pineal of a teleost fish, the european sea bass *Dicentrarchus labrax. Endocrinology,151*, 2265-2275.

Shahjahan, M., Motohashi, E., Doi, H. and Ando, H. (2010). Elevation of Kiss2 and its receptor gene expression in the brain and pituitary of grass puffer during the spawning season. *Gen. Comp. Endocrinol.,169*, 48-57.

Sherwood, N., Eiden, L., Brownstein, M., Spiess, J., Rivier, J. and Vale, W. (1983). Characterization of a teleost gonadotropin-releasing hormone. *Proc. Natl. Acad. Sci. USA,* 80, 2794-2798.

Shi, Y., Zhang, Y., Li, S., Liu, Q., Lu, D., Liu, M., Meng, Z., Cheng, C. H., Liu, X. and Lin, H. (2010). Molecular identification of the Kiss2/Kiss1ra system and its potential function during 17alpha-methyltestosterone-induced sex reversal in the orange-spotted grouper, *Epinephelus coioides. Biol. Reprod., 83*, 63-74.

So, W. K., Kwok, H. F. and Ge, W. (2005). Zebrafish gonadotropins and their receptors: II. Cloning and characterization of zebrafish follicle-stimulating hormone and luteinizing hormone subunits--their spatial-temporal expression patterns and receptor specificity. *Biol. Reprod.,* 72, 1382-1396.

Suzuki, K., Kanamori, A., Nagahama, Y. and Kawauchi, H. (1988). Development of salmon GTH I and GTH II radioimmunoassays. *Gen. Comp. Endocrinol.,* 71, 459-467.

Swanson, P., Suzuki, K., Kawauchi, H., and Dickhoff, W. W. (1991). Isolation and characterization of two coho salmon gonadotropins, GTH I and GTH II. *Biol. Reprod.* 44, 29-38.

Swapna, I., Sudhakumari, C. C., Sakai, F., Sreenivasulu, G., Kobayashi, T., Kagawa, H., Nagahama, Y. and Senthilkumaran, B. (2008). Seabream GnRH immunoreactivity in brain and pituitary of XX and XY Nile tilapia, *Oreochromis niloticus* during early development. *J. Exp. Zool A Ecol. Genet. Physiol., 309*, 419-426.

Szarek, E., Farrand, K., McMillen, I. C., Young, I. R., Houghton, D. and Schwartz, J. (2008). Hypothalamic input is required for development of normal numbers of thyrotrophs and gonadotrophs, but not other anterior pituitary cells in late gestation sheep. *J. Physiol., 586*, 1185-1194.

Temple, J. L., Millar, R. P. and Rissman, E. F. (2003). An evolutionarily conserved form of gonadotropin-releasing hormone coordinates energy and reproductive behavior. *Endocrinology,144*, 13-19.

van, A. R., Kille, P., Lange, A. and Tyler, C. R. (2008). Evidence for the existence of a functional Kiss1/Kiss1 receptor pathway in fish. *Peptides,* 29, 57-64.

Vischer, H. F. and Bogerd, J. (2003). Cloning and functional characterization of a gonadal luteinizing hormone receptor complementary DNA from the African catfish (*Clarias gariepinus*). *Biol. Reprod.,68*, 262-271.

Volkoff, H. and Peter, R. E. (1999). Actions of two forms of gonadotropin releasing hormone and a GnRH antagonist on spawning behavior of the goldfish *Carassius auratus*. *Gen. Comp. Endocrinol.,116*, 347-355.

Weltzien, F. A., Norberg, B., Helvik, J. V., Andersen, O., Swanson, P. and Andersson, E. (2003). Identification and localization of eight distinct hormone-producing cell types in the pituitary of male Atlantic halibut (*Hippoglossus hippoglossus* L.). *Comp. Biochem. Physiol. A Mol. Integr. Physiol.,134*, 315-327.

Whitlock, K. E., Smith, K. M., Kim, H. and Harden, M. V. (2005). A role for foxd3 and sox10 in the differentiation of gonadotropin-releasing hormone (GnRH) cells in the zebrafish *Danio rerio*. *Development,132*, 5491-5502.

Whitlock, K. E., Wolf, C. D. and Boyce, M. L. (2003). Gonadotropin-releasing hormone (GnRH) cells arise from cranial neural crest and adenohypophyseal regions of the neural plate in the zebrafish, *Danio rerio*. *Dev. Biol., 257*, 140-152.

Wong, A. C. and Van Eenennaam, A. L. (2004). Gonadotropin hormone and receptor sequences from model teleost species. *Zebrafish,1*, 203-221.

Wong, T. T., Gothilf, Y., Zmora, N., Kight, K. E., Meiri, I., Elizur, A. and Zohar,Y. (2004). Developmental expression of three forms of gonadotropin-releasing hormone and ontogeny of the hypothalamic-pituitary-gonadal axis in gilthead seabream (*Sparus aurata*). *Biol. Reprod., 71*, 1026-1035.

Yamamoto, N., Oka, Y. and Kawashima, S. (1997). Lesions of gonadotropin-releasing hormone-immunoreactive terminal nerve cells: effects on the reproductive behavior of male dwarf gouramis. *Neuroendocrinology,* 65, 403-412.

Yan, L., Swanson, P. and Dickhoff, W. W. (1992). A two-receptor model for salmon gonadotropins (GTH I and GTH II). *Biol. Reprod., 47*, 418-427.

Yang, B., Jiang, Q., Chan,T., Ko, W. K. and Wong, A. O. (2010). Goldfish kisspeptin: molecular cloning, tissue distribution of transcript expression, and stimulatory effects on prolactin, growth hormone and luteinizing hormone secretion and gene expression via direct actions at the pituitary level. *Gen. Comp. Endocrinol.,*165, 60-71.

Zhang, Y., Li, S., Liu, Y., Lu, D., Chen, H., Huang, X., Liu, X., Meng, Z., Lin, H. and Cheng, C. H. (2010). Structural diversity of the GnIH/GnIH receptor system in teleost: its involvement in early development and the negative control of LH release. *Peptides,31*, 1034-1043.

Zmora, N., Stubblefield, J., Zulperi, Z., Biran, J., Levavi-Sivan, B., Munoz-Cueto, J. A. and Zohar, Y. (2012). Differential and Gonad Stage-Dependent Roles of Kisspeptin1 and Kisspeptin2 in Reproduction in the Modern Teleosts, Morone Species. *Biol. Reprod., 86*, 177.

In: Sexual Plasticity and Gametogenesis in Fishes
Editor: Balasubramanian Senthilkumaran
ISBN: 978-1-62618-848-8

Chapter VII

Dimorphisms in the Brain and Involvement of Various Neurosubstances in Fish Reproduction: An Overview

Arun G. Jadhao
Department of Zoology, RTM Nagpur University Campus,
Nagpur – (M.S.), India

Abstract

Structural variations among the brains of different species are commonly seen. Most prominently anatomical changes are seen in the neuronal populations of the preoptic area (POA) and nucleus lateralis tuberis (NLT) of the hypothalamus of the brain. The POA and the hypothalamus regulate both endocrine and behavioural components of reproduction i.e., the hypophyseal secretion of gonadotrophins. Role of the preoptic nucleus (NPO) in neurosecretion and hormonal regulations is well known. Many studies in mammals and other vertebrates have focused on the problem of sexual dimorphism in brain morphology. Within the POA, sex differences have been identified in neuronal connectivity, cell nucleus size, volume of groups of cells and differences in the distribution of various peptides/neurotransmitters. Inter-sexual dimorphic changes have been reported in various groups of vertebrate animals, but there are no studies which show the anatomical alterations in the POA within the same gender (intra-sexual) of the species, except in fish. Hence, to understand the functional correlations of intra-sexual dimorphism further studies are required in other vertebrate animals.

In last few decades several neuropeptides/ neurotransmitters, proteins and free radical NO have been identified in the hypothalamic nuclei and particularly in the NPO which regulates the hypothalamo-pituitary-gonadal axis in teleosts. Earlier the concept was that only the main neuropeptide gonadotropin-releasing hormone (GnRH) is governing the hypophyseal secretion, but with new findings of several other peptides and other factors, it seems that the hormonal regulation of pituitary is under the control of several neurohormones, peptides and other chemicals. Hence, the fish reproduction is governed by multiple factors together and is more complex.

Abbreviations

AC: anterior commissure
HB: habenula
HC: horizontal commissure
LFB: lateral forebrain bundle
NH: neurohypophysis
NLT: nucleus lateralis tuberis
NLTl: nucleus lateralis tuberis, pars lateralis
NLTp: nucleus lateralis tuberis, pars posterioris
NLTv: nucleus lateralis tuberis, pars ventralis
NPO: preoptic nucleus
NPOp: parvocellular division of the preoptic nucleus
NPOm: magnocellular division of the preoptic nucleus
ON: optic nerve
OT: optic tract
OTec: optic tectum
PC: posterior commissure
PI: pars intermedia
PPD: proximal pars distalis
PR: preoptic recess
RL: recessus lateralis
RPD: rostral pars distalis
TC: toral commissure
TecV: tectal ventricle
TEL: telencephalon
TLo: torus longitudinalis
V: third ventricle
VC: valvula cerebelli

Introduction

Variations in the brain have always aroused interest and the comparative anatomy has revealed many specializations of the central nervous system (CNS). Structural variations among the brains of different species are commonly seen (Ariëns Kappers et al., 1936). Sexual differentiation of the vertebrate nervous system is well studied (Morris et al., 2004). In many cases, these differences can be correlated with behavioral adaptations of the particular species. In the CNS regions, during the past few decades, many studies have demonstrated sexual dimorphism of brain structures in a rapidly increasing number of vertebrates. Some of the morphological differences between males and females appear to be related to the control of gender-specific behavior. However, in addition to the sexual differences in the brain areas, some studies have also reported the intra-sexual dimorphism in the CNS. The functional correlation of such dimorphism is unknown. The hypothalamus contributes greatly to reproductive function and behaviour in vertebrates as part of the hypothalamic-gonadal-pituitary axis. In higher vertebrate classes, the hypothalamus has been found to be the most sexually dimorphic region of the brain. However, to date, not much work has been focused on gender differences in hypothalamic neuronal structure in fishes. Sexual dimorphism of the CNS has been well described in all vertebrate classes and the mammalian brain is the most extensively studied. During development, steroid hormones from the differentiated gonads organize the brain into a masculine or feminine form. Sex differentiation is manifested partly through morphological sex differences in the ultrastructure of organelles, dendritic organization and the volume of distinct cell groups within the brain (Arnold and Gorski, 1984; De Vries, 2004; MacLusky and Naftolin, 1981). A variety of neuroanatomical, morphometric sex differences have been described which include differences in the size of entire brain regions, volume of distinct sub-nuclei or projections, sex differences in soma size, dendritic length, branch number, spine synapses and total dendrite surface area.

Many studies have shown sexual dimorphism in the preoptic area (POA) of mammals including humans, birds, amphibians and fishes. The morphometric sex differences of the largest magnitude mostly occur in the hypothalamic nuclei of mammals (Ayoub et al., 1983; Gorski et al., 1980) and bird (Balthazart and Adkins-Regan, 2002) or in the song control nuclei of birds (Schlinger and Brenowitz, 2002; Wade and Arnold, 2004). Sexual dimorphism of avian species are found in the nuclei related to song control and are significantly larger in the male than in the female, with an increase in number and size of neurons (Arnold and Gorski, 1984). Gender specific differences in the volume of preoptic nuclei of amphibians and reptiles have also been reported. In toads (*Bufo japonicus*), the regions of the brain involved in mate calling, the anterior portion of the preoptic nucleus and the amygdala pars medialis, are significantly larger in males than in females (Takami and Urano, 1984). In contrast to these studies on sexual dimorphism, very little is known on the morphometric variation in the brain showing intra-sexual dimorphism in the fishes. In the last few decades, several peptides, lectins and certain proteins have been identified in several hypothalamic nuclei of vertebrate and invertebrate CNS and many of them are involved in the neuroendocrine control of reproductive processes. The distribution of neuropeptides/ neurotransmitters like gonadotropin-releasing hormone (GnRH), molluscan cardioexcitatory peptide (FMRF-amide), neuropeptide-Y, neuronal nitric oxide synthase (nNOS), galanin, some calcium binding proteins and many more playing an important role in hormonal regulation via neuroendocrine mechanisms. These are located in the hypothalamic and extra-hypothalamic areas of the brain and also in the pituitary have been well demonstrated in many vertebrate species (D´Aniello et al., 1999; Jadhao, 2001; Jadhao and Malz, 2007; Jadhao and Pinelli, 2001; Jadhao et al., 1999; Meyer et al., 1997; Palande et al., 2011). Few workers have also reported the sexually dimorphic distributions of some of these neuropeptides in the teleosts brain and their involvement in the hormonal control.

Dimorphisms in the Fish Brain

Like that of other vetebrate animals, in fish the neuroanatomical variations are noted in the brain, which are of both inter-sexual and intra-sexual types. The differences in the brain structure of male versus female are commonly referred as sexual differences (i.e., inter-sexual dimorphism). However, the changes in the brain within the same gender i.e., male versus male or female versus female are known as intra-sexual differences (i.e., intra-sexual dimorphism). In both inter-sexual as well as intra-sexual dimorphisms, the nucleus preopticus (NPO) is a prime target for studies on the variability of CNS structures (Charlton, 1932). It contains neurosecretory cells (Scharrer, 1941) and is involved in the control of hormonal regulation by the pituitary, and it influences gonadal maturation in all vertebrates, including teleosts (Peter and Fryer, 1983). The NPO is known to innervate directly in to the pituitary gland (Prasada Rao et al., 1993). The inter-sexual and intra-sexual differences in the POA of the hypothalamus have been mainly seen i) in the gross volume of defined cell groups of the POA, ii) in the shape of hypothalamic nuclei, iii) in the dendritic organization, iv) ultrastructural differences in synaptic organization in the preoptic region and v) in the distribution of various peptidergic or neurotransmitter systems. In cichlid, *Haplochromis burtoni*, the dominant male shows significantly larger gonadotropin releasing hormone (GnRH) positive perikarya than that of subdominant males (Davis and Fernald, 1990). In another species, grass goby (*Zosterisessor ophiocephalus*), GnRH intra-sexual dimorphism in GnRH expression is related to the occurrence of alternative male mating tactics has been reported (Scaggiante et al., 2006).

Inter-Sexual Dimorphism

In vertebrate animals, sexual dimorphism is maintained by the counteracting pressures of natural selection and sexual selection. Inter-sexual variations in the preoptic nucleus are commonly reported in different groups of fish. During last few decades, studies have shown such morphometric sex differences in the brain of many fishes (Cornbrooks and Parsons, 1991; Davis and Fernald, 1990; Elofsson et al., 1997; Foran and Bass, 1998; Grober et al., 1991; Grober et al., 1994; Jadhao et al., 2001; Jadhao and Meyer, 2000; Kotrschal et al., 2012; Lauer et al., 2006; Prasada Rao et al., 1996; Raghuveer et al., 2011; Sudhakumari et al., 2010; Vizzianao-Cantonnet et al., 2011). Although, in fish brain, morphometric sex differences have been reported, but sometimes it is complicated to interpret because of multiple reproductive morphs of one sex or the ability to change sexes over the lifespan (Grober and Bass, 2002). However, a variety of sex differences in fish brains have been reported, including differences in the GnRH/LHRH neurons of goldfish (Parhar et al., 2001), GABA and glutamic acid decarboxylase (Bosma et al., 2001). Other differences between the male and female fish brain include the distribution of galanin neurons (Cornbrooks and Parsons, 1991; Jadhao and Meyer, 2000; Prasada Rao et al., 1996), preprotachykinin (Peyon et al., 2000) and aromatase (Melo and Ramsdell, 2001). There have also been reports of sex differences in the position of the preoptic area and other brain regions of the red salmon (Jadhao et al., 2001). Moreover, in the killifish brain, analysis of dendrite characteristics demonstrated that sex differences in hypothalamic neuronal morphology exist in teleosts (Lauer et al., 2006). Recent studies have reported the gender differences in tryptophan hydroxylase-2 mRNA, serotonin, and 5-hydroxytryptophan levels in the brain of catfish, *Clarias gariepinus*, during sex differentiation (Raghuveer et al., 2011; Sudhakumari et al., 2010).

Although, several studies have been carried out on the presence of GnRH isoforms in the teleosts brain by using immunohistochemistry (more details are given under the section neurohormonal control of reproduction), but not much has been reported on its sexually dimorphic localization in the hypothalamic nuclei. However, few workers have reported sexual-dimorphic distribution of the GnRH in the preoptic region of the fish brain. An interesting finding has been made in the cichlid, which shows that the neuronal size of the GnRH labeled soma was larger in sexually mature males than females. On the other hand in goldfish (Parhar et al., 2001) the numbers of GnRH positive cells were noted fewer in males than females and similar results have been reported for black goby (Scaggiante et al., 2006). Another important neuropeptide, galanin (GAL) has been well reported in mammals and also teleosts, which is well distributed in the preoptic region of the brain and involved in the hormonal regulation of the pituitary. Many studies have demonstrated GAL-immunoreactive neuronal organization in the brain of the teleost (Anglade et al., 1994; Batten et al., 1990; Cornbrooks and Parsons, 1991; Jadhao and Pinelli, 2001; Olivereau and Olivereau, 1991a; Olivereau and Olivereau, 1991b; Prasada Rao et al., 1996; Yamamoto et al., 1992). Although, localization of GAL in the brain has been described in the variety of teleosts species, sexually dimorphic distribution of GAL-like immunoreactivity been reported only in three species that is in molly (Cornbrooks and Parsons, 1991), goldfish (Prasada Rao et al., 1996) and red salmon (Jadhao and Meyer, 2000). It has been reported that in molly (Cornbrooks and Parsons, 1991) and goldfish (Prasada Rao et al., 1996) absence of GAL positive cells in the NPO of female, but present in male. On the contrary, it has been found that GAL is absent in male and present in female NPO of salmon (Jadhao and Meyer, 2000).

In mammals and other vertebrates, it is now well established that the adult brains can synthesize steroids from cholesterol (Do Rego et al., 2009). There are evidences indicating that neurosteroids might modulate neurogenesis in the developing or adult CNS. This can notably lead to permanent sexual differentiation of certain structures involved in sexual behavior and the neuroendocrine control of reproduction (Morris et al., 2004). However, according to the hypothesis of aromatization (MacLusky and Naftolin, 1981) many of the androgen effects on the differentiation of male-specific brain circuits and functions are mediated by their conversion to estrogens in localized areas of the CNS (Forlano et al., 2001; Lauber et al., 1997). Thus, the enzyme cytochrome p450 (cyp19) aromatase, which converts androgens into estrogens, is considered as a key element for the brain sexual differentiation of vertebrates (MacLusky and Naftolin, 1981; Vizziano-Cantonnet et al., 2011). In some species, the sex differences in aromatase activity in some brain regions have been reported (Gonzalez and Piferrer, 2003). Recently, potential sexual differences in the expression and activity of the brain aromatase, estrogen receptors (Nagler et al., 2007) in the brain of developing fish were made by using genetic monosex male and female rainbow trout populations (Guiguen et al., 1999). Sex-specific behavior may originate from differences in brain structure or function.

Intra-Sexual Dimorphism

In fish NPO, three types of magnocellular neuroendocrine cells have been distinguished (Hayward, 1974). Since the diameter of these cells can vary with the annual season (Gregory and Tweedle, 1983), a study of intra-sexual dimorphism requires the selection of an animal model in which all specimens are in identical reproductive stages.

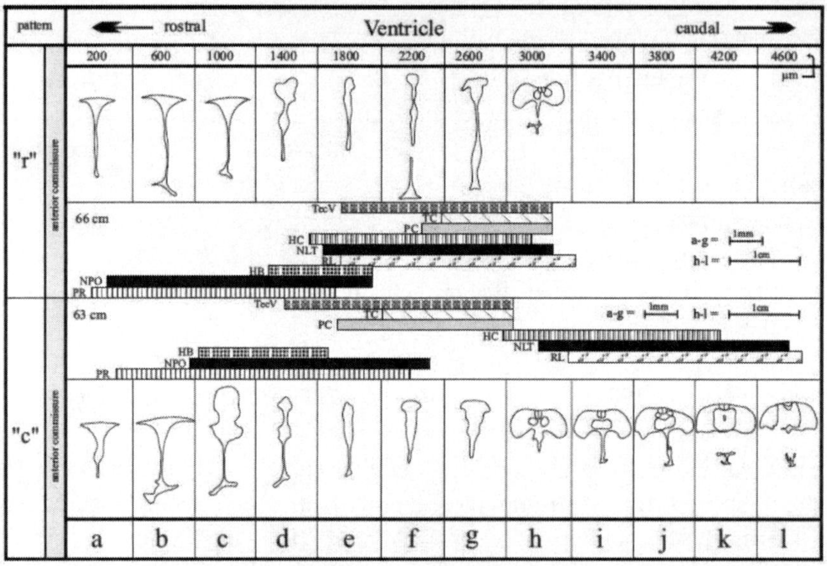

Figure 1. Scheme of the rostrocaudal extension of NPO, NLT, HB, TecV, PR, RL, HC, PC, and TC in an "r" (rostral)- and in a "c" (caudal)-pattern red salmon (for both sexes). The sketch is based on data from "r"-pattern (66cm) and from "c"-pattern (63cm) fish. For both patterns, the variability of the shape of the ventricular wall is illustrated at several rostrocaudal levels. The outline of the TecV wall is shown from its first contacting level with V to the end of RL, although, TecV first appears further rostral in both "r"- and "c"-morphs (modified from Jadhao et al., 2001).

Adult Pacific salmon are a suitable model for investigations of the (structural) variability of the preoptic region since i) many individuals simultaneously arrive at the spawning sites, ii) they spawn only once in their lives, and iii) the sexual maturation of the fish is in synchrony. Hence, comparison of brain structure between specimens is possible without interference from ontogenetic or seasonal factors. A novel finding has been made by showing two anatomically different brain patterns in the brain (Jadhao et al., 2001) same gender of red salmon (which breeds once in life time). This has shown for the first time of any vertebrate, the variations in the number of different structures in the forebrain varies within males and within females (intra-sexual dimorphic changes). This intra-sexual dimorphism are seen with the position of the preoptic nucleus, the nucleus lateralis tuberis, the habenula, the third ventricle, the tectal ventricles, the preoptic recess, the recessus lateralis, the horizontal commissure, the posterior commissure, and the toral commissure (Figure 1).

Further, it has been noted that these intra-sexual difference also show significant changes amongst the different sexes (inter-sexual dimorphism) in the teleost fish (Jadhao et al., 2001). In some cases, intra-sexual dimorphisms are related to different strategies of reproductive behavior (for example two mating strategies). This relationship is well demonstrated in fishes (Henson and Warner, 1997; Moore, 1991) and dependent on such behavior, two alternative reproductive types have been found in many teleosts (Bass, 1992; Bass and Baker, 1990). Besides the presence of alternative reproductive phenotypes, some fishes are known to change their sex (Aldenhoven, 1986). Of the many fish species that are capable of socially mediated sex reversal, the marine goby (*Trimma okinawae*) is known to change sex more than once and in either direction (Sunobe and Nakazono, 1993), whereas the bluehead wrasse (Caribbean reef fish) exhibits only unidirectional sex change, either induced naturally or experimentally (Grober et al., 1991). Although alternative sexual morphs are present in a number of species, the neuroendocrine mechanism of such sexual variability is not clearly understood.

Neuropeptides/ Neurotransmitters and Other Factors in Fish Reproduction

Of all the vertebrate classes, teleosts are the most numerous and large number of species exists in nature. They occupy a variety of aquatic environments and show an incredible diversity in reproductive strategies. Many of them go for seasonal reproduction to ensure that breeding coincides with optimal water temperatures and food supplies for developing offspring. The annual cycle in gonadal growth and function is controlled by circulating concentrations of the gonadotrophic hormones (GTH) released from the pituitary. On several occasions, the morphology and the localizations of peptides and hormones of the teleost hypothalamo-pituitary axis has been reviewed (Gorbman, 1995; Peter et al., 1990; Trudeau, 1997; Zohar et al., 2010). The teleost pituitary consists of glandular part, adenohypophysis and non-glandular portion, neurohypophysis. The adenohypophysis has three distinct lobes, i) the rostral pars distalis, ii) the proximal pars distalis and iii) the neurointermediate lobe, which is composed of the pars intermedia. There are two major GTH molecules secreted from different cells in the proximal pars distalis in fish, GTH-I and GTH-II, which are structurally similar to tetrapode FSH and LH, respectively (Zohar et al., 2010).

The neuroendocrinology of reproduction in vertebrate animals has been well studied, particularly it is well established in mammals compared to other groups of animals and it is still young for teleosts. Although, it is well known that the secretion and regulation of pituitary hormones are under the hypothalamic control of brain, the physiological significance of this multifactorial control of the gonadotropic activity is poorly understood. In fish, many studies have been carried out from last few decades and several workers have shown different neuropeptides, neurotransmitters and other factors in the brain particularly in the NPO and other hypothalamic nuclei such as tyrosine-hydroxylase, gamma-aminobutyric acid, neurophysin, serotonin, arginine vasotocin, enkephalin, angiotensin II, somatostatin, corticotropin-releasing hormone, gonadotropin-releasing hormone, gonadotropin-inhibiting hormone, substance-P, FMRF-amide, neuropeptide Y, galanin, cocaine- and amphetamine-regulated transcript, kisspeptins, aromatase, nitric oxide synthase and calcium binding proteins that stimulate gonadotropin release but only some are playing a very crucial role in controlling gonadotropic activity. Some of these factors having a clearly identified functional significance are discussed below.

Gonadotropin Releasing Hormone (GnRH)

During last few decades, a total number of 24 GnRH variants found in protochordates, vertebrates and invertebrates (Kah et al., 2007; Okubo and Nagahama, 2008; Zohar et al., 2010). The phylogenetic analysis shows the existence of three main GnRH forms. One mainly expressed in the hypothalamus, named as GnRH-I, second present in the synencephalon/

mesencephalon, referred as GnRH-II and third mainly seen in the rostral forebrain, named as GnRH-III. It is interesting to note that GnRH-I and II contain sequences from fish through terrestrial vertebrates, indicating that these forms are ancient and emerged before the divergence of these groups, but GnRH III includes only salmon GnRH (fish) sequences. Among vertebrates, teleost fish represent the group with the highest number of different GnRH isoforms. After the identification of salmon GnRH (Sherwood et al., 1983), seven other GnRH forms have been purified and sequenced in teleosts (Guilgur et al., 2006). GnRH variants were named after the species in which they were first found, despite the fact that they can be present in other species. A considerable amount of work has been done for the identification and localization of GnRH-expressing neurons in the brain of fish using immunohistochemistry, in situ hybridization and transgenesis (Abraham et al., 2008; Guilgur et al., 2006; Kah et al., 2007; Okubo and Nagahama, 2008). Most studies performed earlier suggested the existence of two segregated GnRH systems in fish brain: an anterior system extending from the olfactory bulbs to the pituitary through the ventral telencephalon, preoptic area and ventromedial hypothalamus, and a posterior system restricted to the synencephalon/midbrain tegmentum. The anterior system expressed one variable GnRH form, whereas the posterior system consistently expressed GnRH-II (Kah et al., 1986; Lepretre et al., 1993; Montero et al., 1995; Okuzawa et al., 1990; Okuzawa et al., 2003). The existence of three GnRH forms was confirmed, either by cDNA sequencing or biochemical characterization, in other perciform species (Senthilkumaran et al., 1999; Zmora et al., 2002) and now there is evidence on the existence of three different GnRH forms in most teleost orders (Adams et al., 2002; Amano et al., 2002). The distribution of the cells expressing the different GnRHs was reported in seabream (Gothilf et al., 1996) and European sea bass (Gonzalez-Martinez et al., 2004). The GnRH-II cells appeared restricted to the dorsal synencephalon but GnRH-I and GnRH-III cell distribution overlapped in the olfactory bulbs, ventral telencephalon and preoptic area (Gonzalez-Martinez et al., 2004). Similar results were later obtained in other fish species (Okubo et al., 2006; Vickers et al., 2004; Zohar et al., 2010). In zebrafish, GnRH-III neurons are found in the olfactory bulbs, ventral telencephalon, preoptic area and ventromedial hypothalamus (Abraham et al., 2008; Palevitch et al., 2009).

Gonadotropin-Inhibitory Hormone (GnIH)

Recently, a new hormone has been identified in the brain of birds known as gonadotropin-inhibitory hormone (Tsutsui et al., 2007). This peptide is mainly expressed in hypothalamic and septal neurons and inhibits the synthesis and release of gonadotropins acting directly at the pituitary level (Tsutsui and Ukena, 2006; Tsutsui et al., 2007; Yin et al., 2005). Furthermore, GnIH seems to exert neuromodulatory actions on GnRH cells because preoptic GnRH neurons and GnRH fibers from the median eminence receive a conspicuous GnIH innervation and GnRH cells exhibit GnIH-binding sites (Bentley et al., 2006; Bentley et al., 2008), GnIH inhibits steroidogenesis and development in avian gonad indicating that this neuropeptide could act at different levels in the reproductive axis (Tsutsui et al., 2007). The synthesis and secretion of hypothalamic GnIH is modulated by melatonin, suggesting that GnIH could be implicated in the transduction of photoperiod information to other endocrine centers involved in the control of reproduction (Tsutsui et al., 2007).

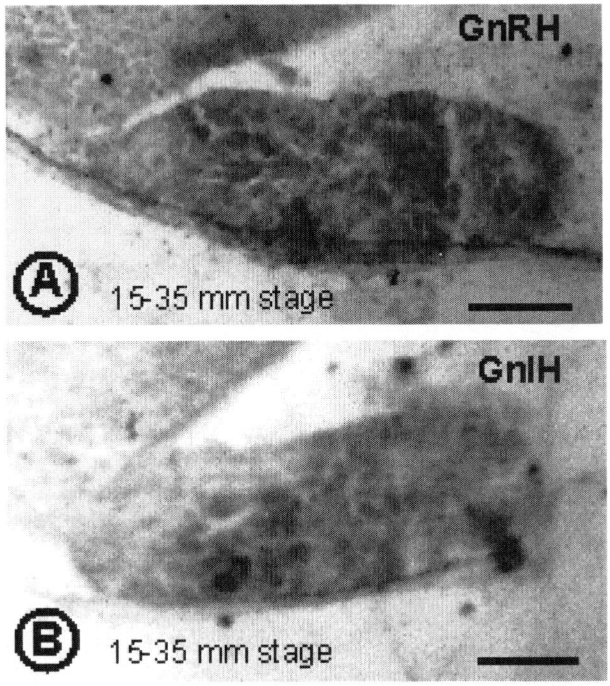

Figure 2. Photomicrograph shows gonadotropin-releasing hormone (GnRH) (A) and gonadotropin-inhibiting hormone (GnIH) (B) immunoreactivity in the pituitary gland of 15-35mm stage larvae of *Labeo rohita*. *Scale bars* 40 μm (A, B).

There are no studies which show GnIH expression in the fish brain, except in goldfish showing putative GnIH in the cells, fibers in the brain and pituitary (Sawada et al., 2002). However, we have seen GnIH expression in the olfactory epithelium and bulb, optic nerve, optic tectum and pituitary in the early developmental stages of carp *Labeo rohita* (Jadhao A. G., unpublished results, Figure 2). However, further studies are needed in order to understand the role of GnIH in fish reproduction.

Dopamine (DA)

An important neurotransmitter has been identified in the brain of teleosts, which is known to inhibit the gonadotropin release in some but not in all teleosts species (Zohar et al., 2010). DA is a small neurotransmitter that is synthesized from tyrosine. DA is known for exerting a wide range of effects in the brain and on several pituitary functions of vertebrates. The distribution of DA in the brain of fish has been extensively studied and shown well developed dopaminergic system (Nieuwenhuys et al., 1998). There are two main classes of DA receptors that differ in their ability to activate (D1) or to inhibit (D2) the enzyme adenylyl cyclase, with each class containing various subtypes (Kebabian and Calne, 1979). The rainbow trout is the only species in which DA is known to inhibit both LH and FSH release acting through D2 receptors (Vacher et al., 2002; Vacher et al., 2000). DA also acts to inhibit GnRH release from GnRH neurons as shown in goldfish. This effect would involve D2 receptors on GnRH pituitary terminals and D1 receptors onto GnRH cell bodies (Yu et al., 1991). Thus, DA in goldfish has a double effect, inhibiting gonadotropin release by direct action on the gonadotrophs and reducing GnRH secretion in the vicinity of the gonadotrophs (Trudeau, 1997). In contrast, DA inhibition of gonadotropin release was not observed in many other teleost fish, particularly in marine species (Copeland and Thomas, 1989).

Gamma-Aminobutyric Acid (GABA)

An inhibitory neurotransmitter GABA is widely distributed and abundantly seen in the brain of all vertebrates. Heavy localization of GABA in the goldfish has been reported (Popesku et al., 2008) and to a lesser extent in trout. Initially, GABA was shown to stimulate LH secretion in goldfish through effects that include stimulation of GnRH release (Kah et al., 1992; Sloley et al., 1992) and inhibition of DA (Trudeau, 1997). In the goldfish and rainbow trout effects on GTH secretion were well studied. It has been seen that after injecting GABA, there is an increase of serum GTH levels in regressed or early maturing fish, but not in late maturing animals (Kah et al., 1992). Moreover, injection of a GABA transaminase inhibitor caused a significant increase of GABA within the hypothalamus and pituitary, and a dose-dependent increase in serum GTH levels (Sloley et al., 1992). The gonadal steroids also affect GABA synthesis in both brain and pituitary (Trudeau, 1997). It has been found that GABA neurons in goldfish are very sensitive to changes in circulating sex steroid levels. However, it is still unknown if GABA neurons express estrogen receptor in the goldfish. However, GABA has an overall stimulatory action on FSH and LH secretion in rainbow trout, which depends on the sex and the reproductive phase of the fish. The stimulatory action of GABA may be exerted directly onto the gonadotrophs and also by GnRH release through GABA A receptors (Kah et al., 1992; Sloley et al., 1992; Mananos et al., 1999; Senthilkumaran et al., 2001).

Neuropeptide Y (NPY)

NPY is another potential molecule has been reported in the regulation of growth, feeding and reproduction in fishes. NPY is a 36 amino acid peptide of the pancreatic peptide family. Though, NPY is stimulating growth hormone (GH) in the goldfish (Peng et al., 1993), it also involved in the regulation of GTH release (Breton et al., 1991; Kah et al., 1989). It is considered that NPY is most important factor linking the growth, feeding and reproductive axes. Many fish species reduce food consumption during the reproductive period. Therefore, the increased expression of NPY in the preoptic area during fasting has been reported in fish (Silverstein et al., 1998). Not only that the stimulatory effects of NPY on LH secretion are also greater in fasted animals (Cerda-Reverter et al., 1999).

NPY in goldfish was shown to be directly involved in the release of LH at the level of the pituitary. However NPY is also able to release GnRH from GnRH nerve terminals in the pituitary or preoptic hypothalamus region slices through NPY Y1 and Y2 Receptor subtypes (Peng et al., 1993; Trudeau, 1997; Senthilkumaran et al., 2001). Immunohistochemical study has shown the presence of NPY in the olfactory bulbs, the nucleus entopeduncularis of the ventral telencephalon, the preoptic area, and various thalamic regions of the goldfish (Kah et al., 1989).

Galanin

Galanin is a 29 amino acid sequence peptide, widely expressed throughout brain and predominantly seen in the preoptic area of the fish. In the preoptic region of the fish, it shows sexually dimorphic distribution (more details are given in the earlier section on inter-sexual dimorphism in fish).

Nitric Oxide (NO)

NO, a free radical gas producing neurons has been localized in the fish brain by using immunohistochemistry. Although, three forms of nitric oxide synthase (NOS) are existing but only neuronal NOS (nNOS) is very much specific to CNS. In mammals, role of NO in the reproduction is clearly understood. The presence of nNOS in the brain and its functional significance in other vertebrates like fish is unclear. However, nNOS has been well demonstrated in the preoptic area and other regions of the brain in fish (Jadhao and Malz, 2003; Jadhao et al., 1999). Based on its immunohistochemical localization it has been argued that NO may play an important role hormonal regulation of pituitary gland in fish (Jadhao and Malz, 2003; Jadhao et al., 1999), like that of mammals. However, more studies are required in fish to clearly understand its functional significance in reproduction of fishes.

Kisspeptins

A new molecule, kisspeptins has been identified in brain which is playing an important role in neuroendocrinology. The kisspeptins are the products of the KiSS1 gene, which encodes for a 145 amino acid precursor protein (Kotani et al., 2001). Recent discovery of the KiSS1/GPR54 system has revolutionized our understanding of the regulation of reproduction and puberty in vertebrate animals. Pioneering studies revealed that inactivating mutations in the G-coupled protein receptor 54 (GPR54, KiSS1r) resulted in idiopathic hypogonadotrophic hypogonadism, where the affected individuals did not undergo puberty (de Roux et al., 2003; Seminara et al., 2003). This phenotype is associated with reduced circulating LH levels, yet maintains the ability to respond to GnRH administration. Recently, a number of studies have reported the identification and characterization of a KiSS1 gene from zebrafish, (van Aerle et al., 2008) and medaka, (Kanda et al., 2008). In fish, the anatomical distribution of KiSS1 mRNA-expressing neurons was examined in the medaka (Kanda et al., 2008), where two populations of cells expressing KiSS1 were identified in the hypothalamic nuclei. In the zebrafish, there is preliminary evidence for the presence of a KiSS-immunoreactive system with perikarya in many hypothalamic nuclei including preoptic area and pituitary.

Cocaine - and Amphetamine-Regulated Transcript (CART)

A recently discovered neuropeptide, CART has been reported to be involved in the neural regulation of reproduction. CART is expressed in the hypothalamus, the key area involved in the central regulation of reproductive processes (Valera et al., 2006; Vrang, 2006). In mammals, role of CART has been suggested as a neuroendocrine-releasing factor (Larsen et al., 2003), which regulates FSH and LH secretion from the pituitary (Baranowska et al., 2003). There is little direct information about the role of CART in the central regulation of reproduction in fish. However, CART-containing neurons and fibers were observed in various areas in the forebrain and pituitary of teleosts (Singru et al., 2007). Moreover, NPY, a neuromodulator of GnRH-LH cells axis in the brain and pituitary, has been found closely associated with the CART in the brain and pituitary (Singru et al., 2008). Recently, in female catfish, it has been shown that the reproductive phase-related variations of CART in the CNS and pituitary (Barsagade et al., 2010). As there are only few studies available on CART in fish, further investigations are required to understand the role of CART in neuroendocrine control of fish reproduction.

Calcium-Binding Proteins (CaBPs)

In the CNS of vertebrates, the intracellular messengers mediate the effects of neurotransmitters on intracellular events and the targets of these messengers include certain intracellular CaBPs (Cheung, 1980). In CNS, CaBPs like calretinin (CR), calbindin (CB), and parvalbumin (PV) are more specifically seen in the nervous tissue (Andressen et al., 1993) and indirectly related to neurotransmission.

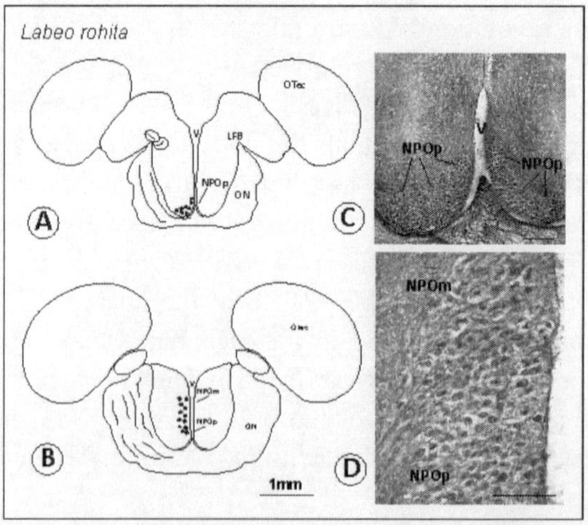

Figure 3. A-B. Schematic representaion of transverse section of the brain showing calretinin (CR) immunostained cells (black circles) in the various subdivisions of nucleus preopticus of the fish *Labeo rohita*. C. A high magnification photomicrograph shows CR-immunoreactivity in NPOp. D. Several CR-ir cells are present in NPOp and NPOm. *Scale bars* 40 μm (C), 80 μm (D).

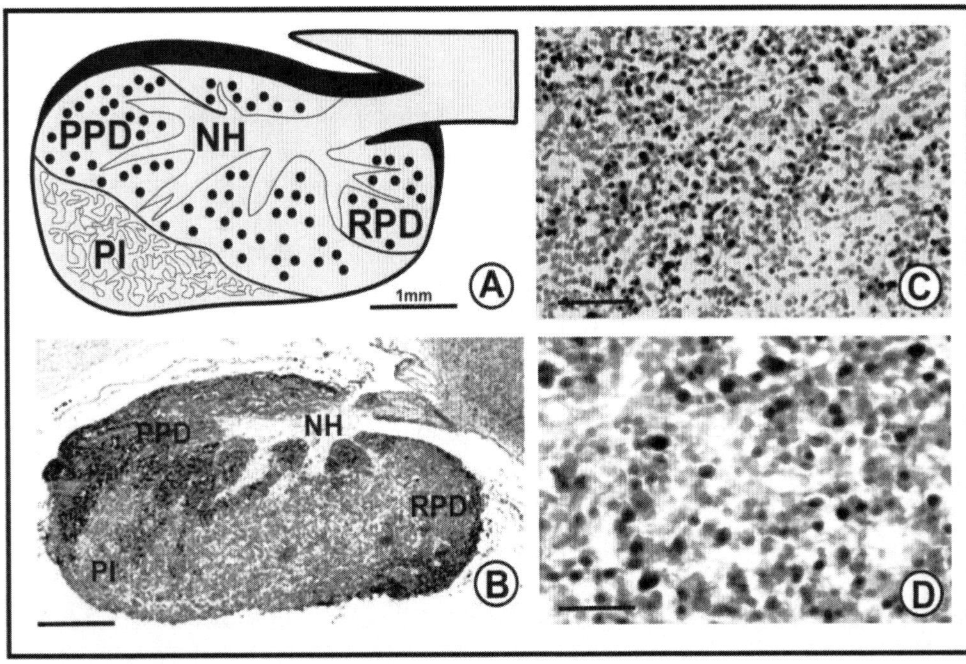

Figure 4. A Schematic representation of saggital section of pituitary gland showing different subdivisions of pituitary and localization of CR (black circles). B A low magnification photomicrograph shows CR-immunoreactivity in the pituitary gland. All subdivisions of the pituitary gland are clearly visible. C Several CR-ir cells are present in PPD of the adenohypophysis. D High magnification photomicrograph showing CR stained cells in PPD. *Scale bars* 40 μm (B), 80 μm (D), 150 μm (C) (modified from Jadhao and Malz, 2007).

However, CR is less widely distributed and principally present in the nerve cells compared to other CaBPs. It has a 29-kDa molecular weight and was identified by analysis of cDNA from chick retina (Rogers, 1987).

CR has been reported by many workers in CNS and pituitary gland (PIT) mammals and varies significantly between species (Andressen et al., 1993). Recent data in amphioxus support the view that CR developed early in vertebrate evolution and represents a phylogenetically old, conserved protein (Castro et al., 2004). Although, it is more rare in non-mammalian vertebrates and amphioxus, its presence has been reported in few teleosts species like grey mullet (Díaz-Regueira and Anadón, 2000), zebra fish (Castro et al., 2006a; Castro et al., 2006b) and catfish Jadhao and Malz, 2007). The presence of CR in the NPO of fish has not been reported before, however, we have seen its presence in the NPO of cypriniformis fish, *Labeo rohita* by using SDS page, western blot and immunohistochemistry (Jadhao A. G., unpublished results, Figure 3). For the first time in catfish localization of CR has been demonstrated in the hypophysis (Figure 4) and suggested that CR is involved in the hormonal regulation by the pituitary in teleost (Jadhao and Malz, 2007). Therefore, on the basis of CR-expressing cells in the pituitary gland (Jadhao and Malz, 2007) and in the NPO, CaBP may be involved in the neurohormonal regulation and hence, intracellular CR might play a significant role in governing hypophyseal functions in fishes, which might resemble its function in the adult rat (Cimini et al., 1997; Miyata et al., 2000). Hence, in fishes more studies are essential to know the exact role of CaBPs in the neuroendocrine regulation.

Conclusion

It is apparent from the existing studies on the several species of teleosts, the POA and other hypothalamic nuclei in the brain are sexually dimorphic and some species also show dimorphism within the same gender i.e., intra-sexual dimorphism. Mainly in the POA, sex differences have been identified in neuronal connectivity, cell nucleus size, volume of groups of cells and differences in the distribution of various peptides/neurotransmitters. Although there are several studies focused on inter-sexual variations in the POA but so far anatomical alterations in the entire forebrain within the same gender (intra-sexual) of the fish has been reported only in salmon brain. Therefore, functional correlations of such intra-sexual dimorphism are unclear and many more investigations are needed to resolve this issue. It is very clear from the available studies that many neuropeptides/ neurotransmitters, proteins and free radical NO have been identified in the hypothalamic nuclei and particularly in the NPO which regulates the hypothalamo-pituitary-gonadal axis in teleosts. Many studies have greatly contributed to understand the neuroendocrine function of the brain and have shown that the hormonal regulation of pituitary is under the control of several neurohormones, peptides and other chemicals. In fish, reproduction is governed by multiple factors singly or together and it seems to be more complex as it is regulated by many factors.

Acknowledgments

The grants received from the Department of Science and Technology [SR/SO/AS-09/2006] and University Grant Commission [32-512/2006(SR)] and [41-22/2012(SR)] Government of India, and Alexander von Humboldt Foundation, and DFG Germany are acknowledged. I am thankful to Biswas, Nikhil and Ketki for the technical assistance.

References

Abraham, E., Palevitch, O., Ijiri, S., Du, S.J., Gothilf, Y. and Zohar, Y. (2008). Early development of forebrain gonadotrophin-releasing hormone (GnRH) neurons and the role of GnRH as an autocrine migration factor. *J. Neuroendocrinol., 20*, 394–405.

Adams, B. A., Vickers, E. D., Warby, C., Park, M., Fischer, W. H., Grey Craig, A., Rivier, J. E. and Sherwood, N. M. (2002). Three forms of gonadotropin-releasing hormone, including a novel form, in a basal salmonid, *Coregonus clupeaformis*. *Biol. Reprod., 67*, 232-239.

Aldenhoven, J. M. (1986). Different reproductive strategies in a sex-changing coral reef fish *Centropyge bicolor* (Pomacanthidae). *Aust. J. Mar. Freshwater Res., 37*, 353-360.

Amano, M., Oka, Y., Yamanome, T., Okuzawa, K. and Yamamori, K. (2002). Three GnRH systems in the brain and pituitary of a pleuronectiform fish, the barfin flounder *Verasper moseri*. *Cell Tissue Res., 309*, 323–329.

Anglade, I., Wang, Y., Jensen, J., Tramu, G., Kah, O. and Conlon, J. M. (1994). Characterization of trout galanin and its distribution in trout brain and pituitary. *J. Comp. Neurol., 350*, 63-74.

Andressen, C., Blümcke, I. and Celio, M. R. (1993). Calcium-binding proteins: selective markers of nerve cells. *Cell Tissue Res., 271*, 181-208.

Ariëns Kappers, C. U., Huber, G. C. and Crosby, E. C. (1936). The comparative anatomy of the nervous system of vertebrates, including man. Vol. II. Hafner Publication Co, New York, pp 606-120.

Arnold, A. P. and Gorski, R. A. (1984). Gonadal steroid induction of structural sex differences in the central nervous system. *Annu. Rev. Neurosci., 7*, 413-442.

Ayoub, D. M., Greenough, W. T. and Juraska, J. M. (1983). Sex differences in dendritic structure in the preoptic area of the juvenile macaque monkey brain. *Science, 219*, 197-198.

Balthazart, J. and Adkins-Regan, E. (2002). Sexual differentiation of brain and behavior in birds. In: D.W. Pfaff, (Eds) *Hormones Brain and Behavior*, Vol. IV. Elsevier, Amsterdam, pp 223–301.

Baranowska, B., Wolińska-Witort, E., Chmielowska, M., Martyñska, L. and Baranowska-Bik, A. (2003). Direct effects of cocaine-amphetamine- regulated transcript (CART) on pituitary hormone release in pituitary cell culture. *Neuroendocrinol. Lett., 24*, 224–226.

Barsagade, V. G., Mazumdar, M., Singru, P. S., Thim, L., Clausen, J. T. and Subhedar, N. (2010) Reproductive phase-related variations in cocaineand amphetamine-regulated transcript (CART) in the olfactory system, forebrain, and pituitary of the female catfish, *Clarias batrachus* (Linn.). *J. Comp. Neurol., 518*, 2503-2524.

Bass, A. H. (1992). Dimorphic male brains and alternative reproductive tactics in a vocalizing fish. *TINS*, 15, 139-145.

Bass, A. H and Baker, R. (1990) Sexual dimorphisms in the vocal control system of a teleost fish: morphology of physiologically identified neurons. *J. Neurobiol., 21*, 1155-1168.

Batten, T. F, Moons, L., Cambre, M. and Vandesande, F. (1990) Anatomical distribution of galanin- like immunoreactivity in the brain and pituitary of teleost fishes. *Neurosci. Lett., 111*, 131-143.

Bentley, G. E., Kriegsfeld, L. J., Osugi, T., Ukena, K., O'Brien, S., Perfito, N., Moore, I. T., Tsutsui, K. and Wingfield, J. C. (2006). Interactions of gonadotropin-releasing hormone (GnRH) and gonadotropin-inhibitory hormone (GnIH) in birds and mammals. *J. Exp. Zool. A. Comp. Exp. Biol., 305*, 807-814.

Bentley, G. E., Ubuka, T., McGuire, N. L., Chowdhury, V. S., Morita, Y., Yano, T., Hasunuma, I., Binns, M., Wingfield, J. C. and Tsutsui, K. (2008). Gonadotropin inhibitory hormone and its receptor in the avian reproductive system. *Gen. Comp. Endocrinol., 156*, 34-43.

Bosma, P. T., Blazquez, M., Fraser, E. J., Schulz, R. W., Docherty, K. and Trudeau, V. L. (2001). Sex steroid regulation of glutamate decarboxylase mRNA expression in goldfish brain is sexually dimorphic. *J. Neurochem., 76*, 945-956.

Breton, B., Mikolajczyk, T., Popek, W., Bieniarz, K., Epler, P. (1991). Neuropeptide Y stimulates in vivo gonadotropin secretion in teleost fish. *Gen. Comp. Endocrinol., 84*, 277-283.

Castro, A., Becerra, M., Manso, M. J. and Anadon, R. (2004). Somatomotor system of the adult amphioxus (*Branchiostoma lanceolatum*) revealed by an anticalretinin antiserum: An immunocytochemical study. *J. Comp. Neurol., 477*, 161-171.

Castro, A., Becerra, M., Manso, M. J. and Anadon, R. (2006a). Calretinin immunoreactivity in the brain of the zebrafish, *Danio rerio*: distribution and comparison with some neuropeptides and neurotransmitter-synthesizing enzymes. I. Olfactory organ and forebrain. *J. Comp. Neurol., 494*, 435-459.

Castro, A., Becerra, M, Manso, M. J. and Anadon, R. (2006b). Calretinin immunoreactivity in the brain of the zebrafish, *Danio rerio*: distribution and comparison with some neuropeptides and neurotransmitter-synthesizing enzymes. II. Midbrain, hindbrain, and rostral spinal cord. *J. Comp. Neurol., 494*, 792-881.

Cerda-Reverter, J. M., Sorbera, L. A., Carrillo, M. and Zanuy, S. (1999). Energetic dependence of NPY-induced LH secretion in a teleost fish (*Dicentrarchus labrax*). *Am. J. Physiol., 277*, R1627-R1634.

Charlton, H. H. (1932). Comparative studies on the nucleus preopticus pars magnocellularis and the nucleus lateralis tuberis in fishes. *J. Comp. Neurol., 54*, 237-275.

Cheung, W. Y. (1980) Calmodulin plays a pivotal role in cellular regulation. *Science, 207*, 19-27.

Cimini, V., Isaacs, K. R. and Jacobowitz, D. M. (1997). Calretinin in the rat pituitary: colocalization with thyroid-stimulating hormone. *Neuroendocrinology, 65*, 179-188.

Copeland, P. A. and Thomas, P. (1989). Control of gonadotropin release in the Atlantic croaker (*Micropogonias undulatus*): evidence for lack of dopaminergic inhibition. *Gen. Comp. Endocrinol., 74*, 474–483.

Cornbrooks, E. B. and Parsons, R. L. (1991) Sexually dimorphic distribution of a galanin-like peptide in the central nervous system of the teleost fish *Poecilia latipinna. J. Comp. Neurol.*, 304, 639-657.

D´Aniello, B., Pinelli, C., Jadhao, A. G., Rastogi, R. K. and Meyer, D. L. (1999). Comparative analysis of FMRFamide-like immunoreactivity in the brain of the caiman (*Caiman crocodilus*) and the turtle (*Trachemys scripta elegans*). *Cell Tissue Res., 298*, 549-559.

de Roux, N., Genin, E., Carel, J. C., Matsuda, F., Chaussain, J. L. and Milgrom, E. (2003). Hypogonadotropic hypogonadism due to loss of function of the KiSS1-derived peptide receptor GPR54. *Proc. Natl. Acad. Sci. USA, 100*, 10972-10976.

De Vries, G. J. (2004). Minireview: sex differences in adult and developing brains: compensation, compensation, compensation. *Endocrinology, 145*, 1063-1068.

Do Rego, J. L., Seong, J. Y., Burel, D., Leprince, J., Luu-The, V., Tsutsui, K., Tonon, M. C., Pelletier, G. and Vaudry, H. (2009). Neurosteroid biosynthesis: enzymatic pathways and neuroendocrine regulation by neurotransmitters and neuropeptides. *Front Neuroendocrinol., 30*, 259-301.

Davis, M. R. and Fernald, R. D. (1990). Social control of neuronal soma size. *J. Neurobiol., 21*, 1180-1188.

Díaz-Regueira, S. and Anadón, R. (2000). Calretinin expression in specific neuronal systems in the brain of an advanced teleost, the grey mullet (*Chelon labrosus*). *J. Comp. Neurol., 426*, 81-105.

Elofsson, U., Winburg, S. and Francis, R. C. (1997). Number of preoptic GnRH-immunoreactive cells correlates with sexual phase in a protandrously hermaphroditic fish (*Amphiprion melanopus*). *J. Comp. Physiol., 181*, 484-492.

Foran, C. M. and Bass, A.H. (1998). Preoptic AVT immunoreactive neurons of a teleost fish with alternative reproductive tactics. *Gen. Comp. Endocrinol., 111*, 271-282.

Forlano, P.M., Deitcher, D.L., Myers, D.A. and Bass, A.H. (2001). Anatomical distribution and cellular basis for high levels of aromatase activity in the brain of teleost fish:

aromatase enzyme and mRNA expression identify glia as source. *J. Neurosci., 21*, 8943-8955.

Gonzalez, A. and Piferrer, F. (2003). Aromatase activity in the European sea bass (*Dicentrarchus labrax* L.) brain. Distribution and changes in relation to age, sex, and the annual reproductive cycle. *Gen. Comp. Endocrinol., 132*, 223-230.

Gonzalez-Martinez, D., Zmora, N., Saligaut, D., Zanuy, S., Elizur, A., Kah, O. and Munoz-Cueto, J. A. (2004). New insights in developmental origins of different GnRH (gonadotrophin-releasing hormone) systems in perciform fish: an immunohistochemical study in the European sea bass (*Dicentrarchus labrax*). *J. Chem. Neuroanat., 28*, 1-15.

Gothilf, Y., Munoz-Cueto, J. A., Sagrillo, C. A., Selmanoff, M., Chen, T.T., Kah, O., Elizur, A. and Zohar, Y. (1996). Three forms of gonadotropin-releasing hormone in a perciform fish (*Sparus aurata*): complementary deoxyribonucleic acid characterization and brain localization. *Biol. Reprod., 55*, 636–645.

Gorbman, A. (1995). Olfactory origins and evolution of the brain–pituitary endocrine system: facts and speculation. *Gen. Comp. Endocrinol., 97*, 171–178.

Gorski, R. A., Csernus, V. J., Jacobson, C. D. Sexual dimorphism in the preoptic area. In: Flerko´, B, Setalo´, G, Tima, L. editors. Advances in physiological sciences. Reproduction and development. Vol. 15. Pergamon Press and Akadémia Kiadó Press, Budapest, 1980, pp 121-130.

Gregory, W.A. and Tweedle, C. D. (1983). Maturation of the goldfish preoptic area: ultrastructural correlates of position and maturation of neurosecretory cells. *Neuroscience, 10*, 425-448.

Grober, M. S. and Bass, A. H. (2002). Life history, neuroendocrinology and behavior in fish. In: Pfaff, DW, editor. *Hormones Brain and Behavior.* (Vol. II. Elsevier, 331-347), Amsterdam.

Grober, M. S., Jackson, I. M. D. and Bass, A. H. (1991). Gonadal steroids affect LHRH preoptic cell number in sex/role changing fish. *J. Neurobiol., 22*, 734-741.

Grober, M. S., Fox, S. H., Laughlin, C. and Bass, A. H. (1994). GnRH cell size and number in a teleost fish with two male reproductive morphs: sexual maturation, final sexual status and body size allometry. *Brain Behav. Evol., 43*, 61-78.

Guiguen, Y., Baroiller, J. F., Ricordel, M. J., Iseki, K., McMeel, O. M., Martin, S. A. and Fostier, A. (1999). Involvement of estrogens in the process of sex differentiation in two fish species: the rainbow trout (*Oncorhynchus mykiss*) and a tilapia (*Oreochromis niloticus*). *Mol. Reprod. Dev., 54*, 154–162.

Guilgur, L. G., Moncaut, N. P., Canario, A. V. and Somoza, G. M. (2006). Evolution of GnRH ligands and receptors in gnathostomata. *Comp. Biochem. Physiol. A Mol. Integr. Physiol., 144*, 272–283.

Hayward, J. N. (1974). Physiological and morphological identification of hypothalamic magnocellular neuroendocrine cells in goldfish preoptic nucleus. *J. Physiol., 239*, 103-124.

Henson, S. A. and Warner, R. R. (1997). Male and female alternative reproductive behaviors in fishes: A new approach using intersexual dynamics. *Annu. Rev. Ecol. Syst., 28*, 571-592.

Henson, S. A. and Warner, R. R. (1997). Male and female alternative reproductive behaviors in fishes: A new approach using intersexual dynamics. *Annu. Rev. Ecol. Syst., 28*, 571-592.

Jadhao, A. G. (2001). Localization of molluscan cardioexcitatory tetrapeptide immunoreactivity in the brain of African Cichlid fish, *Haplochromis burtoni*. *Neurosci. Lett., 303*, 103-106.

Jadhao, A. G., D´Aniello, B., Malz, C. R., Pinelli, C., Meyer, D. L. (2001). Intra-sexual and inter-sexual dimorphisms of the red salmon prosencephalon. *Cell Tissue Res., 304*, 121-140.

Jadhao, A. G. and Malz, C. R. (2003). Localization of NOS-I isoform in the diencephalon and pituitary gland of the catfish, *Synodontis multipunctatus*: An Immunocytochemical Study. *Gen. Comp. Endocrinol., 132*, 278-283.

Jadhao, A.G. and Malz, C. R. (2007). Localization of calcium-binding protein (calretinin, 29kD) in the brain and pituitary gland of teleost fish: An immunohistochemical study. *Neurosci. Res., 59*, 265-276.

Jadhao, A.G. and Meyer, D. L. (2000). Sexually dimorphic distribution of galanin in the preoptic area of red salmon, *Oncorhynchus nerka*. *Cell Tissue Res., 302*, 199-203.

Jadhao, A.G. and Pinelli, C. (2001). Galanin-like Immunoreactivity in the brain and pituitary gland of the 'four-eyed' fish, *Anableps anableps*. *Cell Tissue Res., 306*, 309-318.

Jadhao, A. G., Wilke, I. and Meyer, D. L. (1999). NADPH-d expression in the hypothalamo-hypophysial system of different catfish. *J. Brain Res., 39*, 511-521.

Kah, O., Danger, J. M., Dubourg, P., Pelletier, G., Vaudry, H. and Calas, A. (1989). Characterization, cerebral distribution and gonadotrophin-release activity of neuropeptide Y (NPY) in the goldfish. *Fish Physiol. Biochem., 7*, 69-76.

Kah, O., Breton, B., Dulka, J. G., Nunez-Rodriguez, J., Peter, R. E., Corrigan, A., Rivier, J. E. and Vale, W. W. (1986). A reinvestigation of the Gn-RH (gonadotrophin releasing hormone) systems in the goldfish brain using antibodies to salmon Gn-RH. *Cell Tissue Res., 244*, 327-337.

Kah, O., Lethimonier, C., Somoza, G., Guilgur, L. G., Vaillant, C. and Lareyre, J. J. (2007). GnRH and GnRH receptors in metazoa: a historical, comparative, and evolutive perspective. *Gen. Comp. Endocrinol., 153*, 346-364.

Kah, O., Trudeau, V. L., Sloley, B. D., Chang, J. P., Dubourg, P., Yu, K. L. and Peter, R. E. (1992). Influence of GABA on gonadotrophin release in the goldfish. *Neuroendocrinology, 55*, 396-404.

Kanda, S., Akazome, Y., Matsunaga, T., Yamamoto, N., Yamada, S., Tsukamura, H., Maeda, K., Oka, Y. (2008). Identification of KiSS-1 product kisspeptin and steroidsensitive sexually dimorphic kisspeptin neurons in medaka (*Oryzias latipes*). *Endocrinology, 149*, 2467–2476.

Kebabian, J. W. and Calne, D. B. (1979). Multiple receptors for dopamine. *Nature, 277*, 93-96.

Kotani, M., Detheux, M., Vandenbogaerde, A., Communi, D., Vanderwinden, J. M., Le Poul, E., Brezillon, S., Tyldesley, R., Suarez-Huerta, N., Vandeput, F., Blanpain, C., Schiffmann, S. N., Vassart, G. and Parmentier, M. (2001). The metastasis suppressor gene KiSS-1 encodes kisspeptins, the natural ligands of the orphan G proteincoupled receptor GPR54. *J. Biol. Chem., 276*, 34631-34636.

Kotrschal, A., Räsänen, K., Kristjánsson, B. K., Senn, M. and Kolm, N. (2012). Extreme Sexual Brain Size Dimorphism in Sticklebacks: A Consequence of the Cognitive Challenges of Sex and Parenting. *PLoS ONE, 7*, e30055.

Larsen, P. J., Seier, V., Fink-Jensen, A., Holst, J. J., Warberg, J. and Vrang, N. (2003). Cocaine- and amphetamine-regulated transcript is present in hypothalamic neuroendocrine neurones and is released to the hypothalamic-pituitary portal circuit. *J. Neuroendocrinol., 15*, 219-226.

Lauber, M. E., Sarasin, A. and Lichtensteiger, W. (1997). Transient sex differences of aromatase (CYP19) mRNA expression in the developing rat brain. *Neuroendocrinology, 66*, 173-180.

Lauer, L. E., McCarthy, M. M., Mong, J. and Kane, A. S. (2006). Sex differences in neuronal morphology in the killifish hypothalamus. *Brain Res., 1070*, 145-149.

Lepretre, E., Anglade, I., Williot, P., Vandesande, F., Tramu, G. and Kah, O. (1993). Comparative distribution of mammalian GnRH (gonadotrophin-releasing hormone) and chicken GnRH-II in the brain of the immature Siberian sturgeon (*Acipenser baeri*). *J. Comp. Neurol., 337*, 568-583.

MacLusky, N. J. and Naftolin, F. (1981). Sexual differentiation of the central nervous system. *Science, 211*, 1294-1302.

Mananos, E. L., Anglade, I., Chyb, J., Saligaut, C., Breton, B. and Kah, O. (1999). Involvement of gamma-aminobutyric acid in the control of GTH-1 and GTH-2 secretion in male and female rainbow trout. *Neuroendocrinology, 69*, 269-280.

Melo, A. C. and Ramsdell, J. S. (2001). Sexual dimorphism of brain aromatase activity in medaka: induction of a female phenotype by estradiol. *Environ. Health Perspect., 109*, 257-264.

Meyer, D. L., Fackler, I. R., Jadhao, A. G., D´Aniello, B. and Kicliter, E. (1997). Differential labelling of primary olfactory system subcomponents by SBA (lectin) and NADPH-d histochemistry in the frog *Pipa*. *Brain Res., 762*, 275-280.

Miyata, S., Nakai, S., Kiyohara, T. and Hatton, G. I. (2000). Calbindin-D28k and calretinin in the rat posterior pituitary, light and electron microscopic localization and upregulation with dehydration. *J. Neurocytol., 29*, 5-17.

Montero, M., Le Belle, N., King, J. A., Millar, R. P. and Dufour, S. (1995). Differential regulation of the two forms of gonadotropin-releasing hormone (mGnRH and cGnRH-II) by sex steroids in the European female silver eel (*Anguilla anguilla*). *Neuroendocrinology, 61*, 525-535.

Morris, J. A., Jordan, C. L. and Breedlove, S. M. (2004). Sexual differentiation of the vertebrate nervous system. *Nat. Neurosci., 7*, 1034-1039.

Moore, M. C. (1991). Application of organization-activation theory to alternative male reproductive strategies: a review. *Horm. Behav., 25*, 154-179.

Nagler, J. J., Cavileer, T., Sullivan, J., Cyr, D. G. and Rexroad, C. 3rd (2007), The complete nuclear estrogen receptor family in the rainbow trout: discovery of the novel ERα2 and both ERβ isoforms. *Gene, 392*, 164-173.

Nieuwenhuys, R., Donkelaar, H. J. T. and Nicholson, C. (1998). *The Central Nervous System of Vertebrates Springer*, Berlin, New York.

Okubo, K. and Nagahama, Y. (2008). Structural and functional evolution of gonadotropin-releasing hormone in vertebrates. *Acta. Physiol. (Oxf), 193*, 3–15.

Okubo, K., Sakai, F., Lau, E. L., Yoshizaki, G., Takeuchi, Y., Naruse, K., Aida, K. and Nagahama, Y. (2006). Forebrain gonadotropin-releasing hormone neuronal development:

insights from transgenic medaka and the relevance to X-linked Kallmann syndrome. *Endocrinology, 147*, 1076-1084.

Okuzawa, K., Amano, M., Kobayashi, M., Aida, K., Hanyu, I., Hasegawa, Y. and Miyamoto, K. (1990). Differences in salmon GnRH and chicken GnRH-II contents in discrete brain areas of male and female rainbow trout according to age and stage of maturity. *Gen. Comp. Endocrinol., 80*, 116-126.

Okuzawa, K., Gen, K., Bruysters, M., Bogerd, J., Gothilf, Y., Zohar, Y. and Kagawa, H. (2003). Seasonal variation of the three native gonadotropin-releasing hormone messenger ribonucleic acids levels in the brain of female red seabream. *Gen. Comp. Endocrinol., 130*, 324–332.

Olivereau, M. and Olivereau, J. M. (1991a). Immunocytochemical localization of galanin-like peptidergic system in the brain and pituitary of some teleost fish. *Histochemistry, 96*, 343-354.

Olivereau, M. and Olivereau, J. M. (1991b). Galanin-like immunoreactivity is increased in the brain of estradiol- and methyltestosterone-treated eels. *Histochemistry, 96*, 487-497.

Palande, N. V., Biswas, S. and Jadhao, A. G. (2011). The ontogeny of Mauthner cells in the brain of *Labeo rohita* as revealed by NADPH-d and NOS immunohistochemistry. *Brain Str. Func.*, 216, 67-75.

Palevitch, O., Abraham, E., Borodovsky, N., Levkowitz, G., Zohar, Y. and Gothilf, Y. (2009). Nasal embryonic LHRH factor plays a role in the developmental migration and projection of gonadotropin-releasing hormone 3 neurons in zebrafish. *Dev. Dyn., 238*, 66-75.

Parhar, I. S., Tosaki, H., Sakuma, Y. and Kobayashi, M. (2001).Sex differences in the brain of goldfish: gonadotropin-releasing hormone and vasotocinergic neurons. *Neuroscience, 104*, 1099–1110.

Peng, C., Trudeau, V. L. and Peter, R. E. (1993). Seasonal variation of neuropeptide Y actions on growth hormone and gonadotropin-II secretion in the goldfish: effects of sex steroids. *J. Neuroendocrinol., 5*, 273-280.

Peter, R. E., Fryer, J. N. (1983). Endocrine functions of the hypothalamus of actinopterygians. In: Davis RE, Northcutt RG Editors. *Fish Neurobiology*. Vol. II. Ann Arbor: University of Michigan Press., pp 165-201.

Peter, R. E., Yu, K. L., Marchant, T. A. and Rosenblum, P. M. (1990). Direct neural regulation of the teleost adenohypophysis. *J. Exp. Zoo, 4*, 84-89.

Peyon, P., Saied, H., Lin, X. and Peter, R. E. (2000). Preprotachykinin gene expression in goldfish brain: sexual, seasonal, and postprandial variations. *Peptides, 21*, 225-231.

Popesku, J. T., Martyniuk, C. J., Mennigen, J., Xiong, H., Zhang, D., Xia, X., Cossins, A. R. and Trudeau, V. L. (2008). The goldfish (*Carassius auratus*) as a model for neuroendocrine signaling. *Mol. Cell Endocrinol., 293*, 43-56.

Prasada Rao, P. D., Job, T. C. and Schreibman, M. P. (1993). Hypophysiotropic neurons in the hypothalamus of the catfish *Clarias batrachus*: a cobaltous lysine and HRP study. *Brain Behav. Evol.*, 42, 24-38.

Prasada Rao, P. D., Murthy, C. K., Cook, H. and Peter, R. E. (1996). Sexual dimorphism of galanin-like immunoreactivity in the brain and pituitary of goldfish, *Carassius auratus*. *J. Chem. Neuroanat., 10*, 119-135.

Raghuveer, K., Sudhakumari, C. C., Senthilkumaran, B., Kagawa, H., Dutta-Gupta, A. and Nagahama, Y. (2011). Gender differences in tryptophan hydroxylase-2 mRNA, serotonin, and 5-hydroxytryptophan levels in the brain of catfish, *Clarias gariepinus*, during sex differentiation. *Gen Comp Endocrinol, 171*, 94-104.

Rogers, J. H. (1987). Calretinin: a gene for a novel calcium-binding protein expressed principally in neurons. *J Cell Biol, 105*, 1343-1353.

Sawada, K., Ukena, K., Satake, H., Iwakoshi, E., Minakata, H. and Tsutsui, K. (2002). Novel fish hypothalamic neuropeptide. *Eur. J. Biochem., 269*, 6000-6008.

Scaggiante, M., Grober, M. S., Lorenzi, V. and Rasotto, M. B. (2006). Variability of GnRH secretion in two goby species with socially controlled alternative male mating tactics. *Hormo. Behavior, 50*, 107–117.

Scharrer, E. and Neurosecretion I. (1941). The nucleus preopticus of *Fundulus heteroclitus* L. *J Comp Neurol, 74*, 81-92.

Schlinger, B. A. and Brenowitz, E. A. (2002). Neural and hormonal control of birdsong. In: Pfaff, D.W. editor. *Hormones Brain and Behavior*. Vol. II. Elsevier, 799-839, Amsterdam.

Seminara, S. B., Messager, S., Chatzidaki, E. E., Thresher, R. R., Acierno Jr, J. S., Shagoury, J. K., Bo-Abbas, Y., Kuohung, W., Schwinof, K. M., Hendrick, A.G., Zahn, D., Dixon, J., Kaiser, U. B., Slaugenhaupt, S. A., Gusella, J. F., O'Rahilly, S., Carlton, M. B., Crowley Jr, W. F., Aparicio, S. A. and Colledge, W. H. (2003). The GPR54 gene as a regulator of puberty. *N. Engl. J. Med., 349*, 1614-1627.

Senthilkumaran, B., Okuzawa, K., Gen, K., Ookura, T. and Kagawa, H. (1999). Distribution and seasonal variations in levels of three native GnRHs in the brain and pituitary of perciform fish. *J. Neuroendocrinol., 11*, 181-186.

Sherwood, N., Eiden, L., Brownstein, M., Spiess, J., Rivier, J. and Vale, W. (1983). Characterization of a teleost gonadotropin-releasing hormone. *Proc. Natl. Acad. Sci. USA, 80*, 2794-2798.

Silverstein, J. T, Breininger, J., Baskin, D. G. and Plisetskaya, E. M. (1998). Neuropeptide Ylike gene expression in the salmon brain increases with fasting. *Gen. Comp. Endocrinol., 110*, 157-165.

Singru, P. S., Mazumdar, M., Sakharkar, A. M., Lechan, R. M., Thim, L., Clausen, T. J., Subhedar, N. K. (2007). Immunohistochemical localization of cocaine- and amphetamine-regulated transcript peptide in the brain of the catfish, *Clarias batrachus* (Linn.). *J. Comp. Neurol., 502*, 215-235.

Singru, P. S., Mazumdar, M., Barsagade, V., Lechan, R. M., Thim, L., Clausen, J. S. and Subhedar, N. K. (2008). Association of cocaine and amphetamine-regulated transcript and neuropeptide Y in the forebrain and pituitary of the catfish, *Clarias batrachus*: a double immunofluorescent labeling study. *J. Chem. Neuroanat., 36*, 239-250.

Sloley, B. D., Kah, O., Trudeau, V. L., Dulka, J. G. and Peter, R. E. (1992). Amino acid neurotransmitters and dopamine in brain and pituitary of the goldfish: involvement in the regulation of gonadotropin secretion. *J. Neurochem., 58*, 2254-2262.

Sudhakumari, C. C., Senthilkumaran, B., Raghuveer, K., Wang, D. S., Kobayashi, T., Kagawa, H., Krishnaiah, C. H., Dutta-Gupta, A. and Nagahama, Y. (2010). Dimorphic expression of tryptophan hydroxylase in the brain of XX and XY Nile tilapia during early development. *Gen. Comp. Endocrinol., 166*, 320-329.

Sunobe, T. and Nakazono, A. (1993). Sex change in both directions by alteration of social dominance in *Trimma okinawae* (Pices: Gobiidae). *Ethology, 94*, 339-345.

Takami, S. and Urano, A. (1984). The volume of the toad medial amygdala–anterior preoptic complex is sexually dimorphic and seasonally variable. *Neurosci. Lett., 44*, 253-258.

Trudeau, V. L. (1997). Neuroendocrine regulation of gonadotrophin II release and gonadal growth in the goldfish, *Carassius auratus. Rev. Reprod., 2*, 55-68.

Tsutsui, K. and Ukena, K. (2006). Hypothalamic LPXRF-amide peptides in vertebrates: identification, localization and hypophysiotropic activity. *Peptides, 27*, 1121-1129.

Tsutsui, K., Bentley, G. E., Ubuka, T., Saigoh, E., Yin, H., Osugi, T., Inoue, K., Chowdhury, V. S., Ukena, K., Ciccone, N., Sharp, P. J. and Wingfield, J. C. (2007). The general and comparative biology of gonadotropin-inhibitory hormone (GnIH). *Gen. Comp. Endocrinol., 153*, 365–370.

Vacher, C., Ferriere, F., Marmignon, M. H., Pellegrini, E. and Saligaut, C. (2002). Dopamine D2 receptors and secretion of FSH and LH: role of sexual steroids on the pituitary of the female rainbow trout. *Gen. Comp. Endocrinol., 127*, 198-206.

Vacher, C., Mananos, E. L., Breton, B., Marmignon, M. H. and, Saligaut, C. (2000). Modulation of pituitary dopamine D1 or D2 receptors and secretion of follicle stimulating hormone and luteinizing hormone during the annual reproductive cycle of female rainbow trout. *J. Neuroendocrinol., 12*, 1219-1226.

van Aerle, R., Kille, P., Lange, A. and Tyler, C. R. (2008). Evidence for the existence of a functional Kiss1/Kiss1 receptor pathway in fish. *Peptides, 29*, 57-64.

Valera, A. G., Cavalcante, J. C., Elias, C. F., Felício, L. F. (2006). Cocaineand amphetamine-regulated transcript is overexpressed in the anteroventral periventricular nucleus of pregnant rats. *J. Neuroendocrinol., 18*, 711-714.

Vickers, E. D., Laberge, F., Adams, B. A., Hara, T. J. and Sherwood, N. M. (2004). Cloning and localization of three forms of gonadotropin-releasing hormone, including the novel whitefish form, in a salmonid, *Coregonus clupeaformis. Biol. Reprod., 70*, 1136-1146.

Vizziano-Cantonnet, D., Anglade, I., Pellegrini, E., Gueguen, M., Fostier, A., Guiguen, Y. and Kah, O. (2011). Sexual dimorphism in the brain aromatase expression and activity, and in the central expression of other steroidogenic enzymes during the period of sex differentiation in monosex rainbow trout populations. *Gen. Com.p Endocrinol., 170*, 346-355.

Vrang, N. (2006). Anatomy of hypothalamic CART neurons. *Peptides, 27*, 1970–1980.

Wade, J. and Arnold, A. P. (2004). Sexual differentiation of the zebra finch song system. *Ann. N. Y. Acad. Sci., 1016*, 540-559.

Yamamoto, T., Maler, L. and Nagy, J. I. (1992). Organization of galanin-like immunoreactive neuronal systems in weakly electric fish (*Apteronotus leptorhynchus*). *J. Chem. Neuroanat., 5*, 19-38.

Yin, H., Ukena, K., Ubuka, T. and Tsutsui, K. (2005). A novel G protein-coupled receptor for gonadotropin-inhibitory hormone in the Japanese quail (*Coturnix japonica*): identification, expression and binding activity. *J. Endocrinol., 184*, 257-266.

Yu, K. L., Rosenblum, P. M. and Peter, R. E. (1991). In vitro release of gonadotropinreleasing hormone from the brain preoptic-anterior hypothalamic region and pituitary of female goldfish. *Gen. Comp. Endocrinol., 81*, 256-267.

Zmora, N., Gonzalez-Martinez, D., Munoz-Cueto, J. A., Madigou, T., Mananos-Sanchez, E., Doste, S. Z., Zohar, Y., Kah, O. and Elizur, A. (2002). The GnRH system in the European sea bass (*Dicentrarchus labrax*). *J. Endocrinol., 172*, 105-116.

Zohar, Y., Muñoz-Cueto, J. A., Abigail Elizur, E. and Kah, O. (2010). Neuroendocrinology of reproduction in teleost fish. *Gen. Comp. Endocrinol., 165*, 438- 455.

In: Sexual Plasticity and Gametogenesis in Fishes
Editor: Balasubramanian Senthilkumaran
ISBN: 978-1-62618-848-8

Chapter VIII

Ovarian Catecholestrogens: Role in Final Oocyte Maturation

***K. P. Joy*[1,*] *and R. Chaube*[2,#]**

[1]Department of Zoology, Centre of Advanced Study, Banaras Hindu University, Varanasi, India

[2]Zoology Department, Mahila Mahavidhyalaya, Banaras Hindu University, Varanasi, India

Abstract

Catecholestrogens are naturally–occurring estrogen metabolites meant for excretion. The wide- spread distribution in tissues other than liver, the site of estrogen catabolism, has rekindled interests in these molecules. Even then research in catecholestrogens, primarily in mammals and humans has a chequered history, perhaps due to their very labile and short half-life, and equivocal physiological response in experimental studies. Over the last one decade we focused our research work on ovarian catecholestrogens in catfish and the results are reviewed to stimulate further interests in lower vertebrates. Ovarian follicular layer (theca and granulose) is responsible for the production of catecholestrogens. Estrogen hydroxylase, and 2-hydroxylated and methlylated estradiol-17β elicited significant seasonality and periovulatory changes. Hydroxyestrogens (2- and 4-hydroxylatedE_2 and estrone) stimulated final oocyte maturation. 2-hydroxy E_2 ($2HE_2$) acts as the local signal for induction of steroidogenic shift, inhibiting C19 - C18 pathway and concurrently stimulating C21 progestin pathway resulting in the secretion of the maturation-inducing steroid (MIS). $2HE_2$ is a potent inhibitor of P_{450} aromatase resulting in suppression of E_2 synthesis. These effects are mediated through both E_2 and adrenergic receptors. The post-receptor signaling mechanisms involve suppression of cAMP-protein kinase A, and up regulation of MAP kinase and protein kinase C pathways. $2HE_2$ stimulates prostaglandin ($PGF_{2\alpha}$ and PGE_2) secretion. Our results show that ovarian follicles are good model systems for catecholestrogen research.

* Corresponding author. E-mail: kpjoybhu@gmail.com.
# E-mail: chauberadha@rediffmail.com.

Introduction

In vertebrates, the brain-pituitary-gonad (BPG) - endocrine axis is the key regulator of reproductive activity. Gonadotropin-releasing hormone (GnRH), gonadotropins (FSH and LH) and gonadal steroids are thought to be the key players in the BPG axis, acting in a cascade manner in the regulation of gametogenesis, final oocyte maturation (FOM) and ovulation. A number of recent studies implicate other endocrine and paracrine/autocrine factors in the regulation of gonadal functions (Tyler and Sumpter, 1996). Involvement of growth hormone (GH), insulin, insulin-like growth factor I and II (IGF-I and IGF-II) has been demonstrated in follicular development, acquisition of maturational competence and maturation directly or synergistically by influencing steroidogenic activity (Van der Kraak et al., 1990; Kagawa et al., 1994; Srivastava and Van der Kraak, 1994; Maestro et al., 1997; Negatu et al., 1998; Weber et al., 2007). Our recent research has shown that ovarian catecholestrogens (CEs) and arginine vasotocin act as paracrine/autocrine factors participating in this hormonal cascade. The involvement of paracrine/autocrine factors strongly indicate that control of FOM and ovulation involves more complex co-lateral cascades than previously thought to be. In this chapter, we review the progress made on the role of CEs, as potential modulators of steroidogenic shift during FOM and ovulation as well as the molecular mechanisms underlying the process.

Catecholestrogens: An Overview

The major estrogens estradiol and estrone are removed from circulation through stepwise enzymatic transformations involving hydroxylation, methylation and conjugation to form water-soluble compounds for excretion, collectively called as catecholestrogens (Fishman, 1983; Merriam and Lipsett Eds. 1983; Ball and Knuppen, 1990; Martucci and Fishman, 1993). The hydroxylations involve estrogen-2/4-hydroxylases (CYP_{450} 1A1 and CYP_{450}1B1), a superfamily of heme-containing monooxygenases (Figure 1). The hydroxylated estrogens at 2- or 4-carbon positions are subsequently degraded by catechol-o-methyltransferase (COMT) to form the respective methoxyestrogens. O-methylation is a common biochemical link in estrogen and catecholamine metabolism so that each one can influence the activity of the other. The extensive tissue distribution of CEs particularly in estrogen-synthesizing and target organs indicates that they are not necessarily inactive intermediates of the detoxification pathway but implicates them in a variety of tissue-specific functions (Merriam and Lipsett Eds. 1983). CEs influence catecholamine metabolism and activity by inhibiting tyrosine hydroxylase, the rate-limiting enzyme in the biosynthesis of catecholamines (Saligaut et al., 1993; Chaube and Joy, 2003) or as a high affinity substrate of COMT (Timmers and Lambert, 1989). CEs participate in embryo implantation, gonadotropin release, parturition, uterine weight, egg transport in cycling rats, ovarian steroidogenesis, angiogenesis in ovarian follicles and prostaglandin production (Spicer and Hammond, 1989; Tekpetey and Armstrong, 1991, 1994; Philips et al., 2004; Parada-Bustamante et al., 2007; Basini et al., 2008). CEs may also elicit non-hormonal effects: 2-hydroxyE_2 (2HE_2) and 2-methoxyE_2 (2ME_2) have a higher antioxidant activity than E_2 and protective in neuronal and cardiovascular tissue (Picazo et al., 2003).

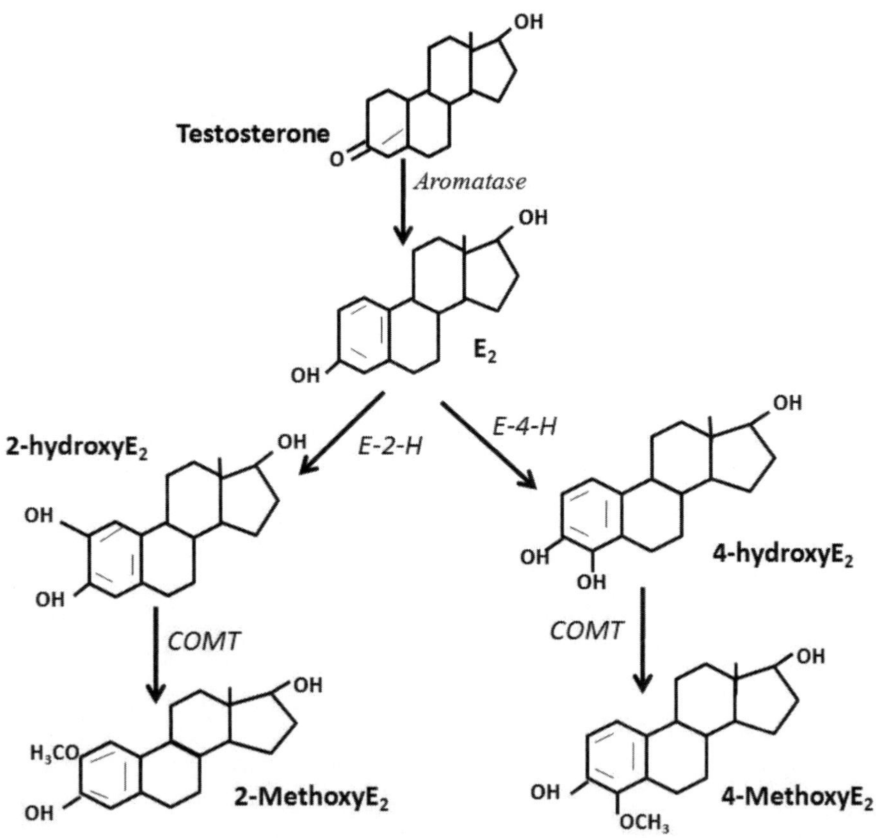

Figure 1. A Schematic diagram showing catecholestrogen formation from estradiol-17β (E_2). E-2-H, Estrogen-2-hydroxylase; E-4-H, Estrogen-4-hydroxylase; COMT, Catechol-o-methyl transferase.

In contrast, hydroxyestrogens generate free radicals from reductive-oxidative cycling (oxidization to ortho-quinone derivatives with concomitant formation of reactive oxygen species, ROS), which cause DNA damage (Nutter et al., 1991). Methylation by COMT is, therefore, physiologically significant. $2HE_2$ is methylated by COMT at a faster rate than $4HE_2$ and free radicals are not so easily generated compared with $4HE_2$ (Tsuchiya et al., 2005). For these reasons, in mammals, E_2 hydroxylation has been investigated for its role in local homeostasis of estrogens, and CEs have been implicated for their involvement in breast and endometrial cancers (Pasqualini, 2004; Salama et al., 2008). $2ME_2$ has been shown to have antitumor and anti-angiogenic effects (Pribluda et al., 2000; Lakhani et al., 2003; Mooberry, 2003; Muek and Seeger, 2010). Therefore, the naturally- occurring CEs have a critical role in both physiological and pathophysiological processes.

Ovarian Catecholestrogen: Distribution, Dynamics and Regulation

In mammalian ovary, follicular layer (theca and granulose) is the major source of CEs with the granulose cells as the main contributor (Spicer and Hammond, 1989). Ovarian activity of the CE synthesizing (EH) and degrading (COMT) enzymes was reported for the first time in a lower vertebrate (*Clarias batrachus*) by us (Senthilkumaran and Joy, 2001). The enzymes showed periovulatory changes with highest EH activity around the time when the eggs were readily strippable and COMT activity increased towards post stripping period. This study gave the first indication of the occurrence of CEs in fish ovary and provided the impetus to take up extensive research in another catfish (*Heteropneustes fossilis*). EH activity is the highest in the liver, followed by brain and ovary (Chourasia and Joy, 2010). Subcellular fraction studies have shown that the enzyme is distributed largely in the microsomal fraction of the tissues. The follicular layer is the main site of EH activity, as in mammals and the denuded oocytes (after hyalouronidase digestion) exhibit very low enzyme activity. Ovarian EH activity showed significant seasonal variation with peak activity in spawning phase as well as periovulatory changes with high activity when the eggs are readily strippable, as in *C. batrachus*. A direct measurement of CEs was achieved with the help of high performance liquid chromatography with electrochemical detector (Mishra and Joy, 2006a). In this study, both $2HE_2$ and $2ME_2$ were characterized and the steroids showed differential and significant seasonal variation. The steroids were not detected in resting phase ovaries but appeared with the start of recrudescence. During recrudescence, both E_2 and CEs showed inverse relationship. Ovarian E_2 level peaked up in preparatory phase fish with peak vitellogenic activity. In the prespawning phase with postvitellogenic follicles, E_2 level declined and $2HE_2$ appeared, which peaked in the spawning phase. $2ME_2$ was not detected in the seasonal samples but after gonadotropin (hCG) challenge, it was measured after egg stripping. These observations indicate that E_2 is hydroxylated and methylated in the ovary and CE formation is a means of removal of the parent steroid. The temporal relationship of the steroids is adapted to their physiological roles. E_2 is well known for its role in the induction of hepatic vitellogenin during follicular growth (vitellogenesis). It maintains postvitellogenic follicles in meiotic arrest by up regulating cAMP- protein kinase A (PKA) signaling mechanism (Jalabert et al., 1991). Hydroxyestrogens, on the other hand, elicit antiestrogenic activity just as synthetic antiestrogens like tamoxifen and clomiphene citrate that compete with estrogens for the receptors, induce FOM and ovulation in fishes (Mishra and Joy, 2006b).

The seasonality in EH activity and CE levels indicate tight regulation of CE metabolism by factors originating from external (photoperiod and temperature) and internal (BPG-endocrine axis) environments. The intrinsic factors controlling EH activity were investigated under both *in vivo* and *in vitro* conditions. Human CG, GH, insulin and IGF-1, which are known to induce FOM in teleosts, including the catfish, stimulate EH activity. IGF-1 and hCG appear to be highly effective in this regard (Chourasia and Joy, 2008a). These hormones are implicated in steroid shift during the FOM (Kagawa et al., 1994; Maestro et al., 1997; Weber and Sullivan, 2000; Mishra and Joy, 2006c). This positive relationship implies that CEs are important links in the regulatory cascade of FOM and ovulation. On the other hand, steroids such as progestins, estrogens, androgens and corticosteroids are known to suppress EH activity (Chourasia and Joy, 2010). These steroids are functionally active hormones, precursors or metabolites and many are detected or identified in fish ovary. The suppressing activity of the steroids varied greatly depending on their chemical nature. Progesterone (P_4) was found to be the strongest inhibitor of EH activity, as was reported in mammalian tissues like liver and ovarian follicles (Brueggemeier, 1983; Mondschein et al., 1987). $2HE_2$ is a strong inhibitor of EH activity, next to P_4. The suppressing effect may be explained by feedback inhibition of the product. P_4 can be used as an effective inhibitor to intervene CE metabolism and function.

Catecholestrogens and Final Oocyte Maturation

Estrogens are generally ineffective or inhibit FOM *in vitro or in vivo* in several teleosts, including catfish (Nagahama, 1987; Jalabert et al., 1991; Mishra and Joy, 2006b;). On the other hand, antiestrogens such as tamoxifen and clomiphene citrate that compete with estrogens for the receptors induce FOM and ovulation in fishes (Donaldson and Hunter, 1983; Mishra and Joy, 2006b). CEs are naturally–occurring antiestrogens (Merriam and Lipsett Eds. 1983), and this property varies with the nature of the CEs. 2-hydroxyestrogens are more abundant with low receptor affinity and dissociate rapidly. 4-hydroxyestrogens, on the other hand, have higher affinity and longer receptor association. Hence 2-hydroxyestrogens are more antiestrogenic than 4-hydroxyestrogens. Involvement of CEs on FOM was reported for the first time from our laboratory (Senthilkumaran and Joy, 2001). The hydroxyestrogens tested in the *in vitro* study stimulated germinal vesicle breakdown (GVBD) in both *C. batrachus* and *H. fossilis* but the response varied with the nature of the steroid and the species. In the GVBD assay, 2-hydroxyestrogens ($2HE_2$ and $2HE_1$) were more effective than the 4-hydroxyestrogens ($4HE_2$ and $4HE_1$). $2HE_2$ was the most effective and $4HE_1$ the least effective. The GVBD response was higher in *C. batrachus* than *H. fossilis*. The GVBD response was routinely studied in *H. fossilis* in various experiments with CEs (Mishra and Joy 2006b, c; Chourasia and Joy, 2012a). Hydroxyestrogen are strong inducers of GVBD but methoxyestrogens are poor inducers, most of the follicles were dead during the incubation (Mishra and Joy, 2006b). This suggests that methoxyestrogens may be apoptotic in nature. The hydroxyestrogen-induced GVBD response was inhibited to varying degrees by actinomycin D, a transcriptional inhibitor and cycloheximide, a translational inhibitor (Senthilkumaran and Joy, 2001). The inhibition by actinomycin D was higher than that of cycloheximide. It is suggested that both transcriptional activity (gene expression) and protein synthesis related to the synthesis of MIS, maturation-promoting factor (MPF) and other upstream or downstream regulators, are involved during the CE-induced GVBD. This inhibition pattern is similar to that elicited by gonadotropin-induced GVBD (Kagawa et al., 1994; Nagahama, 1994). Like gonadotropin, the CE-induced GVBD involves the synthesis of MIS (see below). A direct involvement of CE in ovulation process is not evident in our studies. Only a small proportion of the follicles (about 20%) elicited ovulation just as with other maturation –inducing steroid i.e., MIS (Mishra and Joy, 2006c).

Catecholestrogen Induces MIS Secretion and Steroidogenic Shift

In several species like salmonids and catfishes, estrogen level decreases with the completion of vitellogenesis and at the onset of FOM and ovulation, it further diminishes coincident with an increase in the secretion of MIS, which is generally the progesterone derivative 17, 20β-dihydroxy-4-pregnen-3-one (17, 20β-DP). The presence of E_2 maintains meiotic arrest by up regulating cAMP-PKA pathway (Jalabert et al., 1991; Nagahama, 1994). E_2 lowers MIS secretion through the inhibition of 20β-hydroxysteroid dehydrogenase, the enzyme that converts 17-hydroxyprogesterone (17P) into 17, 20β-DP (Fostier and Baek, 1994). The switch over from estrogenic to progestational phase in steroidogenesis (shift) is crucial for the resumption of meiosis but the mechanisms triggering the shift are not clearly understood (Lubzens et al., 2010). The suggested mechanisms include an increase in E_2 clearance (Baroiller et al., 1987; Young et al., 1997), decreased aromatase activity and expression (Nagahama et al., 1994; Senthilkumaran et al., 2004; Bobe et al., 2006), or low lyase activity of cytochrome P450 17α-hydroxylase/c17–20 lyase (Sreenivasulu and Senthilkumaran, 2009). In our initial study, cyanoketone and epostane (inhibitors of 3β-hydroxysteroid dehydrogenase) and aminoglutethimide (a broad inhibitor of steroidogenesis), did not show clearly any change in the CE-induced GVBD, which led us to believe that CEs may act directly without involving steroidogenesis (Senthilkumaran and Joy, 2001). To verify it, in our subsequent studies, follicular steroid profile was measured during the CE-induced GVBD (Mishra and Joy, 2006b, c; Chourasia and Joy, 2012a). These *in vitro* investigations have shown clearly, i) a down regulation C19- C18 pathway, ii) concurrently an up-regulation of progestin pathway, leading to the synthesis of 17, 20β-DP, and iii) stimulation of corticosteroids such as cortisol that has maturational activity. These changes are similar to those produced by hCG *in vivo* or *in vitro*, suggesting a similarity or continuity in the mode of action of these two hormonal agents. Furthermore, $2HE_2$ failed to induce GVBD in denuded oocytes (without follicular layer) at the same time stimulated steroidogenesis in follicular preparations (both theca and granulose), similar to intact follicles (Mishra and Joy, 2006c). Investigating further on the mode of action, the suppression of estrogen pathway was shown to be due to inhibition of aromatase activity (Chourasia and Joy, 2008b). $2HE_2$ is more potent than fadrozole (a known aromatase inhibitor) to inhibit follicular aromatase activity. The AI_{50} (50% inhibition of aromatase activity) for fadrozole is 4.4 nM and for $2HE_2$ is 0.864 nM in vitellogenic phase. Our unpublished data show that $2HE_2$ stimulates 20β-hydroxysteroid dehydrogenase (20β-HSD), which catalyzes the synthesis of 17,20β-DP.Thus, at the biochemical level, $2HE_2$ inhibits aromatase and stimulates 20β-HSD activities. Figure 2 summarizes the potential sites of $2HE_2$ action on C19 - C18 and C21 pathways. Under the influence of LH surge, E_2 is catabolised by EH to form $2HE_2$, which inhibits aromatase activity (product inhibition). In this manner, the readily available pool of E_2 is removed and its further synthesis is blocked. The hydroxyestrogen concurrently stimulates 20β-HSD to convert 17P to 17, 20β-DP. The hydroxyestrogen may act at other enzyme-controlled steps in steroidogenesis, which needs to be investigated. It is summed up that the surgic secretion of LH stimulates CE synthesis which, in turn, causes the steroidogenic shift to induce MIS secretion. The CE-triggered events- E_2 removal and MIS production- downregulate cAMP-PKA activity, lifting meiotic arrest (Mishra and Joy, 2006d). In this model, CE is considered the proximate link in the LH cascade of regulation of FOM and ovulation, and the local signal that triggers downregulation of aromatase synthesis and function. This model is not different from the earlier ones proposed around P_{450} aromatase but pinpoints the trigger of steroidogenic shift (Nagahama, 1994; Senthilkumaran et al., 2004). It will be interesting to know the role and regulation of CE in species having dual gonadotropins (FSH and LH).

Catecholestrogens and Prostaglandin Secretion

Prostaglandins (PGs) are known to act as local regulators in various reproductive processes such as ovulation, fertilization, implantation and parturition (Goetz et al., 1991; Patino and Sullivan, 2002). Both $PGF_{2\alpha}$ and PGE have been described in teleost ovary, which show differential distribution. PGE level declined prior to, and during ovulation but $PGF_{2\alpha}$ level remained high through the completion of ovulation or even after ovulation up to 24 hr (Cetta and Goetz, 1982). In catfish, both the PGs showed significant seasonal variation with peak levels in spawning phase and periovulatory changes after hCG treatment (Chourasia and Joy, 2012b). $PGF_{2\alpha}$ is more potent to stimulate GVBD than PGE_2.

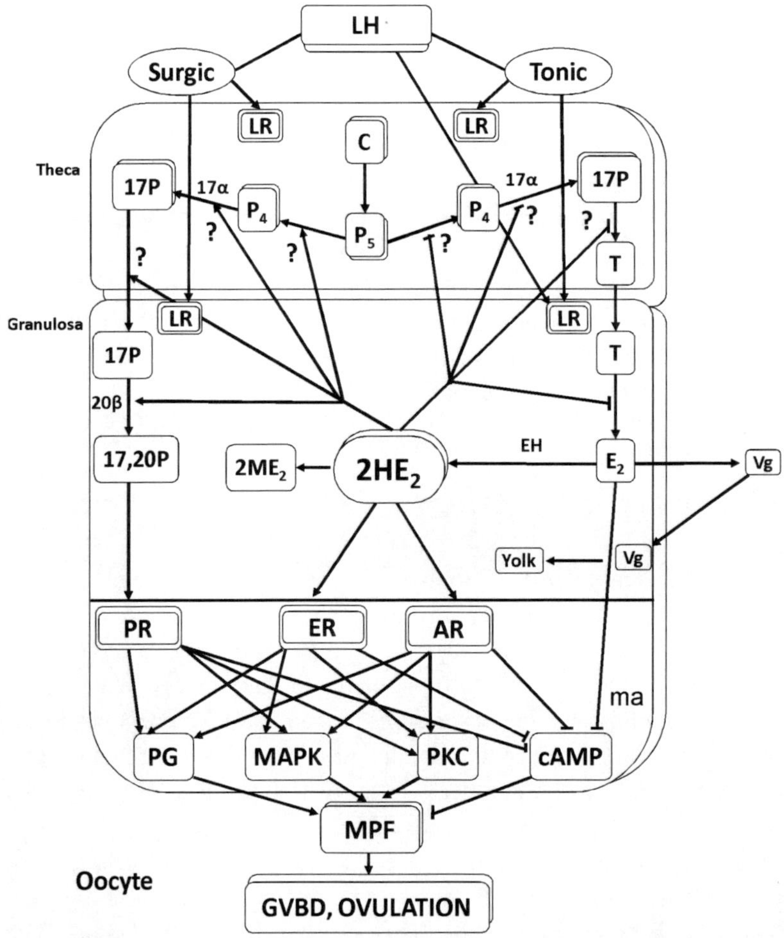

Figure 2. A schematic diagram showing potential sites of action of 2-hydroxyestradiol-17β ($2HE_2$) during final oocyte maturation in catfish. A- aromatase, AR- adrenoceptor, cAMP- cyclic adenosine monophosphate, E_2- estradiol-17β, ER- estrogen receptor, EH- estrogen-2-hydroxylase, GVBD- germinal vesicle breakdown, , LR- LH receptor, ma- meiotic arrest, $2ME_2$- 2-Methoxyestradiol-17β, MAPK- mitogen-activated protein kinase kinase, MPF- maturation-promoting factor, PG- prostaglandins, PR- progestin receptor, PKC-protein kinase C, T- testosterone, Vg- vitellogenin, 17α- 17α-hydroxylase, 20β- 20β-hydroxysteroid dehydrogenase, 17P- 17-hydroxyprogesterone, 17, 20P- 17, 20β-dihydroxy-4-pregnen-3-one.

A functional relation between CEs and prostaglandins has been demonstrated in mammals. In uterine homogenates of rat and human, $2HE_2$ stimulated PGF production more than E_2 (Kelly et al., 1983). The authors showed that $2HE_2$ stimulated PGF production 23 times more than E_2 (5 times) from the basal (control) level in rat uterine homogenates incubated with ^{14}C labeled arachidonic acid. The enhanced activity has been attributed to the presence of a catechol structure, which confers increased chemical reactivity. The catecholamines, epinephrine and norepinepherine also elicit stimulatory effects on $PGF_{2\alpha}$ (Kelly et al., 1983; Skarzynski et al., 1999). In view of the significance of PGs in FOM and ovulation, we investigated the role of CEs on PG secretion *in vitro* (Chourasia and Joy, 2012b). The results show that $2HE_2$ is more effective, followed by $2ME_2$ and E_2. The stimulatory effect of E_2 may be mediated through its conversion into CEs. $2HE_2$ stimulates PGs differentially: PGE_2 level has increased about 2-3 times higher than $PGF_{2\alpha}$, though the latter is more effective in inducing ovulation than the former. The functional significance of $2HE_2$–induced stimulation of both PGE_2 and $PGF_{2\alpha}$ can be correlated with their stimulatory role in FOM and ovulation. It is believed that CEs may act through PGs to stimulate these processes. The functional significance of $2ME_2$ stimulation of PGs in the catfish ovary may be related to a different function. $2ME_2$ was detected in spawning phase fish ovaries filled with postovulatory follicles. Since $2ME_2$ is an apoptotic and antiangiogenic agent and its analogues have been used in cancer treatment (Lakhani et al., 2003; Mooberry, 2003), it is very likely that the methoxyestrogen-stimulated PGs may be involved in follicular regression and atresia, a process active in ovaries soon after spawning. PGs are known luteolysins in mammals (Silvia et al., 1991). $2ME_2$ may act through PGs to induce luteolysis of the transient 'corpus luteum' or postovulatory follicle mass.

Mechanism of Action of Catechoestrogens

Receptors involved: The hydroxyl group at C-2 or C-4 of the Ring A structure of CEs alters its properties as an estrogen and makes it superficially resembling catecholamine neurotransmitters (MacLusky et al., 1983). This property of both a steroid and catechol renders the molecules interaction with both estrogen and catecholamine binding sites (receptors). However, binding and receptor studies in mammalian models have shown that the interaction is varied and complex depending on the tissues and species, chemical nature of the CEs, parameters of the study and a number of other variables. The interactions may involve only classical estrogen receptors (ERα and ERβ) or both estrogen and adrenergic (α-adrenoceptor and β-adrenoceptor) receptors, or even novel putative catecholestrogen binding sites located in cytosol or membrane, which may function as intracellular transcription factors potentially related to ERs (Merriam and Lipsett Eds. 1983; Etchegoyen et al., 1986; Vandewalle et al., 1988; Spicer and Hammond, 1989; van Aswegen et al., 1989; Tekpetey and Armstrong, 1994; Das et al., 1997; Philips et al., 2004; Chiesa et al., 2008). The receptors mediating CE effects have been shown to vary with the nature of hydroxylation. In human breast cell lines, $4HE_2$ was similar to E_2 in its ability to bind to, and activate the classical ERα, but did not show any interaction with α_2- adrenoceptor (Chiesa et al., 2008). On the other hand, $2HE_2$, E_1 and $4HE_1$ competed with both the α_2-adernoceptor and estrogen receptor. In earlier studies, $2HE_2$ has been shown to interact with β-adrenoceptor but not with α-adrenoceptor (Etchegoyen et al., 1986; Spicer and Hammond, 1988).

Estrogens *per se* do not support final maturational activity (Jalabert et al., 1991) and inhibit the stimulatory effect of gonadotropin, pituitary extract, hydrocortisone and $2HE_2$ (Mishra and Joy, 2006b). The study by Mishra and Joy (2006b) indicates an apparent competition between E_2 and $2HE_2$ for the same sites during FOM. These workers have shown that clomiphene citrate stimulated GVBD though moderately in a concentration-dependent manner. In co-incubation and pre-incubation studies with catfish ovarian follicles, tamoxifen inhibited the stimulatory effect of 2-OHE_2 on C_{21} steroids (P_4, 17P, 17, 20α-DP and 17, 20β-DP) but restored the E_2 level (Chourasia and Joy, 2012a). Similarly, tamoxifen inhibited the stimulatory effect of $2HE_2$ on PG levels (Chourasia and Joy, 2012b). These results suggest direct antagonism between E_2 and $2HE_2$ on the receptors. It has been demonstrated that antiestrogens such as tamoxifen and clomiphene citrate that compete with estrogens for the receptors induce ovulation in fishes (Donaldson and Hunter, 1983).

To ascertain the intervention of adrenergic receptors in the mediation of $2HE_2$ effects on steroid profile and PGs, we used phentolamine and propranolol (α- and β- adrenoceptor blockers, respectively) in incubation studies (Chourasia and Joy, 2012a,b). The adrenoceptor blockers produced differential effects on the steroids: the stimulatory effect of $2HE_2$ on C_{21} steroids was suppressed and the inhibitory effect on E_2 was reversed. The α-blocker was found more effective to alter steroid production than the β-blocker. A similar effect was found with PG secretion but $PGF_{2\alpha}$ was inhibited more than PGE_2. Previously, using GVBD as the end point assay, we found that the β-adrenoceptor blocker inhibited only mildly the $2HE_2$ effect but not the α–adrenoceptor blocker (Senthilkumaran and Joy, 2001). It is to be noted that only partial reversal effect is seen with each of the receptor type blockade, the other receptors are open for interaction. Blocking both receptor types simultaneously may give a complete inhibition of the CE effects.

Cell signaling pathways: Multiple cell signaling mechanisms are involved in teleost gonadotropin-induced steroidogenesis. These can be divided into the cAMP-PKA-dependent pathway, and independent pathways such as PKC/Ca^{2+}, arachidonic or mitogen-activated protein kinase kinase-MAP kinase (Tan et al., 1986; Van der Kraak and Wade, 1994; Planas et al., 1997; Rosenfeld et al., 2007; Mazon et al., 2011). On the other hand, activation of PKC pathway and elevation of Ca^{++} level stimulate, and activation of the cAMP pathway inhibits PG secretion in fish (Kellner and Van der Kraak, 1992). Synthesis of PGs by frog ovarian follicles or components is suppressed by exogenous cAMP (Kong et al., 1999). It has been reported that phorbolesters (analogues of diacylglycerol) and calcium ionophore stimulate both PKC activity and PG synthesis in mammalian granulosa cells and fish ovarian follicles (Veldhuis and Demers, 1989; Wang and Leung, 1989; Goetz, 1991; Kellner and Van der Kraak, 1992). Effects of CEs on cAMP are at variance with different studies. CEs significantly induce β-adrenoceptor activity, which is a known stimulator of cAMP (Spicer and Hammond, 1988). $2HE_2$ was shown to enhance cAMP production stimulated by gonadotropin or epinephrine, and enhanced progesterone secretion (Spicer and Hammond, 1989). Steegborn et al. (2005) reported that CEs inhibit adenylyl cyclase through a novel mechanism by forming a complex with the enzyme active center Mg^{++} ions. In agreement with this, Mishra and Joy (2006d) demonstrated that total follicular cAMP was inhibited by $2HE_2$. The functional relationship between CEs and other signaling mechanisms is less known. Therefore, we used specific blockers of cell signaling pathways to demonstrate their involvement in CE-induced GVBD (Mishra and Joy, 2006d, e, f). There is an inverse relation between follicular cAMP level and MIS secretion. Cyclic AMP or cAMP-elevating drugs (3-isobutyl-1-methylxanthine-IBMX, theophylline and caffeine) inhibited the $2HE_2$-induced GVBD response. The PKC inhibitor calphostin C at higher concentrations inhibited GVBD. MAP kinase inhibitor (PD098059) inhibited the steroid-induced MAP kinase expression and GVBD. These findings were further correlated with steroid changes during the $2HE_2$- induced GVBD (Chourasia and Joy, 2012a). Incubations of follicles with IBMX, chelerythrine (a PKC inhibitor) or PD098059 reversed the effect of $2HE_2$ differentially to varying degrees: C_{21} steroids (P_4, 17P, 17, 20α-DP and 17, 20β-DP) were inhibited and E_2 was stimulated. In other words, the blockade of the signaling pathways tends to reverse the $2HE_2$-induced steroidogenic pattern to the meiotic arrest conditions. During the $2HE_2$-GVBD, the parallel increase in MAP kinase expression was suppressed by the MAP kinase inhibitor (Mishra and Joy, 2006e). Considering this fact and the relatively low inhibitions produced by the blockers of other cell signaling pathways, the MAP kinase pathway is prominently involved in the $2HE_2$- stimulation of FOM. An interaction of protein phosphatases and the cell signaling mechanisms was also evident during the $2HE_2$-induced GVBD. Inhibition of protein phosphatases by okadaic acid stimulated $2HE_2$-induced GVBD, and reversed the inhibitory effects of IBMX, calphostin C and PD098059 (Mishra and Joy, 2006d,e,f). The stimulatory effect of $2HE_2$ on PGs was inhibited by the specific blockers of the cell signaling pathways (Chourasia and Joy, 2012b). The PKC inhibitor chelerythrine is the most potent, followed by PD098059 and IBMX, implying that CE acts through multiple cell signaling pathways to modulate PG secretion. The results of the inhibition studies on steroid profile and PG secretion suggest cross-talks at different levels of the cell signaling pathways.

Conclusion

In this chapter, we reviewed our work on ovarian catecholestrogens in catfish in relation to their distribution, seasonal and periovulatory dynamics and functional involvement in steroidogenic shift and FOM. Hydroxyestrogen is an important local paracrine/autocrine link in the LH-mediated regulatory cascade of FOM and ovulation. Catecholestrogens act through both estrogen and adrenergic receptors in this intervention. The post-receptor mechanisms include down regulation of cAMP-PKA pathway, and up regulation of prostaglandin, protein kinase C and MAP kinase pathways. Dynamics of MPF at molecular level is to be correlated in future studies. Extension of research in other fishes and studies on intervention of catecholestrogens in vitellogenesis and follicular atresia are other areas of potential research interests.

Acknowledgment

The research work on catfish catecholestrogens was supported by a project grant from DST, New Delhi (SP/SO/C-13/2001) to KPJ.

References

Ball, P. and Knuppen, R. (1990). Formation, metabolism, and physiologic importance of catecholestrogens. *Am. J. Obstet. Gynecol.,* 163, 2163-2170.

Baroiller, J. F., Fostier, A., Zohar, Y. and Marcuzzi, O. (1987). The metabolic clearance rate of estradiol-17 β in rainbow trout, *Salmo gairdneri* R., estimated by both single injection and constant infusion methods: increase during oocyte maturation. *Gen. Comp. Endocrinol.,* 66, 85-94.

Basini, G., Bussolati, S., Santini, S. E., Bianchi, F., Careri, M., Mangia, A., Musci, M. and Grasselli, F. (2008). Hydroxyestrogens inhibit angiogenesis in swine ovarian follicles. *J. Endocrinol.,* 199, 127-135.

Bobe, J., Montfort, J., Nguyen, T. and Fostier, A. (2006). Identification of new participants in the rainbow trout (*Oncorhynchus mykiss*) oocyte maturation and ovulation processes using cDNA microarrays. *Reprod. Biol. Endocrinol.,* 4, 39.

Brueggemeier, R. W., (1983). Kinetics of rat liver microsomal estrogen 2-hydroxylase: evidence for sex difference at initial velocity condition. *J. Biol. Chem.,* 369, 10239-10242.

Cetta, F. and Goetz, F. W. (1982). Ovarian and plasma prostaglandin E and F levels in brook trout (*Salvelinus fontinalis*) during pituitary-induced ovulation. 27, 1216-1221.

Chaube, R. and Joy, K. P. (2003). *In vitro* effects of catecholamines and catecholestrogens on brain tyrosine hydroxylase activity and kinetics in the female catfish *Heteropneustes fossilis*. *J. Neuroendocrinol.,* 15, 273-279.

Chiesa, I. J., Castillo, L. F. and Lüthy I. A. (2008). Contribution of α-2-adrenoceptors to the mitogenic effect of catecholestrogen in human breast cancer MCF-7 cells. *J. Steroid Biochem. Mol. Biol.,* 110, 170-175.

Chourasia, T. K. and Joy, K. P. (2008a). Estrogen-2/4-hydroxylase activity is stimulated during germinal vesicle breakdown induced by hCG, IGF-1, GH and insulin in the catfish *Heteropneustes fossilis*. *Gen. Comp. Endocrinol.,* 155, 413–421.

Chourasia, T. K. and Joy, K. P. (2008b). Ovarian P450 aromatase activity in the catfish *Heteropneustes fossilis*: Seasonal changes and effects of catecholestrogens. *Gen. Comp. Endocrinol.,* 156, 537-543.

Chourasia, T. K. and Joy, K. P. (2010). Seasonal variation in tissue estrogen-2/4-hydroxylases (EH) and in vitro effects of steroids on ovarian EH activity in the catfish *Heteropneustes fossilis*. *Steroids,* 75, 1097-1105.

Chourasia, T. K. and Joy, K. P. (2012a). *In vitro* effects of 2-hydroxyestradiol-17β on ovarian follicular steroid secretion in the catfish *Heteropneustes fossilis* and identification of the receptor and signaling mechanisms. *Gen. Comp. Endocrinol.,* 175, 500–513.

Chourasia, T. K. and Joy, K. P. (2012b). Role of catecholestrogens on ovarian prostaglandin secretion *in vitro* in the catfish *Heteropneustes fossilis* and possible mechanism of regulation. *Gen. Comp. Endocrinol.,* 177, 128-142.

Das, S. K. Taylor, J. A. Korach, K. S. Paria, B. C. Dey, S. K. and Lubahn, D. B. (1997). Estrogenic responses in estrogen receptor-α deficient mice reveal a distinct estrogen signaling pathway. *Proc. Natl. Acad. Sci. USA,* 94, 12786-12791.

Donaldson, E. M. and Hunter, G. A. (1983). Induced final maturation, ovulation and spermiation in cultured fish. In: W.S. Hoar and D. J. Randall (Eds.), *Fish physiology,* Vol IX B (pp. 347-506). New York, Academic Press.

Etchegoyen, G. S., Cardinali, D. P., Pérez, A. E. Tamayo, J. and Pérez-Palacios, G. (1986). Binding and effects of catecholestrogens on adenylate cyclase activity, and adrenoceptors, benzodiazepine and GABA receptors in guinea-pig hypothalamic membranes. *Eur. J. Pharmacol.,* 129, 1-10.

Fishman, J. (1983). Aromatic hydroxylation of estrogens. *Ann. Rev. Physiol.,* 45, 61-72.

Fostier, A. and Baek, H. (1994). Induction of maturation inducing steroid production in rainbow trout granulose cells: effect of estradiol on gonadotropin stimulated 20β-hydroxysteroid dehydrogenase activity. *Reprod. Nutr. Develop.,* 33, 81-82.

Goetz, F. W. (1991). The role of signal transduction in the control of fish reproduction. In: A. P. Scott, J. P. Sumpter, D. E. Kime and M. S. Rolfe (Eds.), *Proceedings of the 4th International symposium on the reproductive physiology of fish*, (pp. 210-214), Sheffield, FishSymp 91.

Goetz, F. W., Berndtson, A. K. and Ranjan, M. (1991). Ovulation: mediators at the ovarian level. In: P. K. T. Pang and M. P. Schreibman (Eds.), *Vertebrate endocrinology: fundamentals* and *biomedical implications,* Vol. 4 A (pp. 127-203). New York, Academic Press.

Jalabert, B., Fostier, A., Breton, B. And Weil, C. (1991). Oocyte maturation in vertebrates. In: P. K. T. Pang and M. P. Schreibman (Eds.), *Vertebrate endocrinology: fundamentals* and *biomedical implications,* Vol. 4 A (pp 23-90). New York, Academic Press.

Kagawa, H., Kobayashi, M., Hasegawa, Y. and Aida, K. (1994). Insulin and insulin-like growth factors I and II induce final maturation of oocytes of red seabream, *Pagrus major, in vitro*. *Gen. Comp. Endocrinol.,* 95, 293-300.

Kellner, R. G. and Van Der Kraak, G. (1992). Multifactorial regulation of prostaglandin synthesis in preovulatory goldfish ovarian follicles. *Biol. Reprod.,* 46, 630-635.

Kelly, R. W., Abel, M. H., Kennedy, R. L. and Taylor, P. L. (1983). Effects of catechol estrogens on the synthesis and metabolism of prostaglandins, In: G.R. Merriam and M. B. Lipsett (Eds.), *Catechol Estrogens* (pp. 203-214), New York, Raven Press.

Kong, K. Y., Chang, K., Im, W. B. and Kwon, H. B. (1999). *In vitro* ovulation and prostaglandin synthesis by ovarian follicles of *Rana dybowskii*. *Korean J. Biol. Sci.,* 3, 385-391.

Lakhani, N. J., Sarkar, M. A., Venitz, J. and Figg, D. (2003). 2-Methoxyestradiol, a promising anticancer agent. *Pharmacotherapy,* 23,165-172.

Lubzens, E., Young, G., Bobe, J. and Cerda, J. (2010). Oogenesis in teleosts: how eggs are formed. *Gen. Comp. Endocrinol.,* 165, 367-389.

Maestro, M. A., Mendez, E., Pairrizas M. and Gutierrez, J. (1997). Characterization of insulin and insulin-like growth factor-I ovarian receptors during the reproductive cycle of carp *(Cyprinus carpio)*. *Biol. Reprod.,* 56, 1126-1132.

MacLusky, N. J., Barnea, E. R., Clark, C. R. and Naftolin, F. (1983). Catechol estrogens and estrogen receptors. In: G. R. Merriam and M. B. Lipsett (Eds.), *Catechol Estrogens* (pp. 151-165), New York : Raven Press.

Martucci, C. P. and Fishman, J. (1993). P450 enzymes of estrogen metabolism. *Pharmacol. Ther.,* 57, 237-257.

Mazon, M. J., Gomez, A. and Zanuy, S. (2011). Follicle stimulating hormone activates the steroidogenic process via cAMP/PKA and MAP-kinase pathways in the ovary of European sea bass (*Dicentrarchus labrax*), In: K.P. Joy, R. M. Inbaraj, R. Kirubagaran and R. Chaube, (Eds.), *Procedings of the 9th international symposium on reproductive Physiology of fish* (pp.190-191). Cochin, *Indian J. Sci. Technol.*

Merriam, G. R. and Lipsett, M. B. (Eds.). (1983). *Catechol. Estrogens.* New York, Raven Press.

Mishra, A. and Joy, K. P. (2006a). HPLC-electrochemical detection of ovarian estradiol-17β and cateholestrogens in the catfish *Heteropneustes fossilis*: seasonal and periovulatory changes. *Gen. Comp. Endocrinol.,* 145, 84-91.

Mishra, A. and Joy, K. P. (2006b). Relative effects of estradiol-17β (E_2), catecholestrogens and clomiphene citrate on *in vitro* oocyte maturation in the catfish *Heteropneustes fossilis* (Bloch) and E_2 inhibition of 2-hydroxyestradiol-induced maturation. *Gen. Comp. Endocrinol.,* 147, 141-149.

Mishra, A. and Joy, K. P. (2006c). Effects of gonadotrophin *in vivo* and 2-hydroxyestradiol-17β *in vitro* on follicular steroid hormone profile associated with oocyte maturation in the catfish *Heteropneustes fossilis. J. Endocrinol.,* 189, 341-353.

Mishra, A. and Joy, K. P. (2006d). 2-hydroxyestradiol-17β-induced oocyte maturation: involvement of cAMP-protein kinase A and okadaic acid-sensitive protein phosphatases and their interplay in oocyte maturation in the catfish *Heteropneustes fossilis. J. Exp. Biol.,* 209, 2567-2575.

Mishra, A. and Joy, K. P. (2006e). Involvement of mitogen-activated protein kinase in 2-hydroxyestradiol-17β-induced oocyte maturation in the catfish *Heteropneustes fossilis* and a note on possible interaction with protein phosphatases, *Gen. Comp. Endocrinol., 147*, 32- 335.

Mishra, A. and Joy, K. P. (2006f). 2-Hydroxyestradiol-17β-induced oocyte maturation in catfish (*Heteropneustes fossilis*) involves protein kinase C and its interaction with protein phosphatases, *Comp. Biochem. Physiol. A Mol. Integr. Physiol.,* 144, 416-422.

Mondschein, J. S., Hammond, J. M. and Weisz, J. (1987). Characteristics of estrogen-2/4-hydroxylase of porcine ovarian follicles: influence of steroidal and nonsteroidal agents on the activity of the enzymes *in vitro. J. Steroid Biochem.,* 26, 121-124.

Mooberry, S. L. (2003). Mechanism of action of 2-methoxyestradiol: new developments. *Drug Resist. Updat.,* 6, 355-361.

Mueck, A. O. and Seeger, H. (2010). 2-Methoxyestradiol-biology and mechanism of action. *Steroids,* 75, 625-631.

Nagahama, Y. (1994). Endocrine regulation of gametogenesis in fish. *Inter. J. Dev. Biol.,* 38, 217-229.

Nagahama, Y. (1987). Endocrine control of oocyte maturation. In: D. O. Norris and R. E. Jones, R.E. (Eds.), *Hormones* and *reproduction in fishes, amphibians,* and *reptiles* (pp.171-202), New York : Plenum Press.

Nagahama, Y., Yoshikuni, M., Yamashita, M., and Tanaka, M. (1994). Regulation of oocyte maturation in fish. In: N. M. Sherwood and C. L. Hew (Eds.), *Fish physiology,* Vol. XIII, Molecular endocrinology of fish (pp. 393-439). New York: Academic Press.

Negatu, Z., Hsiao, S. M., and Wallace, R. A. (1998). Effects of insulin-like growth factor-I on final oocyte maturation and steroid production in *Fundulus heteroclitus. Fish Physiol. Biochem.,* 19, 13–21.

Nutter, L. M., Ngo, E. O. and Abul-Hajj, Y. J. 1991. Characterization of DNA damage induced by 3, 4-estrone-o-quinone in human cells. *J. Biol. Chem.,* 266, 16380-16386.

Parada-Bustamante, A., Orihuela, P. A., Ríos, M., Navarrete-Gómez, P. A., Cuevas, C. A., Velasquez, L. A., Villalón, M. J. and Croxatto, H. B. (2007). Catechol-o-methyltransferase and methoxyestradiol participate in the intraoviductal nongenomic pathway through which estradiol accelerates egg transport in cycling rats. *Biol. Reprod.,* 77, 934-941.

Pasqualini, J. R. (2004). The selective estrogen enzyme modulators in breast cancer: a review. *Biochem. Biophys. Acta,* 1654, 123-143.

Patino, R. and Sullivan, C. (2002). Ovarian follicle growth, maturation, and ovulation in teleost fish. *Fish Physiol. Biochem.,* 26, 57-70.

Philips, B. J., Ansell, P. J., Newton, L. G., Harada, N., Honda, S., Ganjam, V. K., Rottinghaus, G. E., Welshons, W. V. and Lubahn, D. B. (2004). Estrogen receptor-independent catechol estrogen binding activity: protein binding studies in wild-type, Estrogen receptor-α KO, and aromatase KO mice tissues. *Biochemistry,* 43, 6698-6708.

Picazo, O., Azcoitia, I. and Garcia-Segura, L. M. (2003). Neuroprotective and neurotoxic effects of estrogens. *Brain Res.,* 990, 20-27.

Pribluda, V. S., Gubish, E. R., LaVallee, T. M., Treston, A., Swartz, G. M. and Green, S. J. (2000). 2-Methoxyestradiol: an endogenous antiangiogenic and antiproliferative drug candidate. *Cancer Metastasis Rev.,* 19, 173-179.

Planas, J. V., Goetz, F. W. and Swanson, P. (1997). Stimulation of brook trout ovarian steroidogenesis by gonadotropins I and II is mediated by the cyclic adenosine 3',5'-monophosphate/protein kinase A pathway. *Biol. Reprod.,* 57, 647-654.

Rosenfeld, H., Meiri, I. and Elizur, A. (2007). Gonadotropic regulation of oocyte development, In: P.J. Babin, J. Cerdà, and E. Lubzens (Eds.), *The fish oocyte: from basic studies to biotechnological applications* (pp. 175-202). The Netherlands, Springer.

Salama, S. A., Kamel, M., Awad, M., Nasser, A. H., Al-Hendy, A., Botting, S. and Arrastia, C. (2008). Catecholestrogens induce oxidative stress and malignant transformation in human endometrial glandular cells: protective effect of catechol-O-methyltransferase. *Int. J. Cancer,* 123, 1246-1254.

Saligaut, C., Bennani, S. and Bailhache, T. (1993). Catecholamine synthesis in the rainbow trout (*Onchorhynchus mykiss*) brain: modulaion of tyrosine hydroxylase activity. *Fish Physiol. Biochem.,* 11, 139-144.

Senthilkumaran, B. and Joy, K. P. 2001. Periovulatory changes in catfish ovarian oestradiol-17β, oestrogen-2-hydroxylase and catechol-O-methyltransferase during GnRH analogue-induced ovulation and *in vitro* induction of oocyte maturation by catecholestrogens. *J. Endocrinol.,* 168, 239-247.

Senthilkumaran, B., Yoshikuni, M. and Nagahama, Y. (2004), A shift in steroidogenesis occurring in ovarian follicles prior to oocyte maturation. *Mol. Cell Endocrinol.,* 215, 11-18.

Silvia, W. J., Lewis, G. S., McCracken, J. A., Thatcher, W. W. and Wilson, L. Jr. (1991). Hormonal regulation of uterine secretion of prostaglandin F2α during luteolysis in ruminants. *Biol. Reprod.* 45, 655-663.

Skarzynski, D. J., Uenoyama, Y., Kotwica, J. and Okuda, K. (1999). Noradrenaline stimulates the production of prostaglandin F2 α in cultured bovine endometrial cells. *Biol. Reprod.,* 60, 277-282.

Spicer, L. J. and Hammond, J. M. (1988). Effects of 2-hydroxyestradiol on the number of granulose cell β-adrenergic receptors. *Endocrinology,* 122, 2710-2717.

Spicer, L. J. and Hammond, J. M. (1989). Regulation of ovarian function by catecholestrogens: current concepts. *J. Steroid Biochem.,* 33, 489-501.

Sreenivasulu, G. and Senthilkumaran, B. (2009). A role for cytochrome P450 17α-hydroxylase/c17- 20 lyase during shift in steroidogenesis occurring in ovarian follicles prior to oocyte maturation, *J. Steroid Biochem. Mol. Biol.,* 115, 77–85.

Srivastava, R. K. and Van der Kraak, G. (1994). Insulin as an amplifier of gonadotropin action on steroid production: mechanisms and sites of action in goldfish prematurational full-grown ovarian follicles. *Gen. Comp. Endocrinol.,* 95, 60–70.

Steegborn, C., Livin, T. N., Hess, K. C., Capper, A. B., Taussig, R., Buck, J., Levin, L. R. and Wu, H. (2005). A novel mechanism for adenylyl cyclase inhibition from the crystal structure of its complex with catechol estrogen. *J. Biol. Chem.,* 280, 31754-31759.

Tan, J. D. Adachi, S. and Nagahama, Y. (1986). The *in vitro* effects of cyclic nucleotides, cyanoketone, and cycloheximide on the production of estradiol-17β by vitellogenic ovarian follicles of goldfish (*Carassius auratus*), *Gen. Comp. Endocrinol.,* 63, 110-116.

Tekpetey, F. R. and Armstrong, D. T. (1991). Steroidogenic response of rat and pig luteal cells to estradiol-17β and catecholestrogens in vitro. *Biol. Reprod.,* 45, 498-505.

Tekpetey, F. R. and Armstrong, D. T. (1994). Catecholestrogen modulation of steroid production by rat luteal cells: mechanism of action. *Mol. Cell Endocrinol.,* 101, 49-57.

Timmers, R. J. M. and Lambert, J. G. D. (1989). Catechol-O-methyltransferase in the brain of the male African catfish, *Clarias gariepinus*: distribution and significance for the metabolism of catecholestrogens and dopamine. *Fish Physiol. Biochem.,* 7, 201-210.

Tsuchiya, Y. Nakajima, M. and Yokoi, T. (2005). Cytochrome P450-mediated metabolism of estrogens and its regulation in human, *Cancer Lett.,* 227, 115-24.

Tyler, C. R. and Sumpter, J. P. (1996). Oocyte growth and development in teleosts. *Rev. Fish Biol. Fish.,* 6, 287–318.

Van Aswegen, C. H., Purdy, R. H. and Wittliff, J. L. (1989). Binding of 2-hydroxyestradiol and 4-hydroxyestradiol to estrogen receptors from human breast cancers, *J. Steroid Biochem.,* 32, 485-492.

Van der Kraak, G., Rosenblum, P. M. and Peter, R. E. (1990). Growth hormone-dependent potentiation of gonadotropin-stimulated steroid production by ovarian follicles of the goldfish. *Gen. Comp. Endocrinol.,* 79, 233–239.

Van Der Kraak, G. J. and Wade, M. G. (1994). A comparison of signal transduction pathways mediating gonadotropin actions in vertebrates. In: K. G. Davey, R. E. Peter and S. S. Tobe (Eds.), *Perspectives in comparative endocrinology*, (pp. 59–63), Ottawa: National Research Council of Canada.

Vandewalle, B., Hornez, L. and Lefebvre, J. (1988). Characterization of catecholestrogen membrane binding sites in estrogen receptor positive and negative human breast cancer cell-lines, *J. Recept Res.,* 8, 699-712.

Veldhuis, J. D. and Demers, L. M. (1989). Mechanism(s) by which activation of protein kinase C is coupled to prostacyclin synthesis in granulosa cells. *Mol. Cell Endocrinol.,* 63, 219- 226.

Wang, J. and Leung, P. C. (1989). Synergistic stimulation of prostaglandin E2 production by calcium ionophore and protein kinase C activator in rat granulosa cells. *Biol. Reprod.,* 40,1000-1006.

Weber, G. M., Moore, A. B. and Sullivan, C. V. (2007). In vitro action of insulin-like growth factor-I on ovarian follicle maturation in white perch (*Morone americana). Gen. Comp. Endocrinol.,* 151, 180-187.

Weber, G. M. and Sullivan, C. V. (2000). Effects of insulin-like growth factor-I on oocyte maturation and ovarian steroidogenesis in striped bass, *Morone saxtilis. Biol. Reprod.,* 63, 1049-1057.

Young, G., Todo, T., Kobayashi, T., Guan, G., Nagahama, Y. (1997). Steroidogenesis by the salmonid ovarian follicle: the two cell type model revisited. In: S. Kawashima and S. Kikuyama (Eds.), *Advances in comparative endocrinology,* Vol. 2, (pp. 1443-1449), Monduzzi Editore, Bologna.

In: Sexual Plasticity and Gametogenesis in Fishes
Editor: Balasubramanian Senthilkumaran ISBN: 978-1-62618-848-8

Chapter IX

Understanding the Mechanism of Oocyte Maturation and Ovulation to Induce Spawning for Seed-Production in the Japanese Eel: Tough and Challenging Task

H. Kagawa[*]
Department of Marine Biology and Environmental Sciences,
Faculty of Agriculture, University of Miyazaki, Miyazaki 889-2192, Japan

Abstract

Reduction in eel resources and the catches of glass eels as seedlings for aquaculture have been a serious concern in recent years in both Europe and East Asia. Thus, technical advancement to produce eel seeds for artificial cultivation is strongly desired. Fundamental information on oocyte maturation and ovulation, and its application to artificial induction of sexual maturation are needed to produce good quality seeds of the Japanese eel. This review introduces hormonal mechanisms of cytoplasmic maturation (such as hydration, lipid coalescence, and clearing of the ooplasm) and the maturational competence (the ability to respond to maturation-inducing steroid) and nuclear maturation (germinal vesicle breakdown, GVBD). In addition, ordinary and newly developed methods for induction of spawning have been described.

Keywords: Eel, Oocyte maturation, Artificial hormonal treatment, Spawning, Seed production

[*] E-mail: kagawa@cc.miyazaki-u.ac.jp.

Introduction

The Japanese eel, *Anguilla japonica*, is an economically important fish for Japanese food culture and for Japanese freshwater fish aquaculturists and scientists. The grilled eel dish called "kabayaki" is one of the representative dishes of traditional Japanese food. However, reduction in eel resources and the catches of glass eels as seedlings for aquaculture have been a serious concern in recent years in both Europe, North America and East Asia. The stock of the European eel is obviously in decline (Dekker, 2003). Recruitment has declined since 1980, to 10% of former levels. Landings of the yellow/silver eel fisheries have been in decline since the mid-1960s. Similar pattern of decrease in eel commercial caches have been found in North America (Casselman, 2003) and Japan (Tatsukawa, 2003). To counteract the decrease in eel resources, technical advancement to produce eel seeds for cultivation and recover of eel resources is very much desired.

However, male and female eels are sexually immature under normal cultivate conditions (Yamamoto et al., 1974; Dufour et al. 1988). Cultivated female eels are sexually immature, and their gonadosomatic index (GSI) values slightly increase to about 1-2% in fall but never grow up. Moreover, silver female eels that migrate down rivers in fall also have GSI values of about 1-4% and their ovaries contain oocytes at the oil droplet stage or the primary yolk globule stage (Uto et al., 2004). Sexually mature male and females have never been obtained from the wild until recently (Tsukamoto et al., 2010). Therefore, artificial induction of sexual maturation in both male and female eels has been examined since the 1930s in some European countries and since 1960s in Japan. Repeated injection of mammalian gonadotropins (Yamamoto et al., 1972; Miura et al. 1991) induced spermatogenesis and spermiogenesis in male eels (Ohta and Unuma, 2003). Repeated injections of carp (Fontaine et al., 1964) and salmon pituitary homogenates or extracts induce oocyte growth (vitellogenesis) and full-grown female eels were obtained (Yamamoto et al., 1974). Using these methods, Yamamoto and his colleagues for the first time succeeded to obtain fertilized eggs and eel larvae (Yamamoto and Yamauchi 1974; Yamauchi et al., 1976). Since then, repeated injections of piscine pituitary extracts are now routinely used for induction of vitellogenesis in female eels (Kagawa, 2003). However, the percentages of ovulated females are low, and even if ovulated eggs are obtained, these eggs show low fertility and hatchability. We developed the method for induced maturation and ovulation in the Japanese eel by using the maturation-inducing steroid (17, 20β-dihydroxy-4-pregnene-3-one, DHP) (Kagawa, 2003). Although 100% of female eels are matured and ovulated in response to DHP injection, egg quality are still low (both fertility and hatching rates are 30 and 20 %, respectively) and vary depending to female used. Therefore, fundamental information on oocyte maturation and ovulation, and its application to artificial induction of sexual maturation are needed to produce good quality seeds of the Japanese eel. In the present review, we illustrate basic knowledge of mechanism of oocyte maturation and ovulation, and its application to artificial induction of sexual maturation of female eels.

Induction of Sexual Maturation in Male and Female Eels by Newly Developed Methods

Cultivated male and female eels are sexually immature in general and are never mature under commercial rearing conditions (Yamamoto et al., 1972). One exceptional report describes spermatogenesis and spermiation of two cultivated male eels which were maintained under natural environmental conditions in freshwater or seawater (Matsubara et al., 2008). Ordinary methods for induction of sexual maturation have been proposed mainly from Japanese scientists. Yamamoto and Yamauchi (1974) first succeeded to obtain fertilized eggs and larvae from sexually mature female and male Japanese eel by injection of salmon pituitary extracts (SPE) for female and of human chorionic gonadotropin (HCG) for male. Afterwards, weekly repeated injections of SPE or HCG have been used as ordinary methods for induction of vitellogenesis and of spermatogenesis in female and male eels respectively (Kagawa, 2003; Kagawa et al., 2005; Ohta et al., 1996b; Yamamoto and Yamauchi, 1974; Yamamoto et al., 1974). In male eels, repeated injections of HCG at 1 IU/g body weight once a week over 10 weeks efficiently induced spermatogenesis and spermiation, and 1–2 g of approximately 70% motile spermatozoa was obtained (Ohta et al., 1996b). However, weekly injection of HCG requires repetitive handling of the broodstock and substantial labor, time, and monitoring, resulting in stress to and increased mortality of the fish. An osmotic pump is a delivery device for long-term administration of drugs and hormones. Implantation of a single osmotic pump loaded with a gonadotropic hormone (e.g., pregnant mare serum gonadotropin or HCG) has been carried out experimentally in mammals (Gibson et al., 1994; Patton et al., 1990). We applied the osmotic pump first time on sexual maturation in both male and female eels (Figure 1). A newly introduced osmotic pump (Osmotic PumpType 2002; Alzet Osmotic Pumps Co., Cupertino, CA; diameter=7 mm, length=30 mm, reservoir volume=approximately 200 μl) that can release constant amounts of hormones for a long period was loaded with various amounts of GnRHa, HCG, or SPE. According to the manufacturer's instruction manual, this osmotic pump can release 5 μl of a solution per day for approximately 45–50 days when the fish are maintained at a water temperature of 20 °C.

Implantation of an HCG-loaded osmotic pump significantly increased the mean GSI value in a dose-dependent manner (1-50 IU/day)(Kagawa et al., 2009a). Spermiation was observed in all the HCG-administered experimental groups; the highest amount (1.32±0.30 g) was obtained from the fish administered with the 50 IU/day. Percent of sperm motilities are relatively low from 25-50% in the groups administered 5, 10, 25, and 50 IU/ day of HCG. Spermatocrit values were high in all the treatment groups, and no significant differences were observed among them. This is the first study demonstrating that implantation of a single osmotic pump loaded with HCG stimulates spermatogenesis and spermiation in sexually immature male Japanese eels. These data are comparable with those of the previous study showing that repeated weekly injection of HCG (~1 IU/g body weight) in immature male Japanese eels induces spermiation after the sixth injection (Ohta et al., 1996b).

Implantation of a SPE-loaded osmotic pump (0.5-2.2 mg/day) also stimulated spermatogenesis, although the GSI values were lower than those of the fish implanted with the HCG loaded osmotic pump. Implantation of a single osmotic pump loaded with various doses of GnRHa (0.94, 1.86, or 3.75 μg/day) did not stimulate spermatogenesis.

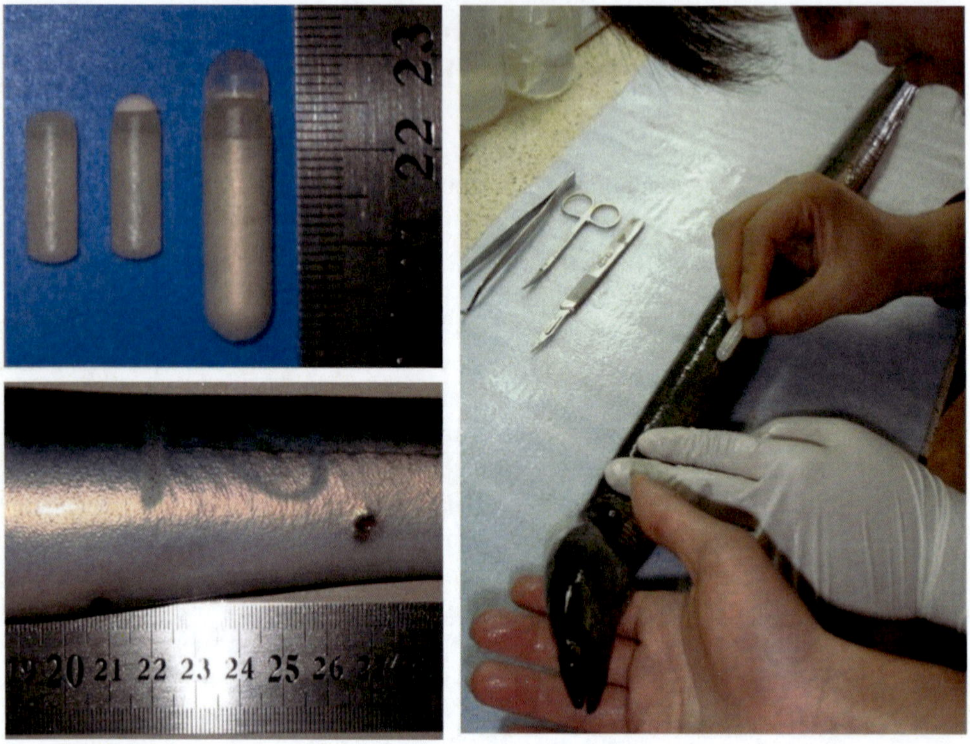

Figure 1. Photographs of different type of osmotic pumps (A) and after cutting the abdomen with a scalpel approximately 8mm (B) osmotic pump are implanted (C).

In the previous study, the samples were obtained after more than 10 weeks of injection, indicating that the milt weight and sperm motility gradually increase from 6 weeks onward and reach the maximal levels after 10 weeks, following which they remain constant. Therefore, if HCG is administered via an osmotic pump for more than 10 weeks, a useful quantity and good quality milt might be obtained. Recent studies in our laboratory have demonstrated that approximately 1 g of spermatozoa with relatively high sperm motility (40–80%) can be obtained from male Japanese eels receiving prolonged HCG treatment (10 weeks) via an osmotic pump (Kagawa et al., in preparation).

Full-grown cultivated female eels were obtained by intraperitoneal weekly injections of SPE (Kagawa, 2003) or by implantation of a single SPE-loaded osmotic pump with a long-term sustained hormone-release system. Sexually immature female eels for experiments are obtained by feminization using oral administration of estradiol-17β at a concentration of 10 mg/kg diet for 4 months during the juvenile stage, since high percentage of eels (more than 90%) become males under cultivate condition (Chiba et al., 1993). Initially, they have ovaries containing oocytes at the oil droplet stage. After 8 to 13 injections of SPE (20 mg/fish/week), almost all female eels became to posse ovaries containing full-grown oocytes (Kagawa, 2003; Ohta et al., 1996a). Implantation of a single SPE-loaded (1.5-4 mg/day/fish) osmotic pump also significantly stimulated vitellogenesis and subsequent implantation of the same osmotic pump yielded the full-grown female eels (Figure 2) (Kagawa et al., 2013a). Implantation of a HCG-loaded osmotic pump (75 IU/day/fish) also stimulated vitellogenesis, although the GSI values were lower than those of the fish implanted with the SPE-loaded osmotic pump. However, implantation of a single osmotic pump loaded with GnRHa (5.25 g/day/fish) did not stimulate vitellogenesis in the eels. The periods until oocytes of osmotic pump implanted female eels attain at migratory nucleus stage take longer than those of SPE injected female eels and the number of fish attained full-grown are less than those of SPE injected female eels. These results indicate also in female eels that osmotic pump is useful for induction of sexual maturation and for obtaining full-gown experiment fish.

Hydration of Oocyte

After a relatively long period of growth (the vitellogenic phase), oocytes undergo maturation, accompanied by several maturational processes in the cytoplasm (such as hydration, lipid coalescence and clearing of the ooplasm) and in the nucleus (such as germinal vesicle breakdown, GVBD) followed by ovulation (Figure 3).

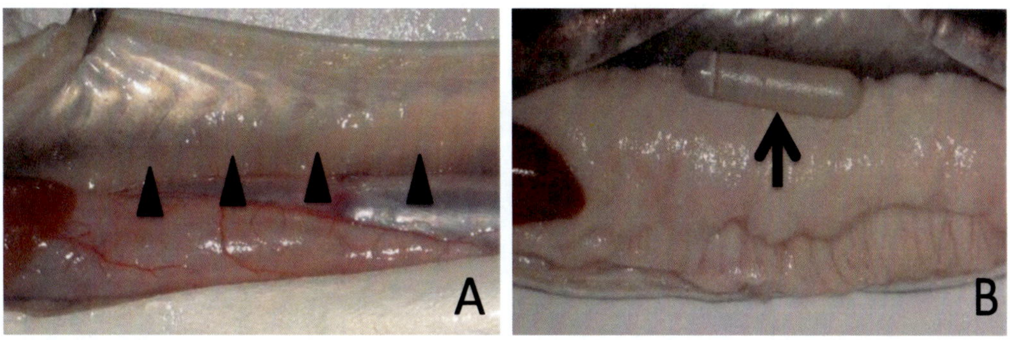

Modified from Kagawa et al. 2013a

Figure 2. Photographs of the ovary of a female Japanese eel before (A) and after (B) implantation with a single osmotic pump loaded with salmon pituitary extract (SPE, 3 mg /day/fish). The arrowheads indicate the immature ovary and the arrow indicates the osmotic pump. Developed ovary possessing full-grown oocytes occupies the abdomen (B).

In particular, in marine teleosts spawning buoyant eggs in seawater, oocytes undergo a significant increase in size because of rapid water uptake during meiosis resumption (Adachi et al., 1988, Fabra et al., 2006, Kagawa et al., 2009). During these processes, the oocytes become buoyant, which is essential for their oceanic survival and dispersal as well as for the initiation of early embryogenesis (Matsubara et al., 1999). In the Japanese eel, we found that the full-grown oocytes underwent more than threefold increase in volume during maturation and ovulation, which was artificially induced by injecting SPE and 17, 20β-dihydroxy-4-pregnen-3-one (DHP) (Kagawa et al. 2009b). Wet and dry weight measurements (Figure 3) indicated that water accumulation during oocyte maturation is the major factor contributing to the follicular diameter increase and oocyte volume. *In vitro* experiments revealed that gonadotropins (HCG, SPE, recombinant eel LH) and DHP stimulated an increase in the diameter of follicle-enclosed oocytes at the migratory nucleus stage.

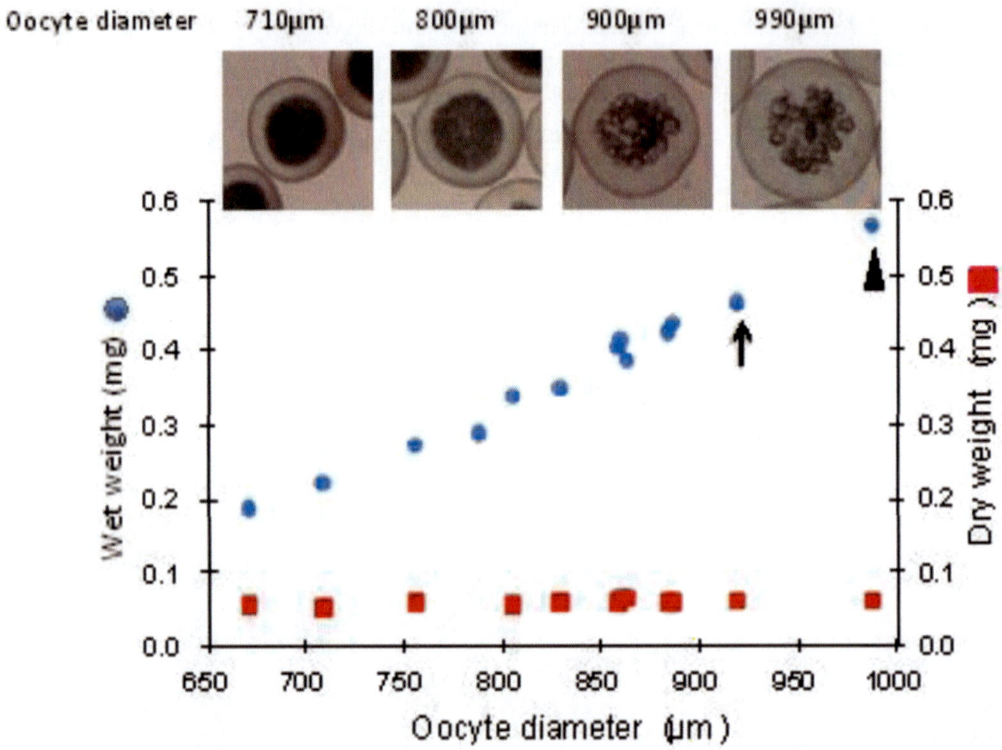

Modified from Kagawa et al. 2009a.

Figure 3. Changes in the wet and dry weights of the follicle-enclosed oocytes during oocyte maturation and hydration induced by salmon pituitary extract (SPE) and 17,20β-dihydroxy-4-pregnene- 3-one (DHP) injections. Arrow: mature oocyte. Arrowhead :ovulated egg. Upper photographs: morphological changes during oocyte maturation and ovulation.

During oocyte maturation, the yolk globules fuse and, concomitantly, pronounced proteolysis of the yolk proteins occurs, generating an increase in free amino acids; the resulting increase in small peptides in the oocytes provides the driving force for the water influx into the oocytes (Fabra et al., 2006; Matsubara et al., 1999; Selman et al., 2001). We found that bafilomycin A1, a specific inhibitor of vacuolar proton-ATPase and acidification of yolk inclusion (Raldu'a et al., 2006), prevented the HCG- and DHP-induced oocyte hydration in a dose-dependent manner. Previous studies indicate that bafilomycin A1 prevents yolk protein hydrolysis and the generation of free amino acids (Selman et al., 2001). Therefore, in the Japanese eel, acidification of the yolk compartments is necessary for appropriate protein hydrolysis, increasing small peptides (such as free amino acids) in the oocytes and providing the driving force for the water influx into the oocytes, as suggested previously (Fabra et al., 2006; Matsubara et al., 1999; Selman et al., 2001). Addition of inhibitors of aquaporin (Aqp) water permeability ($HgCl_2$) to the incubation media inhibited the gonadotropin- and DHP-induced increase in the follicular diameter in a dose-dependent manner, suggesting that aquaporin facilitates water uptake by acting as a water channel, is essential for water influx into oocytes via osmotic mechanisms. However, as per previous report Cerda et al. (2007), it might be possible that $HgCl_2$ also affects other important factors for oocyte hydration, such as K^+ channels. Further studies are necessary to confirm the involvement of Aqp1ab in oocyte hydration of the eel. Recent study by Zapater et al. (2011)

pointed out the fist direct experimental evidence of involvement of Aqp1ab in oocyte hydration of the Atlantic halibut (*Hippoglossus hippoglossus*) by microinjection of an antiserum specific to Aqp1ab.

To elucidate the molecular mechanisms underlying hydration during oocyte maturation, we have cloned novel-water selective aquaporin 1 (*aqp1b*) of the eel (Kagawa et al. 2011). Aquaporin, an open molecular channel transporting water and other solutes along an osmotic gradient (Arge et al., 2002), was found to contribute to the rapid water influx into the oocyte during oocyte maturation in the gilthead seabream (*Sparus auratus*) (Ferbra et al., 2005, 2006). We have isolated and characterized a Japanese eel *aqp1b* cDNA derived from ovary. The predicted amino acid sequences of the cloned Japanese eel ovary-derived *aqp1b* shared 99% overall sequence identity with that of the AQP1 previously reported in the European eel, *Anguilla anguilla* (Martinez et al., 2005) termed AQP1dup. The Japanease eel APQ1b contains three functional domains; an N-terminal extracellular domain, a large transmembrane domain, and a C-terminal cytoplasmic domain. In particular, six potential transmembrane domains and two NPA motifs are conserved. Moreover, amino acids known to be essential for the pore-forming region in human AQP1 (i.e., Phe_{56} , His_{180} , and Arg_{195} Sui et al., 2001) were present in an analogous position in Japanese eel AQP1b. Therefore, these amino acids in Japanese eel AQP1b may be involved in water selective pore formation. Also, a Cys residue located N terminal to the second NPA motif, which may be involved with inhibition of water permeability by mercurial compounds, was identical in Japanese eel AQP1b.

In situ hybridization studies with the eel *aqp1b* cRNA probe revealed intense signals in the oocytes at the perinucleolus stage which became faint during the process of oocyte development. Light microscopic immunocytochemical analysis of ovary using antibody against Japanese eel AQP1b revealed that the reactions was first expressed in the cytoplasm around the yolk globules of oocyte at the primary yolk globule stage and became localized around the large membrane-limited yolk masses which were formed by the fusion of yolk globules during the oocyte maturation phase. These results indicate that during the previtellogenic stage (at the perinucleolus stage), mRNA of eel *aqp1b* are synthesized in the oocytes, perhaps by maternal gene expression and/or from ovarian follicle. Synthesis of Aqp1b protein is stimulated when oocytes begin vitellogenesis. During oocyte maturation of the Japanese eel (meiosis resumption), the yolk granules fused and increased in size to become large yolk masses but did not form a single yolk mass. These morphological changes observed during oocyte maturation is different from those observed in the gilthead seabream (Fabra et al. 2006) in which the yolk granules fuse into a single yolk mass. In the gilthead seabream, during oocyte maturation, Aqp1b translocated towards oocyte periphery and become concentrated within a thin layer just below the oocyte. Further studies are necessary to obtain more conclusive evidence of Aqp1b localization on plasma membrane of the Japanese eel oocytes. In addition, studies on other evolutionally primitive species, such as conger eel *Conger myriaster* and Pike eel *Muraenesox cinereus*, may substantiate this contention.

Maturational Competence and Oocyte Maturation and Ovulation (See Figure 4)

During oocyte maturation phase, full-grown oocytes acquire ability to respond the maturation-inducing steroid (maturational competence). SPE (or DHP) did not induce *in vitro* germinal vesicle breakdown (GVBD) and ovulation in full-grown oocytes at the tertiary yolk globule stage below 700 μm in diameter (Kagawa et al., 2013b). One day after SPE injection, oocytes at the migratory nucleus stage approximately 800 μm in diameter underwent GVBD and ovulation, *in vitro,* in response to SPE and DHP. Oocytes became increasingly sensitive to SPE in association with the increase of oocyte diameter due to everyday SPE-injection, but oocytes obtained from female eels treated with single SPE-injection became less sensitive to SPE and DHP later on. Moreover, pre-incubation with SPE (or recombinant eel LH, but not FSH, which were produced using a HEK 293 cells) and subsequent incubation with DHP induced GVBD. These results indicate that oocytes acquire the ability to respond to DHP at the migratory nucleus stage over 800 μm in diameter and furthermore SPE (maybe LH) has an essential role for initiation and maintenance of the ability of oocytes to respond to gonadotropin and DHP.

After acquisition of maturational competence, oocytes become sensitive to SPE or DHP in association with the continuous increase of oocyte diameter by hydration. Recently, we found that SPE did not stimulate *in vitro* DHP production of ovarian follicles at the tertiary yolk globule stage between 600-700 μm in diameter. However, *in vitro* DHP production was stimulated by SPE in ovarian follicles over 800 μm in diameter and even the production became higher in ovarian follicles with 900 μm in diameter. These results indicate that ovarian follicles acquire the potency to produce maturation-inducing steroid (DHP) at migratory nucleus stage over 800 μm, more specifically at 900 μm in diameter (Kagawa et al., in preparation).

Artificial Induction of Spawning

Ordinary Methods

Ordinary methods for artificial induction of spawning were reviewed earlier (Kagawa 2003; 2005). Briefly, female eels were intraperitoneally injected with SPE (20-30 mg/kg body weight) once a week. After 10-13 injections, females that possessed oocytes over 850 μm in diameter at the migratory nucleus stage were injected with SPE (30 mg/kg body weight) as a priming dose, followed by an intraperitoneal injection of DHP (2 μg/g body weight) 24 h later. Ovulated eggs were obtained 12-18 hr after DHP injection.

Newly Developed Methods (Figure 4)

New methods for artificial induction of spawning have been developed from the basic information obtained by *in vitro* and *in vivo* experiments described above. Eels having ovaries containing full-grown oocytes (approximately 700-750 μm in diameter) were injected with SPE to induce oocyte hydration and maturational competence. Eels possessing competent oocytes (850-900 μm) were injected again (a priming dose) with SPE to induce and maintain maturational competence and ability to respond to gonadotropin. Final treatments of gonadotropin-releasing hormone analog (GnRHa) in combination with SPE given 24 hr after SPE-priming dose to female eels (900-950 μm in diameter) succeeded in induction of spawning in a rearing tank with spermiated male eels injected with hCG. Average fertility and hatchability are approximately 60 and 40%, respectively.

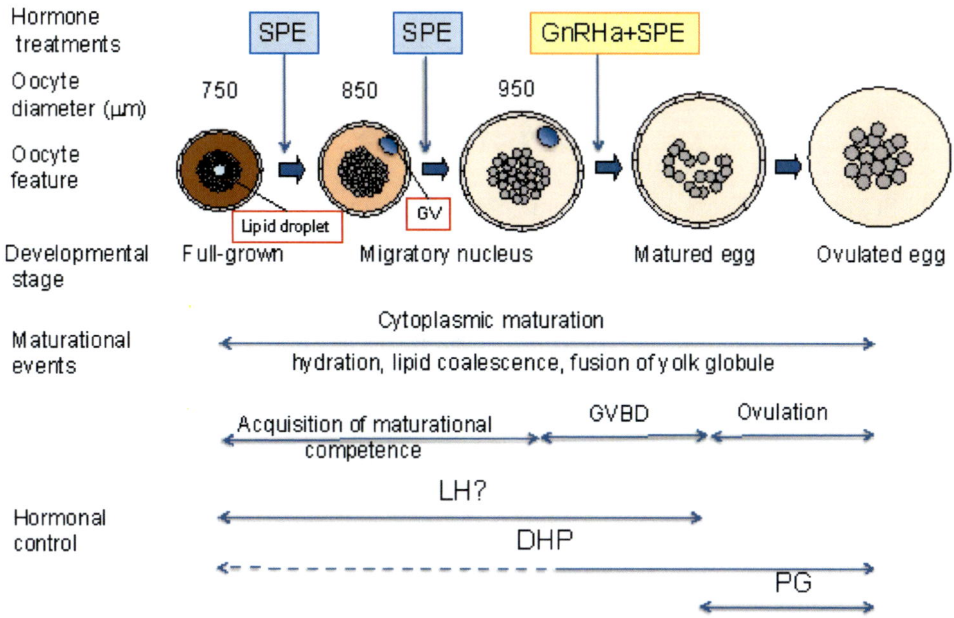

Modified from Kagawa et al. 2013b

Figure 4. Hormonal regulation of oocyte maturation (cytoplasmic and nuclear maturation) and ovulation, and artificial hormone treatments for induction of spawning of the female Japanese eel. DHP: 17,20β-dihydroxy-4-pregnene-3-one, GVBD: germinal vesicle breakdown, PG: prostaglandin, SPE: salmon pituitary extract.

Conclusion

We have developed new methods for induction of spawning in the Japanese eel by improving the ordinary hormonal treatments. The percentage of fertility and hatchability were improved after adopting these methods. However, egg quality obtained from the female eels induced by our hormonal treatment procedures varies among the female eels used. Further studies are necessary to elucidate factors associated with egg quality and improve technical procedures.

Development for artificial induction of maturation and seed production in the Japanese eel started almost half a century ago. Finally, *leptocephalus* larvae and glass eels have been produced first in the world from artificially induced mature male and female eels by elaborate and reasonable hormonal treatments (Kagawa et al. 2005; Tanaka 2003). In 2010, “complete culture" system for the Japanese eel (production of the next generation from bloodstocks which are reared from artificially reproduced eggs) has been developed.

Acknowledgments

This work was financially supported by the Ministry of Education and Science Japan and by the Agriculture, Forestry, and Fisheries Research Council to H.K. We also thank the staffs of Shibushi Branch, National Research Institute of Aquaculture for providing experimental fish and rearing tanks.

References

Adachi S, Ouchi K, Hirose K, Nagahama Y. (1988) Induction of oocyte maturation *in vitro* by steroid hormones in the red sea bream *Pagrus major*. *Nippon Suisan Gakkaishi* 54: 1665.

Casselman, D. J. (2003) Dynatics of resources of the American eel, Anguilla rostrata: Declining abundance in the 199s. In: Aida K, Tsukamoto K, Yamauchi K (eds) *Eel Biology*, Springer, Tokyo, pp 255-274.

Cerda J, Fabra M, Raldua D (2007) Physiological and molecular basis of fish oocyte hydration. In *The Fish Oocyte :From Basic Studies to Biotechnological Applications.* (ed. P. J. Babin, J. Cerda, E. Lubzens), pp. 349–396. Dordrecht, The Netherlands: Springer.

Chiba H, Iwatsuki K. Hayami K, Yamauchi K (1993) Effects of dietary estradiol-17β on feminization, growth and body composition in the Japanese eel (*Anguilla japonica*) *Comp. Biochem. Physiol.* 106: 367–371.

Dekker, W. (2003) Status of the European eel stock and fisheries. In: Aida K, Tsukamoto K, Yamauchi K (eds) *Eel Biology*, Springer, Tokyo, pp 237-254.

Dufour S, Lopes E, Le Menn F, Le Belle N, Baloche S, Fontaine YA (1988) Stomulation of gonadotropin release and of ovarian development by the administration of a gonadoliberin agonist and of dopamine antagonists, in female silver eel pretreated with estratiol. *Gen. Comp. Endocrinol.* 70: 20-30.

Fabra M, Raldua D, Bozzo MG, Deen PMT, Lubzens E, Cerda J (2006) Yolk proteolysis and aquaporin-1o play essential roles to regulate fish oocyte hydration during meiosis resumption. *Dev. Biol.* 295:250-262.

Fontaine M, Bertrand E, Lopes E, Callamand O (1964) Surla maturation des organs genitaux de l'anguiila femelle (Anguilla Anguilla L.) et emission speontanee des oeufs en aquarium C R Acad Sci Paris 259: 2907-2910.

Gibson, M. J., Kasowski, H., Dobrjansky, A., 1994. Continuous gonadotropin-releasing hormone infusion stimulates dramatic gonadal development in hypogonadal female mice. *Biol. Reprod.* 50, 680-685.

Kagawa H. (2003) Artificial induction of oocyte maturation and ovulation. In: Aida K, Tsukamoto K, Yamauchi K (eds) *Eel Biology*, Springer, Tokyo, pp 401–414.

Kagawa H, Tanaka H, Ohta H, Unuma T, Nomura K (2005) The first success of glass eel production in the world: basic biology on fish reproduction advances new applied technology in aquaculture. *Fish Physiol. Biochem.* 2005, 31:193-199.

Kagawa H, KasugaY, Adachi J, Nishi A, Hashimoto H, Imaizumi, H, Kaji S (2009a) Effects of continuous administration of human chorionic gonadotropin, salmon pituitary extract, and gonadotropin-releasing hormone using osmotic pumps on induction of sexual maturation in male Japanese eel, *Anguilla japonica. Aquaculture* 296: 117–122.

Kagawa H, Horiuchi Y, Kasuga Y, Kishi T (2009b) Oocyte hydration in the Japanese eel (*Anguilla japonica*) during meiosis resumption and ovulation. *J. Exp. Zool.* 311A: 752-762.

Kagawa H, Kishi T, Gen K, Kazeto Y, Tosaka R, Matsubara H, Matsubara T, Sawaguchi S (2011) Expression and localization of aquaporin 1b during oocyte development in the Japanese eel (*Anguilla japonica*). *Reprod. Biol. Endocrinol.* 9:71.

Kagawa H, Fujie N, Imaizumi h, Masuda Y, Oda K, Adachi J, Nishi A, Hashimoto H, Teruya K, Kaji S. (2013b). Using osmotic pumps to deliver hormones to induce sexual maturation of female Japanese eel, *Anguilla japonica. Aquaculture* 388-391: 30-34.

Kagawa H, Sakurai Y, Horiuchi R, Kazeto Y, Gen K, Imaizumi H, Masuda Y. (2013b). Mechanism of oocyte maturation and ovulation and its application to seed production inn the Japanese eel. Fish Physiol. Biochem. 39, 13-17.

Martinez AS, Cutler CP, Wilson GD, Phillips C, Hazon N, Cramb G (2005) Cloning and expression of three aquaporin homologues from the European eel (*Anguilla anguilla*): effects of seawater acclimation and cortisol treatment on renal expression. *Biol. Cell* 97: 615-627.

Matsubara, T., Ohkubo, N., Andoh, T., Sullivan, C.V., Hara, A. (1999). Two forms of vitellogenin, yielding two distinct lipovitellins, play different roles during oocyte maturation and early development of barfin flunder, *Verasper moseri*, a marine teleost that spawn pelagic eggs. *Dev. Biol.* 213, 18–32.

Matsubara H, Tanaka H, Nomura K, Kobayashi T, Murashita K, Kurokawa T, Unuma T, Kim SK , Lokman MP, Matsubara T, Kagawa H, Ohta H (2008) Occurrence of spontaneously spermiating eels in captivity. *Cybium* 32 suppl: 174-175.

Miura T, Yamauchi K, Nagahama Y. Takahashi H (1991). Induction of spermatogenesis in male Japanese eel, *Anguilla japonica*, by a single injection of human chorionic gonadotropin. *Zoo. Sci.* 8: 63-73.

Ohta H, Unuma T (2003) Induction of sperm maturation. In: Aida K, Tsukamoto K, Yamauchi Y, editors. *Eel biology*. Tokyo: Springer. pp 415–423.

Ohta, H., Kagawa, H., Tanaka, H., Okuzawa, K., Hirose, K., (1996a.) Changes in fertilization and hatching rates with time after ovulation induced by 17,20β-dihydroxy-4-pregnen-3-one in the Japanese eel, *Anguilla japonica. Aquaculture* 139, 291-301.

Ohta H, Kagawa H, Tanaka H, Okuzawa , Hirose K (1996b) Milt in the Japanese eel Anguilla japonica induced by repeated injections of human chorionic gonadotropin. *Fish Sci.* 62; 44-49.

Patton, P. E., Stouffer, R. L., Zelinski-Wooten, M. B., 1990. The effect of intraluteal infusion of deglycosylated human chorionic gonadotropin on the corpus luteum in rhesus monkeys. *J. Clin. Endocrinol. Metabol.* 70, 1213-1218.

Raldu'a D, Fabra M, Bozzo MG, Weber E, Cerda, J. 2006. Cathepsin B-mediated yolk protein degradation during killifish oocyte maturation is blocked by a H+-ATPase inhibitor: effects on the hydration mechanism. *Am. J. Physiol. Regul. Integr. Comp. Physiol.* 290:R456-R466.

Selman K, Wallace RA, Cerda J (2001) Bafilomycin A1 inhibits proteolytic cleavage and hydration but not yolk crystal disassembly or meiosis during maturation of sea bass oocytes. *J. Exp. Zool.* 290, 265–278.

Sui H, Han BG, Lee JK, Walian P, Jap BK (2001) Structural basis of water specific transport through the AQP1 water channel. *Nature* 414: 872-878.

Tanaka, H (2003) Techniques for larval rearing. In: Aida K, Tsukamoto K, Yamauchi K, (eds), Springer, Tokyo, pp427–434.

Tatsukawa, K. W. (2003) Eel resources in East Asia. In: Aida K, Tsukamoto K, Yamauchi K (eds) *Eel Biology*, Springer, Tokyo, pp 293-300.

Tsukamoto K, Chow S, Otake T, Kurogi H, Mochioka N, Miller M J, Aoyama J, Kimura S, Watanabe S, Yoshinaga T, Shinoda A, Kuroki1 M, Oya M, Watanabe T, Hata K, Ijiri S, Kazeto Y, Nomura K, Tanaka H (2010) Oceanic spawning ecology of freshwater eels in the western north Pacific. Nature Communications 2:179 DOI: 10.1038/ncomms1174 www.nature.com/naturecommunications.

Yamamoto, K, Hiroi, O, Hirano O, Morioka T. (1972) Artificial maturation of cultivated male Japanese eels by synaphorin injection (in Japanese with English abstract). *Bull. Jap. Soc. Sci. Fish* 38: 1083-1090.

Yamamoto K, Morioka T, Hiroi O, Oomori M (1974). Artificial maturation of female Japanese eels by the injection of salmonid pituitary (in Japanese with English abstract). *Nippon Suisan Gakkaishi* 40:1-7.

Yamamoto, K, Yamauchi, K (1974) Sexual maturation of Japanese eel and production of eel larvae in the aquarium. *Nature* (Lond.) 251, 220–222.

Yamauchi K, Nakamura M, Takahashi H, Takano K (1976) Cultivation of larvae of Japanese eel. *Nature* (London). 263: 412.

Uto t, Mikawa N, Okamura Y, Yamada S, Tanaka N, Horie A, Akazawa A, Oka HP. (2004). Ovarian morphology of the Japanese eel in Mikawa Bay. *J. Fish Biol.* 64: 502-513.

Zapater, C, Chauvigne, F, Norberg, B, Finn, RN, Cerda, J. (2011) Dual neofunctionalization of a rapidly evolving aquaporin-1 paralog resulted in constrained and relaxed traits controlling channel function during meiosis resumption in teleosts. *Mol. Biol. Evol.* 28: 3151-3169.

In: Sexual Plasticity and Gametogenesis in Fishes
Editor: Balasubramanian Senthilkumaran

ISBN: 978-1-62618-848-8

Chapter X

Transgenic Medaka Lines with a Visible Marker for Phenotypic Sex: Genetic Analyses of an XY Sex-reversal

Akira Kanamori, Naoya Ichikawa and Michiko Ishikawa
Laboratory of Molecular Genetics, Division of Biological Sciences,
Graduate School of Science, Nagoya University,
Chikusa, Nagoya, Japan

Abstract

Transgenic medaka lines expressing strong ovarian GFP can be potentially utilized in various studies on sex differentiation, of which we present an example here. An XY sex-reversed medaka line found in Sendai, Japan, was genetically analyzed by crossing with the transgenic lines. As a result, presence of the Y chromosome from the original sex-reversal was necessary, but not sufficient, for XY sex-reversal of the progenies. Therefore, cause of this sex-reversal is hypomorphic Y chromosome and likely to be loss of function, including altered expression, of the male-determining gene of the medaka, *dmy*. The transgenic lines enabled us to determine sexual phenotypes of the medaka at one week after hatching, which is substantially earlier than by conventional method with secondary sex characteristics.

Introduction

With the advance of GFP (green fluorescent protein) technologies in recent years (see Miyawaki, 2011a and b; Remington, 2011), transgenic fish with fluorescent proteins has become popular in many aspects of basic and applied biological studies primarily because of easy live imaging without sacrificing fishes (see Driesh III, 2008; Higashijima, 2008; Nakamura et al., 2008; Iwai et al., 2009). Among them, visible sex markers can be potentially useful in the following studies: finding sexual differences in transcriptomes or proteomes, screening and characterization of sex differentiation mutants, drug tests on sex differentiation, and effects of temperature or other environmental factors. Here we present an example of mutant characterization using transgenic medaka lines expressing ovary-specific GFP.

Sexually undifferentiated gonads of vertebrates differentiate into ovaries or testes according to their sex. In this fascinating process, many questions still remain to be answered. The sex-determining system of medaka is XX/XY type (Aida, 1921) and their sex-determining gene was identified, second in the vertebrates after mammalian *sry*, as *dmy/dmrt1bY* (Matsuda et al., 2002; Nanda et al., 2002). The gene was originated about one million years ago by duplication of autosomal *dmrt1* gene (Nanda et al., 2002; Kondo et al., 2006). One of the promising approaches to understand sex differentiation network triggered by *dmy* is to analyze sex differentiation mutants (cf. Morinaga et al., 2007). On the basis of presence or absence of *dmy* in the genome, the sex differentiation mutants have been collected from wild medaka populations (Matsuda et al., 2002; Shinomiya et al., 2004; Otake et al., 2006, 2008). In the present study, we analyzed an XY sex-reversal line found in Sendai, Japan, in 2007 (Sugawara et al., unpublished). The phenotype of medaka sex becomes apparent from outside at about 2 months after hatching (see chapter 5 in Kinoshita et al., 2009). Therefore, we utilized transgenic medaka lines with a GFP driven by *42Sp50* (Kinoshita et al., 2009b), which is highly expressed in the developing young oocytes (Kanamori, 2000). With this transgene, we could determine sex phenotypes from one week after hatching.

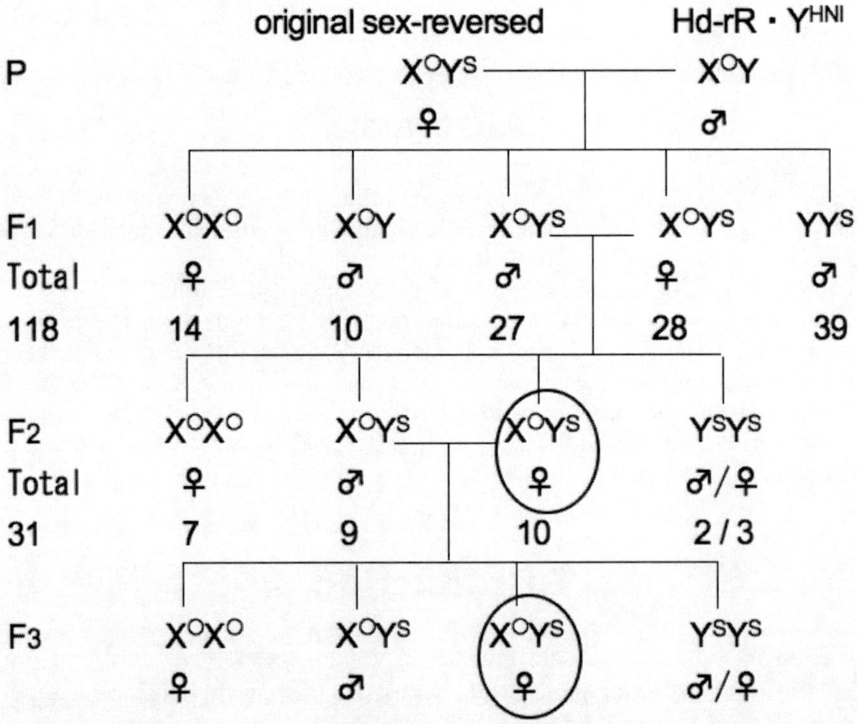

Figure 1. Pedigree of $X^{\circ}Y^{S}$ females used in this study. One F2 and one F3 female (circled) were used in the following experiments.

Materials and Methods

Two XY sex-reversed females were used in this study: one F2 and one F3 progenies from a cross between original XY female (Matsuda et al., 2002) found in Miyagino district of Sendai city, Japan, in 2007, and a Hd-rR • Y^{HNI} male (Figure 1). We analyzed F1 progenies between these XY sex-reversed females and males of two transgenic lines with *42Sp50:EGFP*. Because X chromosomes from either original XY sex-reversed female or Hd-rR • Y^{HNI} showed identical pattern in the following genetic analysis, they are represented here collectively as X°. Y^{S} denotes Y chromosome of the original XY female.

1st Experiment

F1 progenies from a cross between $X^{\circ}Y^{S}$ females (see above) and males ($X^{*}Y^{*}$) homozygous for an oocyte-specific GFP transgene (*Tg (42Sp50:EGFP)*, Kinoshita et al., 2009, Figure 3) were examined. Their genotypes were analyzed by genomic PCR followed by RFLP (restriction fragment length polymorphism) based SNP (single nucleotide polymorphism) detection and phenotypes by ovarian GFP detection. First, we examined fluorescence in the gonads at one week to ten days after hatching under stereo fluorescence microscope (Leica microsystems) and then froze whole body at -70°C.

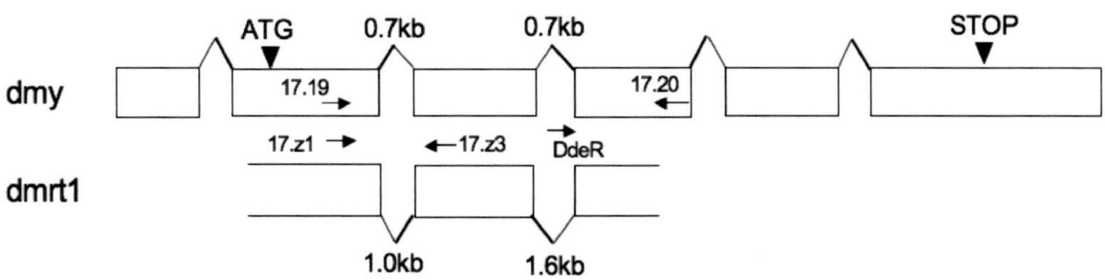

Figure 2. Structure of dmy and dmrt1genes (rectanfgles are exons) with primer locations.

Table 1. Primer sequences

Primer name	sequence (5'-3')
17.19	GAACCACAGCTTGAAGACCCCGCTGA
17.20	GCATCTGCTGGTACTGCTGGTAGTTG
17.z1	CCGCTGAAAGGCCACAAGCGC
17.z3	GCCTGCTGCCTCCTCAAGGCG
DdeR	TCAGGTTTGACTTGGATGCTGACC
2S	CGAGAAATGCACTGGTCAAC
1U	ACCTTTCTGTGTGGAGTTTGC

Their genomic DNA was extracted later with Extract-N-Amp™ Tissue PCR Kit (Sigma Aldrich). PCR were done with a primer pair 17.z1 and 17.z3, which amplifies both *dmy* and *dmrt1* (Table 1 and Figure 2). The fragments from dmy and dmrt1 were 0.8 and 1.1 kb, respectively (Figure 5). Next, Y^S and Y^* were distinguished by a SNP found in their *dmy* genomic sequences between primers, 17.19 and 17.20 (Matsuda et al., 2002). The PCR fragments amplified from DdeR and 17.20 (Table 1 and Figure 2) were digested with *Dde*I. A 350bp fragment were obtained from Y^* and about 315bp and 30bp fragments were obtained from Y^S (Figure 5).

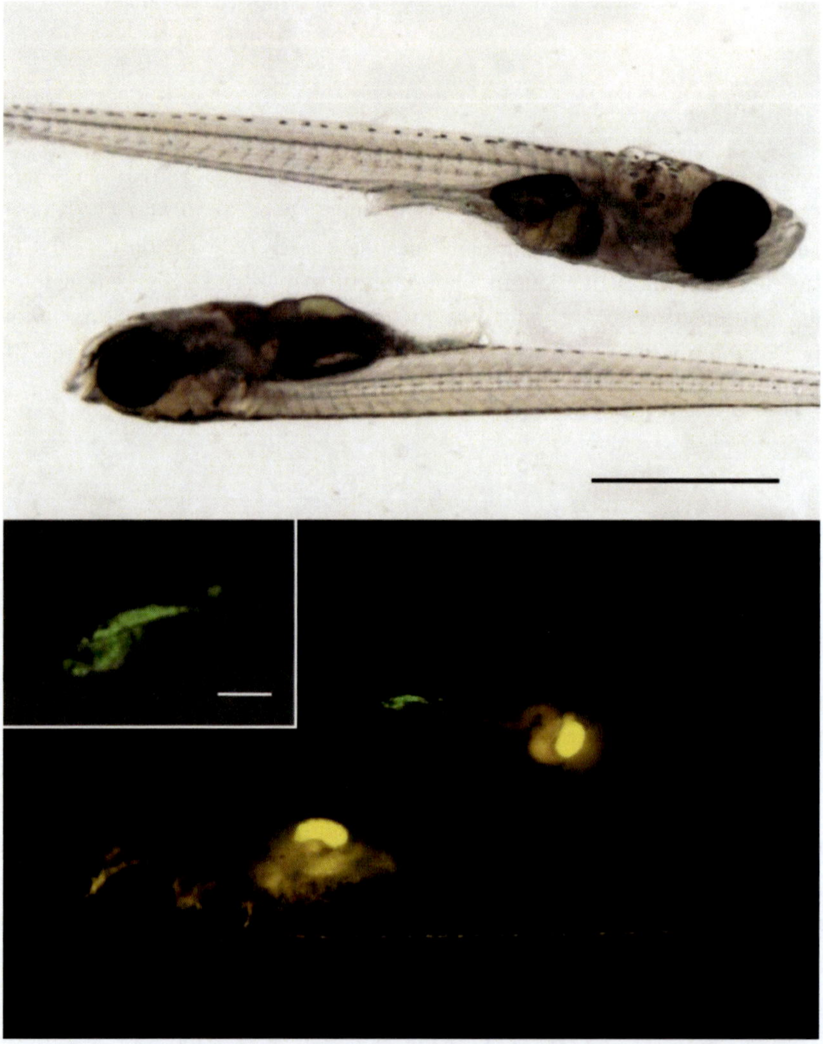

Figure 3. *Tg(42Sp50:GFP)* fish at one week after hatching. upper: with transmitted light (bar=1mm), lower: with excitation light, inset (ovary, bar=0.1mm). The line was based on the mc line described in Kanamori et al., 2008. The wild type allele of lf is on the Y and the mutant allele is on the X. Only XY medaka has leucophores (shown in yellow in the lower panel). The XX (without leucophores) shows ovarian GFP.

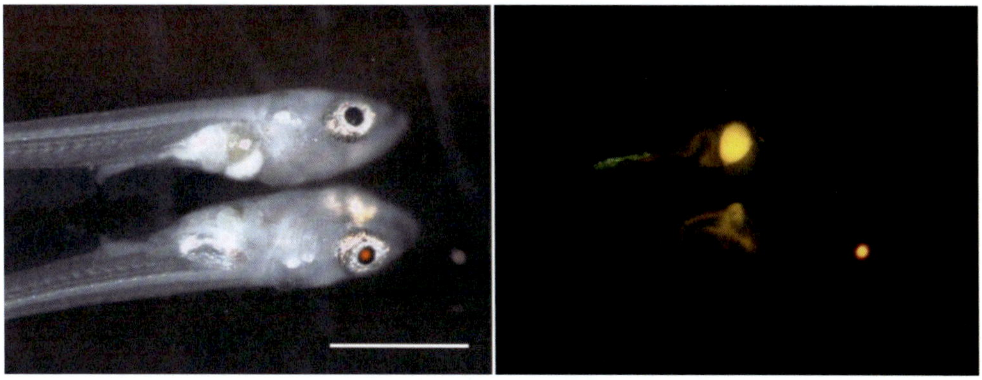

Figure 4. *Tg(dmy/crystalline:DsRed,42Sp50:GFP)* fish at one week after hatching. left: with incident light (bar=1mm), right: with excitation light. The lower fish with a red lens has *dmy* transgene and no fluorescence in the gonad. The upper fish has a black lens (no *dmy* transgene) and strong ovarian fluorescence in the gonad.

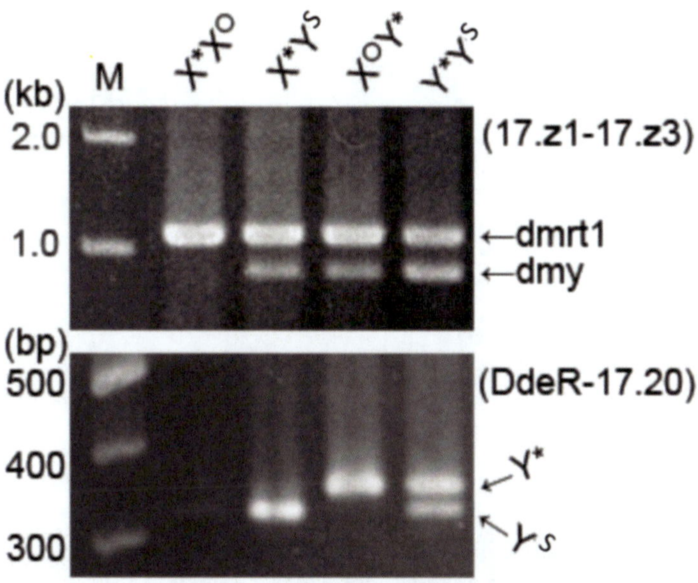

Figure 5. PCR genotyping with a *dmy* fragment (upper) and PCR/RFLP of another *dmy* fragment (lower).

2nd Experiment

We introduced a *dmy* transgene (11 copies, Otake et al., 2010) into *Tg (42Sp50:EGFP)* by crossing (*Tg(dmy/crystalline:DsRed,42Sp50:GFP*, Figure 4). This transgene contains *crystalline*:DsRed that confers red lens penotype. F1 genotypes were analyzed by lens color and genomic PCR, and their phenotypes by ovarian GFP fluorescence as experiment 1. PCR were done with a primer pair 2S and 1U in the PG7 gene nearby *dmy* locus (Matsuda et al., 2002, Table 1). A 800bp fragment were obtained from $X^*$and Y^S and about 500bp from X° (Figure 7).

Results

1st Experiment

The genotypes of sex chromosomes were determined by PCR of the *dmy* fragment and SNP detection by *Dde*I digestion (Figure 5, upper and lower, respectively). The phenotypes of the F1s were determined by presence or absence of ovarian GFP as shown in Figure 3. All X^*X° were females. All X^*Y^* and Y^* Y^S were males. 25 males and 10 females were found from X^*Y^S.

2nd Experiment

The presence of X° was detected by PCR of the PG7 fragment (lower band in Figure 7). The F1 fish with red lenses (with *dmy* transgene) were all males regardless of their genotypes (X^*X° or X^*Y^S). The X^*X° fish with black lenses were all females. 26 males and 29 females were found from the X^*Y^S fish with black lenses (Figure 8).

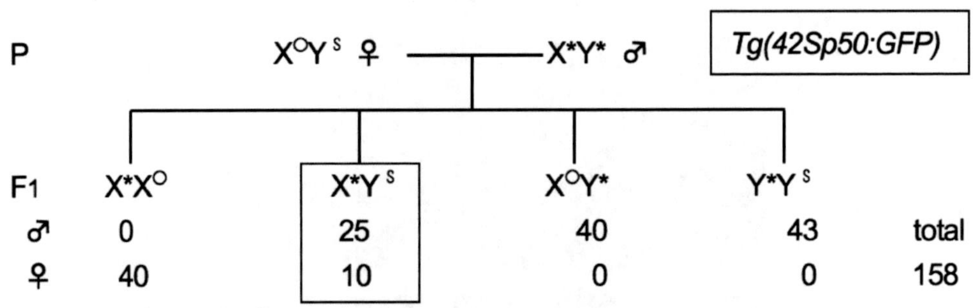

Figure 6. Sex genotypes and phenotypes of F1 (experiment 1).

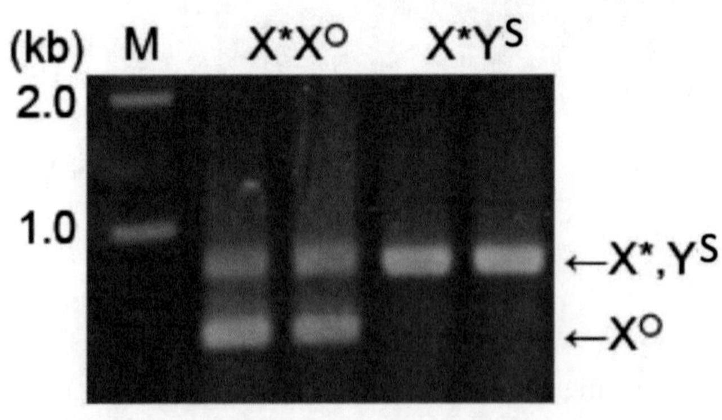

Figure 7. Genotyping by PCR of a PG7 fragment.

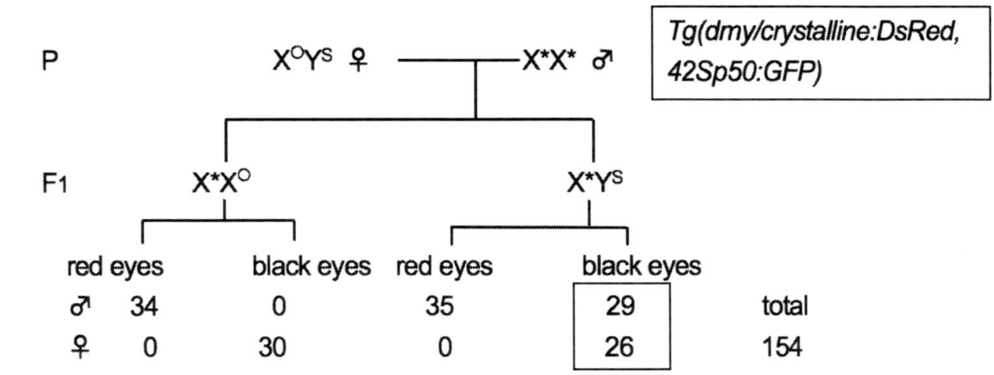

Figure 8. Sex genotypes and phenotypes of F1 (experiment 2).

In both experiments, presence of Y^S was necessary for XY sex-reversal but was not sufficient; presence of Y^* or the *dmy* transgene overrode presence of Y^S.

Discussion

With experiments 1 and 2, we examined sex genotypes and phenotypes of about 300 F1 progenies. As expected, ratio of X^*X°: X^*Y^*: Y^* Y^S: X^*Y^S was 1:1:1:1 in the experiment 1 and ratio of X^*X°: X^*Y^S was 1:1 in the experiment 2. About 30% and 50% of X^*Y^S F1 offspring from experiment 1 and 2, respectively, developed as sex-reversed females. However, those with Y^S developed as males when Y^* or the *dmy* transgene was present. Taken together, these data strongly suggest that the mutations in this strain do not dominantly convert male to female but relative strength of the mutations determine whether they develop as females or males. The results can be most easily explained by presence of a hypomorphic Y choromosome.

The XY sex-reversals from wild-derived medaka populations reported to date are all attributed to the Y chromosomes (Shinomiya et al., 2004; Otake et al., 2006). Their *dmy* alleles showed loss-of-function mutations (Otake et al., 2006). About half contained premature termination of the *dmy* coding sequence and the remaining half showed decreased expression. In an exceptional case, a medaka population in Obu, Aichi, Japan, contained a nonfunctional Y chromosome that behaved as an X chromosome in multiple generations (Otake et al., 2008). Those cases differ from the present mutation in that the *dmy* allele on the Y^S shows no nucleotide mutations in the protein coding region and there is no detectable decrease of the expression level of the *dmy* (Sugawara et al., unpublished).

As the cause of sex reversal in the present case is most likely a hypomorphic Y^* chromosome, we speculate that 1) expression of the *dmy* on the Y^S may be tissue-specifically or developmentally altered or 2) a mutation exists in the Y-specific region of the Y^S outside the *dmy*. The Y-specific region is known to be about 260 kb and does not contain protein coding gene except *dmy* (Nanda et al., 2002; Kondo et al., 2006). We also cannot exclude presence of an autosomal mutation additional to the Y^S, which is responsible for the sex-reversal.

We need about 2 months after hatching to distinguish males and females from their external morphology (see chapter 6 in Kinoshita et al., 2009a). The transgenic medaka used in the study (Kinoshita et al., 2009b) can shorten this period to one week with the strong fluorescence driven by the 42Sp50 promoter and thereby proved to be a useful animal model involved in gonadal sex differentiation.

Conclusion

In the present study, we demonstrated our *42Sp50:GFP* transgenic medaka can be utilized in genetic analyses of a sex-reversal mutant. Noninvasive phenotyping as early as one week after hatching can simplify various assays based on sex differentiation of the medaka.

Acknowledgments

We thank Drs. Mitsuru Sakaizumi, Satoru Hamaguchi, Horoyuki Otake and Ms. Satomi Sugawara for generous gift of the sex-reversed medaka and the *dmy* transgenic fish, and for letting us know the unpublished data.

References

Aida, T. (1921). On the inheritance of color in a fresh-water fish, *Aplocheilus latipes* Temmick and Schlegel, with special reference to sex-linked inheritance. *Genetics*, 6, 554-573.

Higashijima, S. (2008). Transgenic zebrafish expressing fluorescent proteins in central nervous system neurons. *Dev. Growth Differ.*, 50, 407-413.

Detrich III, H. W. (2008). Fluorescent proteins in zebrafish cell and developmental biology. *Method Cell Biol.*, 85, 219-241.

Iwai, T., Inoue, S., Kotani, T. and Yamashita, M. (2009). Production of transgenic medaka fish carrying fluorescent nuclei and chromosomes. *Zool. Sci.*, 26, 9-16.

Kanamori, A. (2000). Systematic identification of genes expressed during early oogenesis in medaka. *Mol. Reprod. Dev.*, 55, 31-36.

Kanamori, A., Toyama, K., Kitagawa, S., Kamehara, A., Higuchi, T., Kinoshita, M. and Hori, H. (2008). Comparative genomics approach to the expression of *figα*, one of the earliest marker genes of oocyte differentiation in medaka (*Oryzias latipes*). *Gene,* 423, 180-187.

Kinoshita, M., Murata, K., Naruse, K. and Tanaka, M. (2009a). *Medaka Biology*. Iowa, USA:Wiley-Blackwell.

Kinoshita, M., Okamoto, G., Hirata, T., Shinomiya, A., Kobayashi, T., Kubo, Y., Hori, H. and Kanamori, A. (2009b). Transgenic medaka enables easy oocytes detection in live fish. *Mol. Reprod. Dev.*, 76, 202-207.

Kondo, M., Hornung, U., Nanda, I., Imai, S., Sasaki, T., Shimizu, A., Asakawa, S., Hori, H., Schmid, M., Shimizu N. and Schartl M. (2006). Genomic organization of the sex-determining and adjacent regions of the sex chromosomes of medaka. *Genome Res.*, 16, 815-826.

Matsuda, M., Nagahama, Y., Shinomiya, A., Sato, T., Matsuda, C., Kobayashi, T., Craig Morrey, E., Shibata, N., Asakawa, S., Shimizu, N., Hori, H., Hamaguchi, S. and Sakaizumi, M. (2002). DMY is a Y-specific DM-domain gene required for male development in the medaka fish. *Nature*, 417, 559-563.

Miyawaki, A. (2011a). Development of probes for cellular functions using fluorescent proteins and fluorescence resonance energy transfer. *Annu. Rev. Biochem.*, 80, 357-373.

Miyawaki, A. (2011b). Proteins on the move: insights gained from fluorescent protein technologies. *Nat. Rev. Mol. Cell Biol.*, 12, 656-668.

Morinaga, C., Saito, D., Nakamura, S., Sasaki, T., Asakawa, S., Shimizu, N., Mitani, H., Furutani-Seiki, M., Tanaka, M. and Kondoh, H. (2007). The hotei mutation of medaka in the anti- Mullerian hormone receptor causes the dysregulation of germ cell and sexual development. *Proc. Natl. Acad. Sci.*, 104, 9691-9696.

Nakamura, S., Aoki, Y., Saito, D., Kuroki, Y., Fujiyama, A., Naruse, K. and Tanaka, M. (2008). *Sox9b/sox9a2*-EGFP transgenic medaka reveals the morphological reorganization of the gonads and a common precursor of both the female and male supporting cells. *Mol. Reprod. Dev.*, 75, 472–476.

Nanda, I., Kondo, M., Hornung, U., Asakawa, S,. Winkler, C., Shimizu, A., Shan, Z., Haaf, T., Shimizu, N., Shima, A., Schmid, M. and Schartl, M. (2002). A duplicated copy of DMRT1 in the sex-determining region of the Y chromosome of the medaka, *Oryzias latipes*. *Proc. Natl. Acad. Sci.*, 99, 11778-11783.

Otake, H., Shinomiya, A., Matsuda, M., Hamaguchi, S. and Sakaizumi, M. (2006). Wild-Derived XY Sex-Reversal mutants in the medaka, *Oryzias latipes*. *Genetics*, 173, 2083-2090.

Otake, H., Hayashi, Y., Hamaguchi, S. and Sakaizumi, M. (2008). The Y Chromosome that lost the male-determining function behaves as an X chromosome in the medaka fish, *Oryzias latipes*. *Genetics*, 179, 2157-2162.

Otake, H., Masuyama, H., Mashima, Y., Shinomiya, A., Myosho, T., Nagahama, Y., Matsuda, M., Hamaguchi, S. and Sakaizumi, M. (2010). Heritable artificial sex chromosomes in the medaka, *Oryzias latipes*. *Heredity*, 105, 247-256.

Remington, S.J. (2001). Green fluorescent protein: a perspective. *Protein Sci.*, 20, 1509-1519.

Shinomiya, A., Otake, H., Togashi, K., Hamaguchi, S. and Sakaizumi, M. (2004). Field survey of sex-reversals in the medaka, *Oryzias latipes*: Genotypic sexing of wild populations. *Zool. Sci.*, 21, 613-619.

In: Sexual Plasticity and Gametogenesis in Fishes
Editor: Balasubramanian Senthilkumaran
ISBN: 978-1-62618-848-8

Chapter XI

Hypothalamic Regulation of Pituitary Gonadotropins

Takashi Kitahashi, Md. Shahjahan and Ishwar S. Parhar
Brain Research Institute, Jeffrey Cheah School of Medicine and Health Sciences, Monash University, Sunway campus, PJ, Malaysia

Abstract

Fish reproduction is regulated by the hypothalamus-pituitary-gonadal (HPG) axis. Within the HPG axis, gonadotropin-releasing hormone (GnRH) has been well recognized as the master molecule of reproduction, which regulates the synthesis and release of gonadotropins, follicle-stimulating hormone (FSH) and luteinizing hormone (LH), from the pituitary. In recent years, new molecules such as kisspeptins and gonadotropin-inhibitory hormone (GnIH) in the HPG axis have become the focus of research in reproductive neuroendocrinology. Each of them is considered to play an important role in the control of the HPG axis, in particular in the positive and negative feedback actions. In this chapter, we summarize the old and new molecules in the HPG axis from anatomical and physiological points of view.

Introduction

As in other vertebrates, the hypothalamus-pituitary-gonadal (HPG) axis predominantly regulates teleost reproduction. The hypothalamus receives a variety of internal and external environmental information and integrates it to regulate the pituitary (hypophysis). The hypothalamus contains many neuropeptides and neurotransmitters that stimulate pituitary gonadotropes (Table 1). Among them, gonadotropin-releasing hormone (GnRH) synthesized in a specific neuronal population of the hypothalamus has been known as the master molecule of reproduction to stimulate the synthesis and release of gonadotropins (GtHs; follicle-stimulating hormone, FSH, and luteinizing hormone, LH) in the pituitary.

Table 1. Hypothalamic hormones and neurotransmitters that regulate gonadotropes

	Hormone/ neurotransmitter	Fish species	Function	Reference
Stimulatory	GnRH1	Blue gourami	Stimulation of FSHβ mRNA expression	(Levy and Degani, 2012)
	GnRH2	Goldfish	Stimulation of LH release	(Chang et al., 1990)
	GnRH3	Coho salmon	Stimulation of FSH and LH release	(Dickey and Swanson, 2000)
		Tilapia	Stimulation of all GtH subunit mRNA expression	(Yaron et al., 2001)
	GnIH (LPXRFa)	Masu salmon	Stimulation of LH, FSH and GH release	(Amano et al., 2006)
		Grass puffer	Stimulation of LH and FSH mRNA expression	(Shahjahan et al., 2011)
		Goldfish	Stimulation of LHβ mRNA expression	(Moussavi et al., 2012)
	PQRFa	Hagfish	Stimulation of GtHβ mRNA expression	(Osugi et al., 2011)
	26RFa	Goldfish	Increase of plasma LH levels	(Liu et al., 2009)
	Kiss1	Sea bass	Increase of plasma LH levels	(Felip et al., 2009)
	Kiss2	Goldfish	Stimulation of LH release	(Yang et al., 2010)
		Zebrafish	Stimulation of FSH and LH mRNA expression	(Kitahashi et al., 2009)
		Sea bass	Increase of plasma FSH and LH levels	(Felip et al., 2009)
	Tachykinin (NKBa, NKF)	Zebrafish	Increase of plasma LH levels	(Biran et al., 2012)
	Serotonin (5-HT)	Goldfish	Increase of plasma LH levels	(Somoza et al., 1988)
		Atlantic croaker	Stimulation of LH release	(Khan and Thomas, 1992)
	NPY	Goldfish	Stimulation of LH release	(Peng et al., 1990)
		Common carp	Stimulation of LH release	(Breton et al., 1991)
		Rainbow trout	Stimulation of LH release	(Cerda-Reverter et al., 1999)
		Sea bass	Stimulation of LH release	(Cerda-Reverter et al., 1999)
		Tilapia	Stimulation of LHβ and GtHα mRNA expression	(Yaron et al., 2001)
	PACAP	Goldfish	Stimulation of LH release	(Wong et al., 1998; Sawisky and Chang, 2005)
		Tilapia	Stimulation of all GtH subunit mRNA expression	(Yaron et al., 2001)
		Blue gourami	Stimulation of FSHβ mRNA expression	(Levy and Degani, 2012)
Inhibitory	GnIH (LPXRFa)	Goldfish	Decrease of serum LH levels	(Moussavi et al., 2012; Zhang et al., 2010)
			Inhibition of FSHβ and LHβ mRNA expression	
	Dopamine	Goldfish	Decrease of plasma LH levels	(Chang et al., 1983)
		Chinese loach	Decrease of plasma LH levels	(Lin et al., 1989)

The release of GnRH and the sensitivity of gonadotropes (FSH and LH cells) to GnRH primarily determine the activity of the HPG axis. GtHs travel throughout the body with the bloodstream and stimulate maturation of the gonads (ovaries and testes) by binding to their respective receptors on them. Matured gonads secrete sex steroid hormones (estrogens and androgens), which negatively regulate the hypothalamus and the pituitary (long- and short-negative feedback loops). This closed-loop system maintains the homeostasis of the reproductive system. On the other hand, in certain situations such as the LH surge in females and the onset of puberty, sex steroids exert stimulatory effects on the hypothalamus and the pituitary (long- and short-positive feedback loops) to initiate the activation of the HPG axis.

Although these mechanisms are very similar in teleosts and tetrapods, there are still significant differences in the roles of individual hypothalamic hormones and the way they are delivered to the pituitary as discussed in later sections. Since GnRH was identified as a neuropeptide that stimulates GtH secretion to play a pivotal role in regulating reproduction, this hypophysiotropic hormone has been recognized as the master molecule of reproduction in all vertebrates.

However, some aspects of the regulatory mechanism of reproduction, especially the sites of negative and positive feedback actions of sex steroids on GnRH secretion, remain unclear. In the last decade, several novel hypothalamic neuropeptides including kisspeptins and gonadotropin-inhibitory hormone (GnIH) have been identified as new players in the control of reproduction. Importantly, these novel neuropeptides fill the "missing link" of the negative and positive feedback actions of sex steroids in mammals. Therefore, the whole picture of the regulatory mechanism of reproduction is more evident now.

In this chapter, we summarize recent knowledge of the hypothalamic neuroendocrine mechanisms in the regulation of pituitary GtHs with specific emphasis on the comparison between teleosts and tetrapods.

Anatomical Characteristics of the Teleost Hypothalamus-Pituitary Connection

The pituitary gland, which consists of the adenohypophysis and the neurohypophysis, is attached to the hypothalamus by a short stalk that contains neurosecretory fibers projecting from the brain to the pituitary in all vertebrates including fishes (Figure 1). The neurohypophysis is a posterior part of the pituitary, with a rich supply of neurosecretory fibers that release vasotocin and isotocin (vasopressin and oxytocin in mammals). On the other hand, the adenohypophysis, an anterior part of the pituitary, contains various pituitary endocrine cells secreting pituitary hormones, which include FSH, LH, thyroid-stimulating hormone (TSH), growth hormone (GH), prolactin (PRL), adrenocorticotropic hormone (ACTH), and somatolactin (SL). The adenohypophysis is further divided into the pars distalis (rostral and proximal pars distalis) and the pars intermedia (Olivereau and Ball, 1964). The dorsal and ventral parts of the proximal pars distalis contain FSH and LH cells (Figure 1b), respectively (Zohar et al., 2010). The relatively segregated cell localization of hormone types to given pituitary regions compared to mammals is one of structural features of the fish pituitary.

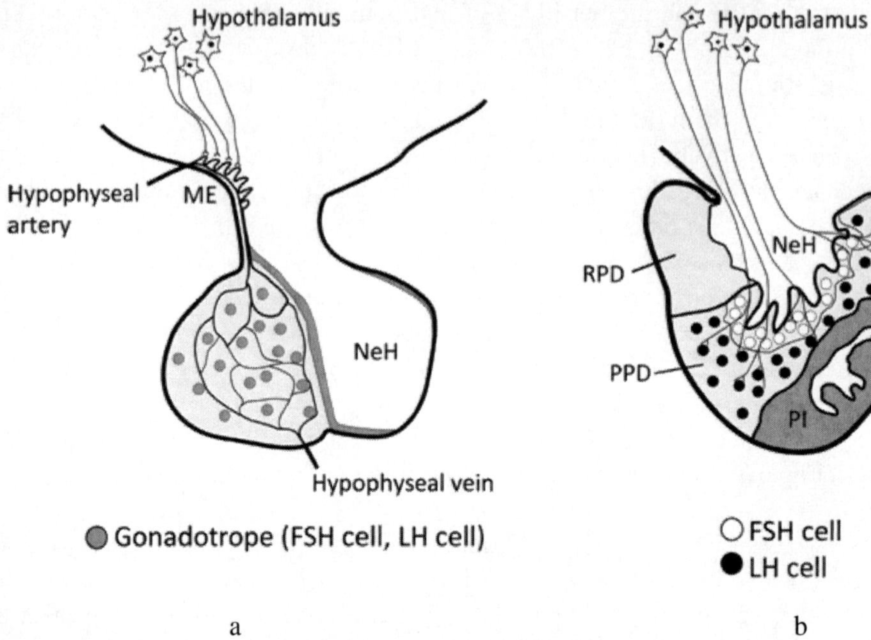

ME, median eminence; NeH, neurohypophysis; PI, pars intermedia; PPD, proximal pars distalis; RPD, rostral pars distalis.

Figure 1. Innervations of hypophysiotropic hypothalamic neurons in mammals (A) and teleosts (B). Innervations of GnRH neurons are shown as an example. Within the adenohypophysis (anterior pituitary), the pars distalis and the pars intermedia are shaded with light and dark gray, respectively. In mammals, the axons of hypophysiotropic neurons terminate at the ME, and released hormones reach target cells via the blood circulation in the hypophyseal portal system. In teleosts, hypophysiotropic neurons directly innervate into the adenohypophysis (the rostral pars distalis and the proximal pars distalis) and terminate near to their target cells, which are located in a cluster manner.

Fish hypothalamus-pituitary connection also has a characteristic anatomical structure when compared to that of mammals. In fish, hypothalamic hormones that control pituitary functions are directly transported to the respective pituitary endocrine cells via neuronal fiber projections to the adenohypophysis (Figure 1b). Indeed, neuronal processes immunoreactive to various hypothalamic hormones project into the pituitary. These hypothalamic hormones are released from the nerve terminals to act directly on nearby target cells. On the other hand, in the mammalian HPG axis, neuronal fibers of the hypophysiotropic hormone cells terminate at the median eminence (ME), which is anatomically located just above the adenohypophysis, and release hormones into the hypophyseal portal system. Released hypothalamic hormones travel with the bloodstream to their target pituitary endocrine cells, which are scattered throughout the adenohypophysis (Figure 1a). Because hypothalamic hormones are delivered throughout the adenohypophysis via the blood circulation, the responsiveness of each pituitary endocrine cell to the hypophysiotropic hormones is determined by the expression of respective receptors in the target pituitary endocrine cells. Direct innervations from the hypothalamus to the adenohypophysis and the segregation of endocrine cells in the pituitary makes teleost a useful model to understand hypothalamic control of the pituitary endocrine cells (Parhar and Iwata, 1994).

In fact, there is good association between the distribution of hypothalamic neuronal fibers and the endocrine target cells in the adenohypophysis of teleosts. For example, fiber terminals of GnRH neurons are distributed in the proximal pars distalis, where LH and FSH cells are located, which justifies the role of GnRH in GtH secretion. Neurons synthesizing several neuropeptides and neurotransmitters, particularly GnRH (Parhar, 1997), GABA (Kah et al., 1987a), neuropeptide Y (Batten et al., 1990), dopamine (Kah et al., 1984), and PACAP (Wong et al., 1998) have been identified to project their fibers to the pars distalis to modulate the release of GtH and other pituitary hormones (Table 1).

Among brain regions, two hypophysiotropic nuclei, namely the nucleus preopticus (NPO) and the nucleus lateralis tuberis (NLT), have been known to control teleostean adenohypophyseal functions. In most teleosts, the NPO and the NLT comprise the major neuronal systems that innervate the pituitary to regulate different endocrine cells (Fryer and Maler, 1981; Peter, 1986). However, some extra hypothalamic areas, especially from the olfactory system and the telencephalon, have also been identified to send neuronal projections to the pituitary (Anglade et al., 1993; Chiba et al., 1996; Rama Krishna et al., 1992). In the POA-hypothalamus, the anterior and posterior subdivisions of the nucleus preopticus periventricularis (NPP) have been identified as the regions that contain a variety of peptide hormones and neurotransmitters, including GnRH (Parhar, 1997) and dopamine (Kah et al., 1984), which are involved in the control of GtH secretion in the pituitary.

Hypothalamic Hormones Regulating GTH Secretion

Gonadotropin-Releasing Hormones (GnRHs)

GnRH was first identified from the ovine and porcine hypothalamus as a peptide hormone that stimulates the release of LH from the adenohypophysis (Amoss et al., 1971; Matsuo et al., 1971), and was called luteinizing hormone-releasing hormone (LHRH). It was identified as a decapeptide with the primary structure of pEHWSYGLRPG-NH_2. Findings that this hypophysiotropic hormone also stimulates the secretion of FSH changed its name to gonadotropin-releasing hormone, GnRH. The first fish GnRH was identified from the salmon by Sherwood and coworkers (Sherwood et al., 1983). To date, 15 forms of GnRH with different amino acid sequences have been identified in vertebrates, including ten teleost-specific forms like salmon, seabream, whitefish, medaka, catfish, herring, dogfish, lamprey I, II and III (Adams et al., 2002; Carolsfeld et al., 2000; Jimenez-Linan et al., 1997; Kavanaugh et al., 2008; Lovejoy et al., 1992; Matsuo et al., 1971; Miyamoto et al., 1982; Miyamoto et al., 1984; Ngamvongchon et al., 1992a; Okubo et al., 2000; Powell et al., 1994; Sherwood et al., 1983; Sower et al., 1993; Yoo et al., 2000). It now appears that most vertebrate species possess two, and some teleost species have three, GnRH forms (Parhar, 1997; Sherwood et al., 1997), although rodents, bovine, and sheep are known to possess only one functional GnRH (Millar, 2003; Morgan et al., 2006). Fish species possess not only the most diverged primary GnRH structure, with 12 in total (10 fish forms, mammalian GnRH, and chicken GnRH-II), but also possess up to 5 GnRH receptor (GnRH-R) variants in some species (Jodo et al., 2003; Kah et al., 2007). Fish GnRH is therefore of interest from an evolutionary point of view.

It has been reported that GnRH and GnRH-R are also expressed in peripheral tissues including the gonads (Lin et al., 2010; Madigou et al., 2002; Pati and Habibi, 1998; Uzbekova et al., 2002; Yu et al., 1998). This gonadal GnRH/GnRH-R system is considered to act in an autocrine or paracrine manner to regulate gonadal steroidogenesis and oocyte maturation (Andreu-Vieyra et al., 2005; Pati and Habibi, 2000). However, in this chapter, we focus only on the central roles of GnRH in the brain.

While GnRH was traditionally named after the species in which it was identified, the same form also exists in other species; salmon GnRH also exists in the herring, goldfish, medaka, seabream, and the tilapia.

Therefore, the current nomenclature uses GnRH1, GnRH2 and GnRH3 based on the phylogenetic sequence analysis and respective location of expression. GnRH1 is the most variable form among the three, and is generally expressed in the ventral forebrain-preoptic area (POA)-basal hypothalamus. GnRH2 is the most evolutionarily conserved GnRH form among vertebrates and expressed in the midbrain. The third type GnRH3 is present in the caudal-most olfactory bulb (Figure 2) along the terminal nerve in certain fish species (Okubo and Nagahama, 2008). GnRH1, of which neurons are located in the POA-basal hypothalamus and send neuronal fibers directly into the adenohypophysis (teleosts) or to the ME (other vertebrates), is considered to be a hypophysiotropic form. In most fish species, GnRH1 neurons are the major source of GnRH fiber projections into the pituitary.

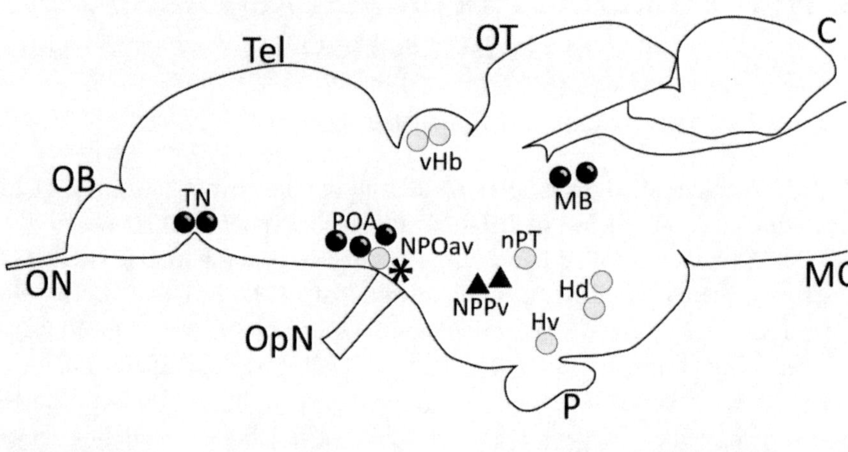

C, cerebellum; Hd, dorsal zone of periventricular hypothalamus; Hv, ventral zone of periventricular hypothalamus; MB, midbrain; MO, medulla oblongata; NPOav, Nucleus preopticus anteroventralis; NPPv, nucleus posterioris periventricularis; nPT, posterior tuberal nucleus; OB, olfactory bulb; ON, olfactory nerve; OpN, optic nerve; OT, optic tectum; P, pituitary; POA, preoptic area; Tel, telencephalon; TN, terminal nerve; vHb, ventral habenula.

Figure 2. A schematic drawing of representative localization of GnRH (black circle), kisspeptin (white circle), GnIH (triangle), and dopamine (asterisk) neurons in a sagittal section of the teleost brain. GnRH neurons in the POA, GnIH neurons in the NPPv, and dopaminergic neurons in the NPOav are shown to innervate the pituitary, while the kisspeptin neuronal population(s) projecting into the pituitary need to be identified.

As in mammals, the LH-releasing role of GnRH has been established in fish species (Trudeau, 1997; Yaron et al., 2003). On the other hand, the lack of quantification method of FSH peptide in most fish species limits information about GnRH effects on FSH release. Some studies in salmonids, however, show that, as in mammals, GnRH also stimulates FSH release from the fish pituitary (Ando et al., 2004; Breton et al., 1998; Dickey and Swanson, 2000; Kawauchi et al., 1989). From these studies, it appears that GnRH1 stimulates the release of FSH and LH during early and late maturational stages, respectively.

Quantitative gene expression analyses of GtH subunits showed the role of GnRH in the regulation of GtH synthesis is dependent on the maturational stage and the fish species (Yaron et al., 2003). For example, in maturing sockeye salmon GnRH analog (GnRHa) significantly increases the amounts of LHβ and GtHα subunit mRNAs in both sexes, while there is no effect on FSHβ subunit mRNA (Kitahashi et al., 1998a).

Similar stimulatory effect on LH subunit genes is observed in reproductively quiescent sea bass *Dicentrarchus labrax* (Mateos et al., 2002). On the other hand, in maturing striped bass and common carp the stimulatory effect of GnRH on gene expression is seen in both FSH and LH subunits in males whereas only LH synthesis is stimulated by GnRH in females (Hassin i, 1998; Kandel-Kfir et al., 2002). *In vitro* studies using a primary culture of pituitary cells from different maturational stages of fish further confirmed that GnRH has direct effects on the synthesis and release of FSH and LH in a maturational stage-dependent manner in different species (Ando and Urano, 2005).

GnRH1 neurons also project into other brain areas including the olfactory bulbs, olfactory nerve, optic tectum, and the spinal cord (Parhar, 1997). The extra-hypothalamic fiber projections of GnRH1 suggest its role in other reproductive functions such as sexual behavior or non-reproductive behavior.

The functions of GnRH2 and GnRH3 are debatable due to limited evidence. Generally, GnRH2 and GnRH3 fibers are observed throughout the brain, but not in the pituitary. Thus, they are thought to have neuromodulatory functions rather than hypophysiotropic functions (Sherwood et al., 1997). This is supported by the spontaneous regular pacemaker activities of GnRH2 and GnRH3 neurons compared to the episodic firing of GnRH1 neurons observed in teleosts (Oka, 2009). In fact, administration of GnRH2 markedly reduced food intake in the goldfish (Matsuda et al., 2008) through an inhibitory effect on orexin neurons (Hoskins et al., 2008). The involvement of GnRH2 in feeding has also been reported in a mammalian species, the musk shrew *Suncus murinus* (Kauffman and Rissman, 2004; Kauffman and Rissman, 2006), suggesting its role in the control of feeding is evolutionally conserved. In addition, a recent study in the sea bass suggested GnRH2 stimulates melatonin secretion from the pineal gland (Servili et al., 2010).

GnRH3 has been suggested to be involved in the control of reproductive behaviors. Immunoneutralization of GnRH3 using a specific antibody significantly decreases aggressive behavior and nest-building ability in the male Nile tilapia (Ogawa et al., 2006). Lesions of GnRH3 neurons cause impairment of nest-building in male dwarf gourami (Yamamoto et al., 1997). Administration of GnRHa shortens the duration of homing migration of the pre-spawning lacustrine sockeye salmon, *Oncorhynchus nerka*, which further provides evidence of its role in stimulating reproductive behavior (Kitahashi et al., 1998b; Sato et al., 1997).

In spite of the similar neuronal localization pattern and function of each GnRH form in different fish species, there are slight variations in the GnRH peptides detected in the pituitary among fish species. For example, GnRH2 is detected in the pituitary of cyprinids (Steven et al., 2003; Yu et al., 1988) and catfish (Ngamvongchon et al., 1992b; Schulz et al., 1993). GnRH3 neurons are also shown to project into the pituitary of the sea bass, but significantly less than GnRH1 neurons (Gonzalez-Martinez et al., 2002a). However the significance of the species differences in pituitary innervation of GnRH neurons is still not clear.

In some fish species including most salmonids and zebrafish, the GnRH1 gene has been lost during evolution. In such species, GnRH3 seems to double as GnRH1 in terms of its function and neuronal localization (Abraham et al., 2009; Amano et al., 1997): i.e. GnRH3 neurons are located in the olfactory bulb, ventral telencephalon, POA, and the ventromedial hypothalamus, and send fibers into the pituitary.

Kisspeptins

Kisspeptin is a member of the RFamide family of peptides, encoded by the KISS1 gene in humans (Kotani et al., 2001). The core 10-amino acid sequence (Kiss1-10) is common among endogenous mature peptides derived from a common precursor peptide. This core sequence is highly conserved during evolution and shows potent biological activity (Bilban et al., 2004). Its important role in reproductive control emerged in 2003 when two independent groups reported inactivating mutations in kisspeptin receptor (kiss1r, G-protein coupled receptor 54, GPR54) result in idiopathic hypogonadotropic hypogonadism in humans and mice (de Roux et al., 2003; Seminara et al., 2003). *Kiss1r* knockout mice also exhibit the failure of pubertal onset (Lapatto et al., 2007; Seminara et al., 2003). Furthermore, similar phenotypes were demonstrated by *Kiss1* knockout in mice, confirming the role of kisspeptin/ kiss1r in the onset of mammalian puberty (d'Anglemont de Tassigny et al., 2007; Lapatto et al., 2007). Kisspeptin administration induces FSH and LH release solely through the activation of GnRH1 release, because treatment with a GnRH antagonist completely blocks the effects of kisspeptin in rodents (Navarro et al., 2005a; Navarro et al., 2005b). Extensive studies revealed that kisspeptin fibers project to GnRH1 neuronal bodies and fiber terminals to regulate GnRH1 secretion at the POA and ME levels, respectively (Parhar et al., 2012).

In rodents, kisspeptin neurons are localized in the arcuate nucleus (Arc) and the anteroventral periventricular nucleus (AVPV) (Clarkson and Herbison, 2006). The two kisspeptin neuronal populations are differently regulated by sex steroids. In the Arc, kisspeptin synthesis is inhibited by sex steroids, while it is stimulated in the AVPV (Smith et al., 2005a; Smith et al., 2005b). Thus, in mammals, kisspeptin neurons in the Arc and the AVPV are considered to mediate the negative and positive feedback actions of sex steroids, respectively.

In teleosts, similar to GnRH, kisspeptins present in multiple forms and multi-functional in one species. The first identification of the cDNAs for two kisspeptin genes, *kiss1* and *kiss2*, was reported in the medaka and the zebrafish (Kitahashi et al., 2009). cDNAs for the two kisspeptin genes have also been cloned in the chub mackerel (Selvaraj et al., 2010), sea bass (Felip et al., 2009), and the goldfish (Li et al., 2009).

Genome comparison suggests that *kiss1* and *kiss2* are paralogs resultant from a whole genome duplication event during vertebrate evolution. Indeed, most vertebrates possess two forms of kisspeptin whereas *kiss2*-homologous sequences are absent in placental mammals.

The absence of *kiss2* could be due to the loss of the *kiss2* gene in this lineage during evolution, because the genomic regions syntenic to where the *kiss2* gene is located in teleost still exist in the mammalian genomes (Kitahashi et al., 2009). In parallel to the multiple kisspeptin forms, there are multiple kisspeptin receptors in teleosts. In the zebrafish, the core sequences of two kisspeptins (Kiss1-10 and Kiss2-10) showed the highest potency for kiss1 receptor (Kiss1R, also called GPR54-1) and kiss2 receptor (Kiss2R, also called GPR54-2), respectively (Lee et al., 2009).

Systemic kisspeptin administrations (Kiss1-10 and Kiss2-10) stimulate the synthesis and release of FSH and LH and gonadal maturation in fish as in mammals. However, the effect of Kiss1-10 and Kiss2-10 in the gonadotrope control varies among species. In sexually mature female zebrafish, intraperitoneal injections of Kiss2-10 but not Kiss1-10 significantly increased the amount of FSHβ and LHβ mRNAs (Kitahashi et al., 2009). The stronger effect of Kiss2-10 compared to Kiss1-10 was also observed in the release of FSH and LH in prepubertal sea bass (Felip et al., 2009). A similar trend of the effects of kisspeptins was observed in the stimulatory effect on gonadal maturation in female sea bass and striped bass (Beck et al., 2012).

On the contrary, in female spotted grouper, an intraperitoneal Kiss2-10 injection stimulated the mRNA expression of FSHβ rather than LHβ (Shi et al., 2010). Furthermore, in the goldfish, an intraperitoneal Kiss1-10 injection, but not Kiss2-10, stimulated LH release in sexually mature female (Li et al., 2009). These results indicate the role of kiss1 and kiss2 in the regulation of GtH is species specific. However, it should be noted that longer kisspeptins, Kiss1-15 and Kiss2-12, are proposed as the endogenous kisspeptin forms in teleost species (Lee et al., 2009).

The anatomical localizations of kisspeptin neurons in the fish brain have been examined by *in situ* hybridization studies. These studies show *kiss1*-expressing neurons in the ventral habenula (vHb) and *kiss2*-expressing neurons in the dorsal zone of the periventricular hypothalamus (Hd; also designated as the nucleus recessus lateralis, nRL) in the zebrafish, medaka, goldfish, and the sea bass (Figure 2; Escobar et al., 2012; Kanda et al., 2012; Kitahashi et al., 2009). *Kiss1*-expressing neuronal populations have also been localized in the nucleus ventral tuberis (nVT) and the nucleus posterior is periventricularis (NPPv) in the medaka (Figure 2; Kanda et al., 2008; Kitahashi et al., 2009).

Interestingly, in the zebrafish and goldfish brains, the *kiss1* gene is not expressed in these hypothalamic neuronal populations. Instead, the *kiss2* gene is expressed in the nVT and NPPv kisspeptin neurons in the goldfish and the zebrafish, respectively (Kanda et al., 2012; Kitahashi et al., 2009; Servili et al., 2011). In addition, there is another *kiss2*-expressing neuronal population in the POA of the zebrafish and goldfish (Kanda et al., 2012; Servili et al., 2011). The variation in the kisspeptin form expressed in hypothalamic nuclei and the different effects of systemic administrations of Kiss1-10 and Kiss2-10 in different fish species raise the possibility that kisspeptin functions diverged during teleost evolution. Similar to the GnRH/ GnRH-R system, kisspeptins and kisspeptin receptors are also expressed in several peripheral tissues including the gonads, suggesting additional local actions of kisspeptins (Biran et al., 2008; Kitahashi et al., 2009; Shahjahan et al., 2010b; Yang et al., 2010).

Fiber projections of kisspeptin neurons in the brain provide important information to understand the endogenous roles of the two kisspeptins in neuroendocrine function. However, information on kisspeptin fiber projections in fish species had been limited by the lack of specific antibodies for fish kisspeptin forms. Recent efforts to develop specific fish kisspeptin antisera and the application of a neuronal tracer revealed that the habenula kiss1 neurons send fibers only to the ventral part of the interpeduncular nucleus, IPN (Ogawa et al., 2012; Servili et al., 2011).

In addition, the expression of Kiss1 receptor is observed only in the habenula and the ventral IPN in the zebrafish (Ogawa et al., 2012). The habenula kiss1 system is thus implicated in the modulation of serotonergic system rather than HPG axis regulation in the zebrafish (Ogawa et al., 2012). On the other hand, Kiss2 fibers are widely distributed in the zebrafish forebrain and midbrain (Servili et al., 2011).

The fact that all three GnRH neuron types express kisspeptin receptors in the Nile tilapia (Parhar et al., 2004) suggests that the role of Kiss2 neurons in the regulation of the HPG axis is via the activation of the GnRH systems.

Gonadotropin-Inhibitory Hormone (GnIH)

Gonadotropin-inhibitory hormone (GnIH) was found in the Japanese quail in 2000 as a hypothalamic neuropeptide that inhibits GtH release from the pituitary (Tsutsui et al., 2000). GnIH was identified as a dodecapeptide SIKPSAYLPLRF-NH_2, and classified as a member of the RFamide family of peptides due to its C-terminal RF-NH_2 motif. GnIH is the first hypothalamic neuropeptide identified as an inhibitor of GtH release in vertebrates. In avian species, GnIH expressed in the paraventricular nucleus (PVN) inhibits the HPG axis at the level of GnRH1 neuronal somata and fiber terminals as well as its direct action on the pituitary gonadotropes (Tsutsui et al., 2012). In the last decade, GnIH and its orthologs have been identified in many vertebrates including birds, fishes, and mammals. As in avian species, mammalian GnIH (RFamide-related peptide-3, RFRP-3) suppresses GtH secretion via the inhibition of GnRH1 neurons, while its direct action on the pituitary is controversial in mammals (Tsutsui et al., 2012).

GnIH has been identified in fish species including the goldfish, zebrafish, medaka, *Takifugu*, grass puffer, and the sea lamprey (Osugi et al., 2006; Sawada et al., 2002; Zhang et al., 2010). Fish GnIH is also referred to as LPXRFa after the sequence of C-terminal motif of its mature peptide. The neurons expressing fish GnIH are localized in the NPPv of the goldfish and the sockeye salmon (Figure 2; Amano et al., 2006; Sawada et al., 2002). As in birds and mammals, fish GnIH neuronal fiber projections are observed in a wide range of brain regions, which include the area where GnRH1 neurons are located. In addition, some GnIH fibers also project to the pituitary in both species (Amano et al., 2006; Sawada et al., 2002), suggesting evolutionally conserved GnIH functions in teleosts.

The physiological function of GnIH in teleosts is complex. As in birds and mammals, intraperitoneal injections of the goldfish and zebrafish GnIH peptides reduce plasma LH levels in adult goldfish (Moussavi et al., 2012; Zhang et al., 2010). However, the inhibitory effect of GnIH injections is not observed in the late recrudescent stage (Moussavi et al., 2012).

In addition, a stimulatory effect of goldfish GnIH peptide on LH release is observed in cultured goldfish and salmon pituitaries (Amano et al., 2006; Moussavi et al., 2012). Furthermore, injections of GnIH drastically increase the levels of FSH and LH mRNAs in the goldfish pituitary in a reproductive stage-dependent manner (Moussavi et al., 2012). *In vitro* studies also show that the goldfish GnIH peptide stimulates GtH subunit gene expression in the goldfish and the grass puffer with clear seasonal differences (Moussavi et al., 2012; Shahjahan et al., 2011). These findings indicate that, in teleosts, the physiological effect of GnIH on the HPG axis differs between GtH synthesis and release and depends on the reproductive stage. To our knowledge, interaction between GnIH and GnRH neurons in teleosts has yet to be reported, while close appositions between GnIH fibers and GnRH1 neuronal somata are evident in birds and mammals (Soga et al., 2010; Tobari et al., 2010; Ubuka et al., 2012; Ubuka et al., 2009). Further studies, especially using other seasonal/non-seasonal breeders are needed, which will provide clearer answers regarding the roles of GnIH in the control of the HPG axis.

Dopamine

Dopamine is a monoamine neurotransmitter that plays a variety of functions in vertebrates. Dopaminergic neuronal distribution in the fish brain has been extensively studied. There are many dopaminergic neuronal populations and wide-spread fiber projections in the brain (Panula et al., 2010). Receptors for dopamine are classified into two major classes that activate (D1-like) or inhibit (D2-like) an enzyme adenylate cyclase (Cardinaud et al., 1998).

Dopamine has an inhibitory role in GtH release in many teleosts (Peter et al., 1986). This inhibitory action has been suggested to be a direct action because dopamine receptors are expressed in gonadotropes (Chang and Peter, 1983). A group of dopaminergic neurons localized in the nucleus preopticus anteroventralis (NPOav) of the ventral preoptic area directly innervate pituitary gonadotropes in the goldfish (Kah I, 1987b) and the European eel (Figure 2; Sébert et al., 2008). Using specific agonists and antagonists *in vivo* and *in vitro*, it was shown that dopamine inhibits GtH release directly through the D2-like receptors expressed in gonadotropes. Studies showed that dopamine inhibits spontaneous and GnRH-stimulated GtH release in the goldfish (Chang and Peter, 1983; Chang et al., 1990), catfish (de Leeuw et al., 1986), Chinese loach (Lin et al., 1989), tilapia (Levavi-Sivan et al., 1995) and the gray mullet (Aizen et al., 2005). In the goldfish, dopamine also acts on GnRH neurons to inhibit GnRH release (Trudeau, 1997). These results indicate that dopaminergic inhibition plays an important role in the control of teleost reproduction. However, the inhibitory role of dopamine in GtH release is not dominant in some fish species (Copeland and Thomas, 1989), suggesting that physiological importance and roles of the dopaminergic inhibition is species-specific in teleosts.

Ontogeny of the GnRH and Kisspeptin Systems

GnRH neurons are known for their migration during embryonic development. In fish, GnRH1 neurons originate in the olfactory placode as in mammals and migrate to their final destination in the POA (Parhar, 2002). In the European sea bass, GnRH1 immunoreactivities start to appear at the olfactory placode on 26 days after hatching (dah). These neurons migrate along the olfactory tract, and are positioned at the POA and the hypothalamus at 45 and 60 dah, respectively (Gonzalez-Martinez et al., 2004; Gonzalez-Martinez et al., 2002b). The migration of GnRH1 neurons during early development, from the olfactory region to the POA, is crucial for the development of the reproductive system in all vertebrates. In patients with Kallmann's syndrome (KS), GnRH1 neurons fail to migrate due to mutations in a gene for anosmin-1, which is responsible for migration of certain nerve cell precursors during embryogenesis, and are trapped at the place where they are originated (Schwanzel-Fukuda et al., 1989; Wray et al., 1989). As a result, patients with KS exhibit hypogonadotropic hypogonadism together with anosmia (partial or complete loss of olfaction) (Naftolin et al., 1971). GnRH3 neurons also originate in the olfactory placode, but at a relatively early stage, around 7 dah, and migrate to the olfactory bulb on 10 dah in the European sea bass, appearing in the ventral telencephalon and the POA from days 15 and 45, respectively (Gonzalez-Martinez et al., 2004; Gonzalez-Martinez et al., 2002b).

In salmonids, immunoreactive GnRH neurons are first detected at 19 day after fertilization (daf) in the olfactory placode, appear in the olfactory nerve by 30 daf, in the olfactory bulb by 51 daf, and appear in the ventral telencephalon and the POA by 462 daf (Parhar et al., 1995). In a fast-developing zebrafish embryo, GnRH3 neurons start appearing in the olfactory placode around 26 hours after fertilization and migrate to the olfactory bulb, POA, and caudal hypothalamus by 4-5 daf (Abraham et al., 2008). Laser ablation of GFP-labeled GnRH3 neurons in the olfactory placode at an early stage of embryonic development completely diminished the number of GnRH3 neurons in the adult stage, indicating all populations of zebrafish GnRH3 neurons originate from the olfactory region (Abraham et al., 2010). On the other hand, GnRH2 neurons originate in the germinal zone of the third ventricle, at a diencephalic/mesencephalic transitional area (Parhar et al., 1996; White and Fernald, 1998). GnRH2 neurons start to be observed on 4 dah in the European sea bass during development (Gonzalez-Martinez et al., 2004; Gonzalez-Martinez et al., 2002b). The earliest expression of the midbrain GnRH2 among the three GnRH forms is also observed in other vertebrate species (Muske and Moore, 1990; White and Fernald, 1998), suggesting an important role of this evolutionally-conserved GnRH form in ontogenesis.

In terms of the roles of each GnRH form during development, the involvement of the GnRH neuronal system(s) in the onset of puberty is well documented in teleosts. Among the three GnRH forms, only hypothalamic GnRH (GnRH1) mRNA and peptide levels show coincident increase with gonadal development (Holland et al., 2001; Okuzawa et al., 2003; Parhar et al., 1996; Senthilkumaran et al., 1999; Shahjahan et al., 2010a), indicating its role in the onset of puberty in fish. On the other hand, during the pubertal stage, three forms of GnRH mRNA expression are significantly high, whereas only GnRH1 shows an increase at the latter stage during the reproduction in grey mullet (Nocillado et al., 2007). These results indicate that, in the species expressing three GnRH forms, GnRH1 has the gonadotropic role, while GnRH2 and GnRH3 have important roles in modulating reproductive behavior (Parhar et al., 2003a; Uchida et al., 2005).

In the last few years, considerable developments have been made in the study of puberty in mammals. It has been demonstrated that kisspeptin and kisspeptin receptor (kissR) play a crucial role in reproduction, mainly by stimulating GnRH release (Messager, 2005), therefore acting as a gatekeeper for puberty.

In the zebrafish, the mRNA levels of kisspeptins (Kiss1 and Kiss2) are increased significantly at the start of the pubertal phase together with hypothalamic GnRH (GnRH3) mRNAs (Kitahashi et al., 2009). The levels of kissR mRNAs in the brain, which are expressed in GnRH neurons (Parhar et al., 2004), are also higher at the start of puberty compared to the pre- or post-pubertal stages in the cobia (Mohamed et al., 2007), grey mullet (Nocillado et al., 2007), fathead minnow (Filby et al., 2008), and the tilapia (Martinez-Chavez et al., 2008). These results indicate that fish kisspeptins most likely control the onset of puberty by stimulating GnRH release, as in mammals (Dungan et al., 2006; Seminara, 2005).

Ontogeny of the Pituitary Gonadotropes

Pituitary GtHs, FSH and LH, directly control gonadal development and reproduction in fish (Lubzens et al., 2010; Yaron et al., 2003). In multiple-spawning fish species, differential roles of FSH and LH in the control of reproductive cycles have been established. In general, FSH is predominant during early gonadal recrudescence and gametogenesis, whereas LH plays a major role in the regulation of final gonadal maturation, ovulation, and spermiation (Swanson et al., 1991; Yaron et al., 2003).

On the other hand, ontogeny studies of FSH and LH cells in the pituitary have suggested additional GtH roles in sex differentiation and pubertal onset. Immunoreactivities of both FSH and LH appear simultaneously in the pituitary; on 14 daf in the Nile tilapia and 51 daf in the sockeye salmon (Parhar et al., 2003b). The pituitary contents of FSH protein and mRNA are much higher than those of LH during early puberty in the masu salmon (Amano et al., 1993) and rainbow trout (Gomez et al., 1999). The earlier appearance of FSH cells compared to LH cells in the pituitary during development commonly occurs in teleost species (Gomez et al., 1999; Guzman et al., 2009; Hassin et al., 1999; Hassin et al., 2000; Moles et al., 2007; Parhar et al., 2003b; Wong et al., 2004). This suggests a role for FSH in sex differentiation and the early stages of gonadal development, whereas LH may play a role mainly in the pubertal stage. However, in some fish species such as the African catfish, pejerrey, and the ricefield eel, LH cells appear earlier than FSH cells (Miranda et al., 2001; Schulz et al., 1997; Wu et al., 2012), suggesting species-specificity in the respective roles of FSH and LH in early gonadal differentiation and development. Immunoreactivities against GnRH receptors (GnRH-R type IA and IB) are first observed in the pituitary on 15 daf in the Nile tilapia (Parhar et al., 2003b). In this species, which undergo sexual differentiation and gonadal steroidogenesis at around 25 daf, immunoreactivity for GnRH-R type IB in the pituitary is decreased at the time of sexual differentiation, while the expression of GnRH-R type IA is significantly increased (Parhar et al., 2002). The presence of GnRH-Rs type IA and IB in gonadotropes suggests that they mediate the different effects of GnRH on the synthesis/release of GtHs from the pituitary. Interestingly, the time of sexual differentiation in the tilapia, salmonids, and the eel coincides with the development of GnRH fibers in the pituitary (Parhar et al., 1995; Swanson et al., 1991). The chronological appearance of gonadotropes (FSH/LH cells), GnRH-Rs, GnRH cells, and GnRH fiber projections into the pituitary provides important insights into the understanding of the complete developmental process of the hypothalamus-pituitary relationship in teleosts (Parhar, 1997).

Feedback Regulations of the HPG Axis by Sex Steroids

The HPG axis forms a closed loop to maintain homeostasis. Alterations in gonadal activity give feedback to the hypothalamus and pituitary through the changes in the circulating plasma levels of sex steroids. The mechanism of the positive and negative actions of sex steroids over the HPG axis, and the GnRH system in particular, has been a long standing question. Despite substantial efforts in the past decades, no concrete evidence of sex steroid receptors in GnRH1 neurons has been shown, although the positive/negative effects of sex steroids on GnRH1 neurons are obvious (Herbison, 2008; Richardson et al., 2004). However, after the findings of kisspeptins and GnIH as new players in the HPG axis, accumulated information suggests that these two hypothalamic neuropeptides mediate the effects of various internal/external environmental signals, including sex steroids, in the HPG axis of vertebrates (Parhar et al., 2012).

Particularly, involvement of kisspeptins and their receptors in the sex steroid feedback mechanism is now evident in fish (Parhar et al., 2012). Ovariectomy diminishes kiss1 gene expression in the Hv neuronal population in the medaka, and E2 replacement rescues it (Kanda et al., 2008; Mitani et al., 2010). Interestingly, in the zebrafish and goldfish, similar positive effects of estrogen are seen in the kiss2 neurons in the ventral hypothalamus and POA populations, respectively (Kanda et al., 2012; Servili et al., 2011). The sex steroid-sensitive kisspeptin neuronal populations express estrogen receptors (ERα in the medaka and ERα, β1 and β2 in the goldfish), and are thus directly regulated by sex steroids.

These data indicate that kisspeptin neurons mediating the positive feedback differ in different species regardless of the kisspeptin form they express. Therefore, information on the anatomy of the kisspeptin systems (Ogawa and Parhar, 2013) and the regulation of each kisspeptin gene (Kitahashi and Parhar, 2013) is important to understand the roles of kisspeptin in a species. Notably, a negative feedback by sex steroids on teleost kisspeptin neurons has not been shown. This is different from the case in mammals, where kisspeptin neurons in the AVPV and Arc mediate the positive and negative feedbacks, respectively (Smith et al., 2006). Therefore, the fish kisspeptin system might receive only positive but not negative feedback by sex steroids (Figure 3).

Although there are no fish studies that report the effects of sex steroids on the GnIH system, mammalian studies have shown possible involvement of the GnIH system in the negative feedback regulation of the HPG axis (Parhar et al., 2012). However the information is still controversial. A subset of GnIH neurons expresses ERα in the Syrian hamster and mice, suggesting a role of E2 in the regulation of GnIH neurons (Kriegsfeld et al., 2006; Molnar et al., 2011). Indeed, treatment with E2 increases c-fos expression in GnIH neurons, indicating the involvement of GnIH in the negative feedback action of sex steroids (Kriegsfeld et al., 2006). Conversely, GnIH mRNA levels in ovariectomized mice are reduced by E2 treatment, suggesting the role of GnIH in positive feedback in mice (Molnar et al., 2011). Furthermore, no effects of E2 treatment on GnIH mRNA levels were observed in ovariectomized rats and ewe (Quennell et al., 2010; Smith, 2008). Therefore, the effects of sex steroids on GnIH neurons might be species-, treatment-, or physiological condition-dependent.

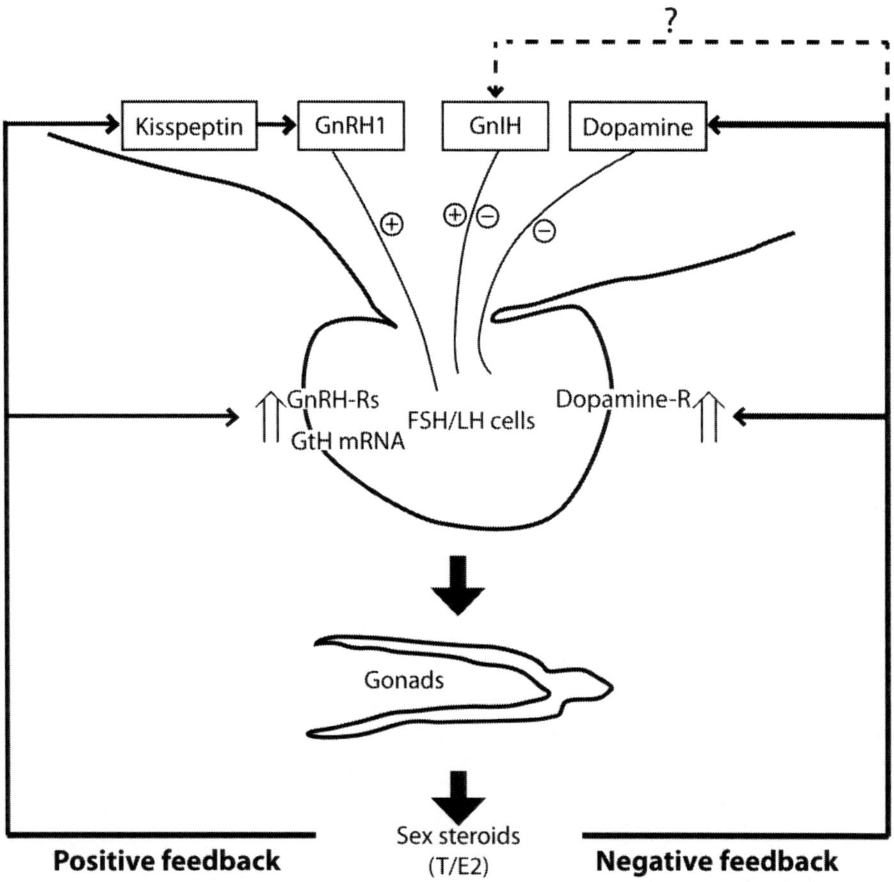

Figure 3. A diagram showing summary of the HPG axis in teleosts. GnRH1 plays a major stimulatory role in GtH secretion, while kisspeptin stimulates GnRH1 release. On the other hand, dopamine plays an important role as an inhibitor of GtH secretion at least in some species. GnIH inhibits GtH release but stimulates GtH synthesis. In terms of sex steroid feedback, the positive feedback seems to be mediated by the stimulatory effects of sex steroids on kisspeptin neurons and the synthesis of GnRH-R in the pituitary. Sex steroids also directly stimulate GtH synthesis at the pituitary level. The negative feedback is mediated by the stimulatory effects of sex steroids on the synthesis of dopamine in the brain and dopamine D2 receptor in the pituitary. Whether GnIH neurons are regulated by sex steroids is unknown. Notably, the strength of each component in the whole system seems to vary depend on species, maturational stage, season, and environmental conditions.

On the other hand, T administrations stimulate the hypophysiotropic population (NPOav) of dopaminergic neurons in the female European eel (Weltzien et al., 2006), suggesting that dopaminergic neurons are involved in the negative feedback regulation of the HPG axis (Figure 3). Further studies in teleosts are required to demonstrate if teleost GnIH neurons also mediate the negative feedback action of sex steroids in the HPG axis.

The pituitary gonadotropes themselves are also the sites of sex steroid feedback. Single-cell real-time PCR combined with laser-capture microdissection showed that individual FSH and LH cells of the male tilapia express multiple types of sex steroid receptors, including ERα, ERβ, ARα, and ARβ (Kitahashi et al., 2007). Interestingly, the combinations of sex steroid receptors detected in individual gonadotropes and the expression levels of each receptor are significantly different between sexually immature and mature fish (Kitahashi et al., 2007). Indeed, the effects of E2 on GtH transcription differ among maturational stages in the masu salmon (Ando et al., 2004), suggesting that changes in the combination of sex steroid receptors decide the response of gonadotropes to circulating sex steroids. Increased levels of GnRH-R and dopamine D2 receptor mRNAs in estradiol-treated pituitary suggest that sex steroids are involved in both positive and negative feedback mechanisms by altering the sensitivity of gonadotropes to GnRH and dopamine (Levavi-Sivan et al., 2006).

Conclusion

A variety of neuropeptides and neurotransmitters, including GnRHs, kisspeptins, and GnIH are secreted from the hypothalamus to regulate the synthesis and release of GtHs in a cooperative manner. Kisspeptins are involved in the positive feedback of sex steroids, and play an important role in the onset of puberty in fish species, as in mammals. The inhibitory role of GnIH on GtH release seems to not be consistent across teleosts and mammals. In fact, *in vivo* administrations of GnIH rather stimulate GtH synthesis in certain maturational stages at least in some species. However, GnIH has an important role in the control of the HPG axis as a member of the inhibitory factors of gonadotropin release. Dopaminergic inhibition of gonadotropes is also largely species-specific, but seems to work as a mediator of the negative feedback action of sex steroids. Therefore, the inhibition of GtH by GnIH and dopamine varies among teleost species. The variation in GtH regulatory mechanism among species is probably the basis of the large variations in reproductive strategies in teleost species. To understand the hypothalamic regulation of pituitary gonadotropes in teleosts, a comparative aspect of studies in different species is very important. Understanding the interactions among hypothalamic neurons is also one of the primary goals of future studies.

References

Abraham, E., Palevitch, O., Gothilf, Y., and Zohar, Y. (2009) The zebrafish as a model system for forebrain GnRH neuronal development. *Gen. Comp. Endocrinol.,* 164, 151-160

Abraham, E., Palevitch, O., Gothilf, Y., and Zohar, Y. (2010) Targeted gonadotropin-releasing hormone-3 neuron ablation in zebrafish: effects on neurogenesis, neuronal migration, and reproduction. *Endocrinology*, 151, 332-340

Abraham, E., Palevitch, O., Ijiri, S., Du, S. J., Gothilf, Y., and Zohar, Y. (2008) Early development of forebrain gonadotrophin-releasing hormone (GnRH) neurones and the role of GnRH as an autocrine migration factor. *J. Neuroendocrinol.,* 20, 394-405

Adams, B. A., Vickers, E. D., Warby, C., Park, M., Fischer, W. H., Grey, C. A., Rivier, J. E., and Sherwood, N. M. (2002) Three forms of gonadotropin-releasing hormone, including a novel form, in a basal salmonid, *Coregonus clupeaformis. Biol. Reprod.,* 67, 232-239

Aizen, J., Meiri, I., Tzchori, I., Levavi-Sivan, B., and Rosenfeld, H. (2005) Enhancing spawning in the grey mullet (*Mugil cephalus*) by removal of dopaminergic inhibition. *Gen. Comp. Endocrinol.,* 142, 212-221

Amano, M., Aida, K., Okumoto, N., and Hasegawa, Y. (1993) Changes in levels of GnRH in the brain and pituitary and GTH in the pituitary in male masu salmon, *Oncorhynchus masou*, from hatching to maturation. *Fish Physiol. Biochem.,* 11, 233-240

Amano, M., Moriyama, S., Iigo, M., Kitamura, S., Amiya, N., Yamamori, K., Ukena, K., and Tsutsui, K. (2006) Novel fish hypothalamic neuropeptides stimulate the release of gonadotrophins and growth hormone from the pituitary of sockeye salmon. *J. Endocrinol.,* 188, 417-423

Amano, M., Urano, A. and Aida, K. (1997) Distribution and function of gonadotropin-releasing hormone (GnRH) in the teleost brain. *Zool. Sci.,* 14, 1-11

Amoss, M., Burgus, R., Blackwell, R., Vale, W., Fellows, R., and Guillemin, R. (1971) Purification, amino acid composition and N-terminus of the hypothalamic luteinizing hormone releasing factor (LRF) of ovine origin. *Biochem. Biophys. Res. Commun.*, 44, 205-210

Ando, H., Swanson, P., Kitani, T., Koide, N., Okada, H., Ueda, H., and Urano, A. (2004) Synergistic effects of salmon gonadotropin-releasing hormone and estradiol-17β on gonadotropin subunit gene expression and release in masu salmon pituitary cells in vitro. *Gen. Comp. Endocrinol.*, 137, 109-121

Ando, H. and Urano, A. (2005) Molecular regulation of gonadotropin secretion by gonadotropin-releasing hormone in salmonid fishes. *Zool. Sci.*, 22, 379-389

Andreu-Vieyra, C. V., Buret, A. G. and Habibi, H. R. (2005) Gonadotropin-releasing hormone induction of apoptosis in the testes of goldfish (*Carassius auratus*). *Endocrinology,* 146, 1588-1596

Anglade, I., Zandbergen, T. and Kah, O. (1993) Origin of the pituitary innervation in the goldfish. *Cell Tissue Res.*, 273, 345-355

Batten, T. F., Cambre, M. L., Moons, L., and Vandesande, F. (1990) Comparative distribution of neuropeptide-immunoreactive systems in the brain of the green molly, *Poecilia latipinna*. *J. Comp. Neurol.*, 302, 893-919

Beck, B. H., Fuller, S. A., Peatman, E., McEntire, M. E., Darwish, A., and Freeman, D. W. (2012) Chronic exogenous kisspeptin administration accelerates gonadal development in basses of the genus Morone. *Comp. Biochem. Physiol. A Mol. Integr. Physiol.*, 162, 265-273

Bilban, M., Ghaffari-Tabrizi, N., Hintermann, E., Bauer, S., Molzer, S., Zoratti, C., Malli, R., Sharabi, A., Hiden, U., Graier, W., Knofler, M., Andreae, F., Wagner, O., Quaranta, V., and Desoye, G. (2004) Kisspeptin-10, a KiSS-1/metastin-derived decapeptide, is a physiological invasion inhibitor of primary human trophoblasts. *J. Cell Sci.,* 117, 1319-1328

Biran, J., Ben-Dor, S. and Levavi-Sivan, B. (2008) Molecular identification and functional characterization of the kisspeptin/kisspeptin receptor system in lower vertebrates. *Biol. Reprod.,* 79, 776-786

Biran, J., Palevitch, O., Ben-Dor, S., and Levavi-Sivan, B. (2012) Neurokinin Bs and neurokinin B receptors in zebrafish-potential role in controlling fish reproduction. *Proc. Natl. Acad. Sci. US,* 109, 10269-10274

Breton, B., Govoroun, M. and Mikolajczyk, T. (1998) GTH I and GTH II secretion profiles during the reproductive cycle in female rainbow trout: relationship with pituitary responsiveness to GnRH-A stimulation. *Gen. Comp. Endocrinol.*, 111, 38-50

Cardinaud, B., Gilbert, J. M., Liu, F., Sugamori, K. S., Vincent, J. D., Niznik, H. B., and Vernier, P. (1998) Evolution and origin of the diversity of dopamine receptors in vertebrates. *Adv. Pharmacol.,* 42, 936-940

Carolsfeld, J., Powell, J. F., Park, M., Fischer, W. H., Craig, A. G., Chang, J. P., Rivier, J. E., and Sherwood, N. M. (2000) Primary structure and function of three gonadotropin-releasing hormones, including a novel form, from an ancient teleost, herring. *Endocrinology*, 141, 505-512

Cerda-Reverter, J. M., Sorbera, L. A., Carrillo, M., and Zanuy, S. (1999) Energetic dependence of NPY-induced LH secretion in a teleost fish (*Dicentrarchus labrax*). *Am. J. Physiol.*, 277, 1627-1634

Chang, J. P. and Peter, R. E. (1983) Effects of dopamine on gonadotropin release in female goldfish, *Carassius auratus*. *Neuroendocrinology*, 36, 351-357

Chang, J. P., Yu, K. L., Wong, A. O., and Peter, R. E. (1990) Differential actions of dopamine receptor subtypes on gonadotropin and growth hormone release in vitro in goldfish. *Neuroendocrinology*, 51, 664-674

Chiba, A., Sohn, Y. C. and Honma, Y. (1996) Distribution of neuropeptide Y and gonadotropin-releasing hormone immunoreactivities in the brain and hypophysis of the ayu, *Plecoglossus altivelis* (Teleostei). *Arch. Histol. Cytol.*, 59, 137-148

Clarkson, J. and Herbison, A. E. (2006) Postnatal development of kisspeptin neurons in mouse hypothalamus; sexual dimorphism and projections to gonadotropin-releasing hormone neurons. *Endocrinology,* 147, 5817-5825

Copeland, P. A. and Thomas, P. (1989) Control of gonadotropin release in the Atlantic croaker (*Micropogonias undulatus*): evidence for lack of dopaminergic inhibition. *Gen. Comp. Endocrinol.*, 74, 474-483

D'Anglemont de Tassigny, X., Fagg, L. A., Dixon, J. P., Day, K., Leitch, H. G., Hendrick, A. G., Zahn, D., Franceschini, I., Caraty, A., Carlton, M. B., Aparicio, S. A., and Colledge, W. H. (2007) Hypogonadotropic hypogonadism in mice lacking a functional Kiss1 gene. *Proc. Natl. Acad. Sci. US*, 104, 10714-10719

De Leeuw, R., Goos, H. J. and van Oordt, P. G. (1986) The dopaminergic inhibition of the gonadotropin-releasing hormone-induced gonadotropin release: an in vitro study with fragments and cell suspensions from pituitaries of the African catfish, *Clarias gariepinus* (Burchell). *Gen. Comp. Endocrinol.*, 63, 171-177

De Roux, N., Genin, E., Carel, J. C., Matsuda, F., Chaussain, J. L., and Milgrom, E. (2003) Hypogonadotropic hypogonadism due to loss of function of the KiSS1-derived peptide receptor GPR54. *Proc. Natl. Acad. Sci. US*, 100, 10972-10976

Dickey, J. T. and Swanson, P. (2000) Effects of salmon gonadotropin-releasing hormone on follicle stimulating hormone secretion and subunit gene expression in coho salmon (*Oncorhynchus kisutch*). *Gen. Comp. Endocrinol.*, 118, 436-449

Dungan, H. M., Clifton, D. K. and Steiner, R. A. (2006) Minireview: kisspeptin neurons as central processors in the regulation of gonadotropin-releasing hormone secretion. *Endocrinology*, 147, 1154-1158

Escobar, S., Felip, A., Gueguen, M. M., Zanuy, S., Carrillo, M., Kah, O., and Servili, A. (2012) Expression of kisspeptinsin the brain and pituitary of the European sea bass (*Dicentrarchus labrax*). *J. Comp. Neurol.*

Felip, A., Zanuy, S., Pineda, R., Pinilla, L., Carrillo, M., Tena-Sempere, M., and Gomez, A. (2009) Evidence for two distinct KiSS genes in non-placental vertebrates that encode kisspeptins with different gonadotropin-releasing activities in fish and mammals. *Mol. Cell. Endocrinol.*, 312, 61-71

Filby, A. L., van Aerle, R., Duitman, J., and Tyler, C. R. (2008) The kisspeptin/gonadotropin-releasing hormone pathway and molecular signaling of puberty in fish. *Biol. Reprod.*, 78, 278-289

Fryer, J. N. and Maler, L. (1981) Hypophysiotropic neurons in the goldfish hypothalamus demonstrated by retrograde transport of horseradish peroxidase. *Cell Tissue Res.,* 218, 93-102

Gomez, J. M., Weil, C., Ollitrault, M., Le Bail, P. Y., Breton, B., and Le Gac, F. (1999) Growth hormone (GH) and gonadotropin subunit gene expression and pituitary and plasma changes during spermatogenesis and oogenesis in rainbow trout (*Oncorhynchus mykiss*). *Gen. Comp. Endocrinol.,* 113, 413-428

Gonzalez-Martinez, D., Zmora, N., Mananos, E., Saligaut, D., Zanuy, S., Zohar, Y., Elizur, A., Kah, O., and Munoz-Cueto, J. A. (2002a) Immunohistochemical localization of three different prepro-GnRHs in the brain and pituitary of the European sea bass (*Dicentrarchus labrax*) using antibodies to the corresponding GnRH-associated peptides. *J. Comp. Neurol.,* 446, 95-113

Gonzalez-Martinez, D., Zmora, N., Saligaut, D., Zanuy, S., Elizur, A., Kah, O., and Munoz-Cueto, J. A. (2004) New insights in developmental origins of different GnRH (gonadotrophin-releasing hormone) systems in perciform fish: an immunohistochemical study in the European sea bass (*Dicentrarchus labrax*). *J. Chem. Neuroanat.*, 28, 1-15

Gonzalez-Martinez, D., Zmora, N., Zanuy, S., Sarasquete, C., Elizur, A., Kah, O., and Munoz-Cueto, J. A. (2002b) Developmental expression of three different prepro-GnRH (gonadotrophin-releasing hormone) messengers in the brain of the European sea bass (*Dicentrarchus labrax*). *J. Chem. Neuroanat.*, 23, 255-267

Guzman, J. M., Rubio, M., Ortiz-Delgado, J. B., Klenke, U., Kight, K., Cross, I., Sanchez-Ramos, I., Riaza, A., Rebordinos, L., Sarasquete, C., Zohar, Y., and Mananos, E. L (2009) Comparative gene expression of gonadotropins (FSH and LH) and peptide levels of gonadotropin-releasing hormones (GnRHs) in the pituitary of wild and cultured Senegalese sole (*Solea senegalensis*) broodstocks. *Comp. Biochem. Physiol. A: Mol. Integr. Physiol.,* 153, 266-277

Hassin, S., Claire, M., Holland, H., and Zohar, Y. (1999) Ontogeny of follicle-stimulating hormone and luteinizing hormone gene expression during pubertal development in the female striped bass, *Morone saxatilis* (Teleostei). *Biol. Reprod.*, 61, 1608-1615

Hassin. S., Gothilf, Y., Blaise, O., and Zohar, Y. (1998) Gonadotropin-I and -II subunit gene expression of male striped bass (*Morone saxatilis*) after gonadotropin-releasing hormone analogue injection: quantitation using an optimized ribonuclease protection assay. *Biol. Reprod.*, 58, 1233-1240

Hassin, S., Holland, M. C. and Zohar, Y. (2000) Early maturity in the male striped bass, Morone saxatilis: follicle-stimulating hormone and luteinizing hormone gene expression and their regulation by gonadotropin-releasing hormone analogue and testosterone. *Biol. Reprod.,* 63, 1691-1697

Herbison, A. E. (2008) Estrogen positive feedback to gonadotropin-releasing hormone (GnRH) neurons in the rodent: The case for the rostral periventricular area of the third ventricle (RP3V). *Brain Res. Rev.*, 57, 277-287

Holland, M. C., Hassin, S. and Zohar, Y. (2001) Seasonal fluctuations in pituitary levels of the three forms of gonadotropin-releasing hormone in striped bass, *Morone saxatilis* (Teleostei), during juvenile and pubertal development. *J. Endocrinol.*, 169, 527-538

Hoskins, L. J., Xu, M. and Volkoff, H. (2008) Interactions between gonadotropin-releasing hormone (GnRH) and orexin in the regulation of feeding and reproduction in goldfish (*Carassius auratus*). *Horm. Behav.,* 54, 379-385

Jimenez-Linan, M., Rubin, B. S. and King, J. C. (1997) Examination of guinea pig luteinizing hormone-releasing hormone gene reveals a unique decapeptide and existence of two transcripts in the brain. *Endocrinology,* 138, 4123-4130

Jodo, A., Ando, H. and Urano, A. (2003) Five different types of putative GnRH receptor gene are expressed in the brain of masu salmon (*Oncorhynchus masou*). *Zool. Sci.,* 20, 1117-1125

Kah, O., Dubourg, P., Chambolle, P., and Calas, A. (1984) Ultrastructural identification of catecholaminergic fibers in the goldfish pituitary. A high-resolution radioautographic study after in vitro 3H-dopamine administration. *Cell Tissue Res.,* 238, 621-626

Kah, O., Dubourg, P., Martinoli, M. G., Geffard, M., and Calas, A. (1987a) Morphological evidence for a direct neuroendocrine GABAergic control of the anterior pituitary in teleosts. *Experientia,* 43, 300-302

Kah, O., Dulka, J. G., Dubourg, P., Thibault, J., and Peter, R. E. (1987b) Neuroanatomical substrate for the inhibition of gonadotrophin secretion in goldfish: existence of a dopaminergic preoptico-hypophyseal pathway. *Neuroendocrinology,* 45, 451-458

Kah, O., Lethimonier, C., Somoza, G., Guilgur, L. G., Vaillant, C., and Lareyre, J. J. (2007) GnRH and GnRH receptors in metazoa: A historical, comparative, and evolutive perspective. *Gen. Comp. Endocrinol.*, 153, 346-364

Kanda, S., Akazome, Y., Matsunaga, T., Yamamoto, N., Yamada, S., Tsukamura, H., Maeda, K., and Oka, Y. (2008) Identification of KiSS-1 product kisspeptin and steroid-sensitive sexually dimorphic kisspeptin neurons in medaka (*Oryzias latipes*). *Endocrinology*, 149, 2467-2476

Kanda, S., Karigo, T. and Oka, Y. (2012) Steroid sensitive kiss2 neurones in the goldfish: evolutionary insights into the duplicate kisspeptin gene-expressing neurones. *J. Neuroendocrinol.*, 24, 897-906

Kandel-Kfir, M., Gur, G., Melamed, P., Zilberstein, Y., Cohen, Y., Zmora, N., Kobayashi, M., Elizur, A., and Yaron, Z. (2002) Gonadotropin response to GnRH during sexual ontogeny in the common carp, *Cyprinus carpio*. *Comp. Biochem. Physiol. B Biochem. Mol. Biol.*, 132, 17-26

Kauffman, A. S. and Rissman, E. F. (2004) A critical role for the evolutionarily conserved gonadotropin-releasing hormone II: mediation of energy status and female sexual behavior. *Endocrinology,* 145, 3639-3646

Kauffman, A. S. and Rissman, E. F. (2006) Role of gonadotropin-releasing hormone II in the mammalian nervous system. *Expert. Rev. Endocrinol. Metab.,* 1, 133-145

Kavanaugh, S. I., Nozaki, M. and Sower, S. A. (2008) Origins of gonadotropin-releasing hormone (GnRH) in vertebrates: identification of a novel GnRH in a basal vertebrate, the sea lamprey. *Endocrinology,* 149, 3860-3869

Kawauchi, H., Suzuki, K., Itoh, H., Swanson, P., Naito, N., Nagahama, Y., Nakai, Y., and Itoh, S. (1989) The duality of teleost gonadotropins. *Fish Physiol. Biochem.,* 7, 29-83

Khan, I. A. and Thomas, P. (1992) Stimulatory effects of serotonin on maturational gonadotropin release in the Atlantic croaker, *Micropogonias undulatus. Gen. Comp. Endocrinol.,* 88, 388-396

Kitahashi, T., Alok, D., Ando, H., Kaeriyama, M., Zohar, Y., Ueda, H., and Urano, A. (1998a) GnRH analog stimulates gonadotropin II gene expression in maturing sockeye salmon. *Zool. Sci.,* 15, 761-765

Kitahashi, T., Ogawa, S. and Parhar, I. S. (2009) Cloning and expression of kiss2 in the zebrafish and medaka. *Endocrinology,* 150, 821-831

Kitahashi, T., Ogawa, S., Soga, T., Sakuma, Y., and Parhar, I. (2007) Sexual maturation modulates expression of nuclear receptor types in laser-captured single cells of the cichlid (*Oreochromis niloticus*) pituitary. *Endocrinology,* 148, 5822-5830

Kitahashi, T. and Parhar, I. S. (2013) Comparative aspects of kisspeptin gene regulation. *Gen. Comp. Endocrinol.,* 181, 197-202

Kitahashi, T., Sato, A., Alok, D., Kaeriyama, M., Zohar, Y., Yamauchi, K., Urano, A., and Ueda, H. (1998b) Gonadotropin-releasing hormone analog and sex steroids shorten homing duration of sockeye salmon in Lake Shikotsu. *Zool. Sci.,* 15, 767-771

Kotani, M., Detheux, M., Vandenbogaerde, A., Communi, D., Vanderwinden, J. M., Le Poul, E., Brezillon, S., Tyldesley, R., Suarez-Huerta, N., Vandeput, F., Blanpain, C., Schiffmann, S. N., Vassart, G., and Parmentier, M. (2001) The metastasis suppressor gene KiSS-1 encodes kisspeptins, the natural ligands of the orphan G protein-coupled receptor GPR54. *J. Biol. Chem.,* 276, 34631-34636

Kriegsfeld, L. J., Mei, D. F., Bentley, G. E., Ubuka, T., Mason, A. O., Inoue, K., Ukena, K., Tsutsui, K., and Silver, R. (2006) Identification and characterization of a gonadotropin-inhibitory system in the brains of mammals. *Proc. Natl. Acad. Sci. US,* 103, 2410-2415

Lapatto, R., Pallais, J. C., Zhang, D., Chan, Y. M., Mahan, A., Cerrato, F., Le, W. W., Hoffman, G. E., and Seminara, S. B. (2007) Kiss1-/- mice exhibit more variable hypogonadism than Gpr54-/- mice. *Endocrinology,* 148, 4927-4936

Lee, Y. R., Tsunekawa, K., Moon, M. J., Um, H. N., Hwang, J. I., Osugi, T., Otaki, N., Sunakawa, Y., Kim, K., Vaudry, H., Kwon, H. B., Seong, J. Y., and Tsutsui, K. (2009) Molecular evolution of multiple forms of kisspeptins and GPR54 receptors in vertebrates. *Endocrinology,* 150, 2837-2846

Levavi-Sivan, B., Biran, J. and Fireman, E. (2006) Sex steroids are involved in the regulation of gonadotropin-releasing hormone and dopamine D2 receptors in female tilapia pituitary. *Biol. Reprod.,* 75, 642-650

Levavi-Sivan, B., Ofir, M. and Yaron, Z. (1995) Possible sites of dopaminergic inhibition of gonadotropin release from the pituitary of a teleost fish, tilapia. *Mol. Cell. Endocrinol.,* 109, 87-95

Levy, G. and Degani, G. (2012) Involvement of GnRH, PACAP and PRP in the Reproduction of Blue Gourami Females (*Trichogaster trichopterus*). *J. Mol. Neurosci.*, 48, 603-616

Li, S., Zhang, Y., Liu, Y., Huang, X., Huang, W., Lu, D., Zhu, P., Shi, Y., Cheng, C. H., Liu, X., and Lin, H. (2009) Structural and functional multiplicity of the kisspeptin/GPR54 system in goldfish (*Carassius auratus*). *J. Endocrinol.,* 201, 407-418

Lin, C. J., Wu, G. C., Lee, M. F., Lau, E. L., Dufour, S., and Chang, C. F. (2010) Regulation of two forms of gonadotropin-releasing hormone receptor gene expression in the protandrous black porgy fish, *Acanthopagrus schlegeli. Mol. Cell. Endocrinol.,* 323, 137-146

Lin, H. R., Peng, C., Van Der Kraak, G., and Peter, R. E. (1989) Dopamine inhibits gonadotropin secretion in the Chinese loach (*Paramisgurnus dabryanus*). *Fish Physiol. Biochem.,* 6, 285-288

Liu, Y., Zhang, Y., Li, S., Huang, W., Liu, X., Lu, D., Meng, Z., and Lin, H. (2009) Molecular cloning and functional characterization of the first non-mammalian 26RFa/QRFP orthologue in Goldfish, *Carassius auratus*. *Mol. Cell. Endocrinol.,* 303, 82-90

Lovejoy, D. A., Fischer, W. H., Ngamvongchon, S., Craig, A. G., Nahorniak, C. S., Peter, R. E., Rivier, J. E., and Sherwood, N. M. (1992) Distinct sequence of gonadotropin-releasing hormone (GnRH) in dogfish brain provides insight into GnRH evolution. *Proc. Natl. Acad. Sci. US,* 89, 6373-6377

Lubzens, E., Young, G., Bobe, J., and Cerda, J. (2010) Oogenesis in teleosts: how eggs are formed. *Gen. Comp. Endocrinol.,* 165, 367-389

Madigou, T., Uzbekova, S., Lareyre, J. J., and Kah, O. (2002) Two messenger RNA isoforms of the gonadotrophin-releasing hormone receptor, generated by alternative splicing and/or promoter usage, are differentially expressed in rainbow trout gonads during gametogenesis. *Mol. Reprod. Dev.,* 63, 151-160

Martinez-Chavez, C. C., Minghetti, M., and Migaud, H. (2008) GPR54 and rGnRH I gene expression during the onset of puberty in Nile tilapia. *Gen. Comp. Endocrinol.,* 156, 224-233

Mateos, J., Mananos, E., Carrillo, M., and Zanuy, S. (2002) Regulation of follicle-stimulating hormone (FSH) and luteinizing hormone (LH) gene expression by gonadotropin-releasing hormone (GnRH) and sexual steroids in the Mediterranean Sea bass. *Comp. Biochem. Physiol. B Biochem. Mol. Biol.,* 132, 75-86

Matsuda, K., Nakamura, K., Shimakura, S., Miura, T., Kageyama, H., Uchiyama, M., Shioda, S., and Ando, H. (2008) Inhibitory effect of chicken gonadotropin-releasing hormone II on food intake in the goldfish, *Carassius auratus*. *Horm. Behav.,* 54, 83-89

Matsuo, H., Baba, Y., Nair, R. M., Arimura, A., and Schally, A. V. (1971) Structure of the porcine LH- and FSH-releasing hormone. I. The proposed amino acid sequence. *Biochem. Biophys. Res. Commun.,* 43, 1334-1339

Messager, S. (2005) Kisspeptin and its receptor: new gatekeepers of puberty. *J. Neuroendocrinol.,* 17, 687-688

Millar, R. P. (2003) GnRH II and type II GnRH receptors. *Trends Endocrinol. Metab.,* 14, 35-43

Miranda, L. A., Strussmann, C. A. and Somoza, G. M. (2001) Immunocytochemical identification of GtH1 and GtH2 cells during the temperature-sensitive period for sex determination in pejerrey, *Odontesthes bonariensis*. *Gen. Comp. Endocrinol.,* 124, 45-52

Mitani, Y., Kanda, S., Akazome, Y., Zempo, B., and Oka, Y. (2010) Hypothalamic Kiss1 but not Kiss2 neurons are involved in estrogen feedback in medaka (*Oryzias latipes*). *Endocrinology,* 151, 1751-1759

Miyamoto, K., Hasegawa, Y., Minegishi, T., Nomura, M., Takahashi, Y., Igarashi, M., Kangawa, K., and Matsuo, H. (1982) Isolation and characterization of chicken hypothalamic luteinizing hormone-releasing hormone. *Biochem. Biophys. Res. Commun.,* 107, 820-827

Miyamoto, K., Hasegawa, Y., Nomura, M., Igarashi, M., Kangawa, K., and Matsuo, H. (1984) Identification of the second gonadotropin-releasing hormone in chicken hypothalamus: evidence that gonadotropin secretion is probably controlled by two distinct gonadotropin-releasing hormones in avian species. *Proc. Natl. Acad. Sci. US,* 81, 3874-3878

Mohamed, J. S., Benninghoff, A. D., Holt, G. J., and Khan, I. A. (2007) Developmental expression of the G protein-coupled receptor 54 and three GnRH mRNAs in the teleost fish cobia. *J. Mol. Endocrinol.,* 38, 235-244

Moles, G., Carrillo, M., Mananos, E., Mylonas, C. C., and Zanuy, S. (2007) Temporal profile of brain and pituitary GnRHs, GnRH-R and gonadotropin mRNA expression and content during early development in European sea bass (*Dicentrarchus labrax L.*). *Gen. Comp. Endocrinol.,* 150, 75-86

Molnar, C. S., Kallo, I., Liposits, Z., and Hrabovszky, E. (2011) Estradiol down-regulates RF-amide-related peptide (RFRP) expression in the mouse hypothalamus. *Endocrinology,* 152, 1684-1690

Morgan, K., Sellar, R., Pawson, A. J., Lu, Z. L., and Millar, R. P. (2006) Bovine and ovine gonadotropin-releasing hormone (GnRH)-II ligand precursors and type II GnRH receptor genes are functionally inactivated. *Endocrinology,* 147, 5041-5051

Moussavi, M., Wlasichuk, M., Chang, J. P., and Habibi, H. R. (2012) Seasonal effect of GnIH on gonadotrope functions in the pituitary of goldfish. *Mol. Cell. Endocrinol.*, 350, 53-60

Muske, L. E. and Moore, F. L. (1990) Ontogeny of immunoreactive gonadotropin-releasing hormone neuronal systems in amphibians. *Brain Res.,* 534, 177-187

Naftolin, F., Harris, G. W. and Bobrow, M. (1971) Effect of purified luteinizing hormone releasing factor on normal and hypogonadotrophic anosmic men. *Nature,* 232, 496-497

Navarro, V. M., Castellano, J. M., Fernandez-Fernandez, R., Tovar, S., Roa, J., Mayen, A., Barreiro, M. L., Casanueva, F. F., Aguilar, E., Dieguez, C., Pinilla, L., and Tena-Sempere, M. (2005a) Effects of KiSS-1 peptide, the natural ligand of GPR54, on follicle-stimulating hormone secretion in the rat. *Endocrinology,* 146, 1689-1697

Navarro, V. M., Castellano, J. M., Fernandez-Fernandez, R., Tovar, S., Roa, J., Mayen, A., Nogueiras, R., Vazquez, M. J., Barreiro, M. L., Magni, P., Aguilar, E., Dieguez, C., Pinilla, L., and Tena-Sempere, M. (2005b) Characterization of the potent luteinizing hormone-releasing activity of KiSS-1 peptide, the natural ligand of GPR54. *Endocrinology,* 146, 156-163

Ngamvongchon, S., Lovejoy, D. A., Fischer, W. H., Craig, A. G., Nahorniak, C. S., Peter, R. E., Rivier, J. E., and Sherwood, N. M. (1992a) Primary structures of two forms of gonadotropin-releasing hormone, one distinct and one conserved, from catfish brain. *Mol. Cell. Neurosci.,* 3, 17-22

Ngamvongchon, S., Sherwood, N. M., Warby, C. M., and Rivier, J. E. (1992b) Gonadotropin-releasing hormone from thai catfish: chromatographic and physiological studies. *Gen. Comp. Endocrinol.,* 87, 266-274

Nocillado, J. N., Levavi-Sivan, B., Carrick, F., and Elizur, A. (2007) Temporal expression of G-protein-coupled receptor 54 (GPR54), gonadotropin-releasing hormones (GnRH), and dopamine receptor D2 (drd2) in pubertal female grey mullet, *Mugil cephalus. Gen. Comp. Endocrinol.,* 150, 278-287

Ogawa, S., Akiyama, G., Kato, S., Soga, T., Sakuma, Y., and Parhar, I. S. (2006) Immunoneutralization of gonadotropin-releasing hormone type-III suppresses male reproductive behavior of cichlids. *Neurosci. Lett.,* 403, 201-205

Ogawa, S., Ng, K. W., Ramadasan, P. N., Nathan, F. M., and Parhar, I. S. (2012) Habenular Kiss1 neurons modulate the serotonergic system in the brain of zebrafish. *Endocrinology,* 153, 2398-2407

Ogawa, S. and Parhar, I. S., (2013) Anatomy of the kisspeptin systems in teleosts. *Gen. Comp. Endocrinol.*, 181, 169-174

Oka, Y. (2009) Three types of gonadotrophin-releasing hormone neurones and steroid-sensitive sexually dimorphic kisspeptin neurones in teleosts. *J. Neuroendocrinol.,* 21, 334-338

Okubo, K., Amano, M., Yoshiura, Y., Suetake, H., and Aida, K. (2000) A novel form of gonadotropin-releasing hormone in the medaka, *Oryzias latipes. Biochem. Biophys. Res. Commun.,* 276, 298-303

Okubo, K. and Nagahama, Y. (2008) Structural and functional evolution of gonadotropin-releasing hormone in vertebrates. *Acta Physiol. (Oxf.)* 193, 3-15

Okuzawa, K., Gen, K., Bruysters, M., Bogerd, J., Gothilf, Y., Zohar, Y., and Kagawa, H. (2003) Seasonal variation of the three native gonadotropin-releasing hormone messenger ribonucleic acids levels in the brain of female red seabream. *Gen. Comp. Endocrinol.,* 130, 324-332

Olivereau, M. and Ball, J. N. (1964) Contribution to the histophysiology of the pituitary gland of teleosts, particularly those of the poecilia species. *Gen. Comp. Endocrinol.,* 47, 523-532

Osugi, T., Uchida, K., Nozaki, M., and Tsutsui, K. (2011) Characterization of novel RFamide peptides in the central nervous system of the brown hagfish: isolation, localization, and functional analysis. *Endocrinology,* 152, 4252-4264

Osugi, T., Ukena, K., Sower, S. A., Kawauchi, H., and Tsutsui, K. (2006) Evolutionary origin and divergence of PQRFamide peptides and LPXRFamide peptides in the RFamide peptide family. Insights from novel lamprey RFamide peptides. *FEBS J.,* 273, 1731-1743

Panula, P., Chen, Y. C., Priyadarshini, M., Kudo, H., Semenova, S., Sundvik, M., and Sallinen, V. (2010) The comparative neuroanatomy and neurochemistry of zebrafish CNS systems of relevance to human neuropsychiatric diseases. *Neurobiol. Dis.,* 40, 46-57

Parhar, I. S., Ogawa, S. and Kitahashi, T. (2012) RFamide peptides as mediators in environmental control of GnRH neurons. *Prog. Neurobiol.,* 98, 176-196

Parhar, I. S. (1997) *GnRH in tilapia: Three genes, three origins and their roles*. In: Parhar, I. S., Sakuma, Y. (eds) GnRH neurons: Gene to Behavior Brain Shuppan, Tokyo, pp 99-122

Parhar, I. S. (2002) Cell migration and evolutionary significance of GnRH subtypes. *Prog. Brain Res.*, 141, 3-17

Parhar, I. S. and Iwata, M. (1994) Gonadotropin releasing hormone (GnRH) neurons project to growth hormone and somatolactin cells in the steelhead trout. *Histochemistry*, 102, 195-203

Parhar, I. S., Iwata, M., Pfaff, D. W., and Schwanzel-Fukuda, M. (1995) Embryonic development of gonadotropin-releasing hormone neurons in the sockeye salmon. *J. Comp. Neurol.,* 362, 256-270

Parhar, I. S., Ogawa, S., Hamada, T., and Sakuma, Y. (2003a) Single-cell real-time quantitative polymerase chain reaction of immunofluorescently identified neurons of gonadotropin-releasing hormone subtypes in cichlid fish. *Endocrinology,* 144, 3297-3300

Parhar, I. S., Ogawa, S., Sakuma, Y. (2004) Laser-captured single digoxigenin-labeled neurons of gonadotropin-releasing hormone types reveal a novel G protein-coupled receptor (Gpr54) during maturation in cichlid fish. *Endocrinology,* 145, 3613-3618

Parhar, I. S., Pfaff, D. W. and Schwanzel-Fukuda, M. (1996) Gonadotropin-releasing hormone gene expression in teleosts. *Brain Res. Mol. Brain Res.,* 41, 216-227

Parhar, I. S., Soga, T, Ogawa, S., and Sakuma, Y. (2003b) FSH and LH-β subunits in the preoptic nucleus: ontogenic expression in teleost. *Gen. Comp. Endocrinol.,* 132, 369-378

Parhar, I. S., Soga, T., Sakuma, Y., and Millar, R. P. (2002) Spatio-temporal expression of gonadotropin-releasing hormone receptor subtypes in gonadotropes, somatotropes and lactotropes in the cichlid fish. *J. Neuroendocrinol.,* 14, 657-665

Pati, D. and Habibi, H. R. (1998) Presence of salmon gonadotropin-releasing hormone (GnRH) and compounds with GnRH-like activity in the ovary of goldfish. *Endocrinology,*139, 2015-2024

Pati, D. and Habibi, H. R. (2000) Direct action of GnRH variants on goldfish oocyte meiosis and follicular steroidogenesis. *Mol. Cell. Endocrinol.,* 160, 75-88

Peng, C., Huang, Y. P. and Peter, R. E. (1990) Neuropeptide Y stimulates growth hormone and gonadotropin release from the goldfish pituitary in vitro. *Neuroendocrinology,* 52, 28-34

Peter, R. E. (1986) Structure-activity studies on gonadotropin-releasing hormone in teleosts, amphibians, reptiles and mammals. *Prog. Clin. Biol. Res.,* 205, 75-93

Peter, R. E., Chang, J. P., Nahorniak, C. S., Omeljaniuk, R. J., Sokolowska, M., Shih, S. H., and Billard, R. (1986) Interactions of catecholamines and GnRH in regulation of gonadotropin secretion in teleost fish. *Recent Prog. Horm. Res.,* 42, 513-548

Powell, J. F., Zohar, Y., Elizur, A., Park, M., Fischer, W. H., Craig, A. G., Rivier, J. E., Lovejoy, D. A., and Sherwood, N. M. (1994) Three forms of gonadotropin-releasing hormone characterized from brains of one species. *Proc. Natl. Acad. Sci. US,* 91, 12081-12085

Quennell, J. H., Rizwan, M. Z., Relf, H. L., and Anderson, G. M. (2010) Developmental and steroidogenic effects on the gene expression of RFamide related peptides and their receptor in the rat brain and pituitary gland. *J. NeuroEndocrinol.,* 22, 309-316

Rama Krishna, N. S., Subhedar, N. and Schreibman, M. P. (1992) FMRFamide-like immunoreactive nervus terminalis innervation to the pituitary in the catfish, *Clarias batrachus* (Linn.): Demonstration by lesion and immunocytochemical techniques. *Gen. Comp. Endocrinol.,* 85, 111-117

Richardson, H. N., Gore, A. C., Venier, J., Romeo, R. D., and Sisk, C. L. (2004) Increased expression of forebrain GnRH mRNA and changes in testosterone negative feedback following pubertal maturation. *Mol. Cell. Endocrinol.,* 214, 63-70

Sébert, M. E., Weltzien, F. A., Moisan, C., Pasqualini, C., and Dufour, S. (2008) Dopaminergic systems in the European eel: characterization, brain distribution, and potential role in migration and reproduction. *Hydrobiologia,* 602, 27-46

Sato, A., Ueda, H., Fukaya, M., Kaeriyama. M., Zohar, Y., Urano, A., and Yamauchi, K. (1997) Sexual differences in homing profiles and shortening of homing duration by gonadotropin-releasing hormone analog implantation in lacustrine sockeye salmon (*Oncorhynchus nerka*) in Lake Shikotsu. *Zool. Sci.,* 14, 1009-1014

Sawada, K., Ukena, K., Satake, H., Iwakoshi, E., Minakata, H., and Tsutsui, K. (2002) Novel fish hypothalamic neuropeptide. *Eur. J. Biochem.,* 269, 6000-6008

Sawisky, G. R. and Chang, J. P. (2005) Intracellular calcium involvement in pituitary adenylate cyclase-activating polypeptide stimulation of growth hormone and gonadotrophin secretion in goldfish pituitary cells. *J. Neuroendocrinol.,* 17, 353-371

Schulz, R. W., Bosma, P. T., Zandbergen, M. A., Van der Sanden, M. C., van Dijk, W., Peute, J., Bogerd, J., and Goos, H. J. (1993) Two gonadotropin-releasing hormones in the African catfish, *Clarias gariepinus*: localization, pituitary receptor binding, and gonadotropin release activity. *Endocrinology*, 133, 1569-1577

Schulz, R. W., Zandbergen, M. A., Peute, J., Bogerd, J., van Dijk, W., and Goos, H. J. (1997) Pituitary gonadotrophs are strongly activated at the beginning of spermatogenesis in African catfish, *Clarias gariepinus*. *Biol. Reprod.*, 57, 139-147

Schwanzel-Fukuda, M., Bick, D. and Pfaff, D. W. (1989) Luteinizing hormone-releasing hormone (LHRH)-expressing cells do not migrate normally in an inherited hypogonadal (Kallmann) syndrome. *Brain Res. Mol. Brain Res.*, 6, 311-326

Selvaraj, S., Kitano, H., Fujinaga, Y., Ohga, H., Yoneda, M., Yamaguchi, A., Shimizu, A., and Matsuyama, M. (2010) Molecular characterization, tissue distribution, and mRNA expression profiles of two Kiss genes in the adult male and female chub mackerel (*Scomber japonicus*) during different gonadal stages. *Gen. Comp. Endocrinol.*, 169, 28-38

Seminara, S. B. (2005) Metastin and its G protein-coupled receptor, GPR54: critical pathway modulating GnRH secretion. *Front Neuroendocrinol.,* 26, 131-138

Seminara, S. B., Messager, S., Chatzidaki, E. E., Thresher, R. R., Acierno, J. S., Shagoury, J. K., Bo-Abbas, Y., Kuohung, W., Schwinof, K. M., Hendrick, A. G., Zahn, D., Dixon, J., Kaiser, U. B., Slaugenhaupt, S. A., Gusella, J. F., O'Rahilly, S., Carlton, M. B., Crowley, W. F., Aparicio, S. A., and Colledge, W. H. (2003) The GPR54 gene as a regulator of puberty. *N Engl. J. Med.,* 349, 1614-1627

Senthilkumaran, B., Okuzawa, K., Gen, K., Ookura, T., and Kagawa, H. (1999) Distribution and seasonal variations in levels of three native GnRHs in the brain and pituitary of perciform fish. *J. Neuroendocrinol.,* 11, 181-186

Servili, A., Le Page, Y., Leprince, J., Caraty, A., Escobar, S., Parhar, I. S., Seong, J. Y., Vaudry, H., and Kah, O. (2011) Organization of two independent kisspeptin systems derived from evolutionary-ancient kiss genes in the brain of zebrafish. *Endocrinology,* 152, 1527-1540

Servili, A., Lethimonier, C., Lareyre, J. J., Lopez-Olmeda, J. F., Sanchez-Vazquez, F. J., Kah, O., and Munoz-Cueto, J. A. (2010) The highly conserved gonadotropin-releasing hormone-2 form acts as a melatonin-releasing factor in the pineal of a teleost fish, the european sea bass *Dicentrarchus labrax*. *Endocrinology,* 151, 2265-2275

Shahjahan, M., Hamabata, T., Motohashi, E., Doi, H., and Ando, H. (2010a) Differential expression of three types of gonadotropin-releasing hormone genes during the spawning season in grass puffer, *Takifugu niphobles*. *Gen. Comp. Endocrinol.,* 167, 153-163

Shahjahan, M., Ikegami, T., Osugi, T., Ukena, K., Doi, H., Hattori, A., Tsutsui, K., and Ando, H. (2011) Synchronised expressions of LPXRFamide peptide and its receptor genes: seasonal, diurnal and circadian changes during spawning period in grass puffer. *J. Neuroendocrinol.,* 23, 39-51

Shahjahan, M., Motohashi, E., Doi, H., and Ando, H. (2010b) Elevation of Kiss2 and its receptor gene expression in the brain and pituitary of grass puffer during the spawning season. *Gen. Comp. Endocrinol.,* 169, 48-57

Sherwood, N. M., Eiden, L., Brownstein, M., Spiess, J., Rivier, J., and Vale, W. (1983) Characterization of a teleost gonadotropin-releasing hormone. *Proc. Natl. Acad. Sci. US*, 80, 2794-2798

Sherwood, N. M., von Schalburg, K. R. and Lescheid, D. W. (1997) Origin and evolution of GnRH in vertebrates and invertebrates. In: Parhar, I. S., Sakuma, Y. (eds) *GnRH Neurons: From Genes to Behavior*. Brain Shuppan, Tokyo, pp 3-25

Shi, Y., Zhang, Y., Li, S., Liu, Q., Lu, D., Liu, M., Meng, Z., Cheng, C. H., Liu, X., and Lin, H. (2010) Molecular identification of the Kiss2/Kiss1ra system and its potential function during 17α-methyltestosterone-induced sex reversal in the orange-spotted grouper, *Epinephelus coioides*. *Biol. Reprod.,* 83, 63-74

Smith, J. T. (2008) Kisspeptin signalling in the brain: steroid regulation in the rodent and ewe. *Brain Res. Rev.,* 57, 288-298

Smith, J. T., Cunningham, M. J., Rissman, E. F., Clifton, D. K., and Steiner, R. A. (2005a) Regulation of Kiss1 gene expression in the brain of the female mouse. *Endocrinology*, 146, 3686-3692

Smith, J. T., Dungan, H. M., Stoll, E. A., Gottsch, M. L., Braun, R. E., Eacker, S. M., Clifton, D. K., and Steiner, R. A. (2005b) Differential Regulation of KiSS-1 Gene Expression by Sex Steroids in the Brain of the Male Mouse. *Endocrinology,* 146, 2976-84

Smith, J. T., Popa, S. M., Clifton, D. K., Hoffman, G. E., and Steiner, R. A. (2006) Kiss1 neurons in the forebrain as central processors for generating the preovulatory luteinizing hormone surge. *J. Neurosci.,* 26, 6687-6694

Soga, T., Wong, D. W., Clarke, I. J., and Parhar, I. S. (2010) Citalopram (antidepressant) administration causes sexual dysfunction in male mice through RF-amide related peptide in the dorsomedial hypothalamus. *Neuropharmacology,* 59, 77-85

Somoza, G. M., Yu, K. L. and Peter, R. E. (1988) Serotonin stimulates gonadotropin release in female and male goldfish, *Carassius auratus L. Gen. Comp. Endocrinol.*, 72, 374-382

Sower, S. A., Chiang, Y. C., Lovas, S., and Conlon, J. M. (1993) Primary structure and biological activity of a third gonadotropin-releasing hormone from lamprey brain. *Endocrinology,* 132, 1125-1131

Steven, C., Lehnen, N., Kight, K., Ijiri, S., Klenke, U., Harris, W. A., and Zohar, Y. (2003) Molecular characterization of the GnRH system in zebrafish (*Danio rerio*): cloning of chicken GnRH-II, adult brain expression patterns and pituitary content of salmon GnRH and chicken GnRH-II. *Gen. Comp. Endocrinol.,* 133, 27-37

Swanson, P., Suzuki, K., Kawauchi, H., and Dickhoff, W. W. (1991) Isolation and characterization of two coho salmon gonadotropins, GTH I and GTH II. *Biol. Reprod.,* 44, 29-38

Tobari, Y., Iijima, N., Tsunekawa, K., Osugi, T., Okanoya, K., Tsutsui, K., and Ozawa, H. (2010) Identification of gonadotropin-inhibitory hormone in the zebra finch (*Taeniopygia guttata*): Peptide isolation, cDNA cloning and brain distribution. *Peptides*, 31, 816-826

Trudeau, V. L. (1997) Neuroendocrine regulation of gonadotrophin II release and gonadal growth in the goldfish, *Carassius auratus*. *Rev. Reprod.,* 2, 55-68

Tsutsui, K., Saigoh, E., Ukena, K., Teranishi, H., Fujisawa, Y., Kikuchi, M., Ishii, S., and Sharp, P. J. (2000) A novel avian hypothalamic peptide inhibiting gonadotropin release. *Biochem. Biophys. Res. Commun.,* 275, 661-667

Tsutsui, K., Ubuka, T., Bentley, G. E., and Kriegsfeld, L. J. (2012) Gonadotropin-inhibitory hormone (GnIH): discovery, progress and prospect. *Gen. Comp. Endocrinol.*, 177, 305-314

Ubuka, T., Inoue, K., Fukuda, Y., Mizuno, T., Ukena, K., Kriegsfeld, L. J., and Tsutsui, K. (2012) Identification, expression, and physiological functions of Siberian hamster gonadotropin-inhibitory hormone. *Endocrinology,* 153, 373-385

Ubuka, T., Morgan, K., Pawson, A. J., Osugi, T., Chowdhury, V. S., Minakata, H., Tsutsui, K., Millar, R. P., and Bentley, G. E. (2009) Identification of Human GnIH Homologs, RFRP-1 and RFRP-3, and the Cognate Receptor, GPR147 in the Human Hypothalamic Pituitary Axis. *PLoS ONE* 4, e8400

Uchida, H., Ogawa, S., Harada, M., Matushita, M., Iwata, M., Sakuma, Y., and Parhar, I. S., (2005) The olfactory organ modulates gonadotropin-releasing hormone types and nest-building behavior in the tilapia *Oreochromis niloticus*. *J. Neurobiol.,* 65, 1-11

Uzbekova, S., Lareyre, J. J., Madigou, T., Davail, B., Jalabert, B., and Breton, B. (2002) Expression of prepro-GnRH and GnRH receptor messengers in rainbow trout ovary depends on the stage of ovarian follicular development. *Mol. Reprod. Dev.,* 62, 47-56

Weltzien, F. A., Pasqualini, C., Sebert, M. E., Vidal, B., Le Belle, N., Kah, O., Vernier, P., and Dufour, S. (2006) Androgen-dependent stimulation of brain dopaminergic systems in the female European eel (*Anguilla anguilla*). *Endocrinology,* 147, 2964-2973

White, R. B. and Fernald, R. D. (1998) Ontogeny of gonadotropin-releasing hormone (GnRH) gene expression reveals a distinct origin for GnRH-containing neurons in the midbrain. *Gen. Comp. Endocrinol.,* 112, 322-329

Wong, A. O., Leung, M. Y., Shea, W. L., Tse, L. Y., Chang, J. P., and Chow, B. K. (1998) Hypophysiotropic action of pituitary adenylate cyclase-activating polypeptide (PACAP) in the goldfish: immunohistochemical demonstration of PACAP in the pituitary, PACAP stimulation of growth hormone release from pituitary cells, and molecular cloning of pituitary type I PACAP receptor. *Endocrinology,* 139, 3465-3479

Wong, T. T., Gothilf, Y., Zmora, N., Kight, K. E., Meiri, I., Elizur, A., and Zohar, Y. (2004) Developmental expression of three forms of gonadotropin-releasing hormone and ontogeny of the hypothalamic-pituitary-gonadal axis in gilthead seabream (*Sparus aurata*). *Biol. Reprod.,* 71, 1026-1035

Wray, S., Grant, P. and Gainer, H. (1989) Evidence that cells expressing luteinizing hormone-releasing hormone mRNA in the mouse are derived from progenitor cells in the olfactory placode. *Proc. Natl. Acad. Sci. US,* 86, 8132-8136

Wu, Y., He, Z., Zhang, L., Jiang, H., and Zhang, W. (2012) Ontogeny of immunoreactive Lh and Fsh cells in relation to early ovarian differentiation and development in protogynous hermaphroditic ricefield Eel *Monopterus albus*. *Biol. Reprod.*, 86:93

Yamamoto, N., Oka, Y. and Kawashima, S. (1997) Lesions of gonadotropin-releasing hormone-immunoreactive terminal nerve cells: effects on the reproductive behavior of male dwarf gouramis. *Neuroendocrinology,* 65, 403-412

Yang, B., Jiang, Q., Chan, T., Ko, W. K. W., and Wong, A. O. L. (2010) Goldfish kisspeptin: Molecular cloning, tissue distribution of transcript expression, and stimulatory effects on prolactin, growth hormone and luteinizing hormone secretion and gene expression via direct actions at the pituitary level. *Gen. Comp. Endocrinol.,* 165, 60-71

Yaron, Z., Gur, G., Melamed, P., Rosenfeld, H., Elizur, A., and Levavi-Sivan, B. (2003) Regulation of fish gonadotropins. *Int. Rev. Cytol.,* 225, 131-185

Yaron, Z., Gur, G., Melamed, P., Rosenfeld, H., Levavi-Sivan, B., and Elizur, A. (2001) Regulation of gonadotropin subunit genes in tilapia. *Comp. Biochem. Physiol. B Biochem. Mol. Biol.,* 129, 489-502

Yoo, M. S., Kang, H. M., Choi, H. S., Kim, J. W., Troskie, B. E., Millar, R. P., and Kwon, H. B. (2000) Molecular cloning, distribution and pharmacological characterization of a novel gonadotropin-releasing hormone ([Trp8] GnRH) in frog brain. *Mol. Cell. Endocrinol.,* 164, 197-204

Yu, K. L., He, M. L., Chik, C. C., Lin, X. W., Chang, J. P. and Peter, R. E. (1998) mRNA expression of gonadotropin-releasing hormones (GnRHs) and GnRH receptor in goldfish. *Gen. Comp. Endocrinol.,* 112, 303-311

Yu, K. L., Sherwood, N. M. and Peter, R. E. (1988) Differential distribution of two molecular forms of gonadotropin-releasing hormone in discrete brain areas of goldfish (*Carassius auratus*). *Peptides,* 9, 625-630

Zhang, Y., Li, S., Liu, Y., Lu, D., Chen, H., Huang, X., Liu, X., Meng, Z., Lin, H., and Cheng, C. H. (2010) Structural diversity of the GnIH/GnIH receptor system in teleost: its involvement in early development and the negative control of LH release. *Peptides*, 31, 1034-1043

Zohar, Y., Munoz-Cueto, J. A., Elizur, A., and Kah, O. (2010) Neuroendocrinology of reproduction in teleost fish. *Gen. Comp. Endocrinol.,* 165, 438-455

In: Sexual Plasticity and Gametogenesis in Fishes
Editor: Balasubramanian Senthilkumaran
ISBN: 978-1-62618-848-8

Chapter XII

Sexual Plasticity of Behavior in Goldfish

Makito Kobayashi,[1] Seiji Saoshiro,[1] Yutaro Kawaguchi,[1] Youichi Hayakawa[1] and Arimune Munakata[2]
[1]Department of Life Science, International Christian University, Mitaka, Tokyo, Japan
[2]Department of Biology, Miyagi University of Education, Sendai, Miyagi, Japan

Abstract

It is known that the sexual differentiation of the brain is induced by *organizational* effects of sex steroids during neonatal development in mammals, and sex steroids exert *activational* effects on the brain of sexually mature individuals for the sexual behavior to occur. Once brain sex is determined, it is irreversible in mammals in the lifetime, and the brain regulates the sexual behavior of respective sexes. However, laboratory and field observations suggest that brain function of some fish species does not conform to the mammalian paradigm. Goldfish, a non-sex changing fish, normally exhibits only gender-typical sexual behavior, but heterotypical (opposite gender) patterns can be induced by hormonal treatments both in males and females even after their sexual maturation. Interestingly, induction of the heterotypical behavior does not inhibit performance of gender-typical behavior. Sex changing fishes in nature are known to perform sexual behavior of both sexes during their lifetime. From this perspective, it is suggested that unlike mammals which have brain sex differentiation, most fish have a sexually bipotential brain that can regulate sexual behavior of both sexes.

Introduction

It is well known that sexual differentiation of the brain in birds and mammals is induced by *organizational* effects of sex steroids during the critical periods of early development. In turn, sex steroids exert *activational* effects on the brain of sexually mature individuals for certain acts of sexual behavior to occur. Once brain sex is determined, it is irreversible in birds and mammals throughout the remainder of their lifetime and the brain regulates the sexual behavior of the respective sexes (Balthazart et al., 2009; Sakuma 2009; Bao and Swaab, 2011; Adkins-Regan, 2012). However, field and laboratory observations suggest that brain function of some teleost fish species does not conform to the paradigm observed in birds and mammals (Kobayashi et al; 2002; Munakata and Kobayashi, 2010; Goncalves and Oliveira, 2011). Sex change is observed in fish species in nature (Helfman et al., 1997; Moyle and Cech, 2000; Godwin, 2010a,b; Larson, 2011). These species change their gonadal sex and also behavioral sex.

Individuals of these species perform sexual behavior of either sex at some point during their lifetime. Non-sex changing fishes normally do not exhibit heterotypical (opposite gender) reproductive functions, but heterotypical sexual behavior can be induced in some species of fishes by hormonal treatments (Munakata and Kobayashi, 2010). The endocrine control of sexual behavior and the induction of heterotypical behavior in goldfish *Carassius auratus*, a non-sex changing fish, are probably most intensively studied among fish species.

From this perspective, it is suggested that unlike mammals which have discrete brain sex differentiation, fish have a sexually bipotential brain that can regulate sexual behavior of both sexes. This chapter is intended to describe sexual behavior of fish in the following aspects: (1) patterns of sex change observed in nature in hermaphroditic fishes, (2) hormonal and pheromonal control of sexual behavior in goldfish, (3) sexual plasticity of behavior in goldfish, (4) sexual bipotentiality of the brain in fish.

Sex Change in Fishes

Among vertebrates, one of the unique characters in fish is that some species experience being both a functional male and a functional female during their lifetime i.e., hermaphroditism (Helfman et al., 1997; Moyle and Cech, 2000).

Therefore, in these species, one individual fish regularly performs both male-typical and female-typical sexual behavior which is rarely observed in other non-sex changing vertebrates (i.e., gonochorism). Hermaphroditism in fish is classified into two categories: sequential hermaphroditism (sex change) and simultaneous hermaphorditism. In sequential hermaphroditism, the ovary and the testis develop at different times resulting in a male phase and female phase in each individual. These sequential hermaphrodites are known to change their gonadal sex and also their behavioral sex at some age or at some social status. Sequential hermaphordites are further classified into three types (Nakashima et al., 1995; Munday et al., 2010) according to the direction of sex change: (1) protandrous hermaphroditism (i.e. male to female sex change); (2) protogynous hermaphroditism (i.e. female to male sex change); (3) bi-directional sex change (serial sex change, both-way sex change, i.e. sex change in either direction).

In these sex-changing fishes, behavioral sex change proceeds to gonadal sex change and some species show rapid sex change in response to social cues (1-2 h in the cleaner wrasse *Labroides dimidiatus*, Nakashima et al., 2000; 1-2 days, in the bluehead wrasse, *Thalassoma bifasciatum*, Godwin et al., 1996). In simultaneous hermaphroditism, both the ovary and the testis develop at the same time in a single individual fish. Two ways of fertilization have been observed in simultaneous hermaphroditism: (1) mating (egg trading) and (2) self-fertilization. In fish which exhibit mating for fertilization, two partners perform the appropriate sexual behavior switching between male and female roles and taking turns fertilizing each other's egg. These species exhibit quick behavioral sex change repeatedly during each spawning bout, releasing eggs and sperm alternately. It is known that two sexually mature belted sandfish *Serranus subligarius*, trades eggs, switching between male and female roles within seconds (Cheek, 1998; Cheek et al., 2000). In the self-fertilizing type, the fish releases fertilized eggs without performing any sexual behavior (Minamimoto et al., 2006).

In these hermaphroditic fishes, except for the self-fertilizing one, individuals will perform both male-typical and female-typical sexual behavior during their lifetime. Although the neuroanatomical basis of behavioral sex change in these fishes is not clearly elucidated, it is suggested that the brain of these fishes can regulate sexual behavior of both sexes unlike mammals and birds.

Regulation of Sexual Behavior in Goldfish

1. Sexual Behavior of Goldfish

Physiological and environmental regulation of sexual behavior has been intensively studied in goldfish. Here we describe sexual behavior (spawning behavior) of goldfish (Kobayashi et al., 2002; Stacey, 2003; Munakata and Kobayashi, 2010). In temperate regions, goldfish spawns several times during their spawning season which occurs during the spring months and when kept under natural environmental conditions. Ovulation of sexually mature post-vitellogenic females is stimulated by environmental cues and spawning behavior of goldfish starts several hours prior to ovulation (Figure 1) in the females (Kobayashi et al., 2002). Females that are in the process of ovulation release a "preovulatory steroid pheromone" (Dulka et al., 1987). Stimulated by this steroid pheromone, males start to chase and sometimes nudges the females (i.e., courtship behavior) normally during the early period of scotophase. Synchronized to photoperiod, ovulation in females occurs in the middle or latter period of scotophase (Kezuka et al., 1989). After ovulation, females produce prostaglandin F2α (PGF) in the ovary which triggers the spawning act in the females (Stacey and Sorensen, 2006; Stacey, 2011). PGF and its metabolites are released into the water as a "postovulatory prostaglandin (PG) pheromone". In turn, the PG pheromone stimulates the males where chasing becomes persistent, increases in intensity and is interspersed with the spawning acts of the female. Spawning acts are initiated by the entry of an ovulated female into the floating aquatic vegetation near the surface of the water and where the male follows the female. The female and the male turn on their sides and swim quickly through the vegetation, releasing eggs (oviposition) and sperm (ejaculation). The male always positions itself underneath and in contact with the female during this act. Then they flip their tails to mix spawned eggs and sperm. Fertilized eggs adhere to the vegetation. Female spawning behavior will continue until most ovulated oocytes are released, and this may involve a hundred or more spawning acts over several hours. In this chapter, the female spawning act is referred to as female sexual behavior, and chasing and the male spawning act is referred to as male sexual behavior in goldfish.

2. Hormonal and Pheromonal Control of Sexual Behavior in Female Goldfish

When sexually mature post-vitellogenic females are exposed to certain external cues, such as a change in water temperature and encountering spawning substrate, they will exhibit release of a large amount of lutenizing hormone (LH) from the pituitary gland (an ovulatory LH surge) that lasts about 15 hours, terminating with ovulation and spawning (Figure 2; Kobayashi et al., 1987a; Kobayashi et al., 2002, Munakata and Kobayashi, 2010). The LH surge in turn stimulates the production of the maturation-inducing steroid, 17α, 20β-dihydroxy-4-pregnen-3-one (17,20-P) in the ovarian follicles, and this ovarian progestin induces oocyte maturation (Kobayashi et al., 1987). 17, 20-P and its conjugates, 17, 20-P sulfate (17, 20-PS) are released into the water as "preovulatory steroid pheromones" (Dulka et al., 1987; Stacey and Sorensen, 2006; Stacey, 2011).

These steroid pheromones induce LH release in male goldfish, i.e., male LH surge. (Kobayashi et al., 1986a, b) and weak chasing by males (Sorensen et al., 1989; Poling et al., 2001).

Increased release of LH further induces milt production in sexually mature males that are ready for spawning. This female to male pheromone interaction is thought to function as a signal to males announcing that ovulation is imminent and stimulates the preparatory steps required for ejaculation.

Ovulation in goldfish occurs at the peak of the LH surge (Kobayashi et al., 1987), and the presence of ovulated oocytes in the oviduct induces the synthesis of PGF (Figure 3; Stacey and Liley, 1974; Stacey, 1976; Sorensen et al., 1995).

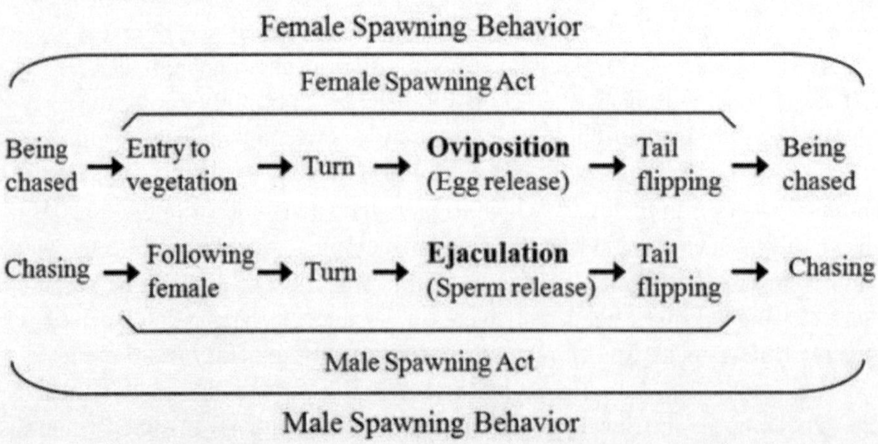

Figure 1. Diagrammatic representation of the spawning behavior of male and female goldfish (Kobayashi et al., 2002). See text for details.

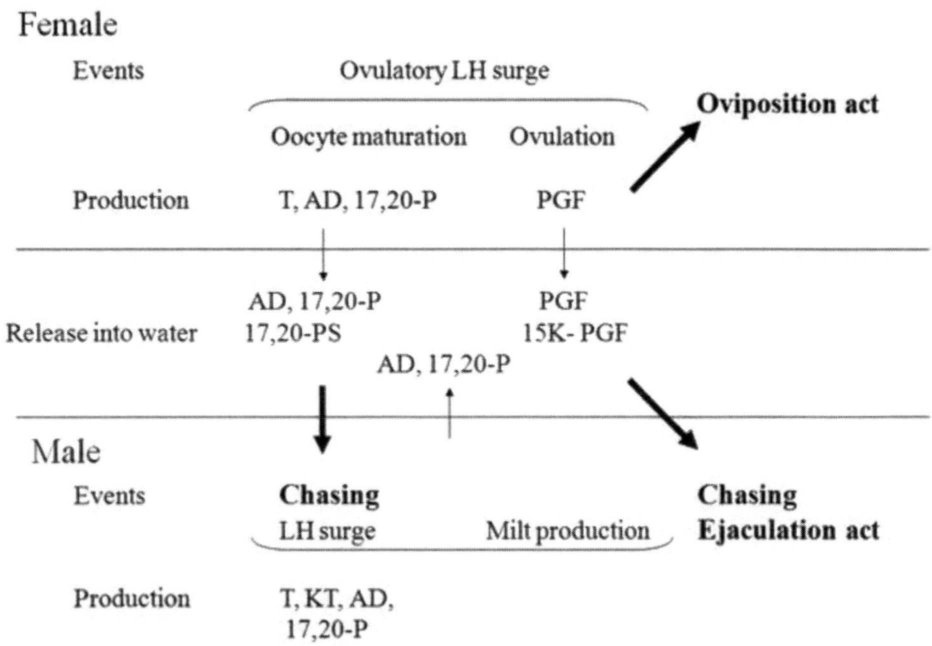

AD, androstenedione; KT, 11-ketotesotosterone; LH, luteinizing hormone; 17,20-P, 17α,20β-dihydroxy-4-pregnen-3-one; 17,20-PS, 17α,20β-dihydroxy-4-pregnen-3-one sulphate; T, testosterone; PGF, prostaglandin F2α; 15K-PGF, 15-keto-prostaglandin F2α.

Figure 2. Diagrammatic representation of the hormonal and pheromonal actions during spawning in goldfish. Thick arrows indicate the action of hormones and pheromones inducing sexual behavior (Munakata and Kobayashi, 2010). See text for details.

PGF is transported to the brain via blood circulation and triggers the female spawning act (Stacey and Peter, 1979). Spawning behavior is terminated when the ovulated oocytes are mostly shed by the spawning act or when manually stripped by researchers (Stacey and Liley, 1974).

It is thought that PGF conveys the information of the presence of ovulated oocytes in the ovarian cavity and that they are ready to be oviposited to the brain of female goldfish.

PGF and its metabolite, 15-keto-prostaglandin F2α (15K-PGF) are released into the water as "postovulatory prostaglandin pheromones" (Stacey and Sorensen, 2006; Stacey, 2011). These pheromones have a strong effect in inducing male sexual behavior in goldfish. When exposed to these PG pheromones, male goldfish exhibit persistent chasing of the ovulated female and perform their spawning act with an ovulated female. The effect of these PG pheromones on LH release in males is very weak compared to that of the preovulatory steroid pheromones (Sorensen et al., 1989). This would indicate that the postovulatory PG pheromone's main function is a signal from female to male announcing that the female has ovulated oocytes and is ready to spawn. Interestingly, when PGF is intramuscularly injected to a non-ovulated female, this female starts to perform the female sexual behavior (i.e., oviposition act) within several minutes although no egg release is accompanied under these conditions (Figure 3; Stacey, 1976; Kobayashi et al., 2002). The injected PGF is released into the water as a pheromone just as PGF is released in ovulated females and stimulates males to perform their sexual behavior (i.e., chasing and ejaculation act with sperm release). The males do not discriminate between ovulated females and PG-injected non-ovulated females in performing male specific sexual behavior, and thus, sexual behaviors of male and female goldfish can be artificially induced all through the year regardless of maturity of females by using the PG-injection method and if the sexual maturity of males can be maintained in the laboratory. It is by the use of this method that study of sexual behavior in goldfish has been greatly advanced.

Estrogens are required for sexual behavior in many female vertebrates (Sakuma, 1997), but ovarian sex steroids are not essential for the occurrence of sexual behavior in female goldfish. Ovariectomized female goldfish exhibited female sexual behavior (i.e., spawning act) after injection of PGF, while treatment with sex steroids (i.e., estradiol-17β, testosterone (T), or 17,20-P) did not trigger any signs of sexual behavior or potentiate the effect of PGF (Kobayashi and Stacey, 1993).

Sexually mature female goldfish exhibit high blood levels of T produced in the ovary. This T is important for the occurrence of the ovulatory LH surge (Kobayashi et al., 1989a,b; 2002) but is not involved in female sexual behavior as shown by the experiments mentioned above. While progesterone is known to complement the expression of proceptive behavior in female rats (Sakuma, 1997), administration of 17,20-P did not enhance the activity of PG-induced spawning behavior in female goldfish (Kobayashi and Stacey, 1993).

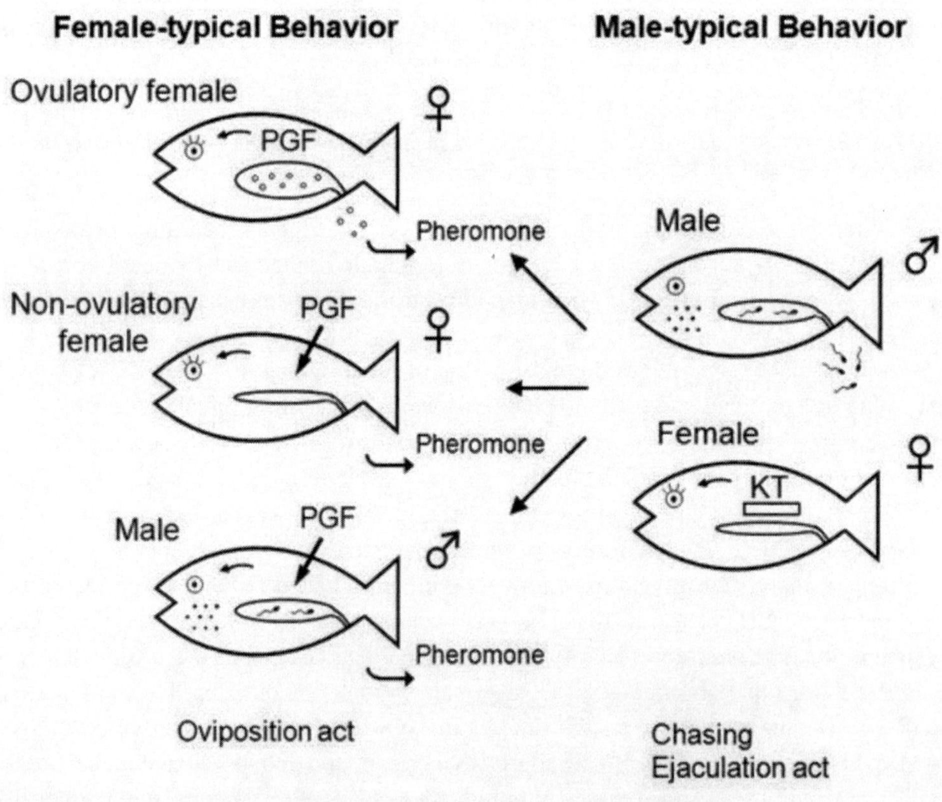

Figure 3. Induction of gender-typical and heterotypical sexual behavior in male and female goldfish by hormonal treatments. See text for details. PGF, prostaglandin F2α; KT, 11-ketotestosterone.

Gonadotropin-releasing hormone (GnRH) shows potentiating effects on female sexual behavior in the goldfish. Intracerebroventricular injection of salmon-type GnRH and chicken-II-type GnRH enhanced PG-induced spawning activity, and GnRH antagonist suppressed the activity (Peter and Volkoff, 1999).

Although goldfish have been reported to have three GnRH neuronal populations in the brain, terminal nerve (TN) GnRH, preoptic GnRH, and midbrain GnRH (Kim et al., 1995a), it is unclear which of the three neuronal GnRH populations is physiologically involved in behavioral facilitation. Since the female goldfish ovulate and spawn even after the axonal transport of GnRH from TN to the other brain areas is blocked by using an olfactory tract section, GnRH of TN origin apparently is not essential for the occurrence of the female sexual behavior (Stacey and Kyle, 1983; Kobayashi et al., 1992, 1994; Kim et al., 1995b, 2001).

3. Hormonal and Pheromonal Control of Male Sexual Behavior in Goldfish

Sexually mature male goldfish follow or chase ovulatory females stimulated by the preovulatory steroid pheromone (Kobayashi et al., 2002; Stacey and Sorensen, 2006; Munakata and Kobayashi, 2010; Stacey, 2011). Preovulatory steroid pheromones (17,20-P and 17,20-PS) stimulate LH release in males and induce weak chasing (Sorensen et al., 1989). The spawning act, however, is performed only after ovulation of females which is the source of the postovulatory PG pheromone that are released into the water column. Androgen(s) is considered to be a requirement for the occurrence of male behavioral responses (Kobayashi and Munakata, 2010). Although complete castration of male goldfish is impractical because of testicular regeneration, involvement of androgen(s) is supported by the fact that male-typical sexual behavior is performed by sexually mature males with secondary sex characteristics and also androgen-treated females (Kobayashi and Stacey, 1996). 11-Ketotestosterone (KT) and T are found in the plasma of male goldfish (Kobayashi et al., 1986a), and KT is the more potent androgen than T (Kobayashi and Stacey, 1996). Based on the studies with KT-implanted female goldfish, KT is considered to be essential for LH release in response to the preovulatory steroid pheromone and male sexual behavior in response to the postovulatory PG pheromones (Kobayashi et al., 1997). In addition to preovulatory pheromones, LH release in males is stimulated by behavioral interactions with ovulated females (Sorensen et al., 1989; Zheng and Stacey, 1997). Increased blood levels of LH stimulate synthesis of 17,20-P, T, and androstendione (AD), and among these steroids, a large amount of AD is released into the water column as male pheromone (Sorensen et al., 2005). Sorensen et al. (2005) provided the following interpretation of the function of AD: This male pheromone acts on other males by the induction of agonistic behavior. This male to male pheromone seems to be an indicator of male sexuality providing a warning to other males and preventing being chased by other males during the pursuit of suitable females. In turn, females release preovulatory and postovulatory pheromones to stimulate physiological readiness for spawning in males.

Since goldfish spawn during late scotophase and sexual dimorphism is not distinct, these pheromones also seem to function as a means to discriminate both the sex and maturational status of individuals amongst the population of fishes maximizing the encounters of appropriate mates.

Although GnRH exhibited a potentiatory effect on sexual behavior in female goldfish, this peptide does not show any clear effect on the sexual behavior in male goldfish (Volkoff and Peter, 1999).

4. Involvement of Olfaction in Sexual Behavior in Goldfish

In male goldfish, olfaction is essential for the occurrence of sexual behavior since PG pheromone from ovulatory females is a trigger of the male sexual behavior. Involvement of olfaction in the male sexual behavior was confirmed by two methods: nasal occlusion and olfactory tract section (Figure 4; Kawaguchi et al., 2011; Partridge et al., 1976; Stacey and Kyle, 1983).

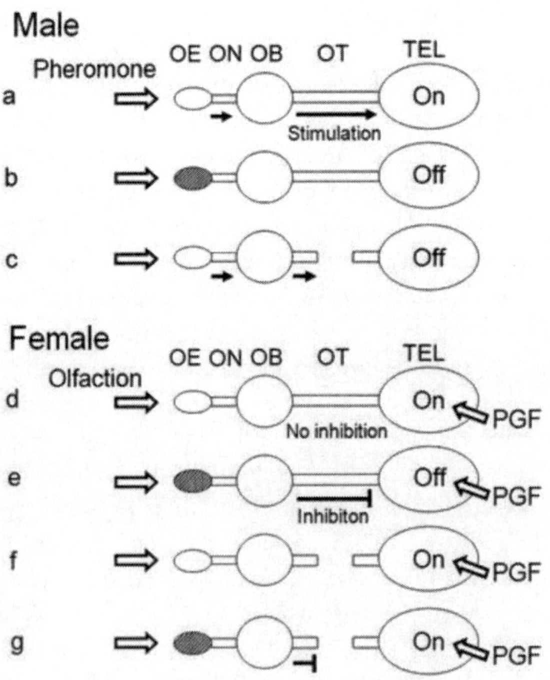

Treatment	Sexual Behavior
Intact	YES
Nasal occlusion	NO
OTX	NO

Treatment	Sexual Behavior
Intact	YES
Nasal occlusion	NO
OTX	YES!
Nasal occlusion + OTX	YES!

Figure 4. Involvement of olfaction in sexual behavior in male and female goldfish. In males, pheromone from females induces male-typical sexual behavior. a. Pheromone is received in the olfactory epithelium (OE) and the signal is transmitted to the telencephalon (TEL) for the regulation of sexual behavior via the olfactory nerve (ON), the olfactory bulb (OB), and the olfactory tract (OT). Since the olfactory bulb of goldfish is of the pedunculated type with elongated olfactory tracts, we can easily section the olfactory tract surgically. b. By nasal cavity occlusion with glue, reception of pheromone is blocked. c. By olfactory tract section (OTX), olfactory information is not transmitted from OB to TEL although pheromone is received in the OE. d. In females, prostaglandin F2α (PGF) triggers female sexual behavior. e. When olfaction is blocked by nasal cavity occlusion with glue, strong inhibition exerts on sexual behavior from OB to TEL even under the stimulation of PGF. f and g. No inhibition occurred from OB by OTX and females resume sexual behavior by the stimulation of PGF although these fish receive no olfactory information. Since unilateral nasal occlusion did not suppress the behavior both in male and female, a possibility that toxicity of glue suppressed the behavior is eliminated (data not shown).

Nasal occlusion was conducted by giving glue into the nasal cavity and this method blocks the reception of pheromones by the olfactory epithelium. The olfactory tract is a nerve bundle connecting the olfactory bulb and the telencephalon. Since the olfactory bulbs of goldfish are of the pedunculated type with elongated olfactory tracts, we can easily sever the tracts surgically (Figure 4). Olfactory tract section (OTX) blocks the transmission of olfactory information from the olfactory bulbs to the telencephalon by which sexual behavior is regulated (Kyle and Peter, 1982; Kyle et al., 1982; Koyama et al., 1984; Koyama et al., 1985). Blockage of olfaction by both of the methods suppressed the occurrence of sexual behavior in males even if the males were mated with PG-injected females.While involvement of olfaction in sexual behavior in male goldfish has been studied, the involvement of olfaction in females has not been studied for a long time. Recently, we have found that olfaction is essential in female goldfish to perform sexual behavior (Kawaguchi et al., 2011). Nasal occlusion suppressed sexual behavior of PG-injected females (Figure 4). However, it is not clarified whether there is any special olfactant which is essential for the occurrence of female sexual behavior or whether whole blockage of olfaction simply causes suppression of female sexual behavior. Surprisingly, when olfactory tract was sectioned in PG-injected females, unlike males these females performed female sexual behavior normally although these females receive no olfactory information (Stacey and Kyle, 1983; Kawaguchi et al., 2011). The effects of nasal occlusion and olfactory tract section seem to be contradictory but we have following interpretation: There is an inhibitory system of female sexual behavior from the olfactory bulb to the telencephalon. When olfaction is blocked, this system exerts suppression on the behavior even under the condition that PGF stimulates the brain.

By OTX, the inhibition is removed and females resume sexual behavior by the simulation of PGF although these fish receive no olfactory information. Male goldfish has a stimulatory system of sexual behavior from the olfactory epithelium to the telencephalon via the olfactory nerve, the olfactory bulb, and the olfactory tract, and this stimulatory system starts to function when olfactory epithelium receives the PG pheromone.

The olfactory tract is subdivided into the medial and the lateral tracts. It is known that stimulatory information is transmitted by the medial olfactory tract, not by lateral olfactory tract (Stacey and Kyle, 1983). As mentioned above, female goldfish has an inhibitory system from the olfactory epithelium to the telencephalon via the olfactory bulb and the olfactory tract. This inhibitory system functions when olfaction is blocked even if the brain is stimulated by PGF. We have recently identified that inhibition of female-typical behavior is mediated not by the lateral olfactory tract but by the medial olfactory tract (Kawaguchi, in preparation). These stimulatory and inhibitory systems of sexual behavior in goldfish are considered to function gender-typically.

Sexual Plasticity of Behavior in Goldfish

Sex change is observed in some hermaphroditic teleosts in nature and these fishes change their gonadal sex, behavioral sex and sex of secondary sexual characteristics (Helfman et al., 1997; Moyle and Cech, 2000). Although gonochoristic (i.e., non-sex changing) teleosts normally do not exhibit heterotypical reproductive functions, heterotypical patterns of behavior can be induced in adult goldfish by hormonal treatments (Kobayashi et al., 2002;

Munakata and Kobayashi, 2010). In males, PGF injection induces female-typical sexual behavior at a rate similar to that seen in females (Figure 3) within several minutes after PG-injection (Kyle and Stacey, 1983). Similarly, androgen capsule implant induces male-typical sexual behaviors in mature females. The shortest latency of induction of male-typical sexual behavior was one day after androgen capsule implantation based on our preliminary experiments (Kobayashi, unpublished data). Interestingly, these hormonal treatments and the performance of heterotypical sexual behavior do not suppress the occurrence of gender-typical sexual behavior. Our recent study showed that male goldfish were shown to retain their ability to perform male-typical sexual behavior within a week after being induced to perform female-typical sexual behavior by PG injection. Likewise, females showed female-typical sexual behavior after being induced to perform male-typical sexual behavior by KT or methyltestosterone (MT) capsule implantation (Stacey and Kobayashi, 1996; Saoshiro et al., 2013).

It would appear that some adult fishes have retain relatively more bisexual potential than in mammals, where the normal processes of male development typically involve inhibition of female-typical potential (Yamanouchi, 1997; Sakuma 2009; Bao and Swaab, 2011). Also, this behavioral plasticity observed in goldfish is comparable to that of bi-directional sex changing fish which change its sex serially.

Furthermore, our recent study (Saoshiro et al., 2013) demonstrated that both male and female goldfish performed male-typical and female-typical sexual behavior alternately and in rapid succession within minutes, and within seconds in some individuals (See Figure 5 for details). It is surprising that non-sex changing goldfish have an ability to switch roles in sexual behavior rapidly similar to what has been observed in egg trading simultaneous hermaphrodite (Cheek, 1998; Cheek et al., 2000). Switch of sexual roles observed in our experiments indicate that the underlying mechanism is context-dependent. That is, experimental fish are ready to perform both male- and female-typical behavior by PG pheromone and PG-injection, respectively and seem to perform appropriate sexual behavior depending on the interaction to the partner fish. It seems unlikely that each behavioral switch is regulated by hormone release, gene expression and/or proliferation of neurons in the brain since sex change occurs in short periods of time (Godwin, 2010a, b; Larson, 2011). Gonochoristic teleosts other than goldfish also exhibit sexual plasticity of behavior as adults. Induction of male-typical sexual behavior was observed in females (See review by Munakata and Kobayashi, 2010): stickleback *Gasterosteus aculeatus* guppy *Poecilia reticulata*, medaka *Oryzias latipes*, mosquitofish *Gambusia affinis holbrooki*, and tilapia *Oreochromis mossambicus* (Kuramochi et al., 2011). Male-typical sexual behavior is also induced in gynogenetic crucian carp *Carassius auratus langsdorfii* (Kobayashi and Nakanishi, 1999). This would mean that a species that has evolved as an all genetically females population can exhibit male-typical behavior. Compared to the induction of male-typical sexual behavior, female-typical sexual behavior in males has been demonstrated only in goldfish by PG injection (Munakata and Kobayashi, 2010).

Sexual Bipotentiality of the Brain in Fish

In mammals and birds, sexual differentiation of the brain occurs during early development and once brain sex is determined, it is irreversible. Androgen can induce male-typical sexual behavior in adult males but not in adult females, and estrogen can induce female-typical behavior in females but not in males (Bao and Swaab, 2011; Sakuma, 2009; Yamanouchi, 1997). This clear sex difference of responsiveness to hormones indicates that male and female have different types of brain in these animals.

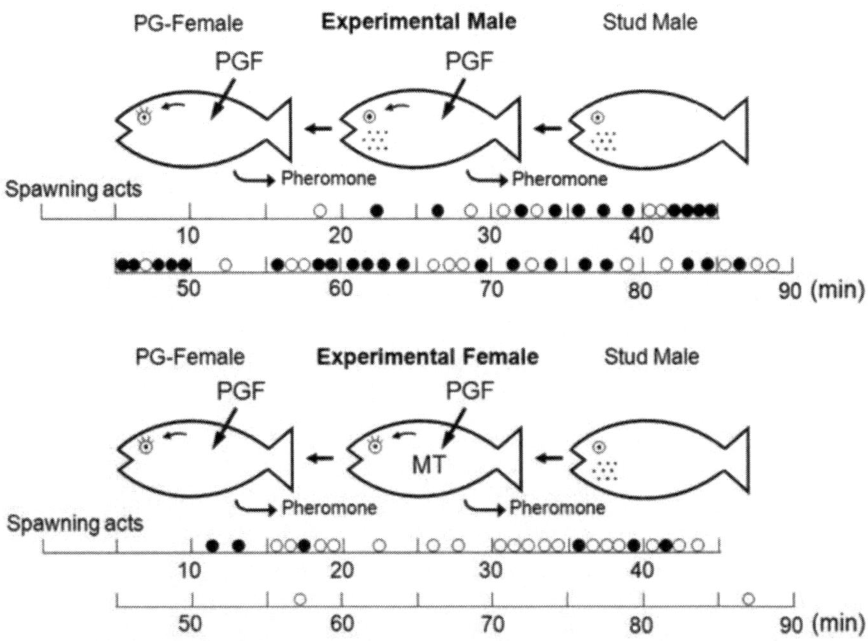

Figure 5. Induction of alternate sexual behavior in male and female goldfish. Upper panel, an experimental male injected with prostaglandin F2α (PGF) was placed with a "PGF-injected" partner female (PG-female) and a "stud" partner male (stud male) in an observation aquarium and behavior was observed for 90 min. The experimental male performed male-typical sex behavior with a PG-female stimulated by pheromone released from the PG-female and female-typical sex behavior with a stud male by the effect of injected PGF. A typical pattern of alternate sex behavior performed by one of experimental males is shown with circles. Solid and open circles indicate performance of male-typical and female-typical behavior, respectively during 90 min observation period. Lower panel, an experimental female implanted with methyltestosterone (MT) capsule and injected with PGF was placed with a PG-female and a stud male in an observation aquarium and behavior was observed for 90 min. The experimental female responded to pheromone released by PG-female by the effect of MT and performed male-typical sex behavior with a PG-female. The experimental female also performed female-typical sex behavior with a stud male by the effect of injected PGF. A typical pattern of alternate sex behavior performed by one of experimental females is shown with circles. Solid and open circles indicate performance of male-typical and female-typical behavior, respectively during 90 min observation period. The sex of the spawning act of the experimental fish can be distinguished by the behavioral pattern and position at spawning. When the fish entered into the aquatic plant ahead of the other fish and took position above the other fish, the spawning act of this fish was judged as a "female-typical spawning act". When the fish followed the fish which went ahead into the aquatic plants and took position underneath the other fish, the spawning act of this fish was judged as a "male-typical spawning act" (Saoshiro *et al.*, 2013).

On the other hand, in some species of fish, male hormone can induce male-typical sexual behavior both in males and females, and female hormone can induce female-typical sexual behavior both in males and females. These results suggest that in some fishes, both males and females possess a sexually bipotential brain which can regulate sexual behavior of both types in response to respective hormones. Existence of sex-changing fishes in nature which perform male-typical and female-typical sexual behavior during their life also supports this idea.

Information of neuroendocrine regulation of gonadal sex change is accumulating (Godwin, 2010a, b; Larson, 2011), but neural control of behavioral sex change remains poorly understood.

It is considered that short latencies of performing heterotypical behavior in response to social cues in sex-changing fish or hormone treatments in non-sex changing fish reflect activation of pre-existing hormone responsive mechanisms in the brain rather than occurrence of neural change for sexual behavior of each sex. In goldfish, injection of PGF induces female-typical sex behavior in males within several minutes, and MT treatment can induce male-typical sex behavior in females in one day (Kobayashi, unpublished data).

Also, both male and female goldfish switch sexual roles within minutes (Figure 5). These results also support the idea of that goldfish have a brain of dual sexual systems.

In addition to the rapid sex change in behavior in goldfish, our recent study on sexual behavior in relation to olfaction demonstrated the sexual bipotentiality of the brain of goldfish. As we described in this chapter, male and female goldfish seem to have sex-typical systems in response to olfaction (Figure 4). Olfactory blockage suppressed male-typical sexual behavior in male goldfish (Kawaguchi et al., 2011), and similarly occurrence of male-typical sexual behavior in MT-treated females was suppressed by nasal occlusion or OTX (Kobayashi, unpublished data). Female-typical sexual behavior was suppressed in PG-injected females by nasal occlusion and resumed by OTX (Kawaguchi et al., 2011). Similarly, female-typical sexual behavior was suppressed in PG-injected males by nasal occlusion and resumed by OTX. These results clearly indicate that male and female goldfish have common bisexual brain systems for the regulation of sexual behavior in response to olfaction. Neuroanatomical basis for the control of gender-typical and heterotypical sexual behavior is not clearly known.

Lesion of the sex steroid-binding area of ventral telencephalon in male goldfish reduced occurrence of male-typical behavior, and interestingly, these males failed to perform female-typical sexual behavior by PG-injection (Kyle et al., 1982). It is of great interest to elucidate whether this area has a single neural unit which regulates both male-typical and female-typical behavior or two discrete neural units which regulate respective sexual behavior. Unlike goldfish, some teleost species seem to have a sexually differentiated brain. Sexually mature males of plainfin midshipman *Porychthys notatus*, generate calls that advertise their presence to attract females, and this vocal activity is induced by KT not by T, whereas agonistic calls of females are induced not by KT but T (Remage-Healy and Bass, 2004, 2007).

In masu salmon *Oncorhynchus masou*, T is a requirement of both males and females for the occurrence of sexual behaviors but T elicits gender-typical behavior in each sex and not heterotypical sexual behavior (Munakata et al., 2001a, b). The different responsiveness to hormones in behavior suggests that the brain of these fishes is sexually differentiated.

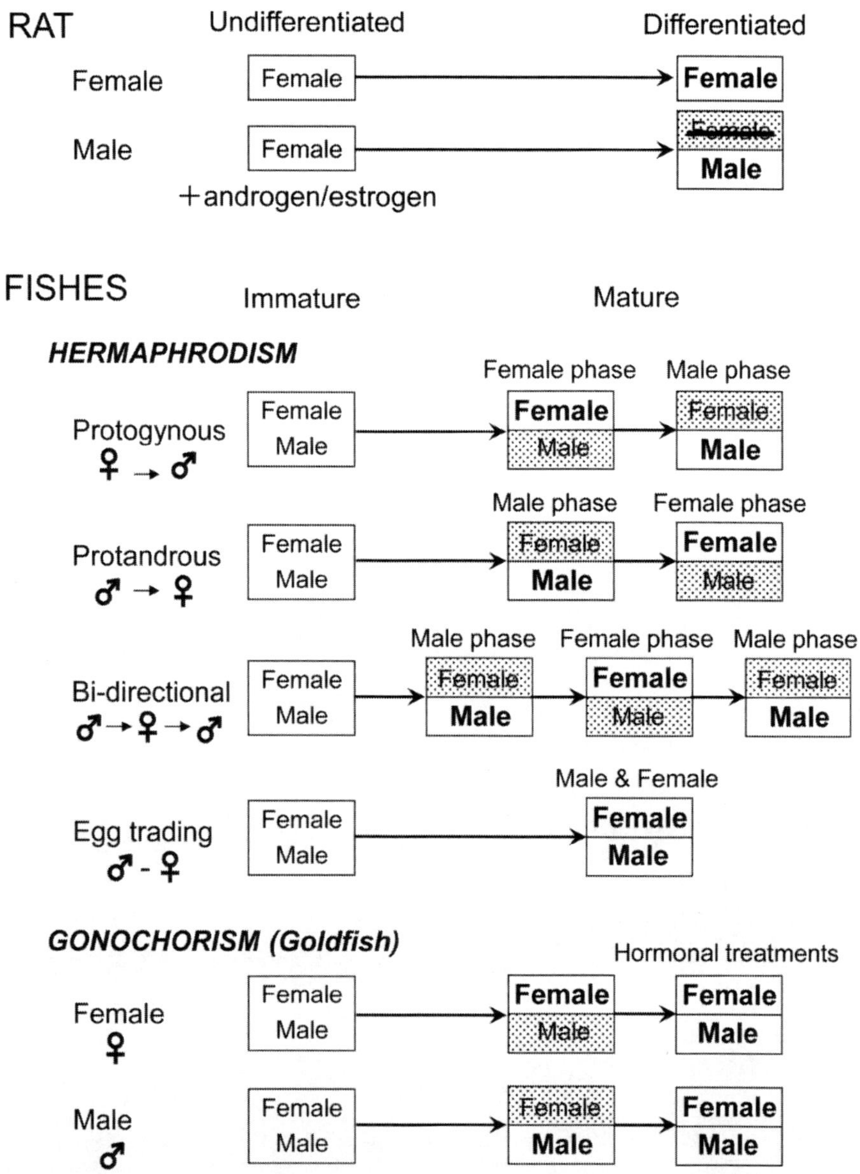

Figure 6. A hypothesis of sexual bipotentiality of the brain in fishes. In the rat, the undifferentiated brain sex is believed to normally be female but in the presence of androgen or estrogen during the perinatal period, the brain develops neural systems which regulate male functions and inhibit female functions (crossed out area). Teleost fish on the other hand, appear to possess a sexually bipotential brain. Among hermaphroditic fishes, when a protogynous sex-changing fish is in the female phase, the female portion of the brain is active and the male portion is quiescent (shaded area). At the time the individual starts to behave like a male, the male portion of the brain is activated and the female portion becomes quiescent (shaded area): Factors (age, social status, etc.) that regulate the sex change vary among species. Protandrous and bi-directional sex-changing fish also use either male portion or female portion of the brain depending on their age or social situation. It seems that egg traders can activate male and female portion of the brain at the same time and can regulate sexual behavior switching between male and female roles rapidly. Although gonochoristic fishes normally use only brain areas controlling gender-typical behaviors during their lifetime, brain areas controlling heterotypical behaviors can be activated by hormonal treatments in the way of sex-changing fish and egg trading fish.

Although examples of adult sexual bipotentiality in teleost behavior admittedly are limited, we suggest that brain bisexuality is typical of most teleosts regardless of their sexual system i.e., hermaphroditic, gonochoristic, gynogenetic (Kobayashi et al., 2002). From this perspective, fish hermaphroditism could be viewed not as the result of specializations that depart dramatically from normal reproductive function, but rather as the opportunistic expression of latent behavioral, endocrine and gonadal bisexuality in ecological situations where hermaphroditism is more adaptive than gonochorism. One of the evolutionary questions that arise during the examination of sexual behavior in fishes is whether only fish obtained the brain bipotentiality in vertebrates during their evolution or have other vertebrates simply lost this bipotentiality during their evolution.

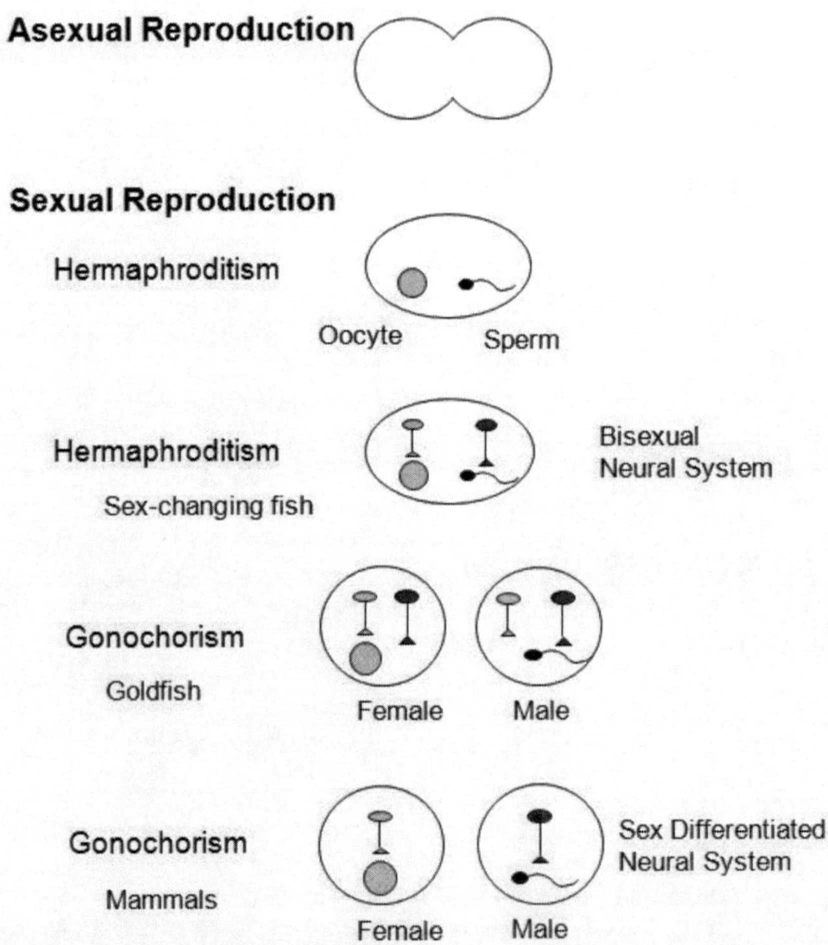

Figure 7. A hypothesis of evolution of sexual bipotentiality of neural system. Appearance of sexual reproduction in primitive animal is considered to be hermaphroditic. If the hermaphroditic animal has neuroendocrine system for the regulation of female function (ovarian development and release of oocytes) and male function (testicular development and release of sperm), this neural system is considered as the prototype of a bisexual neural system. Sex-changing fish can regulate both female and male function by bisexual neural system. Non-sex changing goldfish retains bisexual neural system which enables to perform heterotypical sex behavior. Some species of fish and mammals lost neural bisexuality by brain sex differentiation.

There may be several explanations on these two possibilities, but we propose an explanation for the latter possibility. Considering the fact that the extant animals, including invertebrates, which conduct both asexual and sexual reproduction are mostly hermaphrodites, it is possible that the first animal which reproduced sexually appeared during evolution as hermaphroditic (Brusca and Brusca, 2002). If the primitive hermaphroditic animal has the ovary and the testis and neural systems each for regulation of ovarian development and testicular development, this neuronal system is considered as the prototype of a bisexual neural system (Figure 7). It is not surprising that many invertebrates and some fishes retain this neural bisexuality, and other vertebrates including some fishes, such as plainfin midshipman and masu salmon, have lost this system or have developed an inhibitory system of expression of heterotypical sexual behavior for the gonochoristic reproductive strategy (Yamanouchi, 1997).

Conclusion

Unlike mammals and birds which have sexual differentiation of the brain, field and laboratory observation suggest that some teleost fishes have sexually bipotential brain which can regulate sexual behavior of both sexes. Performance of heterotypical behavior does not suppress the occurrence of gender-typical behavior. Egg trading simultaneous hermaphrodites and hormone treated goldfish can switch sexual behavior between male and female rapidly. Rapid behavioral sex change seems to be regulated context-dependently rather than by hormone release, gene expression and/or proliferation of neurons which need longer time to occur. Neuroanatomical basis of bisexual brain of teleosts which enables gender-typical and heterotypeical sexual behavior remains to be elucidated.

Acknowledgments

We are grateful to Prof. Naoyuki Yamamoto of Nagoya University, for valuable comments for preparing the manuscript.

References

Adkins-Regan, E. (2012). Hormonal organization and activation: Evolutionary implications and questions. *Gen. Comp. Endocrinol.*, 176, 279-285.

Balthazart, J., Cornil, C. A., Charlier, T. D., Taziaux, M., Estradiol, G. F. Ball (2009). A key endocrine signal in the sexual differentiation and activation of reproductive behavior in quail. *J. Exp. Zool.*, 311A, 323-345.

Bao, A. M. and Swaab, D. F. (2011). Sexual differentiation of the human brain: relation to gender identity, sexual orientation and neuropsychiatric disorders. *Frontiers in Neuroendocrinol.*, 32, 214-226.

Brusca, R. C. and Brusca, G. J. (2002). *Invertebrates* (2nd edition). Sunderland, MA: Sinauer Associates.

Cheek, A. O. (1998). Ovulation does not constrain egg parcel size in the simultaneous hermaphrodite *Serranus subligarius*. *Env. Biol. Fishes*, 52, 435-442.

Cheek, A. O., Thomas, P. and Sullivan, C. (2000). Sex steroids relative to alternative mating behaviors in the simultaneous hermaphrodite *Seranus subligarius* (Perciformes: Serranidae). *Horm. Behav.*, 37, 198-211.

Dulka, J. G., Stacey, N. E., Sorensen, P. W., and Van Der Kraak, G. J. (1987). A sex steroid pheromone synchronizes male-female spawning readiness in goldfish. *Nature*, 325, 251-253.

Godwin, J., Crews, D. and Warner, R. R. (1996). Behavioral sex change in the absence of gonads in a coral reef fish. *Proc. R Soc. Lond., B* 263, 1683-1688.

Godwin, J. (2010a). Neuroendocrinology of sexual plasticity in teleost fishes. *Frontier Neuroendocrinol.*, 31, 203-216.

Godwin, J. (2010b). Neuroendocrine regulation of sex change and alternate sexual phenotypes in sex-changing reef fishes. In: K. S. Cole (Ed.) *Reproduction and sexuality in marine fishes* (pp. 273-306). Berkeley and Los Angeles, CA: Univ. of California Press.

Goncalves, D. M. and Oliveira, R. F. (2011). Hormones and sexual behavior of teleost fishes. In: D. O. Norris and K. H. Lopez (Eds.), *Hormones and Reproduction of Vertebrates, Vol. 1, Fishes* (pp. 119-147) New York, NY: Academic Press (Elsevier).

Helfman, G. S., Collete, B. B. and Facey, D. E. (1997). *The Diversity of Fishes.* Malden, MA: Blackwell Science.

Kawaguchi, Y., Mitsuhashi, T., Kitami, A., Nagaoka, A., Hayakawa, Y., and Kobayashi, M. (2011). Involvement of olfaction in sexual behavior in goldfish. *Indian J. Sci. Technol.*, 4, 81.

Kezuka, H., Kobayashi, M., Aida, K., and Hanyu, I. (1989). Effects of photoperiod and pinealectomy on the gonadotropin secretion and ovulation in goldfish *Carassius auratus. Nippon Suisan Gakkaishi* 55, 2099-2103.

Kim, M. H., Oka, Y., Amano, M., Kobayashi, M., Okuzawa, K., Hasegawa, Y., Kawashima, S., Suzuki, Y., and Aida, K. (1995a). Immunohistochemical localization of sGnRH and cGnRH-II in the brain of goldfish, *Carassius auratus. J. Comp. Neurol.*, 356, 72-82.

Kim, M., Kobayashi, M. and Aida, K. (1995b) Changes in brain GnRH after olfactory tract section in female goldfish. *Fish Sci.*, 61, 614-617.

Kim, M. H., Kobayashi, M., Oka, Y., Amano, M., Kawashima, S., and Aida, K. (2001). Effects of olfactory tract section on the immunohistochemical distribution of the brain GnRH fibers in the female goldfish, *Carassius auratus. Zool. Sci.*, 18, 241-248.

Kobayashi, M., Aida, K. and Hanyu, I. (1986a). Gonadotropin surge during spawning in male goldfish. *Gen. Comp. Endocrinol.*, 62, 70-79.

Kobayashi, M., Aida, K. and Hanyu, I. (1986b). Pheromone from ovulatory female goldfish induces gonadotropin surge in males. *Gen. Comp. Endocrinol.*, 63, 451-455.

Kobayashi, M., Aida, K. and Hanyu, I. (1987). Hormone changes during ovulation and effects of steroid hormones on plasma gonadotropin levels and ovulation in goldfish. *Gen. Comp. Endocrinol.*, 67, 24-32.

Kobayashi, M., Aida, K. and Hanyu, I. (1989a). Induction of gonadotropin surge by steroid hormone implantation in ovariectomized and sexually regressed female goldfish. *Gen. Comp. Endocrinol.*, 73, 469-476.

Kobayashi, M., Aida, K. and Hanyu, I. (1989b). Involvement of steroid hormones in the preovulatory gonadotropin surge in female goldfish. *Fish Physiol. Biochem.*, 7, 141-146.

Kobayashi, M., Amano, M., Okuzawa, K., Hasegawa, Y., and Aida, K. (1992) Effects of olfactory tract section on brain GnRH distribution, plasma gonadotropin levels, and gonadal stage in goldfish. *Zool. Sci.*, 9, 765-773.

Kobayashi, M. and Stacey, N. E. (1993). Prostaglandin-induced female spawning behavior in goldfish (*Carassius auratus*) appears independent of ovarian influence. *Horm. Behav.*, 27, 38-55.

Kobayashi, M., Amano, M., Kim, M. H., Furukawa, K., Hasegawa, Y., and Aida, K. (1994). Gonadotropin-releasing hormones of terminal nerve origin are not essential to ovarian development and ovulation in goldfish. *Gen. Comp. Endocrinol.*, 95, 192-200.

Kobayashi, M., Furukawa, K., Kim, M. H., and Aida, K.(1997). Induction of male-type gonadotropin secretion by implantation of 11-ketotestosterone in female goldfish. *Gen. Comp. Endocrinol.*, 108, 434-445.

Kobayashi, M. and Nakanishi, T. (1999). 11-Ketotestosterone induces male-type sexual behavior and gonadotropin secretion in gynogenetic crucian carp, *Carassius auratus langsdorfii*. *Gen. Comp. Endocrinol.*, 115, 178-187.

Kobayashi, M., Sorensen, P. W. and Stacey, N. E. (2002). Hormonal and pheromonal control of spawning behavior in the goldfish. *Fish Physiol. Biochem.*, 26, 71-84.

Koyama, Y., Satou, M., Oka, Y., and Ueda, K. (1984). Involvement of the telencephalic hemispheres and the preoptic area in sexual behavior of the male goldfish, *Carassius auratus*: a brain-lesion study. *Behav. Neural. Biol.*, 40, 70-86.

Koyama, Y., Satou, M. and Ueda, K. (1985). Sexual behavior elicited by electrical stimulation of the telencephalic and preoptic areas in the goldfish, *Carassius auratus*. *Zool. Sci.*, 2, 565-570.

Kuramochi, A., Tsuchiya, A., Kaneko, T., Ohtani-Kaneko, R. (2011) Sexual dimorphism of gonadotropin-releasing hormone type-III (GnRH3) neurons and hormonal sex reversal of male reproductive behavior in Mozanbique tilapia. *Zool. Sci.*, 28, 733-739.

Kyle, A. L. and Peter, R. E. (1982). Effects of forebrain lesions on spawning behaviour in the male goldfish. *Physiol. Behav.*, 28, 1103-1109.

Kyle, A. L, Stacey, N. E. and Peter, R. E. (1982). Ventral telencephalic lesions: Effects on bisexual behavior, activity, and olfaction in the male goldfish. *Behav. Neural Biol.*, 36, 229-241.

Larson, E. T. (2011). Neuroendocrine regulation of sex-changing fishes. In: D. O. Norris and K. H. Lopez (Eds.), *Hormones and Reproduction of Vertebrates, Vol. 1, Fishes* (pp.149-168) New York, NY: Academic Press (Elsevier).

Minamimoto, M., Sakakura, Y., Soyano, K., Akaba, Y., and Hagiwara, A (2006). Plasma sex steroid levels and steroidogenesis in the gonad of the self-fertilizing fish *Rivulus marmoratus*. *Env. Biol. Fishes*, 75, 159-166.

Moyle, P. B. and Cech, J. J. Jr. (2000). *Fishes, An Introduction to Ichthyology* (4th edition). Upper Saddle River, NJ: Prentice Hall.

Munakata, A., Amano, M., Ikuta, K., Kitamura, S., and Aida, K. (2001a). The involvement of sex steroid hormones in downstream and upstream migratory behavior of masu salmon. *Comp. Biochem. Physiol. B.* 129, 661-669.

Munakata, A., Amano, M., Ikuta, K., Kitamura, S., and Aida, K. (2001b). Sex steroids control migration of masu salmon. *Fish Sci.*, 68, *suppl. I*, 49-52.

Munakata, A. and Kobayashi, M. (2010). Endocrine control of sexual behavior in teleost fish. *Gen. Comp. Endocrinol.*, 165, 456-468.

Munday, P. L., Kuwamura., T. and Kroon, F. J. (2010). Bidirectional sex change in marine fishes. In: K. S. Cole (Ed.) *Reproduction and sexuality in marine fishes* (pp. 241-271). Berkeley and Los Angeles, CA: California: Univ. of California Press.

Nakashima, Y., Kuwamura, T. and Yogo, Y. (1995). Why be a both-way sex change? *Ethology*, 177, 1007-1009.

Nakashima, Y., Sakai, Y., Karino, K., and Kuwamura, T. (2000). Female-female spawning and sex change in a haremic coral-reef fish, *Labroides dimidiatus*. *Zool. Sci.*, 17, 967-970.

Partridge, B. L., Liley, N. R. and Stacey, N. E. (1976). The role of pheromones in the sexual behavior of the goldfish. *Anim. Behav.*, 24, 291-299.

Poling, K. R., Fraser, E. J. and Sorensen, P. W. (2001). The three steroidal components of the goldfish preovulatory pheromone signal evoke different behaviors in males. *Comp. Biochem. Physiol. B*, 129, 645-651.

Remage-Healey, L. and Bass, A. H. (2004). Rapid, hierarchical modulation of vocal pattering by steroid hormones. *J. Neurosci.*, 24, 5892-5900.

Remage-Healey, L. and Bass, A. H. (2007). Plasticity in brain sexuality is revealed by the rapid actions of steroid hormones. *J. Neurosci.*, 27, 1114-1122.

Sakuma, Y. (1997). Neural control of reproductive behavior: Bird's eye view. In: K. I. Maeda, H. Tsukamura and A. Yokoyama (Eds.), *Neural Control of Reproduction.* (pp. 155-164). Tokyo, Japan: Japan Sci. Soc. Press.

Sakuma, Y. (2009). Gonadal steroid action and brain sex differentiation in the rat. *J. Neuroendocrinol.*, 21, 410-414.

Saoshiro, S., Kawaguchi, Y., Hayakawa, Y. and Kobayashi, M. (2013). Sexual bipotentiality of behavior in male and female goldfish. *Gen. Comp. Endocrinol.*, 181, 265-270.

Sorensen, P. W., Stacey, N. E. and Chamberlain, K. J. (1989). Differing behavioral and endocrinological effects of two female sex pheromones on male goldfish. *Horm. Behav.*, 23, 317-332.

Sorensen, P. W., Brash, A. R., Goetz, F. W., Kellner, R. G., Bowdin, L., and Vrieze, L. A. (1995). Origins and functions of F prostaglandins as hormones and pheromones in the goldfish. In: F. W. Goetz and P. Thomas (Eds.), *Proceedings of the Fourth International Symposium on the Reproductive Physiology of Fish.* (pp. 252-254). Austin, TX: Fish Symp 95.

Sorensen, P. W., Pinillos, M. and Scott, A. P. (2005). Sexually mature male goldfish release large quantities of androstendione into the water where it functions as a pheromone. *Gen. Comp. Endocrinol.*, 140, 164-175.

Stacey, N. E. and Liley, N. R. (1974). Regulation of spawning behavior in the female goldfish. *Nature*, 247, 71-72.

Stacey, N. E. (1976). Effects of indomethacin and prostaglandins on the spawning behavior of female goldfish. *Prostaglandins*, 12, 113-126.

Stacey, N. E. and Peter, R. E. (1979). Central action of prostaglandins in spawning behaviour of female goldfish. *Physiol. Behav.*, 22, 1191-1196.

Stacey, N. E. and Kyle, A. L. (1983). Effects of olfactory tract lesions on sexual and feeding behavior in the goldfish. *Physiol. Behav.*, 30, 621-628.

Stacey, N. E. and Kobayashi, M. (1996). Androgen induction of male sexual behaviors in female goldfish. *Horm. Behav.*, 30, 434-445.

Stacey, N. E. (2003). Hormones, pheromones and reproductive behavior. *Fish Physiol. Biochem.*, 28, 229-235.

Stacey, N. E. and Sorensen, P. W. (2006). Reproductive pheromones. In: K. A. Sloman, R. W. Wilson, and S. Balshine (Eds), *Fish Physiology, Vol.* 24, *Behaviour and Physiology of Fish.* (pp. 359-412) New York, NY: Elsevier.

Stacey, N. E. (2011). Hormonally derived sex pheromones in fishes. In: D. O. Norris and K. H. Lopez (Eds.), *Hormones and Reproduction of Vertebrates, Vol. 1, Fishes* (pp. 169-192). New York, NY: Academic Press (Elsevier),.

Volkoff, H. and Peter, R. E. (1999). Actions of two forms of gonadotropin releasing hormone and a GnRH antagonist on spawning behavior of the goldfish *Carassius auratus*. *Gen. Comp. Endocrinol.*, 116, 347-355.

Yamanouchi, K. (1997). Brain mechanisms inhibiting the expression of heterotypical sexual behavior in rats. In: K. I. Maeda, H. Tsukamura and A. Yokoyama (Eds.), *Neural Control of Reproduction.* (pp. 291-235) Tokyo, Japan: Japan Sci. Soc. Press.

Zheng, W. and Stacey, N. E. (1997). A steroidal pheromone and spawning stimuli act via different neuroendocrine mechanisms to increase gonadotropin and milt volume in male goldfish (*Carassius auratus*). *Gen. Comp. Endocrinol.*, 105, 228-235.

In: Sexual Plasticity and Gametogenesis in Fishes
Editor: Balasubramanian Senthilkumaran
ISBN: 978-1-62618-848-8

Chapter XIII

Molecular Mechanisms of Gonadal Sex Differentiation and Sex-Reversal in Fish: Tilapia and Medaka

Tohru Kobayashi*
Laboratory of Molecular Reproductive Biology, Institute for Environmental Sciences, University of Shizuoka, Shizuoka, Japan

Abstract

No clear evolutionary conservation exists between phyla for the mechanisms underlying sex determination. In contrast, many studies indicate that various downstream gene products of sex determination genes are functionally similar in different species. In the teleost fish–Nile tilapia, *Oreochromis niloticus*, and medaka, *Oryzias latipes*–numerous conserved genes involving gonadal sex differentiation have been examined. In the latter species, the sex-determining gene, *dmy/dmrt1bY*, has been identified as the second sex-determining gene in vertebrates. Morphogenesis during gonadal differentiation is also conserved, as is evident from the differentiation and development of the parenchyma/germinal epithelium (GE) cells (testis cord) and follicles.

Therefore, it is crucial to understand the molecular mechanisms of gonadal sex differentiation from the perspective of the relationship between conserved gene expression cascades and morphogenesis during gonadal sex differentiation. This article reviews the histogenesis and expression profiles of male and female differentiation-related genes involved in sex differentiation in tilapia and medaka. In both species, the process of gonadal differentiation, comprising histogenesis and gene expression, is almost similar. First, we describe the components of gonadal sex differentiation in each species, including the processes of histogenesis and gene expression. Subsequently, we discuss the functions of the genes during gonadal sex differentiation and propose a concept of a gonadal sex differentiation pathway.

* Corresponding author: Tohru Kobayashi, Ph.D. Laboratory of Molecular Reproductive Biology, Institute for Environmental Sciences, University of Shizuoka, Japan. 52-1 Yada, Suruga-ku, Shizuoka 422-8526, Japan. E-mail: tohruk@u-shizuoka-ken.ac.jp.

Introduction

In most mammals, the male-inducing master gene, SRY (Sex-determining Region Y), is located on the Y chromosome, and is, therefore, absent in XX females. SRY seems to be specific to mammals (Capel, 2000). Recently, the sex-determining gene, *dmy/dmrt1bY* (DM-domain gene on Y chromosome)–a *dmrt1* homolog–has been identified in the teleost fish, medaka (*Oryzias latipes*), as the second sex-determining gene in vertebrates (Matsuda et al., 2002; Nanda et al., 2002).

In addition, *dmy* is only found in 2 subspecies of medaka, within the genus, *Oryzias* (Matsuda et al., 2003). Recent reports suggest that the downstream products of sex determination genes are functionally similar in different species (Capel, 2000; Kobayashi and Nagahama, 2009). Over the past several years, sex-determining genes have been identified from several lower vertebrates, e.g., the amphibian, *Xenopus laevis*: *DM-W* (Yoshimoto et al., 2006); the teleostei, medaka, *Olyzias luzonensis*: *GsdfY* (Myosho et al., 2012); and Patagonian pejerrey, *Odontesthes bonariensis*: *amh-Y* (Hattori et al., 2012). Thus, in contrast to many developmental processes, there is no clear evolutionary conservation for sex-determining genes among phyla.

The undifferentiated gonads of teleosts differentiate into testes or ovaries in a similar way to other vertebrates. Sex differentiation progresses under genetic control together with the effects of various environmental factors. To date, much research has been carried out on gonadal sex differentiation in teleosts. Sexual characteristics and differentiation/development of gonads vary in teleosts, e.g., in the case of gonochorism and hermaphroditism which includes protandry, protogyny, and synchronous hermaphroditism (Atz, 1964; Yamamoto, 1969; Renboth, 1970). In gonochoristic teleost species, gonadal sex differentiation has been examined extensively using medaka and tilapia as model species (Devlin and Nagahama, 2002; Scartl, 2004; Saito and Tanaka, 2009; Kobayashi and Nagahama, 2009).

This review describes a concept in which the developmental sequences of gonadal sex differentiation in both species are comprehensively analyzed.

Formation of Gonadal Primordia

Gonadal primordia have important roles in gonadal sex differentiation. The gonads consist of germinal epithelium and stromal tissue. Historically, undifferentiated gonads consist of 2 somatic components: the cortex and the medulla. It has been suggested that the former originates from the peritoneal wall, or coelomic epithelium, and the latter derives from the mesonephric blastema. A classical concept is that the medulla and cortex participate in the formation of the structure of the gonads during gonadal differentiation in vertebrates (Witschi, 1957).

In this concept, the primordial gonads were thought to consist of the cortex: the coelomic epithelium including germ cells and the medulla: an aggregation of mesonephric blastemal cells. However, during gonadal development in teleosts, the participation of mesonephros-derived cells has been rejected (Lepori, 1980). In mammals and birds, testis cord as medulla is formed by proliferation of coelomic epithelial cells together with cells derived from the mesonephros as interstitial cells (Capel, 2000).

Even in amphibians, it has been reported that the medulla originates from the coelomic epithelium but not from the mesonephros, suggesting that the development is similar to teleost testis differentiation (Merchant-Larios and Villalpando, 1981; Iwasawa and Yamaguchi, 1984; Tanimura and Iwasawa, 1988; 1989; Kumakura and Iwasawa, 1992).

Thus, the gonads in vertebrates, including the teleosts, consist of the parenchyma/germinal epithelium (GE) and interstitium. This indicates that the medulla and cortex in vertebrate gonads are formed by the parenchyma/GE, consisting of germ cells and somatic cells surrounded by a basement membrane, and are named according to their localization in the gonads. In this paper, therefore, we focus on the parenchyma/GE during gonadal sex differentiation (Figure 1).

Overview of Gonadal Sex Differentiation in Tilapia

Sex Differences in Germ Cell Number during Gonadal Differentiation

During gonadal differentiation, morphological sexual dimorphism consists of changes in the germ cell number and histogenesis (Figure 1; Kobayashi et al., 2008). In Nile tilapia, on the day of hatching (4 days post fertilization: 4 dpf at 26°C) primordial germ cells (PGCs), which are morphologically distinguishable from somatic cells, are located in the outer layer of the lateral plate mesoderm around the hindgut. At 3 days post hatching (dph), i.e., 7 dpf, the PGCs are located in the gonadal anlagen after formation of the coelomic cavity in the lateral plate mesoderm, rather than migrating actively (Kobayashi et al., 2000; 2002).

Thereafter, germ cell numbers do not change significantly in either sex from 5 to 8 dph. After 8 dph, in contrast, the XX female germ cells continue to proliferate, whereas the germ cell numbers do not change from 9 to 14 dph in XY male gonads.

As described in previous reports, the number of germ cells in XY male gonads increases again after 14 dph, but spermatogenesis is not observed until 70 dph (Kobayashi et al., 2008; Kobayashi and Nagahama, 2009).

Sex Differences in Histogenesis during Gonadal Differentiation

The morphological sign for a sex difference in gonadal histogenesis is the formation of an intratesticular efferent duct or ovarian cavity (Nakamura, 1998). The formation of the ovarian cavity or intratesticular efferent duct occurs between 20 and 25 dph in the XX and XY gonads, respectively (Kobayashi et al., 2008).

Figure 2 shows the sex differences in histogenesis during gonadal differentiation in tilapia. In XY tilapia gonads, the parenchyma/GE cells including germ cell-surrounding cells developed into the precursor of seminiferous tubule structure and intratesticular efferent duct (Figures 1, 2), which is similar to a cord-like structure. Although no parenchyma cells developed in the XX gonads, an ovarian cavity was formed by the extension of both tips of the gonads (Kobayashi et al., 2008). The development of the parenchyma/GE cells, derived from the GE (coelomic epithelium) during testicular differentiation in tilapia, is similar to testicular differentiation in amphibians (cf. Merchant-Larios and Villalpando, 1981).

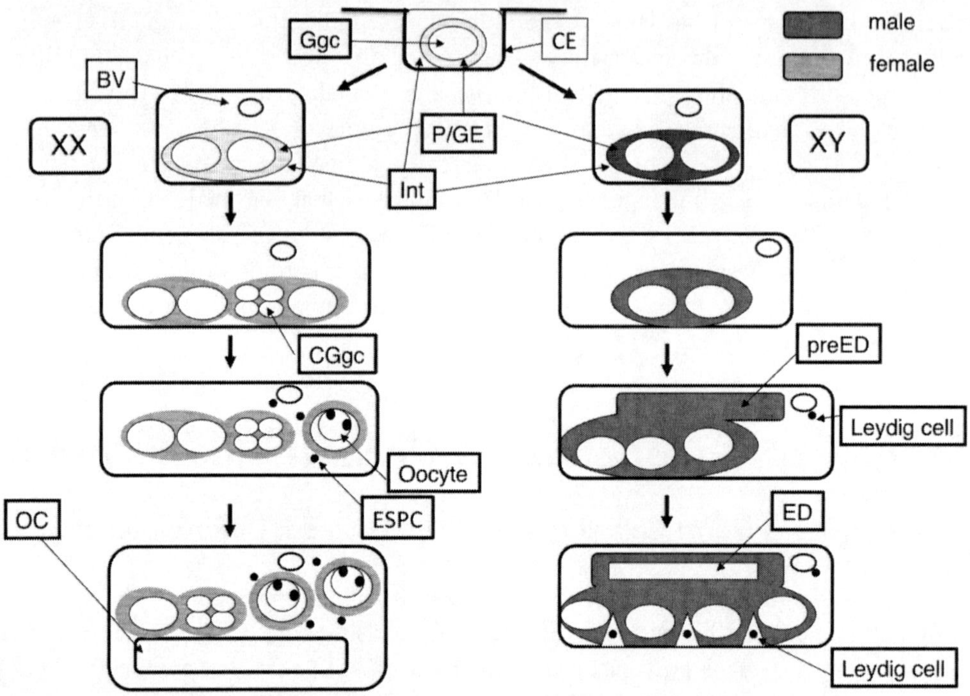

Figure 1. Schematic representation of gonadal sex differentiation in teleost fish. The gonads are consisted of parenchyma/germinal epithelium (GE) and interstitum (stromal tissue). First indication of sex difference is seen in parenchyma/GE, such as sex determining gene expression. Then, morphological sex difference is seen as sex difference in germ cell number in both sexes. Sex differences in histogenesis are seen, such as development for parenchyma/GE (acinus structure or precursor of ED) and ovarian cavity. In XY gonads, parenchyma/GE develops into seminiferous tubule and ED. In XX gonads, ovarian cavity is formed at dorsal or ventral side in medaka and tilapia, respectively. The timing of steroid producing cell differentiation and meiotic transition during gonadal differentiation is different in each species. Ggc, gonial germ cell. CGgc, cystic gonial germ cell. CE, coelomic epithelium. BV, blood vessel. P/GE, parenchyma/germinal epithelium. Int, interstitum. preED, precursor of intratesticular efferent duct. ED, intratesticular efferent duct. OC, ovarian cavity. ESPC, estrogen-synthesizing steroid producing cell.

In addition, gonadal sex differentiation in teleost fish is dependent on the development of the parenchyma cells (parenchyma/GE cells) derived from the GE (cf. Nakamura et al., 1998; Figure 1).

Molecular Events during Gonadal Sex Differentiation in Tilapia

In tilapia, *Ad4BP/SF-1 (nr5a1)* was first detected at 3 dpf (1 day before hatching) as clusters in the primordium of interrenal glands. At 3 dph, when the gonadal anlagen formed, *nr5a1* mRNA was expressed in the epithelial cells of the gonadal anlagen but not in the epithelium of the lateral plate mesoderm, which was the origin of the gonadal anlagen.

After 5 dph, *nr5a1* mRNA became localized to the stromal cells in the vicinity of blood vessels and disappeared from the outermost epithelium in the gonads of both sexes.

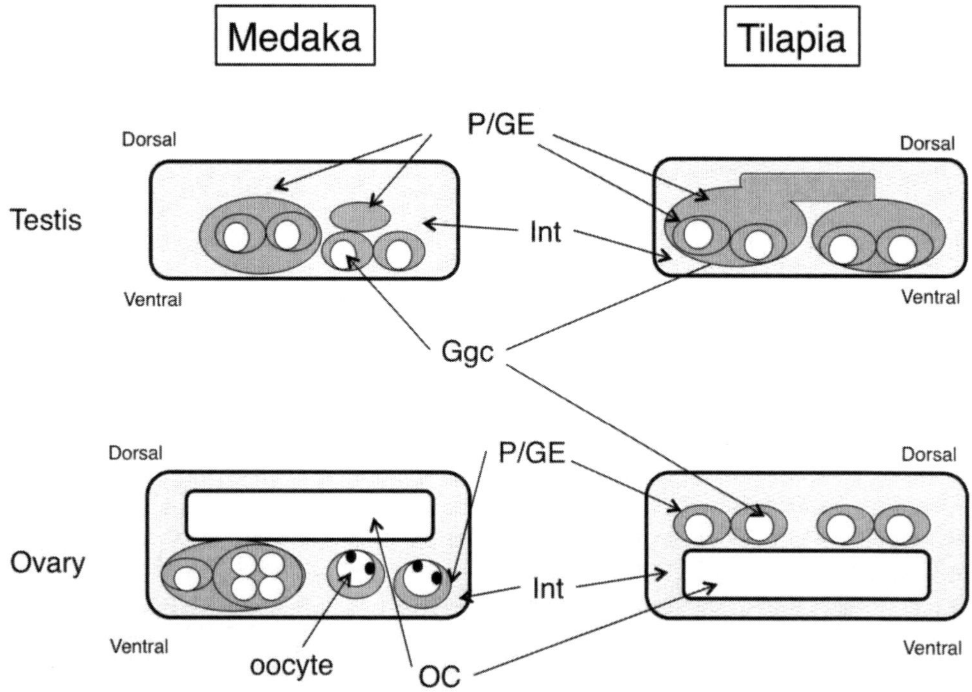

Figure 2. Schematic representation of sex differences in histogenesis in medaka and tilapia. Sex differences in histogenesis become apparent as development of parenchyma/GE and ovarian cavity in XY and XX gonads, respectively. The former is seen as the acinus structure and precursor of intratesticular efferent duct in medaka and tilapia, respectively. The later is seen at dorsal and ventral side in medaka and tilapia, respectively. In medaka, ovarian cavity is formed after the appearance of oocytes, whereas in tilapia, before the appearance of oocytes.

In contrast, the expression of steroidogenic enzymes, including aromatase (*cyp19a1*), occurred in stromal cells near the blood vessels at 5 dph in XX fry but not in XY fry (Wang et al., 2007).

Therefore, these cyp19a1-positive cells were also immunoreactive for P450scc (*cyp11a1*), P450c17 (*cyp17a1*), and 3β-HSD (*3b-hsd*) (Kobayashi et al., 2003; Wang et al., 2007). This suggests that the expression of *nr5a1* does not induce the expression of steroidogenic enzymes in early gonads by itself. Prior to 25 dph, no difference was observed in the expression pattern of *nr5a1* between XX and XY gonads. After 25 dph, the distribution of *nr5a1* mRNA differed between the sexes following the progression of gonadal histogenesis at the formation of an ovarian cavity in females and the intratesticular efferent duct in males, suggesting that sexual dimorphism of *nr5a1* expression is consistent with the sexual dimorphism of gonadal histogenesis (Wang et al., 2007).

In the female pathway, *foxl2* expression was higher in XX than XY gonads from 5 dph onward, indicating a female-dominant expression of *foxl2* in gonads. The expression profiles also indicated co-localization of the expression of *foxl2*, *nr5a1*, and *cyp19a1* in stromal cells near the blood vessels in XX gonads, suggesting an important role for *foxl2* in the transcriptional regulation of *cyp19a1* (Wang et al., 2007).

Similarly, this relationship for *foxl2*, *nr5a1*, and *cyp19a1* was also observed during sex reversal induced by estrogens in the ovaries of XY fry (Kobayashi et al., 2003; Wang et al., 2007).

Since *foxl2* was localized in both the steroid-producing cells in the interstitium of XX gonads and in the germ cell-surrounding cell lineage, including oogonia and oocytes (Wang et al., 2004; 2007), it was also suggested that *foxl2* had another function other than steroidogenesis involving ovarian differentiation, including estrogen production.

In the male pathway, *dmrt1* had a male-specific expression pattern in XY gonads from 6 dph onward. *dmrt1* was detected in the gonial germ cell-surrounding cells in XY gonads, specifically before the appearance of any signs of morphological sex differentiation, i.e., sex differences in the germ cell number and histogenesis such as differentiation into the intratesticular efferent duct or ovarian cavity.

The signals became localized in the Sertoli and epithelial cells of the efferent duct during testicular differentiation (Kobayashi et al., 2008).

In contrast to *dmrt1*, *sox9a* did not reveal sexual dimorphism before the appearance of sex differences in histogenesis and was not expressed in the efferent duct in the testis (Kobayashi et al., 2008). Molecular events during gonadal sex differentiation in tilapia are summarized in Figure 3.

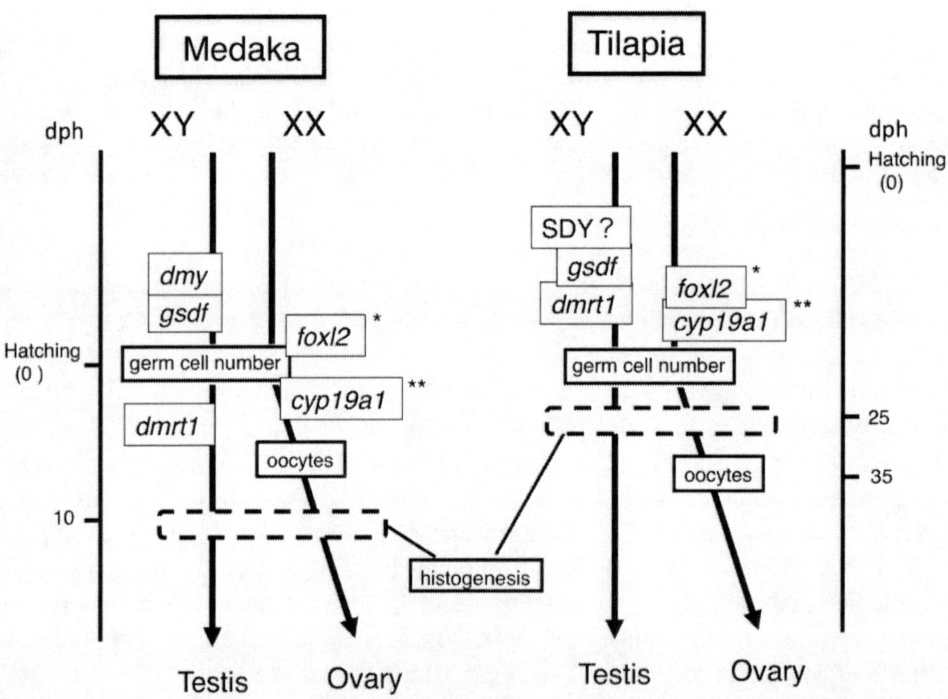

Figure 3. Gene expression during gonadal sex differentiation in medaka and tilapia. dph, days post hatching. 'germ cell number' and 'histogenesis' indicate the developmental stage that the sex difference becomes apparent in each species. * gene expressed in parenchyma/GE and some interstitial cells. ** gene expressed in some interstitial cells.

Overview of Gonadal Sex Differentiation in Medaka

Sex Differences in Germ Cell Number during Gonadal Differentiation

PGCs first appear at the early blastula stage and then migrate to the posterior end of the lateral plate mesoderm. The bilateral clusters of PGCs migrate from both sides up the dorsal surface of the hindgut and are located in the gonadal primodium, which is situated between the hindgut and the coelomic epithelium (Hamaguchi, 1982; Shinomiya et al., 2000). Initially, there are around 10 PGCs, which then undergo 2 or 3 rounds of mitotic division during migration. Finally, the gonadal primordium contains 30–40 PGCs (Hamaguchi, 1982; Kurokawa et al., 2006).

In medaka, previous reports revealed that sex differences in germ cell number were detected at the hatching time (stage 39), with the result that the germ cell number in XX embryos is higher than in XY embryos (Satoh and Egami, 1972; Quirk and Hamilton, 1973; Hamagauchi, 1982). A recent finding using genetic sex markers (*dmy/dmrt1*) clearly indicates that sex differences in germ cell numbers are detected at embryonic stage 38, after the expression of *dmy* (Kobayashi et al., 2004). In early oogenesis, gonial germ cells are divided into type I and type II cells (Saito et al., 2007). The former comprises relatively large germ cells separated by surrounded somatic cells, i.e., isolated gonial germ cells (single cysts). The latter comprises clusters of small and densely-packed germ cells delineated by somatic cells, i.e., cystic gonial germ cells (Shibata and Hamaguchi, 1989). During oogenesis, cystic gonial germ cells had similar morphology to early type B spermatogonia in teleost fish. During early gonadal sex differentiation in medaka, the germ cells in XY gonads are gonium germ cells within a single cyst (Hamaguchi, 1982; Kanamori et al., 1985). Therefore, it is suggested that sex differences in germ cell numbers during early gonadal sex differentiation depend mainly on the proliferation of cystic gonial germ cells in XX gonads (type II division). In XY gonads, in contrast, germ cells undergo several mitotic divisions in a stochastic manner during early gonadal differentiation and then become mitotically quiescent in the post hatching stage (Satoh and Egami, 1972; Hamaguchi, 1982). Gonial germ cells resume mitotic division when spermatogenesis begins around 30 to 50 dph (Kanamori et al., 1985).

Sex Differences in Histogenesis during Gonadal Differentiation

In medaka, sex differences in histogenesis during gonadal differentiation are noticed in the ovarian cavity and acinus in XX and XY gonads, respectively (Kanamori et al., 1985; Shimomiya et al., 2001). In XY gonads, the cluster of gonial germ cells surrounded by somatic cells constitutes the acinus around 5–10 dph, which is the precursor of seminiferous tubules (or seminiferous lobules) and efferent duct in the mature testis. Concerning the parenchyma/GE cells, as mentioned previously, it has been suggested that the acinus is made up of a kind of parenchyma/GE cell structure. In XX gonads, in contrast, the ovarian cavity becomes apparent in the dorsal area of the ovary. It begins as an extension of the cell mass from the medial and lateral regions, which then fuse to cover the dorsal area of the ovary.

As above-mentioned, in contrast, the ovarian cavity is formed on the ventral side in tilapia (Figure 2; Nakamura et al., 1998; Shinomiya et al., 2001).

Molecular Events during Gonadal Sex Differentiation in Medaka

Molecular events during gonadal sex differentiation in medaka are summarized in Figure 3. Sex differentiation in somatic cells becomes apparent during the embryonic period in both sexes, in addition to the expression of the sex-determining gene *dmy* (Kobayashi et al., 2004). A recent finding indicates that the gonadal soma-derived factor (*gsdf*) becomes expressed in germ cell-surrounding cells around stage 38 before hatching and is expressed at a higher level in XY gonads than in XX gonads.

Furthermore, the evidence that *gsdf* is co-expressed in germ cell-surrounding cells with *dmy* in XY gonads suggests that *gsdf* is a downstream gene of *dmy* (Shibata et al., 2010). The expression of *dmrt1* as a gene expressed specifically in males became apparent in the Sertoli cell lineage cells (parenchyma/GE cells) at 10–20 dph, e.g., in the acinus, which include the Sertoli cells and the precursors of the intratesticular efferent duct (Kobayashi et al., 2004; Suzuki et al., 2004). The expression of AMH/MIS, a member of the transforming growth factor-β (TGF-β) superfamily protein, is seen in Sertoli cells during gonadal sex differentiation in mammals. In mammals, *amh* is expressed specifically in Sertoli cells, suggesting that *amh* expression shows male differentiation in mammals. In medaka, it was reported that *amh* was expressed in the developing gonads in both sexes in a similar way to non-mammalian vertebrates, whereas *amh* demonstrated a male-dominant expression profile in developing gonads (Oreal et al., 1998; von Hofsten et al., 2005; Shiraishi et al., 2008; Ijiri et al., 2008).

Sox9 is up-regulated in the male gonads in a number of species in which it is the essential regulator of testis determination. In medaka, *sox9a* is expressed in the brain but not in the gonads and *sox9b* is expressed specifically in the gonads (Yokoi et al., 2002; Nakamura et al., 2012). However, *sox9b* is one of the orthologues of mammalian *sox9* at syntenic and expression levels. Medaka *sox9b* is also expressed in the Sertoli cell lineage cells and in the supporting cells of early oogenesis, consisting of the ovarian cradle (Nakamura et al., 2011; 2012).

Although the dynamics of germ cells in XX individuals showed defined sex differences during early oogenesis, sex differentiation in somatic cells was seen to be obscure. *foxl2,* known as a key gene for ovarian differentiation and development (Pailhoux et al., 2000; 2001; Nakamoto et al., 2006; Wang et al., 2007), is expressed in somatic cells in XX gonads from the day of hatching, suggesting that *foxl2* expression precede the differentiation of the ovarian cavity as the first morphological difference in histogenesis but not sex difference in germ cell numbers (Nakamoto et al., 2006). *cyp19a1* also had a female-specific expression pattern around 5–10 dph in some somatic cells located on the ventral side (Suzuki et al., 2004; Nakamoto et al., 2006). Together with the evidence that *foxl2* and *cyp19a1* are co-expressed in those cells, it is suggested that *foxl2* is involved in the initiation of *cyp19a1* expression during early ovarian differentiation (Nakamoto et al., 2006; Nakamura et al., 2008). As it had been pointed out that not all *foxl2*-positive cells eventually expressed *cyp19a1* in all cases, it was suggested that *foxl2* alone is insufficient for *cyp19a1* expression, indicating that another specific factor(s) might be required (Nakamoto et al., 2006).

Along with the production of steroid hormones, *nr5a1* was expressed in somatic cells surrounding the germ cells in both sexes at stage 36, in which the gonadal primordia were completed. Following this, *nr5a1*was expressed in interstitial cells, which co-expressed *3b-hsd* in XY gonads at 20 dph.

In XX gonads, in contrast, *nr5a1*-expressing cells were also located in interstitial cells, which co-expressed *3b-hsd* at 10 to 20 dph, in addition to the somatic cells surrounding the germ cells, including the pregranulosa cells (Nakamoto et al., 2012).

Recent findings in medaka have demonstrated that *star*, *hsd3b*, and *cyp11a1* were all co-expressed in the identical interstitial cells. The onset of *star* mRNA expression, however, occurred after the expression of *hsd3b* and *cyp11a1* had been initiated (Nakamoto et al., 2010; 2012). In XY medaka, in contrast, *star* is located in interstitial cells around 25 dph, when the appearance of Leydig cells is seen in XY gonads (Kanamori et al., 1985).

Sex Reversal of Tilapia and Medaka

In teleost fishes, sex reversal is well known to occur artificially and naturally. Even in gonochoristic fish species, sex reversal is induced artificially. The 2 species–tilapia and medaka showed gonochoristic sex differentiation as above-mentioned.

In both species, many investigators have reported that sex reversal can be induced easily under artificial conditions, e.g., with sex steroid hormones, temperature, etc (Nakamura et al., 1998; Baroiller et al., 2009). In medaka, the sex-determining gene, *dmy*, determines the sex. Alternatively, the sex of medaka is easily distinguished genetically by the presence or absence of *dmy* (Matsuda et al., 2002; Nanda et al., 2002; Kobayashi et al., 2004).

However, in tilapia, the sex-determining gene has not yet been identified and instead, YY supermales and XX sex-reversed males are able to produce an all-male or all-female population, respectively (Kobayashi et al., 2000; 2008; Kobayashi and Nagahama, 2009).

Therefore, medaka and tilapia are good experimental models for the investigation of mechanisms underlying sex differentiation and sex reversal.

Sex Reversal by Steroid Hormones and Inhibitors

In tilapia, administration of androgens such as 17α-methyltestosterone induced easily masculinization in XX fry (Nakamura et al., 1998; Kobayashi et al., 2008). The treatments of XX fry with the non-steroidal and steroidal aromatase inhibitors (AIs), fadrozole and exemestane, caused complete sex reversal in functional males, avoiding simultaneous estrogen treatment (Nakamura et al., 1999; Afonso et al., 2001; Kwon et al., 2002; Rukusana et al., 2010). Although XY sex reversal by estrogen was not induced as easily as XX sex reversal by androgens and AIs, the treatment with 17α-ethynylestradiol from 4 to 6 dph caused complete XY sex reversal with germ cell proliferation in similar to XX gonads (Kobayashi et al. 2003; 2008).

In medaka, Yamamoto (1953) first reported that sex reversal was induced artificially in genotypic medaka; estrogen treatment in XY males resulted in ovarian development and XX females treated with androgen developed into testes.

Recently, a new method was reported where the immersion of fertilized XY eggs into water containing estrogen, from fertilization to hatching, caused sex reversal (Iwamatsu et al., 2005).

Although this treatment induced XY sex reversal, neither *dmy* expression nor the number of germ cells was affected during early development (Sholz et al., 2003; Suzuki et al., 2005). This suggests that in medaka, early gonadal sex differentiation is not affected by exogenous estrogen.

Sex Reversal by Temperature

In medaka, exposure of embryos to high temperature resulted in female-to-male sex reversal (Sato et al., 2005; Hattori et al., 2007). In the Hd-rR strain, when fertilized eggs were exposed to temperature between 17 and 34°C up until hatching, complete XX sex reversal occurred in accordance with the increasing temperature at 34°C (Hattori et al., 2007).

Interestingly, the effects of temperature were not observed when embryos were exposed to high temperature after stage 36, suggesting that the process of gonadal sex differentiation affected by high temperature is limited to the period before the morphological sex differentiation stage.

In addition, high temperature treatment induced *dmrt1* expression in XX fry after stage 36 (Hattori et al., 2007), whereas in normal XY fry, *dmrt1* became expressed in the XY gonads after 10 to 20 dph (Kobayashi et al., 2004).

In tilapia, elevated temperatures (32–34°C) caused an increase in the number of males in XX progenies and a decrease in the number of females in YY and XY progenies (Baroiller et al., 1995; Abucay et al., 1999; Kwon et al., 2002). It was suggested that the period of thermosensitivity ranged from 10 to 20 or 40 dpf (Baroiller et al., 2009). When XX fry were exposed to high masculinizing temperatures, the expression of *cyp19a1* and *foxl2* mRNA was suppressed (d'Cotta et al., 2001; 2008; Baroiller et al., 2008).

Furthermore, treatment with high masculinizing temperatures caused a rapid upregulation of *dmrt1* in XX fry (d'Cotta et al., 2008). Together with the evidence in medaka, high masculinizing temperatures may cause female-to-male sex reversal through the induction of *dmrt1*.

The Mechanisms of Sex Differentiation: Comparison between Tilapia and Medaka

As mentioned above, it appears that gonadal sex differentiation occurs in a similar manner in both species, though the time scale for gonadal sex differentiation is different (Figure 3). First, a sex-determining gene switches on the male differentiation pathway. In XX embryos, then, gonial germ cells within a single cyst differentiate into cystic gonial germ cells and proliferate, resulting in the gender differences in germ cell numbers in both sexes, which are seen as the first morphological sex differences.

Following this step, sex differences in histogenesis occur, e.g., development of the parenchyma/GE and the ovarian cavity in XY and XX gonads, respectively. Although the molecules involved in the morphogenesis during sex differentiation are similar in both species, it appears that the details differ. Therefore, we will compare the evidence from tilapia and medaka and discuss the similarities and differences.

Estrogen and Gonadal Sex Differentiation

As mentioned previously, in XX fry of tilapia, the expression of steroidogenic enzymes required for estrogen production precedes any morphological gonadal sex differentiation. Furthermore, treatment of AI in XX fry during the gonial germ cells proliferative period caused complete sex reversal, resulting functional males (Nakamura et al., 1999; Afonso et al., 2001; Kwon et al., 2002; Rukusana et al., 2010).

It was also reported that fadrozole treatment for 3 days was sufficient to induce XX sex reversal (Kobayashi et al., 2008). As mentioned previously, it is suggested that estradiol-17β plays a critical role in ovarian differentiation in tilapia. It is also suggested that the concentration of estrogens versus androgens is important in the differentiating gonads (Baroiller and d'Cotta, 2001; d'Cotta et al., 2001). Our recent studies suggest an important role of *foxl2* in the transcriptional regulation of *cyp19a1* but not of *cyp19a2*.

Indeed, our in vitro transcription assay indicates that *foxl2* up-regulated the aromatase expression via specific binding sites for *foxl2* in cooperation with *nr5a1*. When a dominant negative mutant construct of *foxl2* for aromatase transcription was overexpressed in XX fertilized eggs, functional XX males were obtained (Wang et al., 2007). Therefore, it is concluded that *foxl2* regulates ovarian differentiation and development via aromatase (*cyp19a1*) expression and estrogen production in tilapia. Taken together, these results suggest that endogenous estrogens are required for early ovarian differentiation in tilapia.

In addition, *foxl2* is expressed not only in *cyp19a1*-expressing cells, but also in germ cell-surrounding cells during gonadal differentiation (Wang et al., 2007). The localization of *foxl2* mRNA in germ cell-surrounding cells was also noted in medaka (Nakamoto et al., 2006). Since *cyp19a1* is not expressed in oogonia-surrounding cells during gonadal differentiation in these fish species, the possibility remains that *foxl2* is also involved in ovarian differentiation via another pathway, in addition to the regulation of estrogen synthesis.

In other teleosts, it remains unclear whether estrogens are required for early ovarian differentiation. In several teleosts, treatment with an AI caused sex reversal from female to male (Pifferer et al., 1994; Kitano et al., 2000). In medaka, the expression of *star* and all steroidogenic enzymes involved in the estrogenic pathway (*cyp11a1*, *cyp17*, *hsd3b*, *hsd17b*,and *cyp19a1*) are essential for the production of estrogen (Suzuki et al., 2004; Nakamura et al., 2009; Nakamoto et al., 2010).

Previous reports suggested that these enzymes become expressed in the interstitial cells located on the ventral side of the ovary during ovarian development between 5 and 10 dph. In medaka, however, it was reported that expression of *star* became apparent in the interstitial cells in the XX gonads around 20 to 25 dph, suggesting that the onset of *star* expression is delayed compared to the steroidogenic enzymes. As estrogen production depends on *star* expression, it may be impossible to synthesize estrogen in XX medaka until 25 dph (Nakamoto et al., 2012).

Furthermore, XX fry of the *scl* mutant, which lack *cyp17a1*, resulting in no production of estrogen, showed early ovarian differentiation until vitellogenesis. In a similar way to tilapia, it was also reported that *cyp17a2* was cloned and localized in the head kidney, and gonads in medaka (Zhou et al., 2007a, b). XY fry of the *scl* mutant medaka also did not show any male sexual characters depending on androgens. This suggests that, at least in the case of early ovarian differentiation, newly synthesized estrogen is not required in contrast to tilapia (Sato et al., 2008).

Since fertilized medaka eggs contained only a small amount of estrogens (Iwamatsu et al., 2005), however, there is a possibility that a small amount of endogenous estrogen is sufficient for early ovarian differentiation in the absence of the sex-determining gene *dmy*.

In tilapia and medaka, as mentioned above, it seems that the requirements of estrogens for early ovarian differentiation are different. For late ovarian differentiation, however, it appears that estrogens are required in both species. To be precise, the *scl* mutant medaka (XX) shows female-to-male sex reversal after the formation of oocytes (Sato et al., 2008).

In addition, in germ cell-deficient XX medaka, female-specific somatic differentiation marker genes, including *cyp19a1,* are expressed until 20 dph, followed by the expression of male differentiation genes, including *dmrt1*, after down regulation of the female-specific genes (Kurokawa et al., 2006).

The Male Pathway

In medaka, sex differentiation in somatic cells becomes apparent during the embryonic period in both sexes, in addition to the expression of the sex-determining gene, *dmy* (Kobayashi et al., 2004). A recent finding indicates that *gsdf* becomes expressed in germ cell-surrounding cells around stage 38 before hatching, and shows the dominant expression in XY gonads. Furthermore, co-localization of *gsdf* and *dmy* in germ cell-surrounding cells in XY gonads suggests that *gsdf* is a downstream gene of *dmy* (Shibata et al., 2010). Recently, in a subspecies of medaka, *O. luzonensis*, it was reported that *gsdf* was located on the sex chromosome and that *GsdfY* acts as the sex-determining gene. Furthermore, the over expression of *GsdfY* in the XX fry of *O. luzonensis* and *O. latipes* caused XX male differentiation (Myosho et al., 2012).

This suggests that *gsdf* acts as a testis differentiation factor in *dmy* male differentiation cascade in *O. latipes*. The expression of *dmrt1* as a gene expressed specifically in males became apparent in the Sertoli cell lineage cells (such as in the acinus, which include Sertoli cells and the precursors of the intratesticular efferent duct) after the appearance of acinus (Kobayashi et al., 2004; Suzuki et al., 2004).

A recent report indicated that *dmrt1* is required for testis differentiation by the analysis of the *dmrt1* mutant medaka (Masuyama et al., 2011). Thus, sequential evidence strongly suggests that the *dmy/gsdf/dmrt1* cascade plays important roles in testis differentiation in the medaka, *O. latipes*, whereas it remains unclear whether *dmrt1* is part of the *gsdf* downstream cascade. In tilapia, much information has been reported on the expression profiles of genes involved in sex differentiation (Kobayashi and Nagahama, 2009).

Unfortunately, functional analyses of these genes have not been carried out, except for *foxl2*. A recent finding suggested that *gsdf* was involved in testis differentiation as an upstream gene to *dmrt1* (Kaneko et al., unpublished). This suggests that *gsdf/dmrt1* plays important roles in testis differentiation in medaka and tilapia.

Although *gsdf* and *dmrt1* are specifically expressed in Sertoli cell-lineage cells (parenchyma/GE cells) and are key players in testis differentiation, to date, the functions of both genes during testis differentiation remain unclear.

Conclusion

The genetic mechanisms triggering sex determination appear to be diverse in non-mammalian vertebrates. In teleosts, the sex-determining gene, *dmy* (discovered in the medaka, *Oryzias latipes)* exists in only 2 species (Matsuda et al., 2002; 2003). Nonetheless, morphological differentiation and development of the gonads in all vertebrates appear to have been conserved throughout evolution. Many genes involved in gonadal sex differentiation in mammals, e.g., *dmrt1* and *foxl2*, are conserved in other vertebrates. Recently, several reports suggest that in other medaka, e.g., *Oryzias dancena, Oryzias luzonensis*, and *Oryzias minutillus*, a gene other than *dmy* is a candidate for the sex-determining gene (Takehana et al., 2007, Nagai et al., 2008; Myosho et al., 2012). In *O. luzonensis*, *GsdfY* has been identified as the sex-determining gene. Interestingly, in *O. latipes*, it has been suggested that *gsdf* is part of a downstream cascade of *dmy* and that the increase in *GsdfY* in *O. latipes* XX embryos resulted in male differentiation (Shibata et al., 2010; Myosho et al., 2012). In *O. latipes*, *dmrt1* is also required for testis differentiation. Together with the evidence for *gsdf* and *dmrt1* in tilapia, it may be suggested that, at least in teleost fish, testis differentiation progresses via the *gsdf/dmrt1* cascade.

In morphogenesis during gonadal sex differentiation in all vertebrates, original parenchyma/GE cells derived from the germinal epithelium develop into testicular structures, e.g., the testis cord in mammals and the pre-seminiferous tubule structure in amphibians and teleost fish; however, they fail to develop in the ovary. Since *dmrt1* and *foxl2* are expressed in the germ cell-surrounding cell lineages in males and females, respectively, the clarification of the relationship between morphogenesis and these genes products during gonadal sex differentiation is essential for understanding the mechanisms of gonadal sex differentiation. In medaka, *dmrt1*, as a downstream element of the *dmy/gsdf* male differentiation cascade, can induce a bipotential gonad for testis. Since *dmy/gsdf/dmrt1* is expressed in the Sertoli cell lineage (Kobayashi et al., 2004; Shibata et al., 2010), it appears that the function is the proper differentiation and development of the Sertoli cells. As mentioned previously, in tilapia, *foxl2* plays important roles in ovarian differentiation. Together, these facts may suggest that the balance of sexual dimorphism, i.e., sex differentiation, in the parenchyma/GE cells is essential for gonadal sex differentiation, as determined by the expression of *dmrt1* and *foxl2* in the testis and ovary, respectively (male pathway genes vs. female pathway genes). To understand the mechanisms of gonadal sex differentiation in vertebrates, including the teleosts, further study will be necessary to clarify the molecular mechanisms underlying the parenchyma/GE cell differentiation and development together with analyses of the gene cascade from the upstream genes, *dmrt1* and *foxl2,* expressed in germ cell-surrounding cell lineages.

References

Afonso, L. O., Wassermann, G. J. and de Oliveila, R. T. (2001). Sex reversal in Nile tilapia (*Oreochromis niloticus*) using a nonsteroidal aromatase inhibitor. *J. Exp. Zool.*, 290, 177-181.

Atz, J. W. (1964). Intersexuality in fishes. In: *Intersexuality in Vertebrates Including Man.*

Baroiller, J. F. and d'Cotta, H. (2001). Environment and sex determination in farmed fish. *Comp. Biochem. Physiol. C,* 130, 399-409.

Baroiller, J. F., d'Cotta, H. and Saillant, E. (2009). Environmental effects on fish sex determination and differentiation. *Sex Dev.,* 3, 118-135.

Capel, B. (2000). The battle of the sexes. *Mech. Dev.,* 92, 89-103.

Carrasco, L. A. P., Penman, D. J., Villalos, S. A., and Bromage, N. (1999). The effects of oral administration with 17α-methyltestosterone on chromosomal synapsis in *Oreochromisniloticus* (Pisces, Cichlidae). *Mut. Res.,* 430, 87-98.

D'Cotta, H., Fostier, A., Guigen, Y., Govooroun, M., and Baroiller, J. F. (2001) Aromatase plays a key role during normal and temperature-induced sex differentiation of tilapia *Oreochromis niloticus*. *Mol. Reprod. Dev.,* 59, 265-276.

Devlin, R. H. and Nagahama, Y. (2002). Sex determination and sex differentiation in fish: an overview of genetic, physiological, and environmental influences. *Aquaculture,* 208, 191-364.

Hamaguchi, S. (1982). A light- and electron-microscopic study on the migration of primodial germ cells in the teleost, *Oryzias latipes*. *Cell Tissue Res.,* 227, 139-151.

Hattori, R. S., Gould, R. J., Fujioka, T., Saito, T., Kurita, J., Strüssmann, C. A., Yokota, M., Watanabe, S. (2007). Temperature dependent sex determination in Hd-rR medaka *Oryzias latipes*: gender sensitivity, thermal threshold, critical period, and DMRT1 expression profile. *Sex Dev.,* 1, 138-146.

Hattori, R. S., Murai, Y., Oura, M., Masuda, S., Majhi, S. K., Sakamoto, T., Fernandino, J. I., Somoza, G. M., Ykota, M., and Strüssmann, C. A. (2012). A Y-linked anti-Müllerian hormone duplication takes over a critical role in sex determination. *Proc. Natl. Acad. Sci. US,* 109, 2955-2959.

Iwasawa, H. and Yamaguchi, K. (1984). Ultrastructural study of gonadal development in *Xenopus laevis*. *Zool. Sci.,* 1, 591-600.

Iwamatsu, T., Kobayashi, H., Hamaguchi, S., Sagegami, R., and Shuo, T. (2005). Estradiol-17ß content in developing eggs and induced sex reversal of the medaka (*Oryzias latipes*). *J. Exp. Zool.,* 303,161-167.

Ijiri, S., Kaneko, H., Kobayashi, T., Wang, D. S., Sakai, F., Paul-Prasanth, B., Nakamura, M., and Nagahama, Y. (2008). Sexual dimorphic expression of genes in gonads during early differentiation of a teleost fish, the Nile tilapia, *Oreochromis niloticus*. *Biol. Reprod.,* 78, 333-341.

Kanamori, A., Nagahama, Y. and Egami, N. (1985). Development of the tissue architecture in the gonads of the medaka *Oryizias latipes*. *Zool. Sci.,* 2, 695-706.

Kitano, T., Takamune, K., Nagahama, Y., and Abe, S. I. (2000). Aromatase inhibitor and 17α-methyltestosterone cause sex-reversal from genetical females to phenotypic males and suppression of P450 aromatase gene expression in Japanese flounder (*Paralichthys olivaceus*). *Mol. Reprod. Dev.,* 56, 1-5.

Kobayashi, T., Kajiura-Kobayashi, H. and Nagahama, Y. (2000). Differential expression of *vasa* homologue gene in the germ cells during oogenesis and spermatogenesis in a teleost fish, tilapia, *Oreochromis niloticus*. *Mech. Develop.,* 99, 139-142.

Kobayashi, T., Kajiura-Kobayashi, H. and Nagahama, Y. (2002). Two isoforms of *vasa* homologs in a teleost fish: their differential expression during germ cell differentiation. *Mech. Develop.,* 111, 167-171.

Kobayashi, T., Kajiura-Kobayashi, H. and Nagahama, Y. (2003). Induction of XY sex reversal by estrogen involves altered gene expression in a teleost, tilapia. *Cytogenet. Genome Res.,* 101, 289-294.

Kobayashi, T., Matsuda, M., Kajiura-kobayashi, H., Suzuki, A., Saito, N., Nakamoto, M., Shibata, N., and Nagahama, Y. (2004). Two DM domain genes, *DMY* and *DMRT1*, involved in testicular differentiation and development in the medaka, *Oryzias latipes*. *Dev. Dyn.,* 231, 518-526.

Kobayashi, T., Kajiura-Kobayashi, H., Guan, G., and Nagahama, Y. (2008). Sexual dimorphic expression of DMRT1 and Sox9a during gonadal differentiation and hormone-induced sex reversal in the teleost fish Nile tilapia (*Oreochromis niloticus*). *Dev. Dyn.,* 237, 297-306.

Kobayashi, T. and Nagahama, Y. (2009). Molecular aspects of gonadal differentiation in a teleost fish, the Nile tilapia. *Sex Dev.,* 3, 108-117.

Kumakura, M. and Iwasawa, H. (1992). Ultrastructural observations on the development of the gonad in two Species of urodelan amphibians, *Hynobius nigrescens* and *Cynops pyrrhogaster*. *Sci. Rep. Niigata Univ., Ser. D* (Biology), No. 29, 13-26.

Kurokawa, H., Aoki, Y., Nakamura, S. Ebe, Y., Kobayashi, D., and Tanaka, M. (2006). Time-lapse analysis reveals different modes of primordial germ cell migration in the medaka *Oryzias latipes*. *Dev. Growth Differ.,* 48, 209-221.

Kwon, J. Y., McAndrew, B. J. and Penman, D. J. (2002) Treatment with an aromatase inhibitor suppresses high-temperature feminization of genetic male (YY) Nile tilapia. *J. Fish Biol.*, 60, 625-636.

Lepori, N. G. (1980). Sex differentiation, hermaphroditism and intersexuality in Vertebrates including Man. Piccin. Medical Books, Padua.

Merchant-Larios, H. and Villalpando, I. (1981). Ultrastructural events during early gonadal development in *Rana pipiens* and *Xenopus laevis*. *Anat. Rec.*, 199, 349-360.

Matsuda, M., Nagahama, Y., Shinomiya, A., Sato, T., Matsuda, C., Kobayashi, T., Morrey, C. E., Shibata, N., Asakawa, S., Shimizu, N., Hori, H., Hamaguchi, S., and Sakaizumi, M. (2002). DMY is a Y-specific DM-domain gene required for male development in the medaka fish. *Nature,* 417, 559-563.

Matsuda, M., Sato, T., Toyazaki, Y., Nagahama, Y., Hamaguchi, S., and Sakaizumi, M. (2003). *Oryzias curvinotus* has DMY, a gene that is required for male development in the medaka, *O. latipes*. *Zool. Sci.,* 20, 159-161.

Merchant-Larios, H. and Villalpando, I. (1981). Ultrastructural events during gonadal development in *Rana pipiens* and *Xenopus laevis*. *Anat. Rec.*, 199, 349-360.

Myosho, T., Otake, H., Masuyama, H., Matsuda, M., Kuroki, Y., Fujiyama, A., Naruse, K., Hamaguchi, S., and Sakaizumi, M. (2012). Tracing the emergence of a novel sex-determining gene in medaka, *Oryzias luzonensis*. *Genetics,* 191, 163-170.

Nagai, T., Takehana, Y., Hamaguchi, S., and Sakaizumi, M. (2008). Identification of the sex-determining locus in the Thai medaka, *Oryzias minutillus*. *Cytogenet. Genome Res.*, 121, 137-142.

Nakamoto, M., Matsuda, M., Wang, D. S., Nagahama, Y., and Shibata, N. (2006). Molecular cloning and analysis of gonadal expression of Foxl2 in the medaka, *Oryzias latipes*. *Biochem. Biophys. Res. Commun.*, 344, 353–361.

Nakamoto, M., Fukasawa, M., Orii, S., Shimamori, K., Maeda, T., Suzuki, A., Matsuda, M., Kobayashi, T., Nagahama, Y., and Shibata, N. (2010). Cloning and expression of medaka cholesterol side chain cleavage cytochrome P450 during gonadal development. *Dev. Growth Differ.*, 52, 385-395.

Nakamoto, M., Fukasawa, M., Tanaka, S., Shimamori, K., Suzuki, A., Matsuda, M., Kobayashi, T., Nagahama, Y., and Shibata, N. (2012). Expression of 3ß-hydoxysteroid dehydrogenae (hsd3b), star and ad4bp/sf-1 during gonadal development in medaka (*Oryzias latipes*). *Gen. Comp. Endocrinol.*, 176, 222-230.

Nakamura, M., Kobayashi, T., Chang, X. T., and Nagahama, Y. (1998). Gonadal sex differentiation in teleost fish. *J. Exp. Zool.*, 281, 362-372.

Nakamura, M., Kobayashi, T., Yoshiuram, Y., and Nagahama, Y. (1999). Role of endogenous steroid hormones on gonadal sex differentiation in fish. In: *Proceedings of the 6th International Symposium on the Reproductive Physiology of Fish*. eds. by Norberg, B., Kjesbu, O. S., Taranger, G. L., Andersson, E., and Stefansson, S. O. Bergen, pp. 247-249.

Nakamura, S., Kurokawa, H., Asakawa, S., Shimizu, N., and Tanaka, M. (2009). Two distinct types of theca cells in the medaka gonad: germ cell-dependent maintenance of cyp19a1-expressing theca cells. *Dev. Dy.*, 238, 2652-2657.

Nakamura, S., Kobayashi, K., Nishimura, T., Higashijima, S., and Tanaka, M. (2010). Identification of germ line stem cells in the ovary of the teleost medaka. *Science*, 328, 1561-1563.

Nakamura, S., Watakabe, I., Nishimura, T., Toyoda, A., Taniguchi, Y., and Tanaka, M. (2012). Analysis of medaka sox9 orthologue reveals a conserved role in germ cell maintenance. *PLoS One*, 7, e29982.

Nanda, I., Kondo, M., Hornung, U., Asakawa, S., Winkler, C., Shimizu, A., Shan, Z., Haaf, T., Shimizu, N., Shima, A., Schmid, M., and Schartl, M. (2002). A duplicated copy of DMRT1 in the sex-determining region of the Y chromosome of the medaka, *Oryzias latipes*. *Proc. Natl. Acad. Sci. US*, 99, 11778-11783.

Pailhoux, E., Vigier, B., Chaffaux, S., Servel, N., Taourit, S., Furet, J. P., Fellous, M., Grosclaude,F., Cribiu, E. P., Cotinot, C., and Vaiman, D. (2001). A 11.7-kb deletion triggers intersexuality and polledness in goats. *Nat. Genet.*, 29, 453–458.

Pailhoux, E., Vigier, B., Vaiman, D., Servel, N., Chaffaux, S., Cribiu, E. P., and Cotinot C. (2002). Ontogenesis of female-to-male sex-reversal in XX polled goats. *Dev. Dyn.*, 224, 39–50.

Piferrer, F., Baker, I. J. and Donaldson, E. M. (1994). Effects of natural, synthetic, aromatizable, and nonaromatizable androgens in inducing male sex differentiation in genotypic female chinook salmon (*Oncorhynchus tshawytscha*). *Gen. Comp. Endocrinol.*, 91, 59-65.

Oreal, E., Pieau, C., Tattei, M-G., Josso, N., Picard, J-Y., Carre-Eusebe, D., and Marge, S. (1998). Early expression of *AMH* in chicken embryonic gonads precedes testicular *SOX9* expression. *Dev. Dyn.*, 238, 2652-2657.

Quirk, J. G. and Hamilton, J. B. (1973). Number of germ cells in known male and female genotypes of vertebrate embryo (*Oryzias latipes*). *Science*, 180, 963-964.

Reinboth, R. (1970). Intersexuality in fishes. *Mem. Soc. Endocrinol.,* 18, 515-543.

Ruksana, S., Pandit, N. P. and Nakamura, M. (2010). Efficacy of exemestane, a new generation of aromatase inhibitor, on sex differentiation in a gonochoristic fish. *Comp. Biochem. Physiol. C Toxicol. Pharmaco.,* 152, 69-74.

Saito, D., Morinaga, C., Aoki, Y., Nakamura, S., Mitani, H., Furutani-Seiki, M., Kondoh, H., and Tanaka, M. (2007). Proliferation of germ cells during gonadal sex differentiation in medaka: Insights from germ cell-depleted mutant *zenzai*. *Dev. Biol.,* 310, 280-290.

Saito, D. and Tanaka, M. (2008). Comparative aspects of gonadal sex differentiation in medaka: A conserved roles of developing oocytes in sexual canalization. *Sex Dev.,* 3, 99-107.

Satoh, N. and Egami, N. (1972). Sex differentiation of germ cells in the teleost, *Oryzias latipes*, during normal embryonic development. *J. Embryol. Exp. Morphol.,* 28, 385-395.

Satoh, N. (1974). An ultrastructural study of sex differentiation in the teleosts *Oryzias latipes. J. Embryol. Exp. Morph.,* 32, 195-215.

Sato, T., Endo, T., Yamahira, K., Hamaguchi, S., and Sakaizumi, M. (2005). Induction of female-to-male sex reversal by high temperature treatment in Medaka, *Oryzias latipes. Zool. Sci.,* 22, 985-989.

Sato, T., Suzuki. A., Shibata, N., Sakaizumi, M., and Hamaguchi S. (2008). The novel mutant scl of the medaka fish, *Oryzias latipes*, shows no secondary characters. *Zool. Sci.,* 25, 299-306.

Shibata, N. and Hamaguchi, S. (1988). Evidence for the sexual bipotentiality of spermatogonia in the fish, *Oryzias latipes. J. Exp. Zool.,* 245, 71-77.

Shibata, Y., Paul-Prasanth, B., Suzuki, A., Usami, T., Nakamoto, M., Matsuda, M., and Nagahama, Y. (2010). Expression of gonadal soma derived factor (GSDF) is spatially and temporally correlated with early testicular differentiation in medaka. *Gene. Expr. Patterns*, 10, 283-289.

Shiraishi, E., Yoshinaga, N., Miura, T., Yokoi, H., Wakamatsu, Y., Abe, S-I., and Kitano, T. (2008). Müllerian inhibiting substance is required for germ cell proliferation during early gonadal differentiation in medaka (*Oryzias latipes*). *Endocrinology,* 149, 1813-1819.

Shinomiya, A., Tanaka, M., Kobayashi, T., Nagahama, Y., and Hamaguchi, S. (2000). The vasa-like gene, olvas, identifies the migration path of primordial germ cells during embryonic body formation stage in the medaka, *Oryzias latipes. Dev. Growth Differ.,* 42, 317-326.

Shinomiya, A., Hamaguchi, S. and Shibata, N. (2001). Sexual differentiation of germ cell deficient gonads in the medaka, *Oryzias latipes. J. Exp. Zool.,* 290, 402-410.

Scholz, S., Rösler, S., Schäffer, M., Hornung, U., Scartl, M., and Gutzeit, H. O. (2003). Hormonal induction and stability of monosex populations in the medaka (Oryzias latipes): expression of sex-specific marker genes. *Biol. Reprod.,* 69, 673-678.

Suzuki, A., Tanaka, M., Shibata, N., and Nagahama, Y. (2004). Expression of aromatase mRNA and effects of aromatase inhibitor during ovarian development in the medaka, *Oryzias latipes. J. Exp. Zool. A,* 301, 266-273.

Suzuki, A., Nakamoto, M., Kato, Y., and Shibata, N. (2005). Effects of estradiol-17ß on germ cell proliferation and DMY expression during early sexual differentiation of the medaka Oryzias latipes. *Zool. Sci.,* 22, 791-796.

Takehana, Y., Naruse, K., Hamaguchi, S., and Sakaizumi, M. (2007). Evolution of ZZ/ZW and XX/XY sex-determination systems in the closely related medaka species, *Oryzias hubbsi* and *O. dancena*. *Chromosoma,* 116, 463-470.

Tanimura, A. and Iwasawa, H. (1988). Ultrastructural observations on the origin and differentiation of somatic cells during gonadal development in the frog *Rana nigromaculata*. *Dev. Growth Differ.,* 30, 681-691.

Tanimura, A. and Iwasawa, H. (1989). Origin of somatic cells and histogenesis in the primordial gland of the Japanse tree frog *Rhacophorus arboreus*. *Anat. Embryol.* (Berl.), 180, 165-173.

Von Hofsten, J., Larsson, A. Olsson, P. E. (2005). Novel steroidogenic factor (*ff1d*) is coexpressed with anti-Müllerian hormone (*AMH*) in zebrafish. *Dev. Dyn.,* 233, 505-604.

Wang, D. S., Kobayashi, T., Zou, L. Y., and Nagahama, Y. (2004). Molecular cloning and gene expression of *Foxl2* in the Nile tilapia, *Oreochromis niloticus*. *Biochem. Biophys. Res. Commun.,* 320, 83-89.

Wang, D. S., Kobayashi, T., Zhou, L. Y., Paul-Prasanth, B., Ijiri, S., Sakai, F., Okubo, K., Morohashi, K. I., and Nagahama, Y. (2007). Foxl2 up-regulates aromatase gene transcription female-specifically by binding to the promoter as well as interacting with Ad4BP/SF-1. *Mol. Endocrinol.,* 21, 712-725.

Witschi, E. (1957). The inductor theory of sex differentiation. J. Fac. Hokkaido. *Univ. Ser. VI Zool.,* 13, 428-439.

Yamamoto, T. (1953). Artifically induced sex-reversal in genotypic males of the medaka (*Oryzias latipes*). *J. Exp. Zool.,* 123, 571-594.

Yamamoto, T. (1969). Sex differentiation. In: *Fish Physiology*, Vol. III. eds. by Hoar, W. S. and Randall, D. J. Academic Press, New York, pp. 117-175.

Yokoi, H., Kobayashi, T., Tanaka, M., Nagahama, Y., Wakamatsu, Y., Takeda, H., Araki, K., Morohashi, K., and Ozato, K. (2002). Sox9 in a teleost fish, medaka (*Oryzias latipes*): evidence for diversified function of sox9 in gonad differentiation. *Mol. Reprod. Dev.,* 63, 5-16.

Yoshimoto, S. and Ito, M. (2011). A ZZ/ZW-type sex determination in *Xenopus laevis*. *FEBS J.*, 278, 1020-1026.

Zhou, L. Y., Wang, D. S., Kobayashi, T., Yano, A., Paul-Prasanth, B., Suzuki, A., Sakai, F., and Nagahama, Y. (2007a). A novel type of P450c17 lacking the lyase activity is responsible for C21-steroid biosynthesis in the fish ovary and head kidney. *Endocrinology*, 148, 428-491.

Zhou, L. Y., Wang, D. S., Shibata, Y., Paul-Prasanth, B., Suzuki, A., and Nagahama, Y. (2007b). Characterization, expression and transcriptional regulation of P450c17-I and –II in the medaka, *Oryzias latipes*. *Biochem. Biophys. Res. Commun.,* 362, 619-625.

In: Sexual Plasticity and Gametogenesis in Fishes
Editor: Balasubramanian Senthilkumaran
ISBN: 978-1-62618-848-8

Chapter XIV

Physiological and Endocrinological Mechanisms of Sex Change in the Grouper

Yasuhisa Kobayashi, [1,2,3,*] ***Ryosuke Murata*** [1,4]
and Masaru Nakamura [1,5]

[1]Tropical Biosphere Research Center, Sesoko Station, University of the Ryukyus, Sesoko, Motobu-cho, Kunigami-gun, Okinawa, Japan
[2]Okinawa Institute of Advanced Science, Suzaki, Uruma, Okinawa Japan.
[3]Ushimado Marine Laboratory, Okayama University, Kashino, Ushimado, Setouchi, Japan
[4]Shimane Aquarium, Kushirocho Hamada-shi, Shimane, Japan
[5] Okinawa Churashima Foundation, Ishikawa, Motobu-cho, Kunigami-gun, Okinawa, Japan

Abstract

Groupers of the genus *Epinephelus* are one of the most important aquaculture species in the world. The sexuality of groupers is almost protogynous hermaphrodite. Since the gonadal re-constriction occurred in adult, grouper provide a good experiment model to investigate the sexual differentiation in vertebrate. Due to the economical and biological importance of groupers, physiological and endocrinological data are eagerly needed for improvement aquaculture. However, the endocrinological mechanism of sex change is not clearly understood. This review addresses the physiological and endocrinological mechanisms and also explains the process of sex differentiation in groupers. Key regulatory factors underlying sex change has been identified to date include sex steroids and follicle stimulating hormone (FSH).

Keywords: Grouper, Sex change, Protogynous, estrogen, androgen, steroid, gonadotorpin

---

[*] E-mail: yasu_cob@amber.plala.or.jp.

Introduction

Sex determination leads to a binary fate choice of the sexual characteristics in an organism. In almost all vertebrates, sexual reproduction requires two sexes (males and females) to maintain variation and survival. Hence, sex determination and differentiation, which develop either into ovary or testis, is one of the most important processes for species survival. However, sex-determination mechanisms vary greatly among taxa (Desjardins and Fernald, 2009; Devlin and Nagahama, 2002; Sandra and Norma, 2010).

With over 30,000 species, teleost fishes are the largest group of vertebrates (Nelson, 2006). Fishes exhibit a remarkable variety of sexuality (Devlin and Nagahama, 2002). The sexuality can be described either gonochorism or hermaphroditism; and the details are given below (de Mitcheson and Liu, 2008). The only known exception to this is unisexual species, such as *Poecillia formosa* (Dries, 2003). Many teleosts are belonging to gonochoristic, wherein individuals develop as males or females only, and retain the same sex throughout their life span. In many gonochoristic fishes, sex determination is genetic, i.e., males and females have different alleles, or even different genes, that specify their sexual morphology (Devlin and Nagahama, 2002). In other cases, sex is determined by environmental variables, such as temperature, pH, social condition and salinity (Baroiller et al., 2009). Alternatively, many teleosts exhibit natural hermaphroditism, whereby an individual changes from one sex to the other during adulthood. This unique feature of sexuality is limited only to fish among the vertebrates. Several different types of hermaphrodites have been described (de Mitcheson and Liu, 2008; Devlin and Nagahama, 2002; Frisch, 2004). Some teleost have gonads containing both mature ovaries and testes (synchronous hermaphroditism). In sequential hermaphroditism, some fishes change sex from male-to-female (protandrous sex change), others change from female-to-male (protogynous sex change), and a few change sex in both directions multiple times (bi-directional sex change). Thus, the mechanisms of sex determination and differentiation in fishes are highly diverse and plastic, differing from those of other higher vertebrates (Desjardins and Fernald, 2009). Hence, fishes are a very interesting group among the vertebrates for the study of sex determination and differentiation.

Hermaphroditism has been documented in about 2% of all extant fish species (Avise and Mank, 2009). Sequential hermaphrodite (sex change) species have been recorded in 27 of 448 families across 7 orders of fishes, of which the niche of many species is coral reefs, however, most research has been focused only on 5 families (de Mitcheson and Liu, 2008). In these fishes, gonadal sex re-differentiation was observed during sex change in adulthood. Thus, sex-changing fishes are ideal models to investigate gonadal differentiation in vertebrates.

Groupers of the genus Epinephelus are naturally distributed in tropical and subtropical regions (Kitajima et al., 1991; Kohno et al., 1997). It is one of the important aquaculture species and cultured in many countries including Japan (Sakakura et al., 2007). Groupers are also known as the protogynous hermaphrodites (Sadovy and Colin, 1995). They can change sex from female-to-male depending on their size and/or age, or on the basis of their complex social structure (Munday et al., 2006). This kind of sexuality raises several problems for brood stock management. The deflection of sex ratio was observed in the brood stock (Yeh et al., 2003). Mass production of grouper fly from hatcheries under controlled reproduction, will certain be the key to success of industry growth, which is also dependent on simultaneous availability of mature broodstock of both sexes. Thus, artificial sex reversal induced by sex steroids and aromatase inhibitor (AI) have been successfully used in several grouper species (Bhandari et al., 2004a,b. 2006; Yeh et al., 2003; Zhou and Gui, 2010). However, endocrine mechanisms of sex change are not clearly understood.

Our major interest is to investigate the mechanisms that regulate the sex change processes in hermaphrodite groupers from endocrinological viewpoint using morphological, physiological, and molecular techniques. In this chapter, we restrict our discussion to two protogynous grouper species, Malabar grouper (*E. malabaricus*) and Honeycomb grouper (*E. merra*). We first summarize the process of gonadal sex differentiation in malabar grouper. Then, we use honeycomb grouper as an example to describe our recent work on mechanism of sex change. This review intends to highlight sexual plasticity and sex change mechanism in grouper related to sex steroids. For detailed information on other aspects of sex change in grouper, particularly genetics, please refer to the recent review by Zhou and Gui, (2010).

Gonadal Differentiation Process in Grouper

The gonad in groupers is unique because of its bipotential nature. However, there is little information on morphological, sexual and gonadal differentiation in groupers.

Malabar grouper, *E. malabaricus*, is widely distributed and considered as an important species in commercial aquaculture in tropical and subtropical regions including Okinawa, Japan (Yoseda et al., 2006). To obtain basic information on sexual differentiation in grouper, the process of gonadal sex differentiation in this fish was followed by using histology for one year from hatching (Murata et al., 2009). In fish at 11 days post-hatching (dph) large primordial germ cells (PGCs) were evident in the primordial gonad tissue located below the mesonephric ducts at the dorsal side of the intestine (Figure 1.A). Until 47 dph, neither germ cells nor somatic tissue showed any of the morphological characteristics of sexual differentiation.

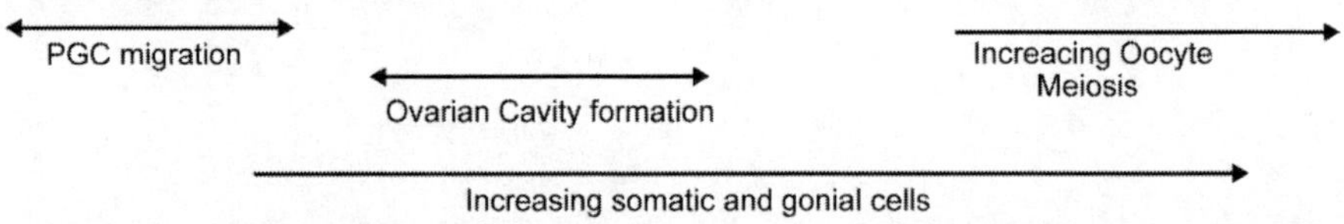

Figure 1. Process of gonadal differentiation in Malabar grouper (*E. malabaricus*). A; The gonad of 11 day-post hatching (dph) larvae. Some primordial germ cells are seen in the area on the dorsal side of the intestine. B; At the 47 dph, two elongations of somatic tissues (arrowheads), indicating ovarian cavity formation, can be seen. C; Completion of ovarian cavity formation are seen at the 75 dph. D; At the 360 dph, some oocyte had already entered into meiosis. Star indicates ovarian cavity. PGC, primordial germ cell; Me, melanophore; BV, blood vessel; Og, oogonium; Oo, oocyte.

By 47 dph, gonads had changed substantially with an increase in the number of somatic cells (Figure 1.B). By 74 dph, the elongations had developed further and were evident in the ovaries of all fish (Figure 1.C). However, ovarian germ cells had not yet begun active division (Figure 1.C). By 360 dph, oocytes were distributed within somatic tissue along the inner periphery of the ovarian cavity (Figure 1.D). These observations suggested that morphological changes associated with ovarian differentiation in the Malabar grouper begin at approximately 75 dph and that in all individuals the gonads differentiate directly into ovaries (Murata, et al., 2009). Additionally, we confirmed this phenomenon by using immunohistochemical observations that estrogen is involved in ovarian differentiation in this fish (Murata et al., 2011).

Gonadal Sex Change in Honeycomb Grouper

Experimentation in the aquaculture groupers may encounter limitations due to the large size of fish. Therefore, a smaller size species (honeycomb grouper, *E. merra*), available in the wild, was chosen as an experimental model. To obtain the basic information about sexuality of this fish, we collected wild honeycomb grouper every month for one year (Bhandari et al., 2003). On the basis of histological observations, gonadal transformation was divided into four phases as described below (Figure 2). Female: mature or immature oocytes were present in the gonad (Figure 2.A).

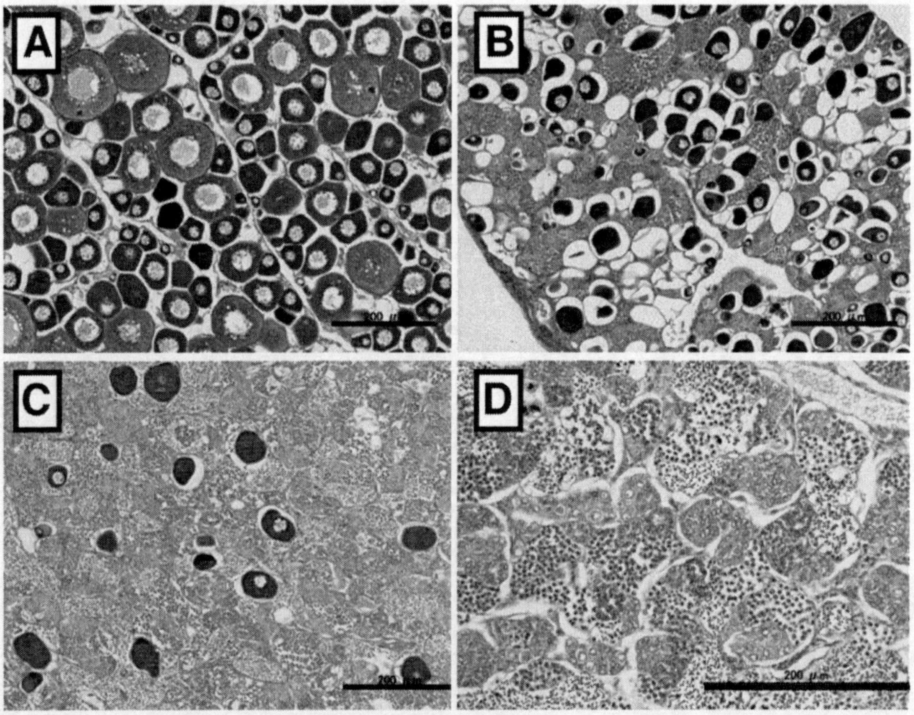

Figure 2. Histological observation of the gonads during natural sex change in honeycomb grouper (*E. merra*). A) Female; B) Early transitional phase (ET); C) Late transitional phase (LT); D) Male. Bars indicate 200 μm.

Early transitional (ET): mixture of male germ cells and primary oocytes were present in the gonad (Figure 2.B). Late transitional (LT): numerous male germ cells with a few primary oocytes were present in the gonad (Figure 2.C). Male: mainly spermatids and some spermatogonia were present in the gonad (Figure 2.D). Female and male fish were respectively found mostly in smaller and larger size weight range, whereas, fish with transitional were observed in intermediate range (Bhandari et al., 2003). In the breeding season, no individuals possessing transitional stage were observed. An overlapping in the sex distribution of transitional individuals in non-breeding season indicates that a sex change occurs during this period. Taken together, in *E. merra*, sex change usually occurs in non-breeding larger female (over 20 cm of total length).

Profiles of Sex Steroids during Natural Sex Change

To investigate the physiological mechanism, especially steroidogenesis, of sex change, we measured sex steroid profiles in grouper during sex change (Figure 3). Plasma estradiol-17ß (E_2) and 11keto-testosterone (11KT) profiles in females were significantly different from males.

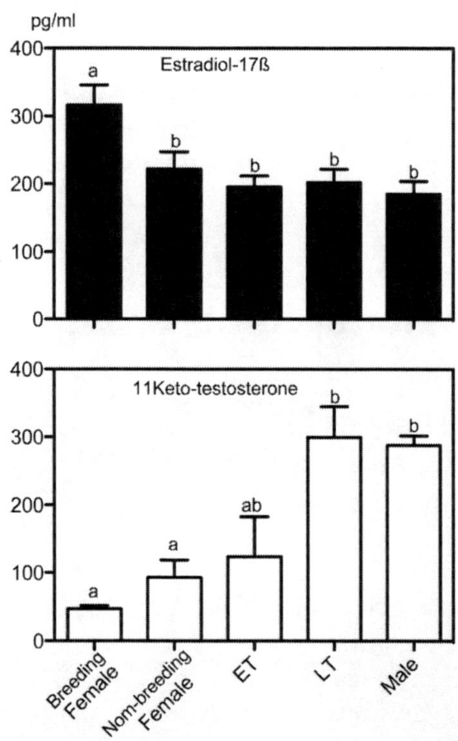

Figure 3. Changes in serum levels of estradiol-17ß (E_2) and 11keto-testosterone (11KT) during natural sex change in honeycomb grouper. Data are shown as means ± SEM. Data points not sharing a letter (a, b, c) are significantly different by Tukey-Kramer multiple comparison test. ET, early transitional; LT, late transitional.

Serum levels of E_2 were high in females in the breeding season, but low in the non-breeding females, transitional and male phase. 11KT was low in females, but not non-detectable, and gradually increased in the transitional and male phase. These results suggest that low serum E_2 levels and degeneration of oocytes accompanied by concomitant increase in the 11KT levels and proliferation of spermatogenic germ cells are probably the events mediating protogynous sex change in this species. In other protogynous fishes, a similar dramatic shift in steroidogenesis from estrogen production to androgen production was occurred at the beginning of sex change (Nakamura et al., 2005).

Effect of Sex Change by Aromatase Inhibitor and Exogenous Hormones

Many attempts have been undertaken on different species to develop standardized methods for sex reversal in captivity by treatment with exogenous sex steroid hormone or aromatase inhibitor (Nakamura et al., 2005; Zhou and Gui, 2010).

Based on the steroid profile during sex change in the honycomb grouper, artificial sex reversal was successesfull in our laboratory. Treatment with androgen (11KT) or aromatase inhibitor (Fadorozole, AI) caused complete masculinization of adult female (Alam et al., 2006; Bhandari et al., 2004a, b). Histological observations revealed that ovaries of the control fish had oocytes at various stages of oogenesis, while in the treated groups, gonads were transformed into testis.

In the sex-reversed fish, plasma E_2 level was significantly low, while androgen increased significantly. These results suggest a balance of estrogens and androgens, which is the most downstream factor in steroidogenesis, is important in sex change.

Androgen Production Cells in Ovary of Honeycomb Grouper

As described before, a constant production of androgen (11KT), albeit at a lower level than in the sexual transitional phase fish and male, was observed in female grouper (Bhandari et al., 2003). To investigate the site of androgen production in ovary, we observed immunopositive cells for cytochrome P450 11ß-hydroxylase (P450 11ß), downstream to 11KT production, throughout the process of sex change in honeycomb grouper (Alam et al., 2006). In female, strong P450 11ß immunopositive cells were observed in the tunica near blood vessel but not in theca layer (Figure 4). In ET and LT phase gonads, P450 11ß localized in remaining follicle layer of degeneration oocyte and tunica near blood vessel. On the other hand, in male, both interstices and tunica showed strong signals against P450 11ß (Alam et al., 2006). Moreover, 11-KT level related with the changes in the nuclei diameter of P450 11ß-positive cells in tunica during sex change. These observations suggest that this special androgen production site in tunica may provide the stimulus for female to undergo oocyte degeneration and initiate gonadal sex change. Additionally, we confirmed that this androgen production site in tunica ovary could be observed in other *Epinephelus* species (Manuscript in preparation).

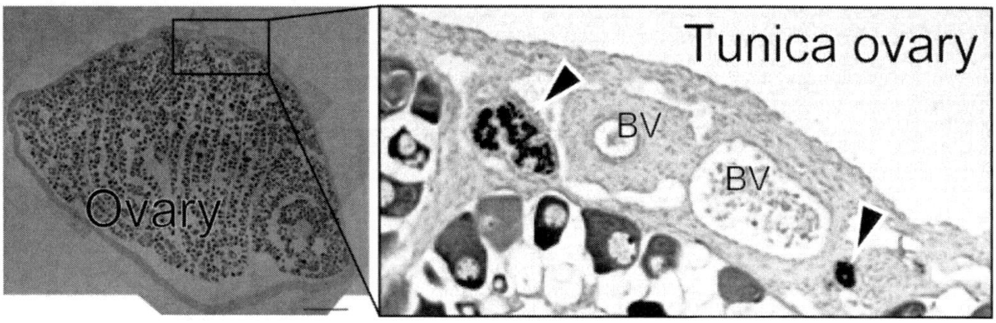

Figure 4. Localization of androgen production cells in the ovary of *E. merra*. P45011ß immunoreactivity (arrowhead) is located in tunica ovary. BV, blood vessel.

Involvement of the Gonadotropin in Gonadal Sex Change

The signals for gonadal sex change probably come from the brain. Therefore, the Brain (hypothalamo) - pituitary- gonad (HPG) axis is involved in the sex change. In this chapter, we focused on the pituitary in the grouper.

In teleosts, like in other vertebrates, gonadal steroidogenesis is largely controlled by pituitary gonadotropins (GTH), FSH and luteinizing hormone (LH). These GTHs contain a common glycoprotein hormone α subunit that forms a heterodimer with unique ß-subunits (Fshß and Lhß). (reviewed in Swanson et al., 2003). GtH actions are mediated by the GtH receptor on the surface of the target cells in the gonads (Oba et al., 2001). Two GtH receptors, which specifically bind FSH and LH, have been cloned from various vertebrates including fishes (Kobayashi et al., 2009; Oba et al., 2001; Swanson et al., 2003). In well-studied salmonids, FSH plays a significant role in puberty and gametogenesis, whereas LH is primarily involved in final maturation of the gametes in both sexes (Swanson, et al., 2003). However, variations in the expression profiles and potential roles of GTHs were reported in other teleost species (Gen et al., 2000; Hellqvist et al., 2006; Kajimura et al., 2001; Mateos et al., 2003). In protogynous wrasse, sexual dimorphic expression patterns of Fshß and Lhß transcripts were observed during the spawning season (Ohta et al., 2008). In protandrous black porgy, plasma LH levels were higher in male than sex-changing fish (Lee et al., 2000). In addition, treatment with exogenous human chorionic gonadotropin (hCG) or LH induced sex change in the protogynous bluehead wrasse and rice-field eel (Koulish and Kramer, 1989). These studies indicate that GTHs participate in regulating sex-changing processes. However, information of GTH expression pattern during sex-changing process is negligible. Therefore, detailed biological functions of GTH during sex change are not clear. Therefore, we quantified the transcripts of GTH subunits in the pituitary at different sexual phases (Figure 5) (Kobayashi et al., 2010). The relative level of *Lhß* mRNA were higher in the breeding season than in the non-breeding season. However, there were no significant differences in these levels among different sexual phases during non-breeding season. In contrast, the expression pattern of *Fshβ* transcripts showed a marked sexual dimorphism. Although *Fshβ* subunit transcripts were low in the breeding and non-breeding female phases, significantly increasing *Fshβ* transcript were observed during the female-to-male sex change, especially in ET stage.

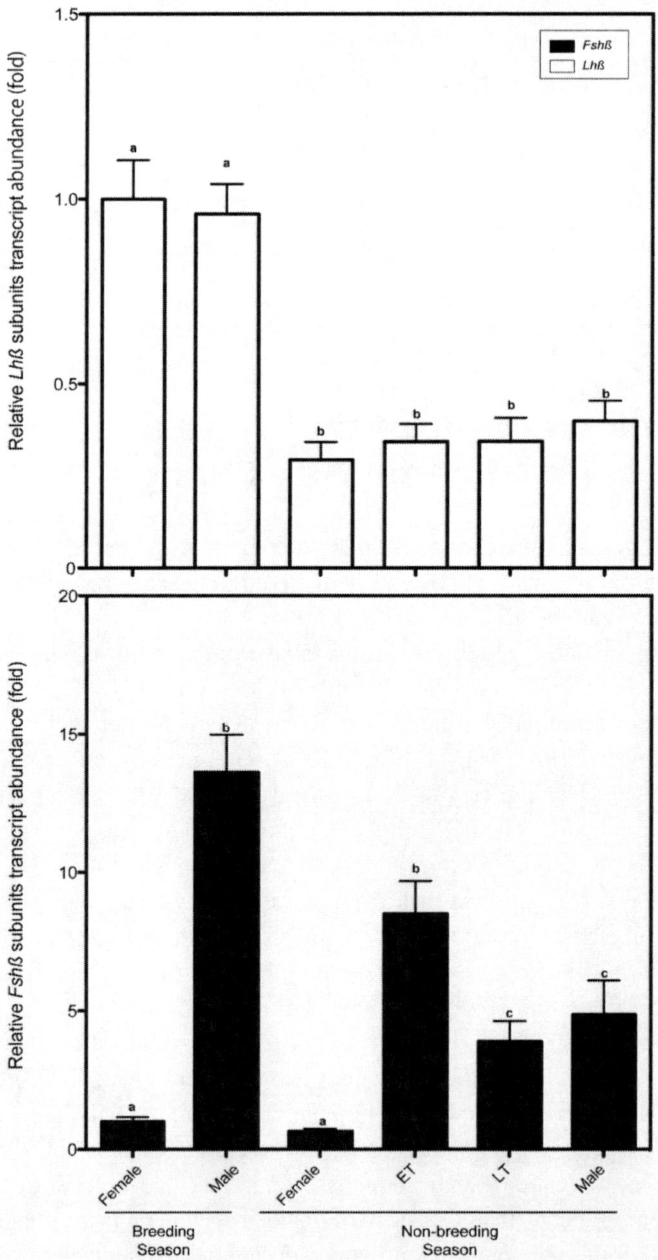

Figure 5. Changes of GTH subunit transcripts in pituitaries of the different sexual phase in honeycomb grouper. The quantification of the transcript abundance of *Lhβ* and *Fshβ* genes in pituitaries were determined by real-time quantitative RT-PCR analysis and normalized to the abundance of *GAPDH* transcript. The values are expressed as a fold change in abundance relative to the means in the breeding

female. Data are shown as means ± SEM. Data points not sharing a letter (a, b, c) are significantly different by Tukey-Kramer multiple comparison test. ET, early transitional; LT, late transitional.

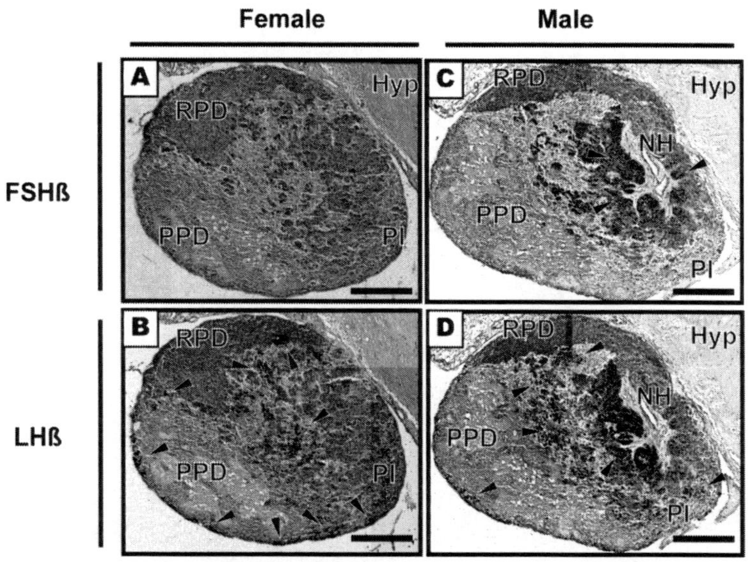

Figure 6. Representative sagittal sections of pituitary gland from non-breeding female (A and B) and male (C and D). Immunostained with anti-Fshß (A and C) and –Lhß (B and D) antibodies. Arrowheads indicate potitive signals. Bars indicate 200 µm. Hyp, hypothalamus; NH, neurohypophysis; PI, pars intemedia; RPD, rostral pars distalis.

To determine if there was sexually dimorphic expression of Fshß protein in the grouper pituitary gland, we examined Fshß and Lhß immunoreactive (ir) cells in pituitary using antisera against conserved epitopes of fish Fshß and Lhß subunits (Figure 6). The Fshß cells could not detect in the female pituitary (Figure 6A). In contrast, appreciable Fshß ir signals were seen around neurohypophysis (NH) of male pituitary (Figure 6C). In female, Lhß ir signals were seen in the proximal par distalis (PPD) and par distalis (PI), not in rostral par distalis (RPD) (Figure 6B). These signals were also seen at same area of male pituitary (Figure 6D). These observations support the result of mRNA expression of GTH subunit in the different sexual phases (Figure 5). Similarly, the up-regulation of FSH receptor in the gonad is associated with sex change (Alam et al., 2010). Interestingly, we confirmed FSH receptor transcripts localized in the androgen production site in tunica ovary (Paper in preparation). In order to identify the role of GTH in gonadal sex change in this fish, we treated female with bovine FSH (50 or 500 ng/fish) or LH (500 ng/fish) *in vivo* (Kobayashi et al., 2010). After 3 weeks, FSH treatment induced female to male sex change and up-regulated endogenous androgen levels and Fshß transcripts, whereas LH treatment had no effect on sex change. Taken together, these results strongly suggest that FSH may trigger the female-to-male sex change in honeycomb grouper.

Conclusion and Future Research

In this chapter, we primarily focus on studies related gonadal sex change. The bisexual gonad in grouper provides a unique model for studying the mechanism of sex change. Although it is not possible to define a hierarchy of regulation during sex change in groupers, a hypothetical model has been proposed, based on our current results (Figure 7).

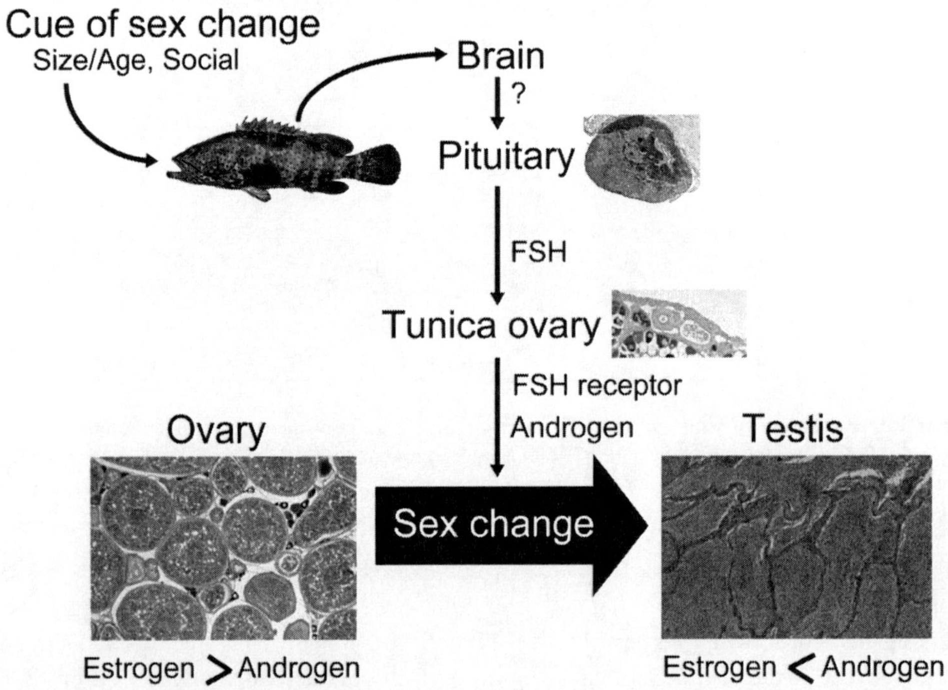

Figure 7. The possible endocrine mechanism of sex change in the honeycomb grouper. FSH produced by pituitary and sex steroid hormones (androgen and estrogen) involved in grouper sex change. Refer to the text for detailed description.

Our data suggest that the endocrine factors (estrogen, androgen, and FSH receptor in gonad; pituitary FSH) may play important roles in the sex change. Future research into the endocrine mechanism of sex change in fishes offers many possibilities. As we mentioned that sex change is a very complex physiological phenomenon, sex changes also occur in organs other than the gonads, such as brain. Thus, there are many unanswered questions about neuroendocrine mechanism of sex change in fishes.

Several neuroendocrine factors (monoamines, gonadotropin-releasing hormone, arginine vasotocin) were involved in brain sex change (reviewed in Godwin, 2010; Sudhakumari et al., 2010; Raghuveer et al., 2011). However the function of these factors has not been analyzed in groupers. Recently, access to detailed genomic information of animals by using next generation sequencer has become easier (Zhang et al., 2011). Therefore, the isolation of additional genes and/or downstream genes related to sex-changing pathway in fishes is expected in the near future.

Acknowledgments

This research work in this review is financially supported by Solution Oriented Research for Science and Technology (SORST), Japan Science and Technology Corporation (JST), Ocean Exposition Commemorative Park Commemorative Management Foundation, Japan Society for Promotion of Science (JSPS) and the Ministry of Education, Culture, Sports, and Technology (Project No.23658166 and 23248034 to MN).

References

Alam, M. A., Bhandari, R. K., Kobayashi, Y., Nakamura, S., Soyano, K. and Nakamura, M. (2006). Changes in androgen-producing cell size and circulating 11-ketotestosterone level during female-male sex change of honeycomb grouper *Epinephelus merra*. *Mol. Reprod. Dev, 73*, 206-214.

Alam, M. A., Bhandari, R. K., Kobayashi, Y., Soyano, K. and Nakamura, M. (2006). Induction of sex change within two full moons during breeding season and spawning in grouper. *Aquaculture, 255*, 532-535.

Alam, M. A., Kobayashi, Y., Hirai, T. and Nakamura, M. (2010). Isolation, characterization and expression analyses of FSH receptor in protogynous grouper. *Comp. Biochem. Physiol. A Mol. Integr. Physiol, 156*, 364-371.

Avise, J. C. and Mank, J. E. (2009). Evolutionary perspectives on hermaphroditism in fishes. *Sex Dev, 3*, 152-163.

Baroiller, J. F., D'Cotta, H. and Saillant, E. (2009). Environmental effects on fish sex determination and differentiation. *Sex Dev, 3*, 118-135.

Bhandari, R. K., Alam, M. A., Soyano, K. and Nakamura, M. (2006). Induction of female-to-male sex change in the honeycomb grouper (*Epinephelus merra*) by 11-ketotestosterone treatments. *Zool. Sci, 23*, 65-69.

Bhandari, R. K., Higa, M., Nakamura, S. and Nakamura, M. (2004a). Aromatase inhibitor induces complete sex change in the protogynous honeycomb grouper (*Epinephelus merra*). *Mol. Reprod. Dev, 67*, 303-307.

Bhandari, R. K., Komuro, H., Higa, M. and Nakamura, M. (2004b). Sex inversion of sexually immature honeycomb grouper (*Epinephelus merra*) by aromatase inhibitor. *Zool. Sci, 21*, 305-310.

Bhandari, R. K., Komuro, H., Nakamura, S., Higa, M. and Nakamura, M. (2003). Gonadal restructuring and correlative steroid hormone profiles during natural sex change in protogynous honeycomb grouper (*Epinephelus merra*). *Zool. Sci, 20*, 1399-1404.

de Mitcheson, Y. S. and Liu, M. (2008). Functional hermaphroditism in teleosts. *Fish Fisheries, 9*, 1-43.

Desjardins, J. K. and Fernald, R. D. (2009). Fish sex: why so diverse? *Curr. Opin. Neurobiol, 19*, 648-653.

Devlin, R. H. and Nagahama, Y. (2002). Sex determination and sex differentiation in fish: an overview of genetic, physiological, and environmental influences. *Aquaculture, 208*, 191-364.

Dries, L. A. (2003). Peering through the looking glass at a sexual parasite: are Amazon mollies red queens? *Evolution, 57*, 1387-1396.

Frisch, A. (2004). Sex-change and gonadal steroids in sequentially-hermaphroditic teleost fish. *Rev. Fish Biol. Fisher, 14*, 481-499.

Gen, K., Okuzawa, K., Senthilkumaran, B., Tanaka, H., Moryama, S. and Kagawa, H. (2000). Unique expression of gonadotropin-I and –II subunit genes in male and female red seabream (*Pagrus major*) during sexual maturation. *Biol. Reprod, 63*, 308-319.

Godwin, J. (2010). Neuroendocrinology of sexual plasticity in teleost fishes. *Front Neuroendocrinol*, *31*, 203-216.

Hellqvist, A., Schmitz, M., Mayer, I. and Borg, B. (2006). Seasonal changes in expression of LH-ß and FSH-ß in male and female three-spined stickleback, *Gasterosteus aculeatus*. *Gen. Comp. Endocrinol, 145*, 263-269.

Kajimura, S., Yoshiura, Y., Suzuki, M. and Aida, K. (2001). cDNA Cloning of Two Gonadotropin ß Subunits (GTH-Iß and -IIß) and Their Expression Profiles during Gametogenesis in the Japanese Flounder (*Paralichthys olivaceus*). *Gen. Comp. Endocrinol, 122*, 117-129.

Kitajima, C., Takaya, M., Tsukashima, Y. and Arakawa, T. (1991). Development of eggs, larvae and juveniles of the grouper, *Epinephelus septemfasciatus*, reared in the laboratory. *Ichthyol Res, 38,* 47-55.

Kobayashi, Y., Alam, M. A., Horiguchi, R., Shimizu, A. and Nakamura, M. (2010). Sexually dimorphic expression of gonadotropin subunits in the pituitary of protogynous honeycomb grouper (*Epinephelus merra*): evidence that follicle-stimulating hormone (FSH) induces gonadal sex change. *Biol. Reprod, 82*, 1030-1036.

Kobayashi, Y., Nakamura, M., Sunobe, T., Usami, T., Kobayashi, T., Manabe, H., et al. (2009). Sex Change in the Gobiid Fish Is Mediated through Rapid Switching of Gonadotropin Receptors from Ovarian to Testicular Portion or Vice Versa. *Endocrinology, 150*, 1503-1511.

Kohno, H., Ordonio-Aguilar, R., Ohno, A. and Taki, Y. (1997). Why is grouper larval rearing difficult?: an approach from the development of the feeding apparatus in early stage larvae of the grouper, *Epinephelus coioides*. *Ichthyol Res, 44*, 267-274.

Koulish, S. and Kramer, C. R. (1989). Human chorionic gonadotropin (hCG) induces gonad reversal in a protogynous fish, the bluehead wrasse, *Thalassoma bifasciatum* (Teleostei, Labridae). *J. Exp. Zool, 252*, 156-168.

Lee, Y. H., Lee, F. Y., Yueh, W. S., Tacon, P., Du, J. L., Chang, C. N., et al. (2000). Profiles of gonadal development, sex steroids, aromatase activity, and gonadotropin II in the controlled sex change of protandrous black porgy, *Acanthopagrus schlegeli Bleeker*. *Gen. Comp. Endocrinol, 119*, 111-120.

Mateos, J., Mañanós, E., Martínez-Rodríguez, G., Carrillo, M., Querat, B. and Zanuy, S. (2003). Molecular characterization of sea bass gonadotropin subunits (alpha, FSHbeta, and LHbeta) and their expression during the reproductive cycle. *Gen. Comp. Endocrinol, 133*, 216-232.

Munday, P. L., Buston, P. M. and Warner, R. R. (2006). Diversity and flexibility of sex-change strategies in animals. *Trends Ecol. Evol, 21,* 89-95.

Murata, R., Karimata, H., Ashraful Alam, M. and Nakamura, M. (2009). Gonadal sex differentiation in the Malabar grouper, *Epinephelus malabaricus*. *Aquaculture, 293*, 286-289.

Murata, R., Karimata, H., Kobayashi, Y., Horiguchi, R., Kishimoto, K., Kimura, M., et al. (2011). Differentiation of steroid-producing cells during ovarian differentiation in the protogynous Malabar grouper, *Epinephelus malabaricus*. *Int. J. Dev. Biol, 55*, 619-625.

Nakamura, M., Kobayashi, Y., Miura, S., Alam, M. A. and Bhandari, R. K. (2005). Sex change in coral reef fish. *Fish Physiol Biochem, 31*, 117-122.

Nelson, J. S. (2006). Fishes of the World, in 4th edn. (John Wiley and Sons, Inc., New Jersey) (2006).

Oba, Y., Hirai, T., Yoshiura, Y., Kobayashi, T. and Nagahama, Y. (2001). Fish gonadotropin and thyrotropin receptors: the evolution of glycoprotein hormone receptors in vertebrates. *Comp. Biochem. Physiol. B, 129*, 441-448.

Ohta, K., Mine, T., Yamaguchi, A. and Matsuyama, M. (2008). Sexually dimorphic expression of pituitary glycoprotein hormones in a sex-changing fish (*Pseudolabrus sieboldi*). *J. Exp. Zool. A, 309*, 534-541.

Raghuveer, K., Sudhakumari, C. C., Senthilkumaran, B., Kagawa, H., Dutta-Gupta, A. and Nagahama, Y. (2011). Gender differences in tryptophan hydroxylase mRNA, serotonin and 5-hydroxytryptophan levels in the brain of catfish, *Clarias gariepinus*, during sex differentiation. *Gen. Comp. Endocrinol, 171,* 94-104.

Sadovy, Y. and Colin, P. L. (1995). Sexual development and sexuality in the Nassau grouper. *J. Fish Biol, 46*, 961-976.

Sakakura, Y., Shiotani, S., Shiozaki, M. and Hagiwara, A. (2007). Larval rearing without aeration: a case study of the seven-band grouper *Epinephelus septemfasciatus*; using a wave maker. *Fish Science, 73*, 1199-1201.

Sandra, G.-E. and Norma, M.-M. (2010). Sexual determination and differentiation in teleost fish. *Rev. Fish Biol. Fisher, 20*, 101-121.

Sudhakumari, C.C., Senthilkumaran, B., Raghuveer, K., Wang, D.S., Kobayashi, T., Kagawa, H., Krishnaiah, Ch., Dutta-Gupta, A. and Nagahama, Y. (2010). Dimorphic expression of tryptophan hydroxylase in the brain of XX and XY Nile tilapia during early development. *Gen. Comp. Endocrinol, 166,* 320-329.

Swanson, P., Dickey, J. T., and Campbell, B. (2003). Biochemistry and physiology of fish gonadotropins. *Fish Physiol. Biochem, 28(1)*, 53-59.

Yeh, S. L., Kuo, C. M., Ting, Y. Y. and Chang, C. F. (2003). Androgens stimulate sex change in protogynous grouper, *Epinephelus coioides*: spawning performance in sex-changed males. *Comp. Biochem. Physiol. C, 135*, 375-382.

Yoseda, K., Dan, S., Sugaya, T., Yokogi, K., Tanaka, M. and Tawada, S. (2006). Effects of temperature and delayed initial feeding on the growth of Malabar grouper (*Epinephelus malabaricus*) larvae. *Aquaculture, 256*, 192-200.

Zhang, J., Chiodini, R., Badr, A. and Zhang, G. (2011). The impact of next-generation sequencing on genomics. *J. Genet. Genomics, 38*, 95-109.

Zhou, L. and Gui, J.-F. (2010). Molecular mechanisms underlying sex change in hermaphroditic groupers. *Fish Physiol. Biochem, 36*, 181-193.

In: Sexual Plasticity and Gametogenesis in Fishes
Editor: Balasubramanian Senthilkumaran

ISBN: 978-1-62618-848-8

Chapter XV

Some Non-Classical Intragonadal Regulators of Fish Reproduction

Bechan Lal*
Department of Zoology, Banaras Hindu University, Varanasi, India

Abstract

It is well established that gonadal activities are essentially regulated by hormonal dialogues between gonad and hyphothalamo-hypophysial axis. However, regulation of gonadal activities is far more complex than the envisaged.Numerous studies on mammals during past decades indicate that the coordination and integration of gonad, which exhibits compartmentalized structural-physiological organization and display spacio-temporal functional heterogenecity, involve local factors too. Gonadal cells not only receive ex-gonad borne chemical inputs via blood but also from different cell types within the gonad

Local regulatory system empowers the individual gonadal cells to respond with greater efficiency than would be possible through hypothalamo-hypophysial regulation alone. The functional state of different type of gonadal cells is, thus, determined by their associations with adjacent cells, which exert local control through paracrine interactions. Present review will highlight these aspects to understand the importance of non-classical intragonadal regulators of reproduction in teleosts.

Introduction

Hypotahlamo-hypophyseal system is the central regulator of reproduction in vertebrates including teleosts. However, recent knowledge provides valid evidences of paracrine interactions among different cell types in the gonad which greatly influence the development and maturation of viable gametes. Such interactions exist between germ cells-Sertoli cells, Sertoli cells-Leydig cells, Leydig cells-macrophages, granulosa cells-oocyte, thecal cells-

* E-mail: lalbhu@yahoo.co.in.

granulosa cells, etc. The Leydig cells fail to develop normally in the absence of macrophages (Hales, 2002). Likewise, morphological and physiological status of macrophages also changes with structure and functions of the Leydig cells (Huston, 2006).

Similarly, germ cells do not proliferate and differentiate normally in the absence of Sertoli cells and granulosa cells (Schulz, et al., 2005; Tesone et al., 2009). An elaborate network of local paracrine interactions between gametogenic and somatic elements is reported to be essential for the generation of normal and viable germ cells.

The bidirectional communication during their interactions takes place through variety of substances produced by somatic cells such as Leydig cells, thecal cells, granulosa cells, macrophages, peritubular cells, fibroblast cells, mast cells, etc., as well as developing germ cells like oocytes and spermatozoa. Many of such substances have been isolated and characterized in past. These substances include growth factors and cytokines [epidermal growth factor- EGFs, fibroblast growth factor- FGFs, transforming growth factors-TGFs, insulin like growth factors- IGFs, platelet derived growth factor- PDGF and tumor necrosis factor-TNF, interleukins], regulatory peptides [endorphin, angiotensin II, arginine-vasopressin, inhibin, activin, endothelin, FSH, LH, GH, GnRH, atrial natriuretic factor-ANF, brain natriuretic peptide- BNP and C-type natriuretic peptide-CNP] etc., and reactive oxygen and nitrogen species [sulphurdioxide-SO_2, hydrogen peroxide- H_2O_2, nitric oxide-NO]. However, most of these studies are limited to mammals.

Despite that fishes exhibit highly diverse modes and mechanisms of reproduction, such studies in fishes are highly scanty. This chapter discusses the current state of art on some intra-gonadal growth factors (EGF, TGF, FGF, IGF) and reactive nitrogen species (NO), which are reported to play very crucial role in regulation of gametogenesis and steroidogenesis in fishes.

The Existence of the NOS/NO System and Its Role in Reproductive Activities

NO, a highly versatile and ubiquitous important gaseous molecule, is produced from L-arginine by catalytic action of one of the three isoforms of NOS: neuronal (nNOS), endothelial (eNOS), and inducible (iNOS). NO is recognized as an important inter- and intra-cellular signaling molecule that regulates diverse reproductive functions such as steroidogenesis, spermatogenesis, penile erection, folliculogenesis and fertilization in mammals (Roselli et al., 1998). In mammals, all forms of NOS are demonstrated in gonadal cells like thecal cells, granulossa cells, oocytes, Leydig cells, endothelial cells, peritubular cells, macrophage, Sertoli cells, spermatogonia, spermatocytes, and spermatids in human, rodents and pig, though the predominant form differs from cell to cell and species to species (Lee and Cheng, 2004; Kim et al., 2006). Expression of NOS stimulates germ cell differentiation (Zini et al., 1996) and activates programmed cell death in mammals by generating NO (Zini et al., 1998; Lee and Cheng, 2004). Involvement of NOS/NO system in oogenesis, ovulation and fertilization are clearly established in mammalian species. Weisman et al. (2005) have reported potent inhibition of basal testosterone production by the rat Leydig cells after treatment with S-itrosoglutathione (GSNO), a NO donor; but have opined that testosterone production by the Leydig cells is sensitive to only exogenous NO (paracrine in nature), not of the Leydig cells (endogenous in nature). Valenti et al. (1999) have, however, established biphasic effects of NO on steroid secretion depending on the chemical nature and concentrations of NO donors such as S-nitroso-N-acetylpenicilamide (SNAP), diethylamine (DEA), and diethylenetriamine (DETA); low doses (10^{-8}–10^{-5}M) stimulate testosterone production, while higher doses (above 10^{-5}M) inhibit testosterone secretion.

In fishes, however, such studies are meager. Lines of evidence from the author's laboratory demonstrate the existence of nNOS and iNOS in germ cells as well as interstitial cells in the recrudescing and fully mature testis of the catfish, *Clarias batrachus*, which differs in intensity and varies with changing reproductive status (Dubey nee Pathak and Lal, 2010). The isolated Leydig cells of the catfish also exhibit distinct NAPDH-diaphorase-positive staining as well as iNOS and nNOS expression (Dubey nee Pathak and Lal, 2008). Thus, fish testis is capable of generating substantial amount of NO. Evidence has also been provided that testicular NO production is under the control of extra-gonadal hormones such as gonadotropin, growth hormone, and insulin. Recently, *Wilson-Leedy and Ingermann (2011) have demonstrated the presence of NO in sperms of Oncorhynchus mykiss, indirectly through DAF-FM technique. Although Creech* et al. *(1998) could not detect NOS activity in sperms but have suggested the presence of trace amount of NO in the sperms with seminal plasma of fathead mi*nnows through ESR signaling (electron spin resonator signals). Like testis, fish ovary is also capable of producing NO. Reproductive stage- and cell-specific expressions of all forms of NOS have been demonstrated in oocyte follicles of the ovary of another catfish, *Heteropeustes fossilis* (Krishna and Tripathi, 2008). NOSs were predominantly localized in pre-vitellogenic follicles; nNOS in oocytes while iNOS and eNOS in granulosa and thecal cells.

NO is shown to have role in regulation of steroidogenesis and spermatogenesis in catfish. Leydig cell- and macrophage- generated NO is reported to inhibit the testosterone production in autocrine and paracrine manner in fish (Dubey nee Pathak and Lal, 2008, 2009). The moderate to intense expressions of nNOS and iNOS in germ cells within the seminiferous tubules of the catfish during the mid- and late-recrudescence phases suggest their role in spermatogenesis too in fishes. *Creech* et al. *(1998) have shown that addition of low dose of exogenous chemical NO (10^{-6}*M SNP derived-NO) enhances the sperm motility of fathead minnows, *Pimephelas promelas as well as velocity parameters such as VCL, VSL and VAP. However, higher dose of SNP (10^{-2}*M) decreased the sperm motility and the velocity parameters. Creech et al. *(1998) have also demonstrated NOS in the chorion around the micropyle of fathead minnows oocytes. They have suggested that extremely local and concentration-dependent stimulatory effect of NO on sperm motility facilitates chemostatic aggregation of sperms towards mic*ropyle, thus ultimately help fertilization of eggs. Wilson-Leedy and Ingermann (2011) have reported no effect of NO on percent motility using a NO donor, SNAP (10^{-9} as well as 10^{-6}M); but they have observed significant increase and decrease in different sperm velocity parameters such as path straightness (STR), lateral head movement (Wobble-Wob) and curvilinear velocity (VCL) depending upon the concentrations of SNAP and a NO scavenger, PTIO (2-phenyl-4,4,5,5-tetramethylimidazoline-1-oxyl-3-oxide). Tripathi and Krishna (2008) have demonstrated the involvement of NO in follicular development and oocyte maturation in *H. fossilis*.

Epidermal Growth Factor Family

The epidermal growth factor (EGF) is a polypeptide of 53 amino acids with three disulfide bonds. This family of growth factors consists of EGF, transforming growth factor-α (TGF-α), epiregulin, amphiregulin, heparin binding EGF (HB-EGF), crypto-1, betacellulin (BTC), Schwannoma-derived growth factor (SDGF), glial-derived growth factor (GGF-II). These proteins are derived from membrane bound precursor molecules. The EGF receptor is a monomeric glycoprotein that binds EGF with high affinity and specificity. Its binding to receptors activates tyrosine kinases, causing both autophosphorylation and phosphorylation of other cellular substrates on their tyrosine residues. It stimulates the proliferation of different cell types in culture for example epidermal cells, fibroblasts, crystalline cells, glial cells, and vascular endothelial cells. Transforming growth factor- α (TGF-α) resembles EGF structurally and functionally. It is composed of 50 amino acids, and binds to a common receptor for both TGF-α and EGF. These growth factors are strong mitogenic as well as angiogenic factors.

EGFs are expressed in the mammalian ovary. The immunoreactive EGF is detected in follicular fluid of human and in other mammalian ovaries, however, the results from different studies in different species are somehow confusing about their cellular origins (Tse and Ge, 2009). But EGF receptor is distinctly shown in thecal cells, granulosa cells and corpus luteum (Maruo et al., 1993). Both EGF and its receptors are also demonstrated in mammalian testis, and their roles in reproductive functions have been thoroughly discussed (see review- Conti et al., 2006; He Junping et al., 2009). The paracrine effects of the EGF/TGF-α secreted by interstitial thecal cells prevent apoptosis of granulosa cells from follicles selected to ovulate (Tilly et al., 1992). In mammals, EGF is shown to stimulate DNA synthesis and granulosa cell proliferation in the ovarian follicles. It also modulates ovarian steroidogenesis and granulosa cell differentiation by increasing progesterone production and inhibiting FSH-induced biosynthesis of estradiol and receptors of FSH and LH in cultured granulosa cells (Tilly et al., 1992). In addition, EGF enhances oocyte maturation in several mammalian species.

In fishes, studies on existence of EGF and its roles in gonadal activities are meager and limited to zebrafish and goldfish only. In contrast to mammals, where EGF is abundantly expressed in the submaxillary gland, the ovary and testis are the predominant sites of EGF expression in the zebrafish, which indicate its potential role in regulation of fish reproduction. Wang and Ge (2004) for the first time cloned full-length cDNA of EGF and its receptors (EGFR) from zebrafish ovary. Subsequently, this group also cloned the other members of EGF family such TGF-α, BTC and HB-EGF (Tse and Ge, 2009, 2010). Although the zebrafish EGF precursor peptide is slightly shorter than its mammalian counterpart, the arrangement of seven EGF-like motifs and mature EGF peptide is similar to the mammals. EGF expresses highly in the immature ovary of zebrafish fish and expression declines with the ovarian development. Previtellogenic and early vitellogenic follicles express relatively high levels of EGF than the more advanced ovarian follicles. EGFR expresses, however, very low in immature ovary, but increases with initiation of vitellogenesis till late stages of ovarian development. In ovarian follicles, EGF expresses in both follicular cells as well as oocytes; more in oocytes than follicular cells. But, EGFR expresses exclusively in the follicular cells and only in traces in oocytes. This spatial expression pattern of EGF and EGFR indicates that the oocytes transmit some signal to follicle cells through EGF. The temporal expressions of TGF- α, BTC and HB-EGF in zebrafish follow the same patterns as that of EGF.

In fish ovary, EGF is reported to play role in regulating follicle survival and prostaglandin synthesis (Srivastava and Van-der Kraak, 1995;Janz and Van-der Kraak, 1997). Pati et al. (1996) have shown that EGF stimulates final oocyte maturation in goldfish. Similarly, TGF-α is reported equally potent in increasing the rate of oocyte maturation in zebrafish. The effects of EGF and TGF-α are found to be mediated by activin (Pang and Ge, 2002). Activin is a member of the transforming growth factor-β (TGFβ) superfamily. Tse and Ge (2009) have provided evidence that EGF signaling network in ovarian follicle of the zebrafish is self-regulated by its own members.

Fibroblast Growth Factor Family

Fibroblast growth factors (FGFs) constitute a family of important regulators of growth and differentiation of various cell types in vertebrates. This family includes acidic and basic FGFs, keratinocyte growth factors, and some other peptides; till date FGF-1 to 22 from mammals and FGF-24 from fish. Out of 23 FGFs, FGF-2 and FGF-9 are implicated in mammalian reproduction. Basic FGF (bFGF) is 146-amino acid polypeptides with mitogenic effects on variety of cells derived from the mesoderm and the neuroectoderm. bFGF also exist in an amino-terminal truncated homolog that lacks the first 15 N-terminal residues and has been detected in the corpus luteum of mammalian ovary. Acidic FGF (aFGF) is a 140-amino acids peptide that can also exist in amino-terminal truncated form, lacking the first six amino acids. The mRNA of both, FGF and its receptor, have been identified in human granulosa cells. It has been suggested that bFGF modulates angiogenesis, cell proliferation, progesterone synthesis, and apoptosis in the ovary (Berisha and Schams, 2005). Like EGF and TGF- α, FGF acts through a tyrosine kinase receptor. Studies conducted on rat granulosa cell culture and pre-ovulatory follicles have shown that bFGF inhibit DNA apoptotic fragmentation (Tilly et al., 1992), which suggest its importance in the modulation of programmed cell death in ovarian follicles. bFGF is also localized in spermatogonia and pachytene spermatocyte during active spermatogenesis, indicating about its role in germ cell proliferation and spermatogenetic progression. There is evidence that bFGF modulates spermatogenesis by acting on Leydig cells through 17β-HSD and aromatase activation (Sordoillet et al., 1992). It also activates Sertoli cells in mammals (Han et al., 1993).

Although, all the mammalian orthologs of FGFs have been identified in zebrafish, FGF-9 is missing from the zebrafish genome. Interestingly, some FGF paralogs, specific to zebrafish, are also detected. But none of them is reported from the gonad (Itoh and Konishi, 2007). In fishes, information on the existence of FGFs in fish gonad and its role in fish reproduction is very limited. The immunoreactive FGF has been demonstrated in the cytoplasm of oocytes in immature ovary of medaka, which disappears before the initiation of vitelllogenesis. But again reappears around the oocytes in the adult ovary at vitellogenic stage. It is reported to act on follicular cells and regulates the initiation of vitellogenesis as well as proliferation and differentiation of granulosa cells in medaka (Watanabe et al., 1998). The tissue distribution analysis of FGFs by RT-PCR in Nile tilapia also reveals complete absence of the FGF-9 expression, but FGF-16 and FGF-20b express in the ovary (Sun et al., 2012).

Russo et al. (2000) have detected the existence of bFGF and FGF receptor-1(FGFR) in Sertoli cells and spermatogonia A and B in Antartic teleosts *Chionodraco hamatus*, immunocytochemically. In medaka also, bFGF is reported in Sertoli cells and spermagonia A, however, its receptors in located only in Sertoli cells (Watbabe et al., 1997). They have suggested its role in Sertoli cell proliferation, which ultimately help spermatogenic progression. FGF is also indentified in immature testis of *Salvelinus Leucomaenis* before spermatogonial proliferation (Watnabe and Onitake , 2005). A week immunostaining against bFGF is also reported in the spermatogonia of *Trematomus bernacchii* (Russo et al., 2000).

Transforming Growth Factor-β Family

TGF-β superfamily is large group of structurally related growth factors which regulate a variety of cellular processes, including cell-cycle progression, development, cell differentiation of somatic and germ cells in gonads, adhesion of cells and immune surveillance. Based on structural and functional characteristics, this superfamily has been classified into transforming growth factor-β (TGF-β), bone morphogenetic protein (BMP), growth and differentiation factors (GDF), inhibins and activins, neurotrophic factors (GDNF) derived from glial cells, and anti-Mullerian hormone (AMH) subfamilies.

TGF-β is synthesized as an inactive latent form, which is converted to its active form enzymatically. In mammals, three biologically active isoforms of TGF-β are known; TGFβ1, TGF-β2 and TGF-β3, while TGFβ4 and TGFβ5 are expressed in bird and *Xenopus laevis*, respectively. TGF-βs are 25 kDa homodimers linked by disulfide bonds, sharing a high level of sequence homology and almost identical tertiary structure. Generally, all isoforms elicit similar biological responses, which are mediated by two different types of receptor; type I and type II. Both receptors have ligand-binding extracellular region and a cytoplasmic serine/threonine kinase domain. A third TGF-β receptor type (III), a proteoglycan, is thought to function as storage and delivery system for TGF-β.

All three isoforms of TGF-β are detected in the theca and granulosa cells of the mammalian ovary, and their intensity varies over the course of gonadotropin-induced follicle maturation (Drummond, 2005). TGF-β is involved in the bidirectional communication between granulosa and thecal cells and also between granulosa cells and the oocyte. TGF-β1 and TGF-β2 act synergistically in the ovary, and is reported to modulate the proliferation and differentiation of granulosa cells in rats (Knight and Glister , 2006). TGF-β is believed to augment several action of FSH including the induction of aromatase activity in murine granulosa cells. Ovarian TGF-β helps folliculogenesis and production of mature oocytes, at least partially by amplifying the effects of FSH and also by stimulating production of estradiol.

In mammalian testis, TGF-β1 is expressed in both somatic cells (Sertoli cells, peritubular myoid cells and macrophages) as well as germ cells, while TGF-β2 and TGF-β3 are expressed only by the somatic cells. There is differential production of TGF-β receptors in testis during development, type-I and -II are highly expressed in the immature testis suggesting that TGF-βs play a crucial role during early testicular development. Differential TGF-β expression also occurs during postnatal development of the testis and onset of spermatogenesis. The actions of TGF-β on testicular target cells are influenced by endocrine hormones and sex steroids. Production of TGF-β1 by rat Sertoli cells are up-regulated by estradiol. TGF-β inhibits Leydig cell steroidogenesis in primary culture. It also influences testicular steroidogenesis depending on the dose; low doses of TGF-β stimulate the testosterone secretion by increasing 3β-hydroxysteroid dehydrogenase activity, while its high doses decrease the secretion of testosterone by inhibiting pregnenolone production.

Activin, the other important member of TGF-β superfamily, is a homo- or heterodimer composed of two β-subunits: βA and βB (βAβB, βAβA, or βBβB, and referred as activin AB, actvin A and activin B, respectively). In humans, activins remain bound to a specific serum binding glycoprotein, follistatin. The activities of activins are modulated by follistatin. High levels of activins have been observed in the mid-cycle and at the end of the luteal phase. Activins are also high during pregnancy. Activins synthesized by the granulosa cells exert autocrine and paracrine effects on the follicle, stimulate the expression of LH receptors in the granulosa cells and inhibit LH-induced synthesis of androgens in thecal cells. In mammals, treatment with activins *in vitro* not only stimulate follicular growth and granulosa cell proliferation, but also increase the production of FSH receptors and FSH-induced luteinizing hormone (LH) receptors in cultured granulosa cells. Activins first bind to a specific receptor type-II, then it recruits a type-I receptor for signaling. Activins play important role in the development of the oocyte and cumulus cells, which express the activin subunits, βA, βB, and follistatin (Silva et al., 2003).

Inhibins, another member of TGF-β superfamily, are glycoproteins consisting of two subunits; α (18 kDa) and β (12 kDa). Inhibin subunits form heterodimers, constituted by a common α-subunit, but with different β-subunits, known as inhibin A (α and βA) and inhibin B (α and βB). These are produced mainly in the gonad. Ovarian inhibins, synthesized in granulosa cells inhibit synthesis of FSH in the pituitary gland. Inhibin B is secreted during the early follicular phase and decreases thereafter. The levels of inhibin A remains low in the follicular phase but high during the luteal phase. It is generally accepted now that inhibins function as a classical endocrine hormone from the gonad to regulate FSH secretion at the pituitary. Contrary to activins, inhibin does not have its own signaling receptor but it largely works as an antagonist to compete with activin for type II activin receptors (Wiater and Vale 2008). The high potency of inhibin in competing for activin receptors is due to the involvement of an inhibin co-receptor, betaglycan (also termed TGF-β type III receptor), which helps to deliver inhibin to activin type II receptors.

Although molecular evidences establish the presence of all three isoforms of TGF-β in many fish tissues, studies on existence of TGFβ in fish gonads and its role in fish reproduction, however, are rare. Presence of TGF-β1 are reported in the kidney of rainbow trout, goldfish, carp and hybrid striped bass, while TGF-β2 is reported from the heart cell of carp. TGF-β3 is demonstrated in blood cell of rainbow trout, European eel and grass carp. Expression of TGF-β1 mRNA has also been demonstrated in thymus, head kidney and spleen of grass carp. Recently, a novel form of transforming growth factor i.e. TGF-β6 is detected in muscle of *Sparus aurata*.

However, there are only two reports on the existence of TGF-β in fish gonad. Calp et al. (2003) have observed expression of TGF-β in the early vitellogenic and pre-maturational full-grown follicles in goldfish ovary through RT-PCR. Parallel to it, Kohli et al. (2003) cloned the TGFβ1 and its receptor type II from the ovary of zebrafish. They have also investigated the down-stream signaling molecules of TGF-β, i.e. Smad 2, 3, and 4 in the ovarian follicles of all developmental stages. They have shown that TGF-β1 inhibits gonadotropin and 17α,

20β-dihydroxyprogesterone (DHP)-induced oocyte maturation in zebrafish (Kohli et al., 2003, 2005). TGF-β has also been reported to enhance the prostaglandin synthesis in the goldfish and rainbow trout ovary. It also suppresses the steroidogenesis in goldfish (Calp et al., 2003). There is no report on the existence of TGF-β and its role in fish testis.

The existence of activin A, activin B, follistatin and inhibins are also demonstrated in fish gonad. Activin βA and βB subunits have been cloned from the goldfish. The evidence is provided that goldfish ovary expresses activins mRNA (Pang and Ge, 2002a). The activin receptors have also been cloned from the ovary of goldfish and zebrafish. Mousa and Mousa (2003) have detected strong immunoreactive activin-β in the cytoplasm of oogonia and primary oocytes of *L. ramada.* They have also observed immunoreactivities for subunits of inhibin and activin (α, βAand βB) in the vitellogenic oocytes. Vitellogenic oocytes exhibit, however, low immunoreactivity as compared to previtellogenic oocytes. Immunoactivities completely disappears when the oocytes become fully mature. Ge et al. (1993) have detected activin A, activin B and follistatin in the granulosa cells of previtellogenic oocytes of the goldfish. Wang and Ge (2003a) have reported the expression of activin βA and βB predominantly in the follicular cells of the full-grown follicles of zebrafish. Similar immunocytochemical staining for activin βA and βB are also demonstrated in rainbow trout ovary. Interestingly, the expression pattern of the activin-binding protein, follistatin, is opposite to that of activin subunits. Its expression level remains significantly high in the oocytes than that in the follicle layers (Wang and Ge , 2004). Expression of activin and follistatin in cultured zebrafish follicle cells is regulated by gonadotropins. Gonadotropin(s) stimulates activin βA and follistatin expression, but suppresses βB expression employing different signaling pathways. The stimulation of activin βA and follistatin is mediated by the cAMP-PKA pathway in cultured zebrafish follicular cells, while inhibition of activin βB expression is achieved through cAMP-dependent but PKA independent pathway (Peng and Ge, 2002b; Wang and Ge, 2003b,c). The temporal expression patterns of these peptides *in vivo* during the development of the ovary and ovarian follicles offer important clues to the differential role of these proteins during the ovarian and follicular development. The presence of activins A and B in the testicular interstitial and Sertoli cells of goldfish has been also established immunocytochemically.

Ovarian activin-follistatin system is reported to play an important role in mediating the effects of gonadotropin(s) on zebrafish oocyte maturation and maturational competence (Pang and Ge, 2002a; Wang and Ge, 2003b). Activin has been shown to stimulate final oocyte maturation (FOM) in zebrafish. Follistatin, however, suppresses gonadotropin-induced oocyte maturation (Kwok et al., 2005). In testis, activin is suggested to augment early spermatogonial proliferation (Miura and Miura, 2003).

Information on gonadal inhibin in fishes is limited to only zebrafish. Dominant expression of inhibin-specific subunit α (inhibin α, known as inha) has been demonstrated in the zebrafish ovary and testis (Poon et al., 2009). Function of inhibin is more restricted to the regulation of reproduction in the zebrafish. Inside the zebrafish ovary, the expression of inha is predominantly localized in the somatic follicular cells, without any expression in the oocyte. In the rainbow trout, inha expression is detected in the granulosa cells through *in situ* hybridization. A remarkable variation in inha expression is reported during folliculogenesis in zebrafish. It is expressed maximally in fully grown follicles, but expression drops significantly in the oocytes follicles undergoing GVBD (Poon et al., 2009). Its expression is regulated though negative feed-back by FSH. In the goldfish, mammalian inhibin stimulates LH secretion *in vitro*. Kwok et al. (2005) have also reported inhibin-induced LH secretion. Taking together the finding of Kwok et al. (2005), Poon et al. (2009) have suggested that the increased production of inhibin by fully grown follicles may serve as an important signal from the ovary to the pituitary for the readiness of the follicles to mature and ovulate. Poon et al. (2009) have shown that recombinant human inhibin A has significant inhibitory effect on the final oocyte maturation induced by DHP. The inhibitory effect of inhibin on DHP induced- final oocyte maturation is also reported in *Fundulus heteroclitus*.

Bone morphogenetic proteins (BMPs) are another member of TGF-β family. Active BMPs are composed of 50-100 amino acids with seven cysteine residues; six of them form three intramolecular disulfide bonds, known as cysteine knots. The seventh cysteine is used for dimerization of monomers through covalent disulfide bond, thus forming the biologically active signaling molecule. There are almost 30 BMPs. Though named initially for their role in bone formation, BMPs play important role in orchestrating tissue architecture throughout the body, and also in many physiological functions, including reproduction. In reproduction, BMP8A is essential for maintenance of spermatogenesis in mammals. BMP-15 is involved in oocyte maturation. BMP-15, also known as growth and differentiation factor (GDF)-9B, which is closely related to another important peptide, GDF-9. The presence of the conserved cysteine knot places GDF-9 and BMP-15 into the TGF-β superfamily, but GDF-9 and BMP-15 lack the seventh cysteine, and GDF-9 and BMP-15 are biologically active as monomer. Considerable progress has been made regarding the function of BMP-15 and GDF-9 in mammalian folliculogenesis and ovulation.

However, very little is known about the structure and function of BMP-15 and GDF-9 in fishes. Recently, BMP-15 and GDF-9 cDNAs have been cloned from the zebrafish (Clelland et al., 2006, 2007; Liu and Ge, 2007) and European sea bass. The fish BMP-15 shares only 45 to 56% of sequence identities in the mature peptide region with their mammalian counterparts (Clelland et al., 2006). Through real-time PCR, BMP-15 is shown to express in ovarian follicles at all the stages of development without any apparent change during folliculogenesis. Using immunocytochemistry and *in situ* hybridization, Clelland et al. (2006) have demonstrated its presence in oocytes as well as follicular cells in ovarian follicles of zebrafish. They have also reported that BMP-15 inhibits oocyte maturation in fish. Tan et al. (2009) have worked out the signaling pathway of BMP-15 action and suggested that activin, TGF-β1 and BMP-15 target common gene(s) to regulate oocyte maturation and amongst them, Smad2 plays crucial role.

Though structure and functions of GDF-9 is known in mammals, its existence and role in fish gonadal activities are largely unknown. GDF-9 cDNAs has been cloned from the zebrafish (Liu and Ge, 2007) and characterized. A 2.1kb cDNA of the zebrafish GDF-9 homolog shares 60% homology with that of mammals in the mature region. It is exclusively expressed in the gonads. In the ovary, GDF-9 is predominantly expressed in the oocytes as demonstrated by both real-time RT-PCR and *in situ* hybridization. Zebrafish GDF-9 expresses maximally in the primary follicles, but expression gradually declines as the follicular development advances (Liu and Ge, 2007). The expression of GDF-9 is down regulated by gonadotropin in both ovarian fragments as well as isolated follicles in dose- and time-dependent manners, and inhibition is stage-dependent; the strongest inhibition in fully grown follicles and no effect on the primary follicles. The information on function of GDF-9 in fishes is still unknown.

Insulin-Like Growth Factor Family

This family has two members; insulin-like growth factor-I (IGF-I) and II (IGF-II). IGF-I is a ubiquitous 70-amino acid polypeptide and regulates variety of functions in different tissues. In the rat ovary, IGF-I mRNA is localized in granulosa cells of developing follicles. IGF-I binds to its specific receptors on the target cells with higher affinity than IGF-II. IGF-I stimulates the production of estradiol and progesterone and expression of LH/hCG receptors. IGF-I is reported to play a crucial role in the selection of preovulatory follicles. Insulin-like growth factor-II (IGF-II) is a 67-amino acid peptide, with 62% homology with IGF-I. IGF-II is expressed in theca cells of antral follicles and granulosa cells of dominant follicles. The switch in the expression of the IGF-II from thecal to granulosa cells during follicle selection is considered as distinct characteristics of the dominant follicle. IGFs are one of the important molecules that play important role in inter-compartmental communications during ovarian follicular development.

Activities of IGFs are modulated by IGF binding proteins (IGF-BPs) which are low molecular weight proteins that bind specifically to IGFs. There are six IGF-BPs that regulate IGF activities in the ovary, either by binding IGFs or by exerting a direct effect on steroidogenesis. FSH regulates the expression of IGF-BPs. Expressions of IGFBP-4 and IGFBP-5 decline in rat granulosa following FSH treatment. *In situ* analysis reveals the presence of IGF-BPs in atretic follicles but not in healthy follicles. IGF-I is reported as survival factor in ovarian follicular cells. Treatment with IGF-I alone or in combination with FSH or hCG suppresses spontaneous apoptosis in rat preovulatory follicles.

Gonadal IGFs system is distinctly established in fish ovary and testis and their role in regulation of different reproductive events are thoroughly studied. Numerous studies suggest that IGF-I and IGF-II are substantially involved in the regulation of growth, differentiation and reproduction by selectively promoting mitogenesis and differentiation and inhibiting apoptosis (Bobe et al., 2003; Wood et al., 2005). Recently, the presence of a third form of IGF encoded by a separate gene has been detected in tilapia and zebrafish (Wang et al., 2008). The presence of this novel form IGF-III appears to be limited to fish. The expression of IGF-III mRNA is gonad specific and restricted to somatic and granulosa cells in the ovary and to interstitial cells in the testis.

IGF-I peptide and/or mRNA have been indentified in granulosa and thecal cells from the onset of their development in red seabream, gilthead seabream and tilapia. In gilthead seabream and tilapia, IGF-I mRNA is detected in young and previtellogenic oocytes of tilapia (Berishvilli et al., 2006; Vinas and Piferrer , 2008). IGF-II mRNA and peptide has also been detected in granulosa cells of the late follicle stages of tilapia (Vinas and Piferrer, 2008). Both the IGFs (IGF-I and IGF-II) have been reported to influence the folliculogenesis, and oocyte maturation in fishes (Bobe et al., 2003; Sarang and Lal, 2005; Lokman et al., 2007). High amount of IGF-I and IGF-IR mRNA is present in previtellogenic follicles of starlet in recrudescing females than in non- recrudescing females, and females entering vitellogenesis exhibit an increase in ovarian IGF-I and IGF-IR mRNA (Wuertz et al., 2007).

The IGF-IR is also reported from the granulosa and thecal cells of coho salmon and in gilthead seabream and sterlet (Wuertz et al., 2007) previtellogenic oocytes, suggesting the importance of local IGF-I in ovary. Reinecke (2010) has recently suggested that IGFs promotes proliferation of follicular cells and oocyte maturation. Ovarian IGF-I increases the production of gap junctions proteins between granulosa cells and oocytes in red seabream (Petino and Kagawa, 1999). IGFs modulate steroidogenesis in fishes. IGF-I inhibits testosterone and 17α-hydroxyprogesterone production by thecal cells and stimulates secretion of 17β-estradiol and 17α, 20β-dihydroxyprogesterone by the granulosa cells in the preovulatory ovary in coho Salmon and stripped bass by inducing both the P450 aromatase gene expression and aromatase activity (Sullivan and Weber, 2000).

In fish testis also, IGFs and their receptors system have been very well demonstrated. IGF-I mRNA and/or peptide have been detected in the spermatogonia, spermatocytes, Sertoli, and Leydig cells of rainbow trout, tilapia, and sea bass (Le Gac et al., 1996; Berishvili et al., 2006; Vinas and Piferrer , 2008) and are reported to influence the testicular activities through autocrine/paracrine mechanism. IGF receptor (IGF-IR) is also demonstrated in spermatogonia and primary spermatocytes of rainbow trout (Le Gac et al., 1996) and gilthead seabream. No study has yet been reported on the cellular sources of IGF-II in fish testis, though its presence has been shown by PCR in carp (Tse et al., 2002) and tilapia (Caelers et al., 2004). Moreover, IGF-I is reported to stimulate spermatogenesis by inducing production of 11-ketotestosterone (Tokalov and Gutzeit, 2005). It also promotes the incorporation of thymidine into spermatogonia and primary spermatocytes from cultured spermatogenic rainbow trout testis. Reinecke (2010) has suggested that IGFs regulate spermatogenesis in fishes in conjunction with 11-ketotestosterone.

Conclusion and Future Perspectives

It can lastly be concluded that intra-gonadal factors effectively regulate follicular growth, gemetogenesis, final maturation of gametes and steroidogenesis in fishes. The existing reports on presence of intra-gonadal growth factors, their receptors and reactive nitrogen species (NO), though meager and fragmentary, undoubtedly lay foundation of a hypothesis that fish gonad is equipped with local network of regulatory paracrine factors, which not only directly regulate different reproductive events but also mediate and/ or modulate the actions of endocrine hormones.

However, many more studies are warranted to strengthen this hypothesis. Because, majority of such evidence are preliminary in nature and rely mostly on the availability of mRNA transcripts of growth factors and their receptors in the ovary of zebrafish and goldfish. Related study in fish testis is highly scanty. Studies focusing on demonstration of related peptides in fish gonad are very few and largely based on the antibodies against mammalian growth factors.

Therefore, the current evidence on paracrine regulation of gonadal activities is only the tip of the iceberg. Continued efforts are undoubtedly needed to characterize such peptides in more and more fish species and also to develop specific analytical tools, as the fishes employ wide variety of mode and regulatory mechanism of reproduction. Spacio-temporal expressions of paracrine factors in somatic and germ cells in fish gonad open new dimensions of future studies on dynamism and regulation of their secretion, site(s) of their action, interactions among themselves as well as with endocrine hormones delivered to gonad to accomplish an intricate regulation of fish reproduction.

References

Berisha, B. and Schams, D. (2005). Ovarian function in ruminants. *Domest. Anim. Endocrinol, 29,* 305-317.

Berishvili, G., D'Cotta, H., Baroiller, J. F., Segner, H. and Reinecke, M. (2006). Differential expression of IGF-I mRNA and peptide in the male and female gonad during early development of a bony fish, the tilapia, *Oreochromis niloticus. Gen. Comp. Endocrinol, 146,* 204-210.

Bobe, J., Maugars, G., Nguyen, T., Rime, H. and Jalabert, B. (2003). Rainbow trout follicular maturational competence acquisition is associated with an increased expression of follicle stimulating hormone receptor and insulin-like growth factor-II messenger RNAs. *Mol. Reprod. Dev, 66,* 46-53.

Caelers, A., Berishvili, G., Meli, M. L., Eppler, E. and Reinecke, M. (2004). Establishment of a real-time RT-PCR for the determination of absolute amounts of IGF-I and IGF-II gene expression in liver and extrahepatic sites of the tilapia. *Gen. Comp. Endocrinol, 137,* 196-204.

Calp, K. M., Matsumoto, J. A. and Van der Kraak, G. (2003). Activin and transforming growth factor-β as local regulators of ovarian steroidogenesis in the goldfish. *Gen. Comp. Endocrinol,* 132, 142-150.

Clelland, E., Kohli, G., Campell, R. K., Sharma, S., Shimasaki, S. and Peng, C. (2006). Bone morphogenetic protein-15 in the zebrafish ovary: complementary deoxyribonucleic acid cloning, genomic organization, tissue distribution, and role in oocyte maturation. *Endocrinol,* 147, 201-209.

Clelland, E., Tan, Q., Balofsky, A., Lacivita, R. and Peng, C. (2007). Inhibition of premature oocyte maturation: a role for bone morphogenetic protein-15 in zebrafish ovarian follicles. *Endocrinol,* 148, 5451-5458.

Conti, M., Hsieh, M., Park, J. and Su, Y. (2006). Role of epidermal growth factor network in ovarian follicles. *Mol Endocrinol, 20,* 715-723.

Creech, M. M., Arnold, E. V., Boyle, B., Muzenich, M. C., Montvtlle, C., Bohle, D. S. and Atherton, R. W. (1998). Sperm motility enhancement by nitric oxide produced by the ocytes of fathead minnows, *Pimephelas promelas. J. Androl, 19,* 667-674.

Drummond, A. E. (2005). TGF-β signaling in the development of ovarian function. *Cell Tissue Res, 322,* 107-115.

Dubey nee Pathak, N. and Lal, B. (2008). Nitric oxide: an autocrine regulator of Leydig cell steroidogenesis in the Asian catfish, *Clarias batrachus. Gen. Comp. Endocrinol, 158,* 161-167.

Dubey nee Pathak, N. and Lal, B. (2009). Paracrine role of macrophage produced-nitric oxide (NO) in Leydig cell steroidogenesis in a teleost, *Clarias batrachus*: impact of gonadotropin, growth hormone and insulin on NO production by testicular macrophages. *Gen. Comp. Endocrinol, 160,* 12-18.

Dubey nee Pathak, N. and Lal, B. (2010). Seasonality in expression and distribution of nitric oxide synthase isoforms in the testis of the catfish, *Clarias batrachus*: role of nitric oxide in testosterone production. *Comp. Biochem. Physiol, 151,* 286-293.

Ge, W., Cook, H., Peter, R. E., Vaughan, J. and Vale, W. (1993). Immunocytochemical evidence for the presence of inhibin and activin-like proteins and their localization in goldfish gonads. *Gen. Comp. Endocrinol, 89,* 333-340.

Hales, D. B. (2002). Testicular macrophages modulation of Leydig cell steroidogenesis. *J. Reprod. Immunol, 57,* 3-18.

Han, I. S., Sylvester, S. R., Kim, K. H., Schelling, M. E., Venteswaran, S., Blanckaert, V. D., McGuinness, M. P. and Griswold, D. (1993). Basic fibroblast growth factor is a testicular germ cell product which regulate Sertoli cell function. *Mol. Endocrinol, 7,* 889-897.

He, J., Dong, C., You, R., Zhu, Z., Lv, L. and Smith, G. W. (2009). Localization of epidermal growth factor (EGF) and its receptor (EGFR) during postnatal testis development in the alpaca, *Lama pacos. Ani. Reprod. Sci, 116,* 155-161.

Huston, J. C. (2006). Physiological interaction between macrophages during development. *Biol. Reprod, 43,* 885-890.

Itoh, N. and Konishi, M. (2007). The zebrafish family. *Zebrafish, 4,* 179-186.

Janz, D. M. and Van Der Kraak, G. (1997). Suppression of apoptosis by gonadotropin, 17β-estradiol, and epidermal growth factor in rainbow trout preovulatory ovarian follicles. *Gen. Comp. Endocrinol, 105,* 186-193.

Kim, H. C., Byun, J., Lee, T. K., Jeong, C. W., Ahn, M. and Shin, T. (2006). Expression of nitric oxide synthase isoforms in the testis of pigs. *Anatomia Histologia Embryologia, 36,* 135-138.

Knight, P. G. and Glister, C. (2006). TGF-β superfamily members and ovarian follicle development. *Reproduction, 132,* 191-206.

Kohli, G., Clelland, E. and Peng, C. (2005). Potential target of transforming growth factor-β1 during inhibition of oocyte maturation in zebrafish. *Reprod. Biol. Endocrinol, 3,* 53-64.

Kohli, G., Hu, S., Clelland, E., Muccio, T. D., Rothenstein, J. and Peng, C. (2003). Cloning of transforming growth factor-β1 (TGF-β1) and its type II receptor from zebrafish ovary and role of TGF-β1 in oocyte maturation. *Endocrinol, 144,* 1931-1941.

Kwok, H. F., So, W. K., Wang, Y. and Ge, W. (2005). Zebrafish gonadotropins and their receptors: I. Cloning and characterization of zebrafish follicle stimulating hormone and luteinizing hormone receptors: evidence for their distinct functions in follicle development. *Biol. Reprod, 72,* 1370-1381.

Le Gac, F., Loir, M., Le Bail, P. Y. and Ollitrault, M. (1996). Insulin-like growth factor-I mRNA and IGF-I receptor in trout testis and in isolated spermatogenic and Sertoli cells. *Mol. Reprod. Dev, 44,* 23-35.

Lee, N. P. Y. and Cheng, C. Y. (2004). Nitric oxide/nitric oxide synthase, spermatogenesis, and tight junction dynamics. *Biol. Reprod, 70,* 267-276.

Liu, L. and Ge, W. (2007). Growth differentiation Factor-9 and its spatiotemporal expression and regulation in the zebrafish ovary. *Biol. Reprod, 76,* 294-302.

Lokman, P.M., George, K.A., Divers, S.L., Algie, M. and Young, G. (2007). 11-Ketotestosterone and IGF-I increase the size of previtellogenic oocytes from shortfinned eel, *Anguilla australis, in vitro. Reproduction, 133,* 955-967.

Maruo, T., Landines-Llave, C. A., Samoto, T., Matsuo, H., Manalo, A. S., Ito, H. and Mochizuki, M. (1993). Expression of epidermal growth factor and its receptor in the human ovary during follicular growth and regression. *Endocrinol, 132,* 924-931.

Miura, T. and Miura, C. I. (2003). Molecular control mechanism of fish spermatogenesis. *Fish Physiol. Biochem, 28,* 181-186.

Mousa, M. A. and Mousa, S. A. (2003). Immunohistochemical localization of inhibin and activin-like proteins in the brain, pituitary gland, and the ovary of thin-lipped grey mullet, *Liza ramada* (Risso). *Gen. Comp. Endocrinol, 132,* 434-443.

Pang, Y. and Ge, W. (2002). Epidermal growth factor and TGFα promote zebrafish oocyte maturation *in vitro*: potential role of the ovarian activin regulatory system. *Endocrinol, 143,* 47-54.

Pang, Y. and Ge, W. (2002a). Activin stimulation of zebrafish oocyte maturation *in vitro* and its potential role in mediating gonadotropin-induced oocyte maturation. *Biol. Reprod, 61,* 987-992.

Pati, D., Balshaw, K., Grinwich, D. L., Hollenberg, M. D. and Habibi, H. R. (1996). Epidermal growth factor receptor binding and biological activity in the ovary of goldfish, *Carassius auratus*. *Am. J. Physiol, 270,* 1065-1072.

Patino, R. and Kagawa, H. (1999). Regulation of gap junctions and oocyte maturational competence by gonadotropin and insulin-like growth factor-I in ovarian follicles of red seabream. *Gen. Comp. Endocrinol, 15,* 454-462.

Peng, Y. and Ge, W. (2002b). Gonadotropin regulation of activin βA and activin type IIA receptor expression in the ovarian follicle cells of the zebrafish, *Danio rerio*. *Mol. Cell Endocrinol, 188,* 195-205.

Poon, S., So, W., Yu, X., Liu, L. and Ge, W. (2009). Characterization of inhibin a subunit (inha) in the zebrafish: evidence for a potential feedback loop between the pituitary and ovary. *Reproduction, 138,* 709-719.

Reinecke, A. (2010). Insulin like growth factors and fish reproduction. *Biol. Reprod, 82,* 656-661.

Roselli, M., Keller, P. J. and Dubey, R. K. (1998). Role of nitric oxide in the biology, physiology and pathophysiology of reproduction. *Hum. Reprod. Update, 4,* 3-24.

Russo, A., Angelini, F., carotenuto, R., Guarino, F. M., Falugi, C. and Campanella, C. (2000). Spermatogenesis in some Antartic teleosts from the Ross Sea: histological organization of the testis and localization of bFGF. *Polar. Biol, 23,* 279-287.

Sarang, M.K. and Lal, B. (2005). Effect of piscine GH/IGF-I on final oocyte maturation in vitro in *Heteropneustes fossilis*. *Fish Phyisol. Biochem, 31,* 231-233.

Schulz, R. W., Menting, S., Bogerd, J., Franca, L. R., Vilela, D. A. R. and Godinho H. P. (2005). Sertoli cells proliferation in the adult testis: evidence from two fish species belonging to different orders. *Biol. Reprod, 73,* 891-898.

Silva, C. C., Groome, N. P. and Knight, P. G. (2003). Immunohistochemical localization of inhibin/activin alpha, beta A and beta B subunits and follistatin in ovine oocytes during *in vitro* maturation and fertlilization. *Reproduction, 125,* 33-42.

Sordoillet, C., Savona, C., Chauvin, M. A., dePertii, E., Feige, J. J., Morera, A. M. and Benahmad, M. (1992). Basic fibroblast growth factor enhances testosterone secretion in cultured porcine Leydig cells: site of action. *Mol. Cell Endocrinol, 89,* 163-171.

Srivastava, R. K. and Van Der Kraak, G. (1995). Multifactorial regulation of DNA synthesis in goldfish ovarian follicles. *Gen. Comp. Endocrinol, 100,* 397-403.

Sun, Y., Zeng, S., Ye, K., Yang, C., Li, M, Huang, B., Sun, L, Zhou, L. and Wang, D. (2012). Involvement of FGF9/16/20 subfamily in female germ cell development of Nile tilapia, *Oreochromis niloticus*. *Fish Physiol. Biochem, 38,* 1427-1439.

Tan, Q., Balofsky, A., Weisz, K. and Peng, C. (2009). Role of activin, transforming growth factor-β and bone morphogenetic protein 15 in regulating zebrafish oocyte maturation. *Comp. Biochem. Physol. Part A, 153,* 18-23.

Tesone, M., Abramovich, D., Irusta, G. and Parvorell, F. (2009). Autocrine and paracrine regulation of the ovary. In P.J. Chedrese (Ed.), *Reproductive Endocrinology: A molecular approach.* (pp. 241-248). London, Springer Science + Business Media, LLC.

Tilly, J. L., Billig, H., Kowalski, K. and Hsueh, A. J. (1992). Epidermal growth factor and basic fibroblast growth factor suppress the spontaneous onset of apoptosis in cultured rat ovarian granulosa cells and follicles by a tyrosine kinase-dependent mechanism. *Mol. Endocrinol, 6,* 1942-1950.

Tokalov, S. V. and Gutzeit, H. O. (2005). Spermatogenesis in testis primary cell cultures of the tilapia, *Oreochromis niloticus. Dev. Dyn, 233,* 238-1247.

Tripathi, V. and Krishna, A. (2008). Changes in nitric oxide (NO) synthase isoforms and NO in the ovary of *Heteropnustes fossilis* (Bloch.) during thereproductive cycle. *J. Endocrinol, 199,* 307-316.

Tse, A. C. and Ge, W. (2009). Differential regulation of betacellulin and heparin-binding EGF-like growth factor in cultured zebrafish ovarian follicle cells by EGF family ligands. *Comp. Biochem. Physiol. Part A, 153,* 13-17.

Tse, A. C. and Ge, W. (2010). Spatial localization of EGF family ligands and receptors in the zebrafish ovarian follicle and their expression profiles during folliculogenesis. *Gen. Comp. Endocrinol, 167,* 397-407.

Tse, M. C. L., Vong, Q. P, Christopher, H.K. and Cheng, K. M. C. (2002). PCR-cloning and gene expression studies in common carp, Cyprinus carpio, insulin-like growth factor-II. *Biochimica et Biophysica Acta*, 1575, 63– 74.

Valenti, S., Cuttica, C. M., Fazzuoli, L., Giordano, G. and Giusti, M. (1999). Biphasic effect of nitric oxide on testosterone and cyclic GMP production by purified rat Leydig cells cultured *in vitro. Int. J. Androl, 22,* 336-341.

Vinas, J. and Piferrer, F. (2008). Stage-specific gene expression during fish spermato-genesis as determined by laser-capture microdissection and quantitative-PCR in sea bass, *Dicentrarchus labrax* gonads. *Biol. Reprod, 79,* 738-747.

Wang, Y. and Ge W. (2003c). Gonadotropin regulation of follistatin expression in the cultured ovarian follicle cells of zebrafish, *Danio rerio. Gen. Comp. Endocrinol, 134,* 308-315.

Wang, D. S., Jiao, B., Hu, C., Huang, X., Liu, Z. and Cheng, C. H. (2008). Discovery of a gonad-specific IGF subtype in teleost. *Biochem. Biophys. Res. Comm, 367,* 336-341.

Wang, Y. and Ge, W. (2003a). Spatial expression patterns of activin and its signaling system in the zebrafish ovarian follicle: evidence for paracrine action of activin on the oocytes. *Biol. Reprod, 69,* 1998-2006.

Wang, Y. and Ge, W. (2003b). Involvement of cyclic adenosine 3',5'-monophosphate in the differential regulation of activin βA and βB expression by gonadotropin in the zebrafish ovarian follicle cells. *Endocrinol, 144,* 491-499.

Wang, Y. and Ge, W. (2004). Developmental profile of activin βA, βB, and follistatin in the zebrafish ovary: evidence for their differential roles during sexual maturation and ovulatory cycle. *Biol. Reprod, 71,* 2056-2064.

Watanabe, A., Kobayashi, E., Ogawa, T. and Onitake, K. (1998). Fibroblast growth factor may regulate the initiation of oocyte growth in the developing ovary of the medaka, *Oryzias latipes*. *Zool. Sci, 15,* 531-536.

Watbabe, A., Hatakeyama, N., Yasuoka, A. and Onikite, K. (1997). The distribution of fibroblast growth factor and the mRNA for its receptor, MFR1, in the developing testis of medaka, *Oryzias latipes*. *J. Expt. Zool, 279,* 177-184.

Watnabe, A. and Onitake, K. (2005). Changes in the distribution of fibroblast growth factor in the teleostean testis during spermatogenesis. *J. Expt. Zool, 272,* 475-483.

Weber, G. M. and Sullivan, C.V. (1999). Effects of insulin-like growth factor-I on in vitro final oocyte maturation and ovarian steroidogenesis in striped bass, *Morone saxatilis*. *Biol. Reprod, 63,* 1049-1057.

Weisman, D. A., Niu, E., Sottas, G. R., Holmes, C. M., Huston, J. C. and Hardy, M. P. (2005). Paracrine modulation of androgen synthesis in rat Leydig cell by nitric oxide. *Int. J. Androl, 26,* 369-378.

Wiater, E. and Vale, W. (2008). Activins and inhibins. In R. Deryck, and K. Miyazono (Eds). *The TGF-β Family* (pp 79-120). New York: Cold Spring Harbor Laboratory Press.

Wilson-Leedy, J. G. and Ingermann, R. L. (2011). Production of nitric oxide by sperm of the steelhead, Oncorhynchus mykiss and its actions on motility and respiration. *Theriogenology, 75,* 144-154.

Wood, A. W., Duan, C. and Bern, H. A. (2005). Insulin-like growth factor signaling in fish. *Int. Rev. Cytol, 243,* 215-285.

Wuertz, S., Gessner, J., Kirschbaum, F. and Kloas, W. (2007). Expression of IGF-I and IGF-I receptor in male and female sterlet, *Acipenser ruthenus*: evidence for an important role in gonad maturation. *Comp. Biochem. Physiol. Part A, 147,* 223-230.

Zini, A., Abitbol, J., Girardi, S. K., Schulsinger, D., Goldstein, M. and Schlegel, P. N. (1998). Germ cell apoptosis and endothelial nitric oxide synthase (eNOS) expression following ischemia-reperfusion injury to testis. *Arch. Androl, 41,* 57-65.

Zini, A., O'Bryan, M. K., Magid, M. S. and Schlegel, P. N. (1996). Immunohistochemical localization of endothelial nitric oxide synthase in human testis, epididymis, and vas deferens suggests a possible role for nitric oxide in spermatogenesis, sperm maturation, and programmed cell death. *Biol. Reprod, 55,* 935-941.

In: Sexual Plasticity and Gametogenesis in Fishes
Editor: Balasubramanian Senthilkumaran
ISBN: 978-1-62618-848-8

Chapter XVI

Involvement of Brain-Pituitary-Gonadal Axis on Regulation of the Reproductive Cycle in Female Chub Mackerel

Michiya Matsuyama, Sethu Selvaraj, Mitsuo Nyuji and Hirofumi Ohga
Laboratory of Marine Biology, Faculty of Agriculture
Kyushu University, Hakozaki, Higashi-ku, Fukuoka, Japan

Abstract

The chub mackerel (*Scomber japonicus*) is one of the commercially important marine pelagic fishery resources of the world, and aquaculture of this species commenced recently in Japan. Captive maintained male fish undergo normal spermatogenesis but female fish after the completion of vitellogenesis fails to undergo final oocyte maturation and ovulation. The Brain-Pituitary-Gonadal (BPG) axis is an indispensable neuroendocrine pathway shown to be involved in the regulation of reproductive cycle of teleosts and understanding their involvement is important for aquaculture and stock management of chub mackerel. This chapter introduces the importance of chub mackerel as an experimental animal and the experimental system, followed by the characterization and analyses of hormonal players of BPG axis namely, kisspeptins, gonadotropin-releasing hormones (GnRHs), gonadotropins (GtHs) and sex steroids during the reproductive cycle.

1. Introduction

In teleosts, like other vertebrates, reproductive processes are regulated by the precise coordination of neuroendocrine hormones acting through the brain–pituitary–gonad (BPG) axis. The neuropeptide gonadotropin-releasing hormone (GnRH) plays a central role by stimulating the synthesis and release of the pituitary gonadotropins (GtHs). These pituitary GtHs, follicle-stimulating hormone (FSH) and luteinizing hormone (LH), act on the gonads to stimulate steroidogenesis, which in turn regulates the ovarian growth and maturation (Nagahama and Yamashita, 2008). Further, gonadal sex steroids exert positive and negative feedback at different levels of the BPG axis. In recent years, kisspeptins (Kiss), a member of the RF-amide family have been shown to act as a critical endogenous regulator of GnRH neurons expressing kisspeptin receptors (Tena-Sempere et al., 2012). In addition, other neural and non-neural factors have been shown to be directly or indirectly involved in the regulation of fish reproduction (Zohar et al., 2010). In this chapter, we will limit ourselves to Kiss–GnRHs–GtHs–sex steroids that have been characterized and analyzed in adult female chub mackerel.

1.1. Experimental System

The chub mackerel, belonging to the order Perciformes and family Scombridae, is a commercial pelagic fish distributed throughout the tropical and temperate waters of the world. In Japan, this fish species is one of the most important commercially exploited fish. Due to unreliable and unpredictable wild catches, high consumer demand, and high early growth potential; this species has been targeted for aquaculture in recent years. The adult chub mackerel reared in outdoor tanks undergo normal spermatogenesis and vitellogenesis (Matsuyama et al., 2005). However, like the hatchery-reared female broodstock of many commercial marine fish species (Zohar and Mylonas, 2001), captive maintained female chub mackerel fails to undergo final oocyte maturation (FOM) and ovulation spontaneously (Shiraishi et al., 2005). These characteristic features make chub mackerel a suitable experimental model for basic research contributing to understand reproductive neuroendocrinology of scombroid fish. The experimental system of chub mackerel comprises of sea pens and outdoor tanks. In the sea pens, wild caught mackerels are reared for one year after capture and sampling are performed at different times of the year corresponding to seasonal reproductive cycle stages (Shiraishi et al., 2009; Figure 1A).

The ovarian developmental stages in the chub mackerel were classified by the most advanced group of oocytes (Shiraishi et al., 2005). Accordingly, the ovarian stages for seasonal reproductive cycle were divided into four stages as follows: IM (immature), EV (early vitellogenesis), LV (late vitellogenesis), and PS (post-spawning). As female chub mackerel fails to undergo FOM and ovulation in captive conditions, our group developed an induced spawning protocol based on sustained release GnRH analogue delivery system. During the spawning season (April-June), fish are transferred to 3-ton outdoor concrete tanks and reared under natural photoperiod and temperature. After acclimatization for a week, fish are anaesthetized with 2-phenoxyethanol (100 mg/l) and screening of females with fully-grown vitellogenic oocytes (600-650 μm in diameter) are performed through ovarian biopsy using a plastic catheter tube. In addition, males oozing milt under gentle abdominal pressure are selected. After selection of required number of females and males, intramuscular injection with the GnRH analogue [des-Gly^{10}-(D-Ala^{6}) LHRH ethylamide, 400 μg/kg BW] mixed with molten cocoa butter was performed. The first ovulation occurs 34-36 h post-injection (Shiraishi et al., 2008), and subsequent daily spawning is observed over a month when the water temperature range between 18-23°C. The fish are fed twice per week during the experiments.

A previous study by Scott et al. (1999) demonstrated significant decline in the plasma concentration of GnRH agonist on day 5 after intramuscular injection with the GnRH agonist suspended in coconut oil in the plaice (*Pleuronectes platessa*). Considering the above, fish sampling are performed after one week of GnRHa injection in chub mackerel based on the presumption that endogenous hormonal pathway of BPG-axis regulates multiple spawning. Based on the information on time course of FOM and ovulation in captive chub mackerel (Shiraishi et al., 2008), sampling are performed at different times of the day, corresponding to spawning cycle stages (Figure 1B). Spawning cycle stages are classified as follows: LV (late vitellogenesis), GVM (germinal vesicle migration), HY (hydration), OV (ovulation), and POV (post-ovulation). In the spawning stock, ovulation occurs synchronously among fish prior to midnight, and spawning is performed immediately following ovulation between 21.00-01.00 h, with peaking around 23.00 h. This experimental system facilitate fish sampling at different ovarian stages to elucidate the role of key endocrine hormones regulating seasonal and spawning cycles in female chub mackerel.

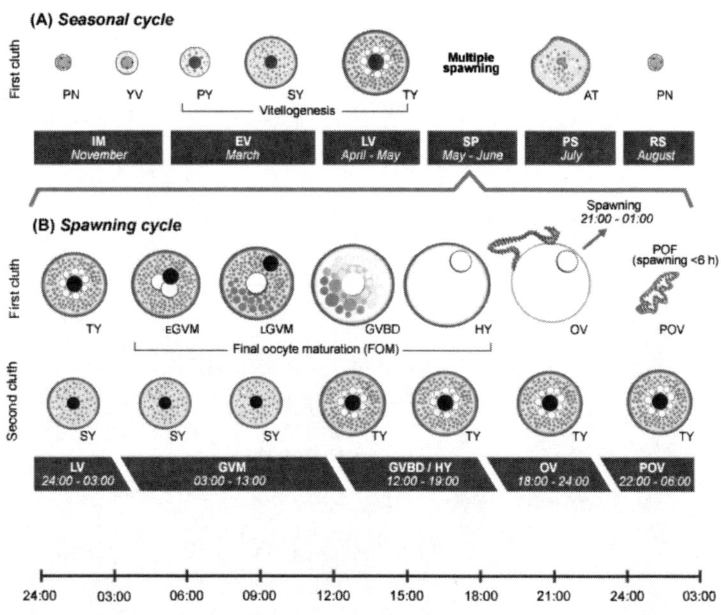

Figure 1. Classification of chub mackerel ovarian stages during the seasonal reproductive and spawning cycles. (A) Ovarian stages during the seasonal cycle were classified based on developmental stages of the most advanced oocytes. IM (immature), oocytes at peri-nucleolus (PN) stage and yolk vesicle (YV) stages; EV (early vitellogenesis), oocytes at primary yolk (PY) and secondary yolk (SY) stages; LV (late vitellogenesis), oocytes at tertiary yolk (TY) stage; SP (spawning); PS (post-spawning), period with atretic oocytes (AT); RS (resting), stage with oocytes at PN stage. (B) Ovarian stages during the spawning cycle were classified based on developmental stages of the most advanced oocytes (first clutch) and the presence of postovulatory follicles (POFs) (Shiraishi et al., 2008; 2009). LV (late vitellogenesis), oocytes at TY stage; GVM (germinal vesicle migration), oocyte at EGVM (early GVM) and LGVM (late GVM) stages; GVBD/HY (germinal vesicle breakdown or hydration), oocytes at germinal vesicle breakdown or hydrated and transparent oocytes after GVBD; OV (ovulated), ovaries just after ovulation; POV (post-ovulation), oocytes at tertiary yolk stage with new POFs (spawning <6h).

2. Kisspeptins and Their Receptors

2.1. Kisspeptins

Kisspeptins are peptide products of the *Kiss1* gene, belonging to the RFamide family, which share a common Arg-Phe-amide motif at their C-termini and are ligands for the G protein–coupled receptor 54 (GPR54 or Kiss1r) (Roseweir and Millar, 2009). In contrast to the placental mammals expressing a single kisspeptin gene, few teleosts express *kiss1* and *kiss2* (reviewed in Tena-Sempere et al., 2012). However, teleost species such as the puffer fish (*Tetraodon nigroviridis)*, stickleback (*Gasterosteus aculeatus*), including some perciform fish express only *kiss2* (reviewed in Kanda and Oka, 2012). In mammals, the Kiss1 precursor protein is cleaved into several mature peptides, including kisspeptin-54, -14, -13 and -10 (Kiss-10) and are reported to activate kisspeptin receptor (Kiss1r) with equal biopotency (Kotani et al., 2001). However, recent studies indicate that teleosts *kiss1* and *kiss2* would produce mature peptides of Kiss1-15 and Kiss2-12, based on the conservation of dibasic amino residues in the N-terminal of Kiss-10 region and ligand selectivity analysis indicating zebrafish (zf) Kiss1-15 and Kiss2-12 exhibiting higher potency for zfKiss1r (GPR54-1) and zfKiss2r (GPR54-2), respectively (Lee et al., 2009; Tena-Sempere et al., 2012).

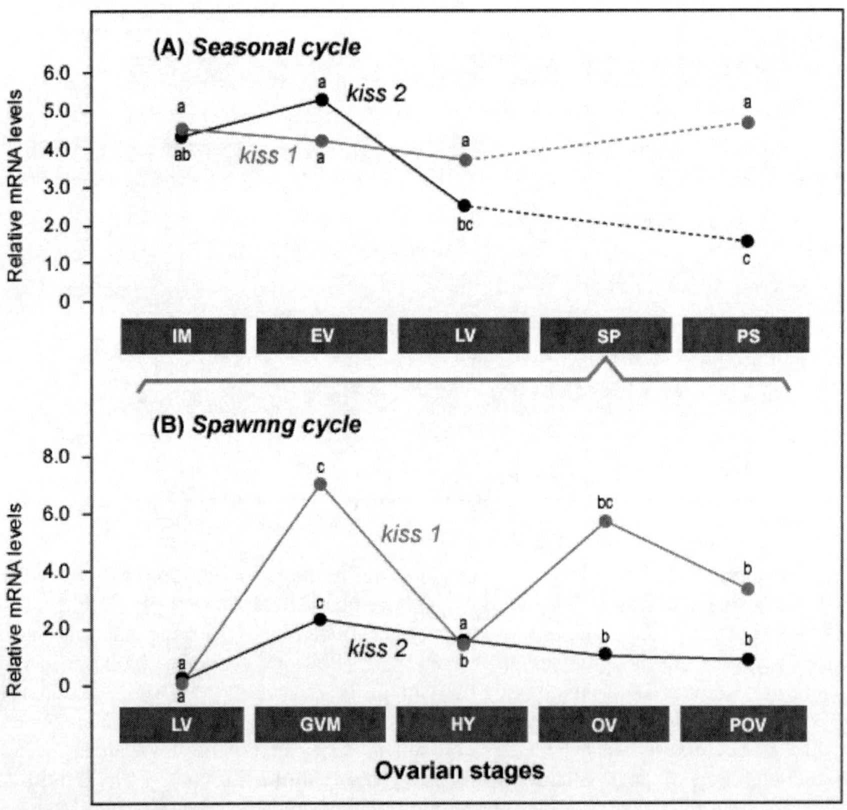

Figure 2. Expression changes of *kiss1* and *kiss2* mRNAs in the brain during the seasonal (A) and spawning (B) cycles. Different letters above and below the circles represent significant differences between ovarian stages. Refer Figure 1 for description of ovarian stages.

The chub mackerel brain found to express both *kiss1* and *kiss2*. Expression analysis indicated increased levels of *kiss2* in the brain during immature and early vitellogenesis and reduced levels during late vitellogenesis and post-spawning periods (Figure 2A). Levels of *kiss1* did not show any significant changes. Interestingly, both *kiss1* and *kiss2* levels fluctuated during the spawning cycle (Selvaraj et al., 2012a). Both *kiss1* and *kiss2* levels in the brain were low at late vitellogenic stage and increased significantly during the germinal vesicle migration (GVM) period. However, *kiss1* levels decreased during hydration before increasing again at ovulatory and post-ovulatory periods. In contrast, *kiss2* levels decreased at ovulatory and post-ovulatory periods (Figure 2B). In Senegalese sole (*Solea senegalensis*), both *kiss2* and *kiss2r* expression was found to be highest either before or during the spawning season (Mechaly et al., 2012). Similarly, in the brain of mature female striped bass (*Morone saxatilis*), both *kiss1* and *kiss2*, including levels of their receptors *gpr54-1* and *gpr54-2*, were found to be significantly higher in comparison to juvenile and prepubertal fish (Zmora et al., 2012). These results suggest an ovarian stage dependent expression of *kiss1* and *kiss2* in teleosts.

In mammals, sex steroids have been demonstrated as key regulator of hypothalamic *Kiss1* expression through negative and positive feedback (Roa and Tena-Sempere, 2007). In medaka (*Oryzias latipes*), *kiss1* neurons of the hypothalamic area co-expressing estrogen receptor-α (ERα) shown to be involved in the positive-feedback effect of ovarian estrogens (Kanda et al., 2008; Mitani et al., 2010). Similarly, in the goldfish (*Carassius auratus*), *kiss2* but not *kiss1* neurons of the preoptic area (POA) co-expressing estrogen receptors shown to be up regulated by ovarian estrogens (Kanda et al., 2012). These studies indicate that ovarian sex steroids are prime regulator of *kiss* expression in the brain.

Neuroanatomical studies have revealed that *kiss1* expressing neurons are mainly localized in the habenula of zebrafish (Kitahashi et al., 2009; Servilli et al., 2011), medaka (Kanda et al., 2008; Mitani et al., 2010), and goldfish (Kanda et al., 2012). However, *kiss2* neurons are distributed in the hypothalamic regions of zebrafish, medaka, and goldfish (Kitahashi et al., 2009; Mitani et al., 2010; Kanda et al., 2012), including the POA of goldfish (Kanda et al., 2012). In red seabream (*Pagrus major*), *kiss2* neurons were localized mainly in the hypothalamic regions with sporadic distribution in the POA (Shimizu et al., 2012). For the first time in a teleost fish, Servili et al. (2011) demonstrated immunocytochemical distribution of kisspeptin peptides in the zebrafish. The authors revealed that habenula Kiss1 neurons only project to the interpeduncular nucleus (IPN) and the raphe regions. Interestingly, Kiss2 neurons of hypothalamic regions project to the POA, where GnRH3 neurons reside in this species. In connection to the above, we recently produced antibodies against chub mackerel Kiss-10 regions of chub mackerel Kiss1 and Kiss2. Future studies on the localization of Kiss neurons will help to clarify its functional role in the reproduction and interaction with GnRH system.

More importantly, recent functional studies indicate that the kisspeptin peptides have the potential to induce gonadal development in teleost fish. In light of the above, our group found that chronic subcutaneous implantation of Kiss1-15 but not Kiss2-12 peptides induced spermiation and vitellogenesis onset in adult immature chub mackerel (Selvaraj et al., 2013). Likewise, peripheral administration of Kiss1-10 and Kiss2-10 peptides found to stimulate testicular development in prepubertal yellowtail kingfish (*Seriola lalandi*) (Elizur et al., 2011). More recently, Beck et al. (2012) found accelerated gonadal development in both males and females of the white bass (*Morone chrysops*), striped bass and their hybrids, when administered with Kiss1-10 and Kiss2-10 peptides.

2.2. Kisspeptin Receptors

Kisspeptins binds to its cognate receptor, G-protein-coupled receptor (GPR54 or Kiss1r) to affect other downstream signalling molecules. Unlike mammals expressing a single *GPR54* gene, teleosts express multiple GPR54 receptor genes (Akazome et al., 2010; Onuma and Duan, 2012). A single form of *GPR54* i.e. *gpr54-2* has been isolated from around twenty different fish species (Tena-Sempere et al., 2012). However, two distinct subtypes, *gpr54-1 and gpr54-2* have been reported in zebrafish (Biran et al., 2008), medaka (Lee et al., 2009), goldfish (Li et al., 2009), European sea bass (Tena-Sempere et al., 2012) and striped bass (Zmora et al., 2012). Our group recently found the presence of *gpr54-1* and *gpr54-2* in the brain of chub mackerel (Ohga et al., 2012a).

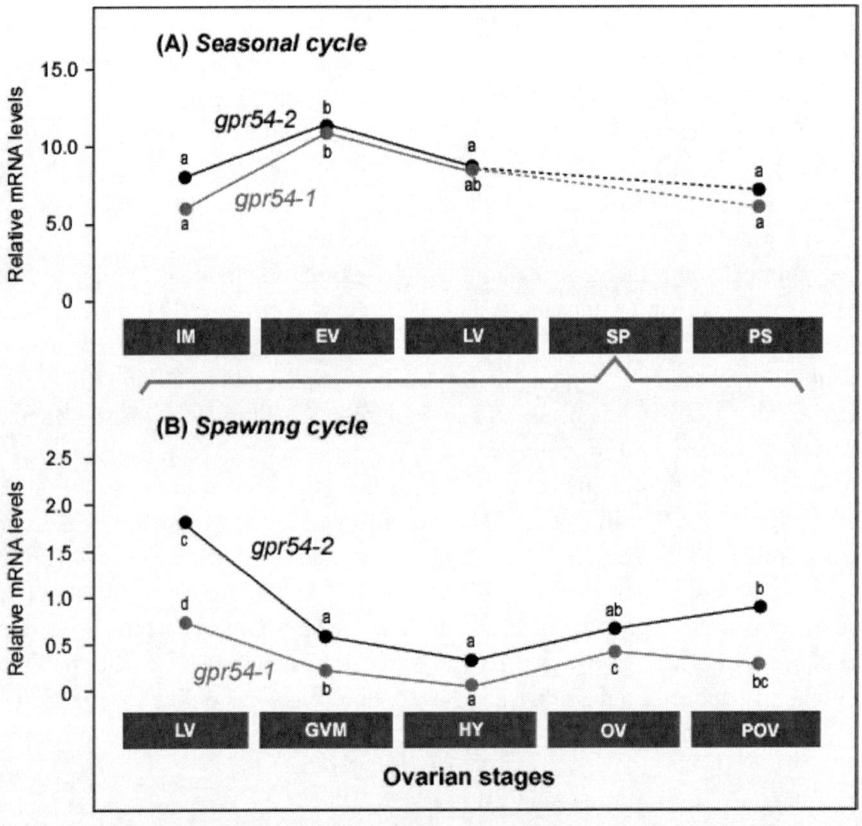

Figure 3. Expression changes of *gpr54-1* and *gpr54-2* mRNAs in the brain during the seasonal (A) and spawning (B) cycles. Different letters above and below the circles represent significant differences between ovarian stages. Refer Figure 1 for description of ovarian stages.

Expression analyses of *gpr54-1* and *gpr54-2* in the brain during seasonal reproductive cycle of female chub mackerel (Figure 3A) indicated that their levels were higher during early vitellogenic period and decreased during late vitellogenic and post-spawning period, in agreement with the expression levels of *kiss2*. Interestingly, *gpr54-1* and *gpr54-2* expression levels were lower during GVM and hydration period and increased during ovulatory and post-ovulatory periods, respectively (Figure 3B; Ohga et al., 2012a). A recent report in zebrafish indicated that the duplicated *gpr54* genes have evolved a distinct gene expression pattern and different Kiss ligand selectivity (Onuma and Duan, 2012). Future studies on Kiss ligands and GPR54 receptors binding assay will help to clarify the physiological significance of Kiss-GPR54 expression profiles in the brain of chub mackerel.

Following the observation of Parhar et al. (2004) demonstrating *GPR54* expression by GnRH1, GnRH2, and GnRH3 neurons in the brain of tilapia (*Oreochromis niloticus*), subsequent studies have clearly indicated GPR54 expression by GnRH neurons, in addition to other brain regions. Similarly, Grone et al. (2010) in another cichlid fish (*Astatotilapia burtoni*) revealed *GPR54* expression in GnRH1 and GnRH3 neurons, including distribution in the olfactory bulb, telencephalon, diencephalon, mesencephalon, and rhombencephalon. In zebrafish, expressing *kiss1r* and *kiss2r*, Servilli et al. (2011) shown that *kiss1r* is distributed in the regions where Kiss1 neurons project i.e. IPN and raphe regions. Similarly, *kiss2r* distribution was found in the regions where Kiss2 neurons project i.e. POA and hypothalamic regions. Recently, we have developed antibodies against chub mackerel GPR54-1 and GPR54-2 and future studies on their distribution and colocalization by GnRH neurons will shed new lights on the interaction of Kisspeptin and GnRH systems.

3. Gonadotropin Releasing Hormone

The gonadotropin releasing hormone (GnRH) forms in different vertebrate species are classified as GnRH1, GnRH2, or GnRH3, based on phylogenetic analysis and neuroanatomical distribution (Fernald and White, 1999). GnRH1 is the hypophysiotropic form, with a distribution in the neuronal population of the preoptic area (POA) and hypothalamus. In fish, the GnRH1 forms include the mammalian form (mGnRH) and various fish-specific peptides such as seabream, medaka, whitefish, catfish, and herring GnRH (sbGnRH, mdGnRH, wfGnRH, cfGnRH, and hrGnRH, respectively) (Okubo and Nagahama, 2008). The GnRH2 form exists in the midbrain tegmentum region, and it is represented in all vertebrates examined to date by chicken GnRH-II (cGnRH-II). On the contrary, GnRH3 is a teleost-specific form (salmon GnRH, sGnRH) that is expressed in neuronal populations in the olfactory bulb, terminal nerve ganglion region, and POA (Lethimonier et al., 2004). For convenience, we refer to the various forms as ‘GnRH forms’ in the present chapter. The presence of three GnRH forms in the brain of the chub mackerel, namely, sbGnRH (GnRH1), cGnRH-II (GnRH2) and sGnRH (GnRH3), was confirmed by combined high performance liquid chromatography (HPLC) and time-resolved fluoroimmunoassay (TR-FIA) (Selvaraj et al., 2009). Immunocytochemical localization of the three GnRH forms in the brain was investigated by using specific antisera and double immunolabeling was used to localize GnRH-ir (immunoreactive) fibers innervating the pituitary.

sbGnRH-ir cell bodies were localized in the nucleus preopticus of the POA, with fibers in the olfactory bulb, POA, and hypothalamus (Figure 4A). cGnRH-II-ir cell bodies were observed only in the midbrain tegmentum region, with a wide distribution of fibers, which were dense in the midbrain tegmentum and spinal cord (Figure 4B). sGnRH-ir neurons were localized in the ventral olfactory bulb and terminal nerve ganglion regions (Figure 4C). Further, sGnRH-ir fibers were found in different regions of the brain, with prominent fibers running in parallel in the preoptic area (POA) without entering the pituitary. Among the three GnRH forms, only sbGnRH-ir fibers innervated the pituitary gland from the preoptic-hypothalamic region, targeting follicle stimulating hormone (FSH) and luteinizing hormone (LH)-producing cells in the proximal pars distalis region.

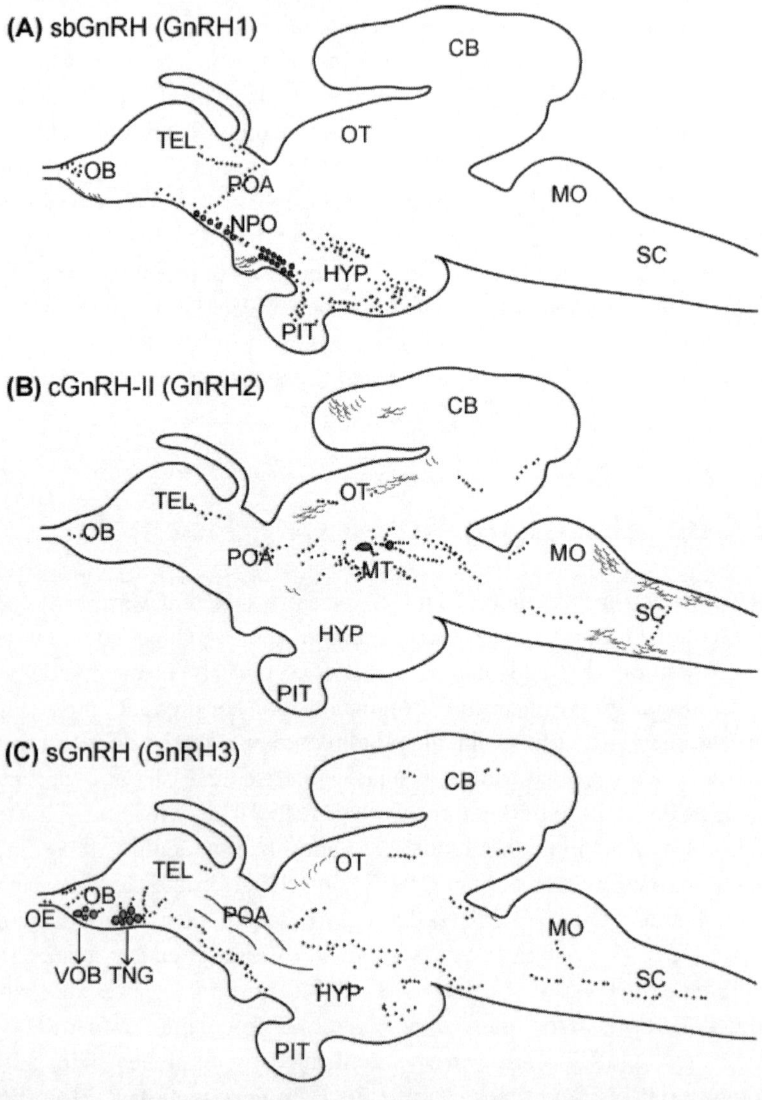

Figure 4. Immunocytochemical distribution of GnRH cell bodies (closed circles) and fibers (dots and lines) in sagittal sections incubated with (A) sbGnRH (GnRH1) antiserum, (B) cGnRH-ll (GnRH2) antiserum, and (C) sGnRH (GnRH3) antiserum. CB, cerebellum; HYP, hypothalamus; MO, medulla oblongata; MT, midbrain tegmentum; NPO, nucleus preopticus; OB, olfactory bulb; OE, olfactory epithelium; OT, optic tectum; PIT, pituitary; POA, preoptic area; SC, spinal cord; TEL, telencephalon; TNG, terminal nerve ganglion; VOB, ventral olfactory bulb. Modified from Selvaraj et al. (2009).

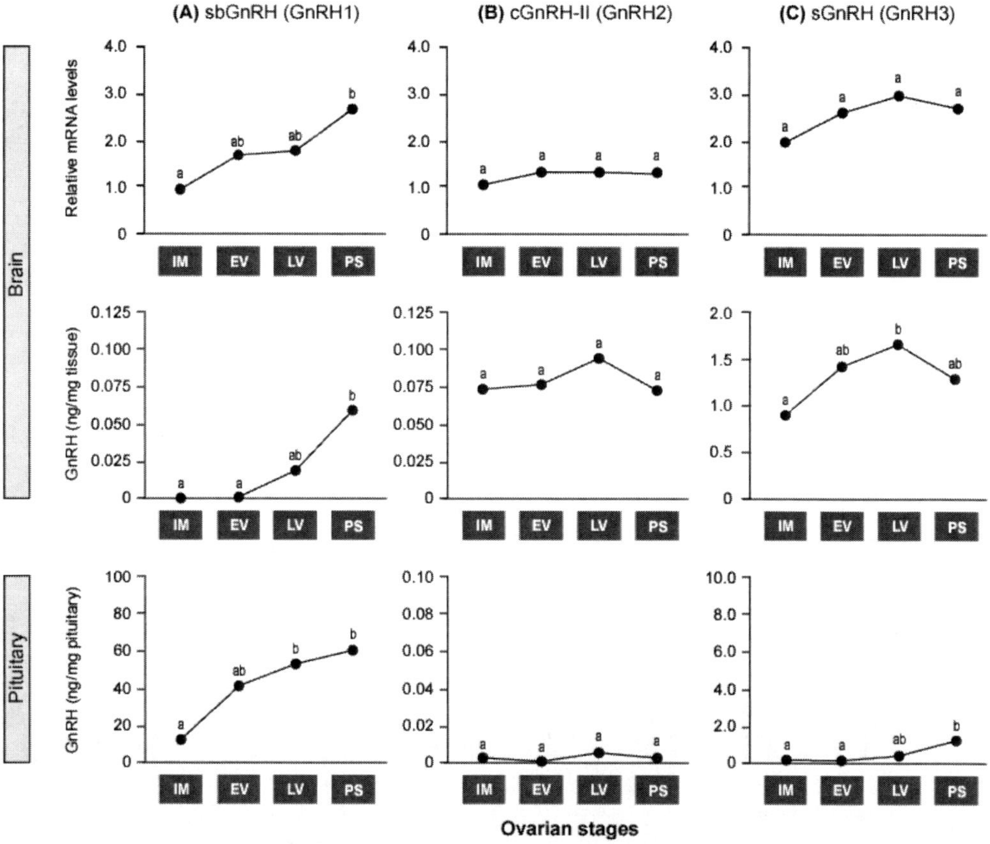

Figure 5. Expression changes of brain GnRH mRNAs and corresponding GnRH peptides in the brain and pituitary during the seasonal reproductive cycle. Different letters above the circles represent significant differences between ovarian stages. Refer Figure 1 for description of ovarian stages. Modified from Selvaraj et al. (2012b).

Recently, we isolated full-length cDNAs encoding three GnRH forms and analyzed seasonal changes in the concentrations of mRNA in the brain and corresponding peptides in the brain and pituitary (Selvaraj et al., 2012b). The results indicated that only sbGnRH mRNA and peptide concentrations in the brain showed close relations to the seasonal ovarian development (Figure 5). During the spawning cycle, brain mRNA and peptide levels of all three GnRH forms showed variations with sbGnRH peptide levels many-fold higher than the cGnRH-II and sGnRH forms (Figure 6). Likewise, pituitary sbGnRH peptide levels were many-fold higher than the other two GnRH forms, with no significant change in their levels. Thus, it was demonstrated that multiple GnRH forms exist in the brain of the chub mackerel and suggest that they serve different functions, with sbGnRH having a significant role in regulating ovarian development via stimulating FSH- and LH-producing cells, and sGnRH and cGnRH-II serving as neurotransmitters or neuromodulators.

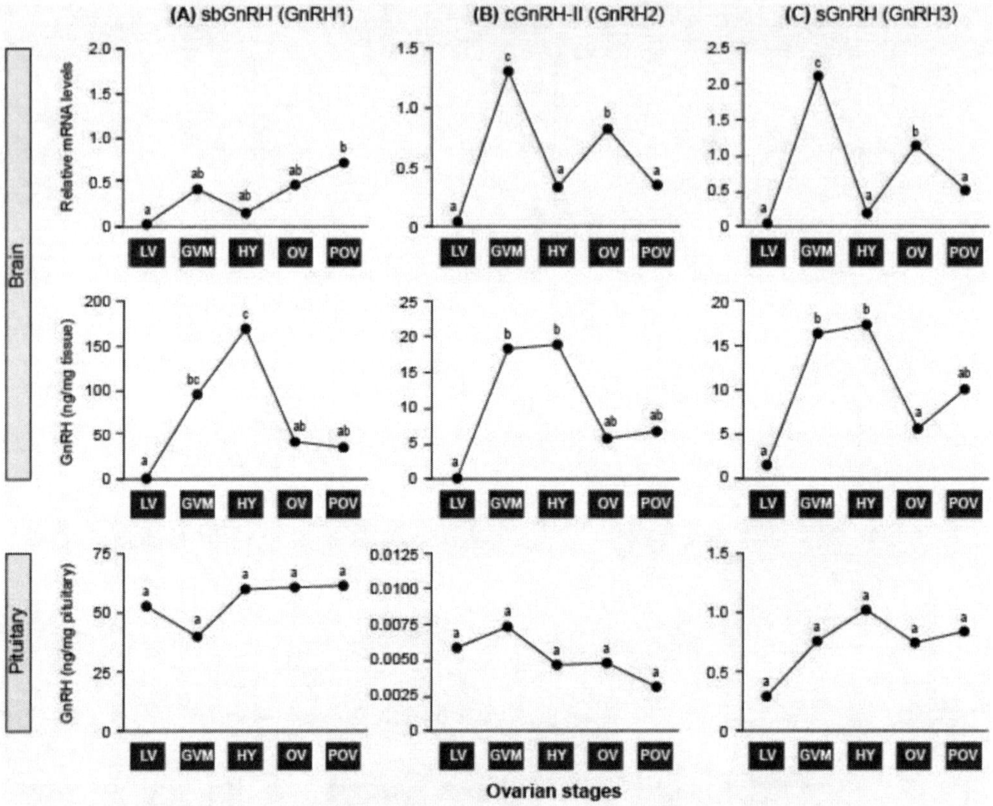

Figure 6. Expression changes of brain GnRH mRNAs and corresponding GnRH peptides in the brain and pituitary during the spawning cycle. Different letters above the circles represent significant differences between ovarian stages. Refer Figure 1 for description of ovarian stages. Modified from Selvaraj et al. (2012a).

GnRH acts via GnRH receptors (GnRHR) present on the pituitary gonadotrophs to activate the synthesis and release of pituitary GtHs (Alok et al., 2000). Teleost fish expresses four or five subtypes of GnRHRs (Moncaut et al., 2005). Even though, there is a difference in the distribution of receptor subtype in the pituitary, only one GnRH receptor subtype in each species shown to regulate pituitary gonadotropin release (reviewed in Chen and Fernald, 2008). Accordingly, few studies indicated expression changes of pituitary GnRHR at different stages of ovarian development. Alok et al. (2000) found a higher expression level of pituitary GnRHR mRNA in female striped bass showing advanced stages of ovarian development. Similarly, in European sea bass, increased expression was noted during late vitellogenesis in comparison with maturation, spawning, and post-spawning periods (González-Martínez et al., 2004). In light of the above, we recently isolated one subtype of GnRH receptor from the pituitary of chub mackerel and future studies on expression analyses and ligand-binding assay will help to further clarify the role of GnRH-GnRH receptor system in the reproductive cycle of chub mackerel.

4. Gonadotropins and Their Receptors

4.1. Gonadotropins

The pituitary of chub mackerel could be separated into four regions as in other teleosts: the rostral pars distalis (RPD), the proximal pars distalis (PPD), the pars intermedia (PI), and the neurohypophysis (NH) (Figure 7A). Immunohistochemical study using anti-FSHβ and LHβ clearly showed that FSH and LH cells were separately localized, indicating that chub mackerel FSH and LH are produced in different cell types (Figure 7B and 7C). Both FSH and LH cells occupied the entire PPD in the pituitaries of chub mackerel; however, FSH cells were localized mainly in the central area of the PPD, whereas LH cells were observed along the border of the PI. Furthermore, LH cells were also observed in a few small clusters of cells in the PI along the external border of the PPD (Nyuji et al., 2011).

In multiple spawning fish, the function of GtHs remains largely unknown, despite numerous studies in different species. Currently, there are few methods available for analyzing the function of GtHs in non-salmonids and little understanding of how they regulate gametogenesis in multiple spawning fish. ELISAs for measuring both FSH and LH have not been developed in multiple spawning fish, except for Nile tilapia using recombinant FSH and LH (Aizen et al., 2007a, b). Thus, the gene expression analyses of GtH subunit mRNAs have predominantly been used to characterize the GtH function in non-salmonids. We cloned GPα, FSHβ and LHβ from chub mackerel and analyzed their pituitary mRNA expression during the reproductive cycle in female chub mackerel.

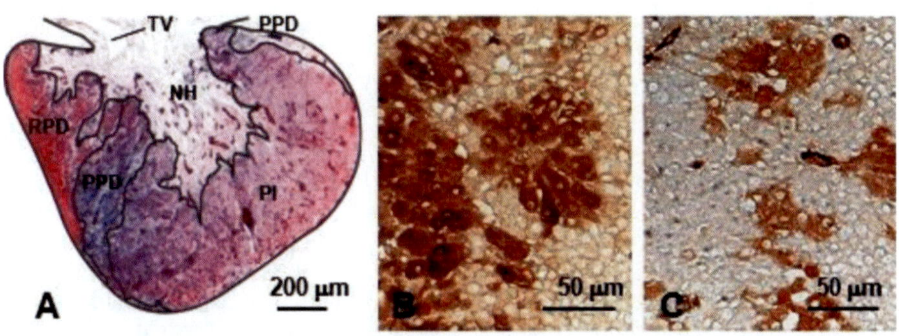

Figure 7. (A) Sagittal section of female chub mackerel pituitary at the early vitellogenic stage stained with Masson's trichrome. Anterior is on the left. (B) and (C) Two adjacent sections through the central part of the proximal pars distalis (PPD) stained with anti-FSHβ (B), and anti-LHβ (C). RPD, rostral pars distalis; PPD, proximal pars distalis; PI, pars intermedia; NH, neurohypophysis; TV, third ventricle. Modified from Nyuji et al. (2011).

During the seasonal reproductive cycle, FSHβ mRNA levels increased during ovarian development and peaked at the end of vitellogenesis (stage LV), whereas LHβ mRNA levels significantly increased during LV compared to immature and early vitellogenesis (Figure 8A). Both FSHβ and LHβ mRNA levels significantly decreased during post-spawning period (Nyuji et al., 2012a). Since pituitary FSHβ mRNA expression was high during vitellogenesis, chub mackerel FSH could be involved in vitellogenesis, similar to salmonids. High LHβ mRNA expression during late vitellogenesis may be important in preparation for FOM in chub mackerel, as has been shown in salmonids (Gomez et al. 1999). During the spawning cycle, FSHβ and LHβ had different mRNA expression patterns. FSHβ mRNA levels were high from FOM to ovulation and constant at the end of vitellogenesis (Figure 8B). Constant high expression levels of FSHβ mRNA may indicate continuous FSH synthesis during the spawning cycle, allowing vitellogenesis to occur in parallel with FOM and spawning. Interestingly, FSHβ mRNA levels increased further just after ovulation and spawning.

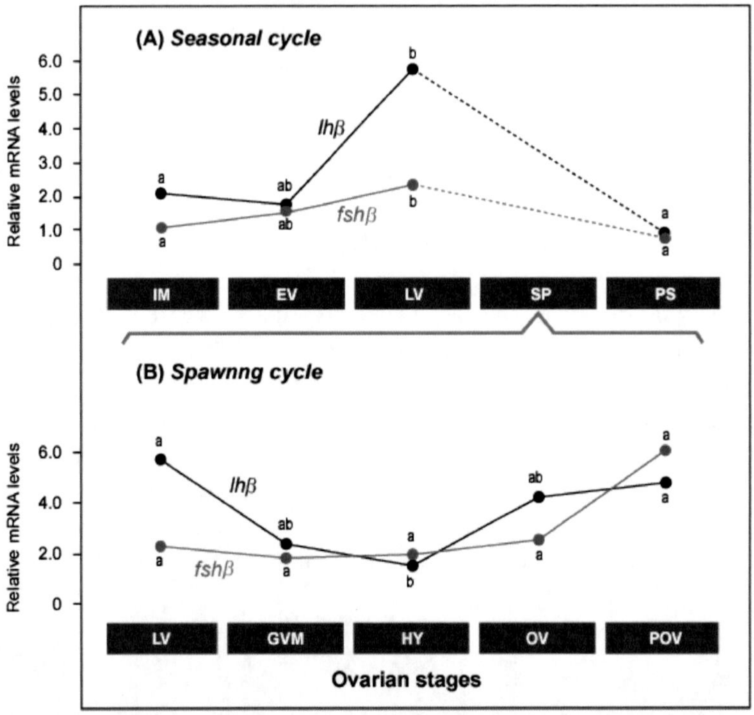

Figure 8. Changes in GtH subunit (*fshβ* and *lhβ*) mRNA levels in female chub mackerel pituitary during the seasonal (A) and spawning (B) cycles. Different letters above and below the circles represent significant differences between ovarian stages. Refer Figure 1 for description of ovarian stages. Modified from Nyuji et al. (2012a).

There are two possible explanations for this transient increase in FSHβ mRNA. First, there may be a diurnal GtH rhythm. Diurnal expression profiles of FSHβ mRNA have been proposed in two daily spawning species: bambooleaf wrasse (*Pseudolabrus sieboldi*), where FSHβ mRNA increased from the end of ovulation to the next initiation of FOM (Ohta et al. 2008), and in gilthead seabream (*Sparus aurata*), FSHβ mRNA levels increased 4 and 8 h prior to spawning during GVBD and were maintained until 8 h after spawning when FOM was initiated (Gothilf et al. 1997). In chub mackerel, it may not be possible to clarify whether the diurnal regulation of GtH subunit transcription plays a role since they do not necessarily undergo daily spawning cycles in the wild (Shiraishi et al. 2009). The second possible reason for transient increase in FSHβ mRNA expression is that ovulation and spawning may be a signal to accelerate vitellogenesis. In female Nile tilapia, plasma FSH increases prior to ovulation, possibly stimulating new generation of follicle for the next cycle (Aizen et al. 2007b).

Recently, we successfully purified native FSH and LH from the pituitaries of adult chub mackerel by anion-exchange chromatography and immunoblotting using specific antisera

(Ohga et al., 2012b). The steroidogenic potency of the intact chub mackerel FSH (cmFSH) and LH (cmLH) were evaluated in vitellogenic follicles by measuring the level of gonadal steroids. In addition, we evaluated the maturation-inducing potency of the GtHs on ovarian fragments containing late vitellogenic follicles. Both cmFSH and cmLH significantly stimulated E2 production in mid-vitellogenic follicles. In contrast, only LH significantly stimulated the production of 17,20β-P in late vitellogenic follicles. Similarly, cmLH induced FOM in ovarian fragments containing late vitellogenic follicles. Results from mRNA expression and bioassay using intact GtHs indicated that both FSH and LH may regulate vitellogenic processes, whereas only LH initiates FOM in chub mackerel (Ohga *et al.*, 2012b). Currently, using the biologically active cmFSH and cmLH purified from pituitaries, we are developing homologous ELISAs to measure plasma FSH and LH concentrations in chub mackerel. The available information regarding FSH and LH plasma profiles in the chub mackerel will confirm the link between GtH synthesis and release during the spawning cycle.

4.2. Gonadotropin Receptors (GtHR)

Multiple spawners like chub mackerel possess numerous developmental stages of oocytes in the ovary simultaneously for achieving multiple spawning episodes in one reproductive season. As vitellogenesis and oocyte maturation occur simultaneously in one ovary, ovarian follicles are likely exposed to both FSH and LH regardless of their developmental status. It is thus of great interest to elucidate how both vitellogenesis and oocyte maturation occur simultaneously and explain their precise control in a single ovary. FSH and LH are generally accepted to act in gametogenesis through their receptors (FSHR and LHR), both being members of the seven transmembrane domain G protein-coupled receptor (GPCR) superfamily (Moyle et al., 1994). To clarify the mode of action of FSH and LH in asynchronous ovarian follicle development in chub mackerel, different developmental stages of ovarian follicles were isolated during the spawning cycle, and mRNA expression levels of FSHR and LHR were measured (Nyuji et al., 2012b). *In situ* hybridization of GtH receptor mRNA in chub mackerel ovarian follicle cells showed FSHR mRNA has already been identified in the follicular layer of the pre-vitellogenic follicles in which LH mRNA has not been observed (Figure 9A). FSHR and LHR mRNAs were found in granulosa cells throughout the vitellogenesis, while both of them were hardly detectable in theca cells (Figure 9A and 9B). FSHR mRNA levels were high during vitellogenesis, while they drastically decreased during FOM. In contrast, LHR mRNA expression showed low levels during early and mid vitellogenesis, then increased and remained high during late vitellogenesis and germinal vesicle migration, but decreased in the hydrated oocyte just after germinal vesicle breakdown (Figure 10). Thus, actions of respective FSH and LH on vitellogenesis and FOM are likely regulated by differential expressions of their receptors which are found to be expressed by granulosa cells of the chub mackerel oocyte. Some *in vitro* studies have reported that both FSH and LH stimulate E2 production in salmonids (Suzuki et al., 1988; Planas et al., 2000), common carp (*Cyprinus carpio*) (Van Der Kraak et al., 1992) and bigeye tuna (*Thunnus obesus*) (Okada et al., 1994). However, the bioactivity of pituitary GtHs is still unclear in teleost fish as a result of promiscuous hormone-receptor (FSHR and LHR) interactions. Several studies have indicated that both FSH and LH bind to the FSHR with similar affinities in salmonids (Yan et al., 1992; Miwa et al., 1994) and African catfish

(*Clarias gariepinus*) (Bogerd et al., 2001; Vischer et al., 2003). Thus, it is possible that the E2 produced by cmLH might be due to cross-activation between cmLH and FSHR. Future studies on the circulating levels of FSH and LH during the spawning cycle and reporter-gene assays of cognate ligand-receptor interactions will help to clarify this possibility.

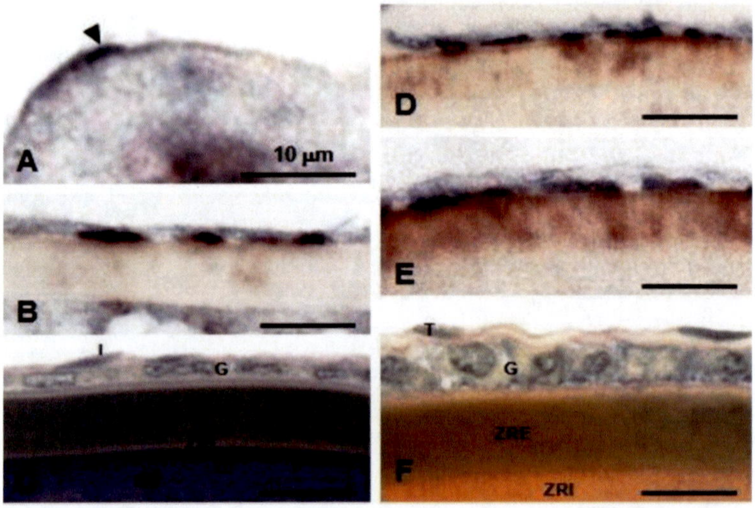

Figure 9. *In situ* hybridization of GtH receptor mRNA in chub mackerel ovarian follicles. (A) Section of peri-nucleolus stage oocyte hybridized with FSHR antisense-strand RNA probe, showing that FSHR mRNA has already been identified in the follicular layer (arrowheads). (B) Section of early vitellogenic stage ovarian follicles hybridized with FSHR antisense-strand probe. (C) Section of early vitellogenic stage ovarian follicles stained with Periodic-acid Schiff (PAS)/Haematoxylin Eosin (HE)/Metanil Yellow (MY). (D) and (E) Sections of late vitellogenic ovarian follicles hybridized with FSHR and LHR antisense-strand RNA probes, respectively. (F) Section of late vitellogenic stage ovarian follicles stained with PAS/HE/MY. Bars indicate 10 μm. G, granulosa cell; OC, oocyte cytoplasm; T, thecal cell; ZRE, zona radiata externa; ZRI, zona radiate interna. Modified from Nyuji et al. (2012b).

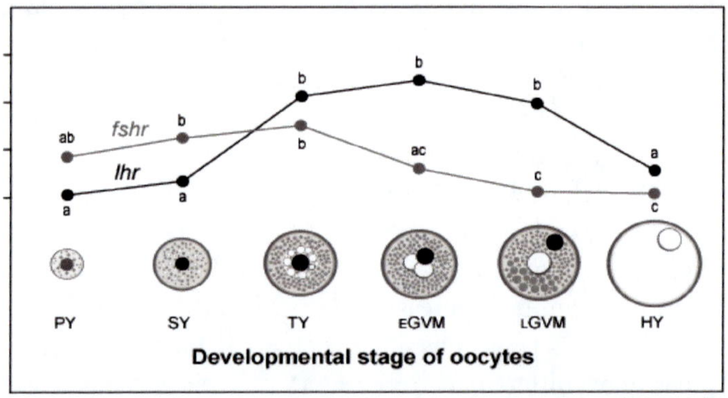

Figure 10. Changes in expression levels of *fshr* and *lhr* mRNAs in the different stage of chub mackerel ovarian follicles. PY, primary yolk stage; SY, secondary yolk stage; TY, tertiary yolk stage; EGVM, early germinal vesicle migration; LGVM, late germinal vesicle migration; HY, hydration. Modified from Nyuji et al. (2012b).

5. Sex Steroid Hormones

Intact chub mackerel ovarian follicles were incubated with different radiolabeled steroid precursors to clarify the synthetic pathways of steroid hormones during vitellogenesis and following FOM. During vitellogenesis, E2 was synthesized from pregnenolone (P5) via 17-hydroxypregnenolone (17-P5), 17-hydroxyprogesterone (17-P), androstenedione (AD), and testosterone (T) (Matsuyama et al., 2005) (Figure 11). After vitellogenesis, the steroidogenic pathway shifts from E2 to maturation-inducing hormone (MIH) production owing to the inactivation of 17,20-lyase and the activation of 20β-hydroxysteroid dehydrogenase (20β-HSD). Of the new steroids produced during FOM, 17,20β-dihydroxy-4-pregnen-3-one (17,20β-P) was most effective at inducing germinal vesicle breakdown *in vitro* (Table 1). Circulating levels of 17,20β-P increased specifically around the time of germinal vesicle migration, while another FOM-specific 20β-hydroxylated progestin, 17,20β,21-trihydroxy-4-pregnen-3-one (20β-S), was present at consistently low levels during FOM (Figure 12). These results indicate that 17,20β-P is the MIH of chub mackerel. During the FOM, the production of 17,20β-P-5β (which is the 5β-reduced form of 17,20β-P) increased with 17,20β-P production. The 17,20β-P-5β was much less effective than 17,20β-P for FOM induction in chub mackerel. The quick conversion of 17,20β-P observed in the chub mackerel probably represents the inactivation process of MIH. Similar 5β-reduction of 17,20β-P has been observed in maturing follicles of medaka (Fukada et al., 1994), Japanese yellowtail (*Seriola quinqueradiata*, Rahman et al., 2001), bambooleaf wrasse (Ohta and Matsuyama, 2002), and red seabream (Ohta et al., 2002a).

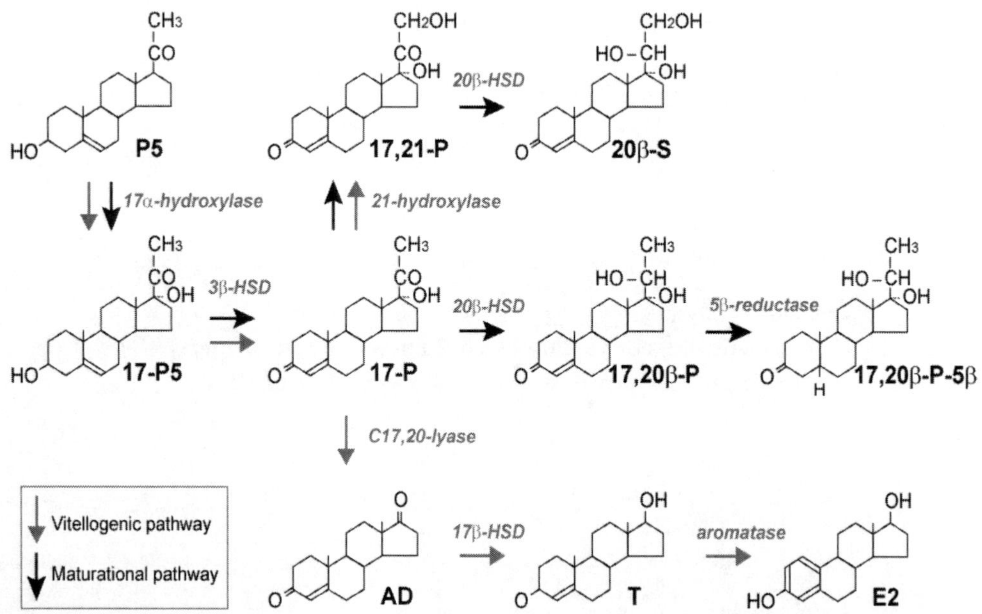

Figure 11. Steroidogenic pathways in the chub mackerel ovarian follicles during vitellogenesis and final oocyte maturation. See Table 1 for steroid abbreviations.

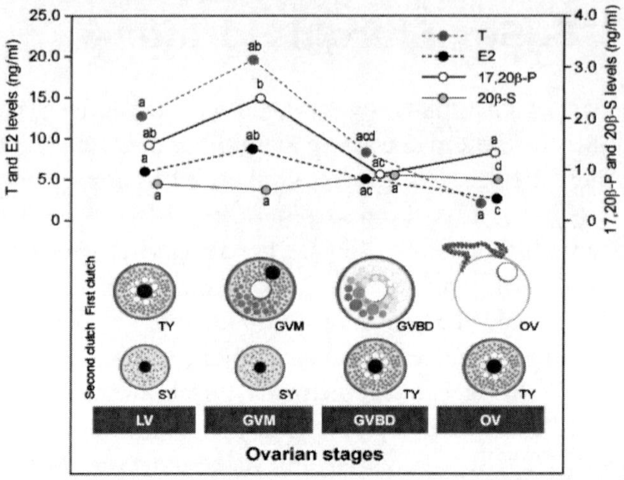

Figure 12. Changes in oocyte composition and serum levels of testosterone (T), estradiol-17β (E2), 17,20β-dihydroxy-4-pregnen-3-one (17,20β-P) and 17,20β,21-tihydroxy-4-pregnen-3-one (20β-S) during the spawning cycle. Different letters above and below the circles represent significant differences between ovarian stages. TY, tertiary yolk stage; GVM, germinal vesicle migration; HY, hydration; OV, ovulated. Refer Figure 1 for description of ovarian stages. Modified from Matsuyama et al. (2005).

Table 1. Nomenclature of steroids

Systematic name	Trivial name	Abbreviation
3β-hydroxy-5-pregnen-20-one	pregnenolone	P5
4-androstene-3,17-dione	androstenedione	AD
17β-hydroxy-4-androsten-3-one	testosterone	T
1,3,5(10)-estratriene-3,17β-diol	estradiol-17β	E2
3β,17-dihydroxy-5-pregnen-20-one	17-hydroxypregnenolone	17-P5
17-hydroxy-4-pregnene-3,20-dione	17-hydroxyprogesterone	17-P
17,20β-dihydroxy-4-pregnen-3-one	—	17,20β-P
17,21-dihydroxy-4-pregnen-3,20-dione	11-deoxycortisol	17,21-P
17,20β,21-trihydroxy-4-pregnen-3-one	—	20β-S
17,20β-dihydroxy-5β-pregnan-3-one	—	17,20β-P-5β

Table 2. In vitro effectiveness of various steroid hormones on the percentage of the chub mackerel ovarian follicles to complete GVBD

Steroids	Dose				
	10^{-5} M	10^{-6} M	10^{-7} M	10^{-8} M	0 M
17,20β-P	50.5±7.7	22.7±2.9	0	0	
17,20β-P-5β	14.7±1.2	10.1±2.7	0	0	
20β-S	15.4±6.0	11.0±2.8	0	0	
17,21-P	0	0	0	0	
Control					0

Control medium contains 1% ethanol. See Table 1 for steroid abbreviations.

Although there are a number of studies on follicular steroidogenesis in various teleost species, the complete steroidogenic pathway from P5 to E2 has been determined in only a few species: medaka (Kobayashi et al., 1996), bambooleaf wrasse (Ohta et al., 2001), red seabream (Ohta et al., 2002b), and yellowtail (Rahman et al., 2002). Chub mackerel, medaka, and yellowtail are gonochoric, while bambooleaf wrasse and red seabream have a bisexual phase in their life history. Interestingly, T is the substrate precursor of E2 in vitellogenic ovarian follicles of gonochoristic species. In contrast, E2 is synthesized via estrone (E1) rather than T in bambooleaf wrasse and red seabream (Ohta et al., 2001; 2002b). We hypothesize that low levels or lack of production of T in ovarian follicles of red seabream and bambooleaf wrasse may be related to sex reversals in their life cycles.

The basic concept of follicular steroidogenesis in teleosts is common between species, where E2 is synthesized during vitellogenesis while MIH is produced during FOM (Nagahama and Yamashita, 2008). In teleosts, two 20β-hydroxylated progestins are known to be MIHs, although a naturally occurring MIH has been conclusively identified in a limited number of species. 17,20β-P has been identified as an MIH in amago salmon (*Oncorhynchus rhodurus*), the first MIH identified in teleosts, and in Indian catfish (*Clarias batrachus*), mummichog (*Fundulus heteroclitus*), medaka, and yellowtail. 20β-S has also been identified as an MIH in Atlantic croaker (*Micropogonias undulatus*), spotted seatrout (*Cynoscion nebulosus*), striped bass, and a puffer fish (reviewed in Matsuyama, 2007). Regardless of the difference in E2 synthetic pathways and whether the MIH is 17,20β-P or 20β-S, there is a common enzymatic kinetics in MIH production; the decrease in C17,20-lyase activity involved in the cessation of E2 production and increase in 20β-HSD activity leads to a dramatic elevation in MIH. In the ovarian follicles of teleost, the steroidogenic enzyme which serves as a branch point for E2 and MIH is cytochrome P-450 17β-hydroxylase/C17,20-lyase (P-450c17). P-450c17 is a single cytochrome P-450 enzyme mediating both 17β-hydroxylase and C17,20-lyase activities (Chung et al., 1987; Hanukoglu, 1992) in which only C17,20-lyase activity is required for the synthesis of C19 steroids such as AD. Therefore, depression of one of two functions of P450c17 is involved in the steroidogenic shift from the pathway leading to E2 to MIH synthesis. Recently, a novel P450c17, named P450c17-II, was cloned and characterized in tilapia, medaka and zebrafish (Zhou et al., 2007a). Unlike the original P450c17, P450c17-II forms of tilapia and medaka possess only 17α-hydroxylase activity, and it has been demonstrated that P450c17-II is predominantly expressed in the ovary at the maturational phase and produces 17,20β-P (Zhou et al., 2007a; 2007b). It should be elucidated whether the P450c17-II gene exists and plays important roles in ovarian 17,20β-P production in the chub mackerel.

Conclusion and Future Perspective

Considerable progress on basic understanding of BPG axis in fish reproduction has been made over the past decade. However, problems that need to be examined are still many. In the classical picture of reproductive BPG axis, GnRH was occupying the central role. However, following the demonstration of kisspeptins role in reproductive function, lately Kisspeptin/GPR54 system has emerged as a critical upstream regulator of GnRH secretion. Presently, several research groups are dedicating efforts to clarify the role of

Kisspeptin/GPR54 system in teleost reproduction. Additionally, functions of GtHs, particularly FSH in multiple spawning fish, are still unclear. Our group has examined the role of BPG axis in chub mackerel reproduction as a model for not only multiple spawning marine species, but also for commercially important fish. Deciphering the endocrinological mechanism of puberty and dysfunction of gametogenic processes in captivity has a strong potential impact on fish farming, as well as enhancing our basic understanding on fish reproduction. More recently, the draft sequence of Pacific bluefin tuna (*Thunnus orientalis*) genome has been determined (Nakamura et al., 2011). Since chub mackerel belong to the family Scombridae, same as that of tunas, analyzing large amounts of data on gene expression dynamics by using tuna genome information will help to screen and analyze specific molecular markers regulating key reproductive processes such as puberty, which takes several years to initiate in tunas. Our chub mackerel experimental system led to major advances in our basic understanding of BPG axis in fish reproduction.

Acknowledgments

We gratefully acknowledge the students in the Laboratory of Marine Biology, Kyushu University, for their assistance during the experiment. We thank support from Grant-in-Aid for Scientific Research from the Ministry of Education, Culture, Sports, Science, and Technology of Japan (MEXT). This research was performed as part of the Recruitment Variability of Small Pelagic Fish Populations Program, which is supported by the Fisheries Research Agency of Japan.

References

Aizen, J., Kasuto, H., Golan, M., Zakay, H. and Levavi-Sivan, B. (2007a). Tilapia follicle-stimulating hormone (FSH): immunochemistry, stimulation by gonadotropin-releasing hormone, and effect of biologically active recombinant FSH on steroid secretion. *Biol. Reprod*, 76, 692–700.

Aizen, J., Kasuto, H. and Levavi-Sivan, B. (2007b). Development of specific enzyme-linked immunosorbent assay for determining LH and FSH levels in tilapia, using recombinant gonadotropins. *Gen. Comp. Endocrinol*, 153, 323–332.

Akazome, Y., Kanda, S., Okubo, K. and Oka, Y. (2010). Functional and evolutionary insights into vertebrate kisspeptin systems from studies of fish brain. *J. Fish. Biol*, 76, 161–182.

Alok, D., Hassin, S., Sampath Kumar R., Trant J. M., Yu, K. and Zohar, Y. (2000). Characterization of a pituitary GnRH-receptor from a perciform fish, *Morone saxatilis*: functional expression in a fish cell line. *Mol. Cell. Endocrinol*, 168, 65–76.

Beck, B. H., Fuller, S. A., Peatman, E., McEntire, M. E., Darwish, A. and Freeman, D. W. (2012). Chronic exogenous kisspeptin administration accelerates gonadal development in basses of the genus *Morone*. *Comp. Biochem. Physiol. A*, 162, 265–273.

Biran, J., Ben-Dor, S. and Levavi-Sivan B. (2008). Molecular identification and functional characterization of the kisspeptin/kisspeptin receptor system in lower vertebrates. *Biol. Reprod*, 79, 776–786.

Bogerd, J., Blomenrohr, M., Andersson, E., van der Putten, H. H., Tensen, C. P., Vischer, H. F., Granneman, J. C., Janssen-Dommerholt, C., Goos, H. J. and Schulz, R. W. (2001). Discrepancy between molecular structure and ligand selectivity of a testicular follicle-stimulating hormone receptor of the African catfish (*Clarias gariepinus*). *Biol. Reprod*, 64, 1633-1643.

Chen, C. C. and Fernald, R. D. (2008). GnRH and GnRH receptors: distribution, function and evolution. *J. Fish Biol*, 73, 1099–1120.

Chung, B. C., Picado-Leonard, J., Haniu, M., Bienkowski, M., Hall, P. F., Shively, J. E. and Miller, W. L. (1987). Cytochrome P450c17 (steroid 17α-hydroxylase/17,20 lyase): cloning of human adrenal and testis cDNAs indicates the same gene is expressed in both tissues. *Proc. Natl. Acad. Sci. USA*, 84, 407-411.

Elizur, A., Nocillado, J. N., Biran, J., Sivan, B. and Zohar, Y. (2011). Advancement of the onset of puberty in *Seriola lalandi* by chronic treatment with Kiss peptides. *Indian Journal of Science and Technology*, 4, 274-275.

Fernald, R. D. and White, R. B. (1999). Gonadotropin-releasing hormone genes: phylogeny, structure, and functions. *Front. Neuroendocrinol*, 20, 224–240.

Fukada, S., Sakai, N., Adachi, S. and Nagahama Y. (1994). Steroidogenesis in the ovarian follicle of medaka (*Oryzias latipes*, a daily spawner) during oocyte maturation. *Develop. Growth Differ,* 36, 81-88.

Gomez, J. M., Weil, C., Ollitrault, M., Le Bail, P. Y., Breton, B. and Le Gac F. (1999). Growth hormone (GH) and gonadotropin subunit gene expression and pituitary and plasma changes during spermatogenesis and oogenesis in rainbow trout (*Oncorhynchus mykiss*). *Gen. Comp. Endocrinol*, 113, 413–428.

Gonzalez-Martinez, D., Madigou, T., Mananos, E., Cerda-Reverter, J. M., Zanuy, S., Kah, O. and Munoz-Cueto, J. A. (2004). Cloning and expression of gonadotropin-releasing hormone receptor in the brain and pituitary of the European sea bass: An *in situ* hybridization study. *Biol. Reprod*, 70, 1380–1391.

Gothilf, Y., Meiri, I., Elizur, A. and Zohar, Y. (1997). Pre-ovulatory changes in the levels of three gonadotropin-releasing hormone- encoding messenger ribonucleic acids (mRNAs), gonadotropin b-subunit mRNAs, plasma gonadotropin, and steroids in the female gilthead seabream, *S. aurata*. *Biol. Reprod*, 57, 1145–1154.

Grone, B. P. Maruska, K. P., Korzan, W. J. and Fernald, R. D. (2010). Social status regulates kisspeptin receptor mRNA in the brain of *Astatotilapia burtoni. Gen. Comp. Endocrinol,* 169, 98–107.

Hanukoglu, I. (1992). Steroidogenic enzymes: structure function, and role in regulation of steroid hormone biosynthesis. *J. Steroid Biochem. Meluc. Biol*, 43, 779–804.

Kanda, S., Akazome, Y., Matsunaga, T., Yamamoto, N., Yamada, S., Tsukamura, H., Maeda, K. and Oka, Y. (2008). Identification of KiSS-1 product kisspeptin and steroid-sensitive sexually dimorphic kisspeptin neurons in medaka (*Oryzias latipes*). *Endocrinology,* **149**, 2467–2476.

Kanda, S., Karigo, T. and Oka, Y. (2012). Steroid sensitive *kiss2* neurones in the goldfish: evolutionary insights into the duplicate kisspeptin gene-expressing neurons. *J. Neuroendocrinol,* 24: 897–906.

Kanda, S. and Oka, Y. (2012). Evolutionary insights into the steroid sensitive kiss1 and kiss2 neurons in the vertebrate brain. *Front. Endocrinol*, 3, 28, 1–10.

Kitahashi, T., Ogawa, S. and Parhar, I. S. (2009). Cloning and expression of kiss2 in the zebrafish and medaka. *Endocrinology,* 150, 821–831.

Kobayashi, D., Tanaka, M., Fukada, S. and Nagahama, Y. (1996). Steroidogenesis in the ovarian follicles of the medaka (*Oryzias latipes*) during vitellogenesis and oocyte matura*tion. Zool. Sci,* 13, 921–927.

Kotani, M., Detheux, M., Vandenbogaerde, A., Communi D., Vanderwinden, J. M., Le Poul, E., Brézillon, S., Tyldesley, R., Suarez-Huerta, N., Vandeput, F., Blanpain, C., Schiffmann, S. N., Vassart, G. and Parmentier, M. (2001). The metastasis suppressor gene KiSS-1 encodes kisspeptins, the natural ligands of the orphan G protein-coupled receptor GPR54. *J. Biol. Chem*, 276, 34631–34636.

Lee, Y. R., Tsunekawa, K., Moon, M. J., Um, H. N., Hwang, J. I., Osugi, T., Otaki, N., Sunakawa, Y., Kim, K., Vaudry, H., Kwon, H. B., Seong, J. Y. and Tsutsui, K. (2009). Molecular evolution of multiple forms of kisspeptins and GPR54 receptors in vertebrates. *Endocrinology*, 150, 2837–2846.

Lethimonier, C., Madigou, T., Munoz-Cueto, J. A., Lareyre, J. J. and Kah O. (2004). Evolutionary aspects of GnRHs, GnRH neuronal systems and GnRH receptors in teleost fish. *Gen. Comp. Endocrinol*, 135, 1–16.

Li, S., Zhang, Y., Liu, Y., Huang, X., Huang, W., Lu, D., Zhu, P., Shi, Y., Cheng, C. H. K., Liu, X. and Lin, H, (2009). Structural and functional multiplicity of the kisspeptin/GPR54 system in goldfish (*Carassius auratus*). *J. Endocrinol*, 201, 407–418.

Matsuyama, M., Shiraishi, T., Sundaray, J. K., Rahman, M. A., Ohta, K. and Yamaguchi, A. (2005). Steroidogenesis in ovarian follicles of chub mackerel, *Scomber japonicus*. *Zool. Sci*, 22, 495, 101–110.

Matsuyama, M. (2007). Follicular Steroidogenesis in Fish. In M. J. Rocha, A. Arukwe, B. G. Kapoor. (Eds.), *Fish Reproduction*. (171-199). Science Publishers (USA).

Mechaly, A. S., Viñas, J. and Piferrer, F. (2012). Sex-specific changes in the expression of kisspeptin, kisspeptin receptor, gonadotropins and gonadotropin receptors in the Senegalese sole (*Solea senegalensis*) during a full reproductive cycle. *Com. Biochem. Physiol*, A, 162, 364-371.

Mitani, Y., Kanda, S., Akazome, Y., Zempo, B. and Oka, Y. (2010). Hypothalamic Kiss1 but not Kiss2 neurons are involved in estrogen feedback in medaka (*Oryzias latipes*). *Endocrinology,* 151, 1751–1759.

Miwa, S., Yan, L. and Swanson, P. (1994). Localization of two gonadotropin receptors in the salmon gonad by *in vitro* ligand autoradiography. *Biol. Reprod*, 50, 629-642.

Moncaut, N., Somoza, G., Power, D. M. and Canario, A. V. (2005). Five gonadotrophin releasing hormone receptors in a teleost fish: isolation, tissue distribution and phylogenetic relationships. *J. Mol. Endocrinol,* 34, 767–779.

Moyle, W. R., Campbell, R. K., Myers, R. V., Bernard, M. P., Han, Y. and Wang, X. Y. (1994). Coevolution of ligand-receptor pairs, *Nature*, 368, 251–255.

Nagahama, Y. and Yamashita, M. (2008). Regulation of oocyte maturation in fish. *Develop. Growth Differ*, 50, S195-S219.

Nakamura, Y., Mori, K., Saitoh, K., Saitoh, K., Sugaya, T., Shigenobu, Y., Ojima, N., Oshima, K., Tashiro, K., Hattori, M., Kuhara, S. and Sano, M. (2011). Whole genome analysis on bluefin tuna: general features of gene set predicted by bioinformatics. *Abstract of Plant and Animal Genomes* XIX Conference, San Diego, p. 625.

Nyuji, M., Kitano, H., Ohga, H., Kodama, R., Selvaraj, S., Yamaguchi, A. and Matsuyama, M. (2012b). Localization and differential expression profiles of two gonadotropin receptors in chub mackerel (*Scomber japonicus*) ovaries. *Proc. 7th Int. Symp. Fish Endocrinol.*, Argentina, 2012, p.181. .

Nyuji, M., Selvaraj, S., Kitano, H., Ohga, H., Yoneda, M., Shimizu, A., Kaneko, K., Yamaguchi, A. and Matsuyama, M. (2012a). Changes in the expression of pituitary gonadotropin subunits during reproductive cycle of multiple spawning female chub mackerel *Scomber japonicas*. *Fish Physiol. Biochem*, 38, 883–897.

Nyuji, M., Shiraishi, T., Selvaraj, S., In, V. V., Kitano, H., Yamaguchi, A., Okamoto, K., Onoue, S., Shimizu, A. and Matsuyama, M. (2011). Immunoreactive changes in pituitary FSH and LH cells during seasonal reproductive and spawning cycles of female chub mackerel *Scomber japonicus*. *Fish. Sci*, 77, 731-739.

Ohga, H., Kaneko, K., Shimizu, A., Kitano, H., Selvaraj, S., Nyuji, M., Yamaguchi, A. and Matsuyama, M. (2012b). Steroidogenic and maturation-inducing potency of native gonadotropic hormones in female chub mackerel, *Scomber japonicas. Reprod. Biol. Endocrinol,* 10:71.

Ohga, H., Selvaraj, S., Fujinaga, Y., Kitano, H., Nyuji, M., Yamaguchi, A. and Matsuyama, M. (2012a). Identification and seasonal expression variations of two subtypes of kisspeptin receptors in a scombroid fish, *Scomber japonicus*. In. Program Abstracts of the 2nd World Conference on Kisspeptin signaling in the brain, Japan, 2012, p. 139.

Ohta, K. and Matsuyama, M. (2002). Steroidogenic pathways to 17,20β-dihydroxy-4-pregnen-3-one and 17,20β,21-trihydroxy-4-pregnen-3-one in the ovarian follicles of the bambooleaf wrasse, *Pseudolabrus sieboldi. Fish. Sci*, 68, 41-50.

Ohta, K., Mine, T., Yamaguchi, A. and Matsuyama, M. (2001). Steroidogenic pathway to estradiol-17β synthesis in the ovarian follicles of the protogynous wrasse, *Pseudolabrus sieboldi. Zool. Sci,* 18, 937–945.

Ohta, K., Mine, T., Yamaguchi, A. and Matsuyama, M. (2008). Sexually dimorphic expression of pituitary glycoprotein hormones in a sex-changing fish (*P. sieboldi*). *J. Exp. Zool*, 309A, 534–541.

Ohta, K., Yamaguchi, S., Yamaguchi, A., Gen, K., Okuzawa, K., Kagawa, H. and Matsuyama, M. (2002a). Biosynthesis of steroids in ovarian follicles of red seabream, *Pagrus major* (Sparidae, Teleostei) during final oocyte maturation and the relative effectiveness of steroid metabolites for germinal vesicle breakdown *in vitro*. *Comp. Biochem. Physiol*, B, 133, 45-54.

Ohta, K., Yamaguchi, S., Yamaguchi, A., Gen, K., Okuzawa, K., Kagawa, H. and Matsuyama, M. (2002b). Biosynthesis of estradiol-17β in the ovarian follicles of the red seabream, *Pagrus major* during vitellogenesis. *Fish. Sci,* 68. 680–687.

Okada, T., Kawazoe, I., Kimura, S., Sasamoto, Y., Aida, K. and Kawauchi, H. (1994). Purification and characterization of gonadotropin I and II from pituitary glands of tuna (*Thunnus obesus*). *Int. J. Peptide Protein Res*, 43, 69-80.

Okubo, K. and Nagahama, Y. (2008). Structural and functional evolution of gonadotropin-releasing hormone in vertebrates. *Acta Physiol*, 193, 3–15.

Onuma, T. A. and Duan, C. (2012). Duplicated Kiss1 receptor genes in zebrafish: distinct gene expression patterns, different ligand selectivity, and a novel nuclear isoform with transactivating activity. *FASEB Journal*, 26, 1-10.

Parhar, I. S., Ogawa, S. and Sakuma, Y. (2004). Laser-captured single digoxigenin-labeled neurons of gonadotropin releasing hormone types reveal a novel G protein-coupled receptor (Gpr54) during maturation in cichlid fish. *Endocrinology,* 145, 3613-3618.

Planas, V. J., Athos, J., Goetz, W. F. and Swanson, P. (2000). Regulation of ovarian steroidogenesis in vitro by follicle-stimulating hormone and luteinizing hormone during sexual maturation in salmonid fish. *Biol. Reprod*, 62, 1262-1269.

Rahman, M. A., Ohta, K., Chuda, H., Nakano, S., Maruyama, K. and Matsuyama, M. (2001). Gonadotropin-induced steroidogenic shift towards maturation-inducing hormone in Japanese yellowtail during final oocyte maturation. *J. Fish Biol*, 58, 462-474.

Rahman, M. A., Ohta, K., Chuda, H., Yamaguchi, A. and Matsuyama, M. (2002). Steroid hormones and their synthetic pathways in the vitellogenic ovarian follicles of yellowtail, *Seriola quinqueradiata*. *J. Fac. Agr. Kyushu Univ,* 46, 311–319.

Roa, J. and Tena-Sempere, M. (2007). KiSS-1 system and reproduction: comparative aspects and roles in the control of female gonadotropic axis in mammals. *Gen. Comp. Endocrinol*, 153, 132–140.

Roseweir, A. K. and Millar, R. P. (2009). The role of kisspeptin in the control of gonadotrophin secretion. *Hum. Reprod. Update*, 15, 203-212.

Scott, A. P., Whitthames, P. R., Vermeirssen, E. L. M. and Carolsfeld, J. (1999). Prolonged-release gonadotropin-releasing hormone analogue implants enhance oocyte final maturation and ovulation, and increase plasma concentration of sulphated C21 steroids in North Sea plaice. *J. Fish Biol*, 55, 316–328.

Selvaraj, S., Kitano, H., Amano, M., Nyuji, M., Kaneko, K., Yamaguchi, A. and Matsuyama, M. (2012b). Molecular characterization and expression profiles of three GnRH forms in the brain and pituitary of adult chub mackerel (*Scomber japonicus*) maintained in captivity. *Aquaculture*,356/357:200-210.

Selvaraj, S., Kitano, H., Amano, M., Ohga, H., Yoneda, M., Yamaguchi, A., Shimizu, A. and Matsuyama, M. (2012a). Increased expression of kisspeptin and GnRH forms in the brain of scombroid fish during final oocyte maturation and ovulation. *Reprod Biol Endocrinol,* 10:64.

Selvaraj, S., Kitano. H., Fujinaga, Y., Amano, M., Takahashi, A., Shimizu, A., Yoneda, M., Yamaguchi, A. and Matsuyama, M. (2009). Immunological characterization and distribution of three GnRH forms in the brain and pituitary gland of chub mackerel (*S. japonicus*). *Zool. Sci, 26,* 828–839.

Selvaraj, S., Kitano, H., Fujinaga, Y., Ohga, H., Yoneda, M., Yamaguchi, A., Shimizu, A. and Matsuyama, M. (2010). Molecular characterization, tissue distribution, and mRNA expression profiles of two Kiss genes in the adult male and female chub mackerel (*S. japonicus*) during different gonadal stages. *Gen. Comp. Endocrinol, 169,* 28–38.

Selvaraj, S., Ohga, H., Kitano, H., Nyuji, M., Yamaguchi, A., Matsuyama, M., 2013. Peripheral administration of Kiss1 pentadecapeptide induces gonadal development in sexually immature adult scombroid fish. *Zool. Sci, 30 (6),* in press.

Servili, A., Page, Y. L., Leprince, J., Caraty, A., Escobar, S., Parhar, I. S., Seong, J. Y., Vaudry, H. and Kah, O. (2011). Organization of two independent kisspeptin systems derived from evolutionary-ancient kiss genes in the brain of zebrafish. *Endocrinology, 152,* 1527–1540.

Shimizu, Y., Tomikawa, J., Hirano, K., Nanikawa, Y., Akazome, Y., Kanda S., Kazeto, Y, Okuzawa, K., Uenoyama, Y., Ohkura, S., Tsukamura, H., Maeda, K. I., Gen, K., Oka, Y. and Yamamoto, N. *(2012).* Central distribution of kiss2 neurons and peri-pubertal changes in their expression in the brain of male and female red seabream *Pagrus major. Gen. Comp. Endocrinol,* 175, 432–442.

Shiraishi, T., Ketkar, S.D., Katoh, Y., Nyuji, M., Yamaguchi, A. and Matsuyama, M (2009) Spawning frequency of the Tsushima Current subpopulation of chub mackerel *Scomber japonicus* off Kyushu, Japan. *Fish. Sci, 75,* 649–655.

Shiraishi, T., Ketkar, S.D., Kitano, H., Nyuji, M., Yamaguchi, A. and Matsuyama, M. (2008). Time course of final oocyte maturation and ovulation in chub mackerel *Scomber japonicus* induced by hCG and GnRHa. *Fish. Sci, 74,* 764–769.

Shiraishi, T., Ohta, K., Yamaguchi, A., Yoda, M., Chuda, H. and Matsuyama, M. (2005). Reproductive parameters of the chub mackerel *Scomber japonicus* estimated from human chorionic gonadotropin induced final oocyte maturation and ovulation in captivity. *Fish. Sci, 71,* 531–542.

Suzuki, K., Kawauchi, H. and Nagahama, Y. (1988). Isolation and characterization of two distinct gonadotropins from chum salmon pituitary glands. *Gen. Comp. Endocrinol, 71,* 292–301.

Tena-Sempere, M., Felip, A., **Gómez,** A., Zanuy, S. and Carillo, M (2012): Comparative insights of the kisspeptin/kisspeptin receptor system: lessons from non-mammalian vertebrates. *Gen. Comp. Endocrinol, 175,* 234-243.

Van Der Kraak, G., Suzuki, K., Peter, R.E., Itoh, H. and Kawauchi, H. (1992). Properties of common carp gonadotropin I and gonadotropin II. *Gen. Comp. Endocrinol, 85,* 217–229.

Vischer, H. F., Granneman, J. C., Linskens, M. H., Schulz, R. W. and Bogerd, J. (2003). Both recombinant African catfish LH and FSH are able to activate the African catfish FSH receptor. *J. Mol. Endocr, 31,* 133-140.

Yan, L., Swanson, P. and Dickhoff, W. (1992). A two-receptor model for salmon gonadotropins (GTH I and GTH II). *Biol. Reprod, 47,* 418-427.

Zhou, L. Y., Wang, D. S., Kobayashi, T., Yano, A., Paul-Prasanth, B., Suzuki, A., Sakai, F. and Nagahama, Y. (2007a). A novel type of P450c17 lacking the lyase activity is responsible for C21 steroid biosynthesis in the fish ovary and head kidney. *Endocrinology, 148,* 4282-4291.

Zhou, L. Y., Wang, D. S., Shibata, Y., Paul-Prasanth, B., Suzuki, A. and Nagahama, Y. (2007b). Characterization, expression and transcriptional regulation of P450c17-I and -II in the medaka *Oryzias latipes. Biochem. Biophys. Res. Commun, 26,* 619–625.

Zmora, N., Stubblefield, J., Zulperi, Z., Biran, J., Levavi-Sivan, B., Muñoz-Cueto, J. A. and Zohar, Y. (2012). Differential and gonad stage-dependent roles of Kisspeptin1 and Kisspeptin2 in reproduction in the modern teleosts, *Morone* Species. *Biol. Reprod, 86,* 177.

Zohar, Y., Muñoz-Cueto, J. A., Elizur, A. and Kah, O. (2010). Neuroendocrinology of reproduction in teleost fish. *Gen. Comp. Endocrinol, 165,* 438–455.

Zohar, Y. and Mylonas, C.C. (2001). Endocrine manipulations of spawning in cultured fish: from hormones to genes. *Aquaculture, 197,* 99–136.

In: Sexual Plasticity and Gametogenesis in Fishes
Editor: Balasubramanian Senthilkumaran
ISBN: 978-1-62618-848-8

Chapter XVII

Teleosts Are Classical Models for the Study of Oogenesis and Shift in Steroidogenesis

Raju Murugananthkumar
and Balasubramanian Senthilkumaran*
Department of Animal Sciences,
School of Life Sciences-Centre for Advanced Studies,
University of Hyderabad, Central University, Hyderabad,
Andhra Pradesh, India

Abstract

Information on the molecular mechanisms of oocyte growth and maturation provides novel inputs to our understanding of teleostian oogenesis. Oogenesis is a complex process primarily coordinated by two steroid hormones, estradiol-17β (E_2) and 17α,20β-dihydroxy-4-pregnen-3-one (17α,20β-DP), under the influence of pituitary gonadotropins, luteinizing hormone (LH) and follicle stimulating hormone (FSH). Oocytes originated from primordial germ cells get arrested in the diplotene stage of prophase I of first meiotic division until they mature. Vitellogenesis is a key event where E_2 stimulates the production of vitellogenin in liver, transport into the developing oocytes where it is selectively taken up into the oocyte by a receptor-mediated process involving specific cell-surface receptors. The events of oogenesis were tightly modulated by several genes which are involved at the level of gonadal steroidogenesis under the influence of FSH. Ovarian aromatase (cyp19a1a) is considered to be one of the rate limiting enzymes for oogenesis. It is well established that production of E_2 and 17α,20β-DP by ovarian follicles occurs via the coordination of two cell layers, the theca and granulosa. A shift in the steroidogenesis pathway from E_2 to 17α,20β-DP occurs in the ovarian follicle layer immediately prior to oocyte maturation. During maturation, gonadotropins primarily regulate the shift, E_2 to 17α,20β-DP in the cholesterol biosynthesis pathway by enhancing the transcription of 20β-hydroxysteroid dehydrogenase (20β-HSD) in granulosa cells. LH

* Corresponding author: bsksl@uohyd.ernet.in/senthilkumaranb@yahoo.com.

enhances the production of 17α,20β-DP by decreasing E_2 levels through the reduction of cyp19a1a and/or P450c17-I in ovarian follicle. Conversely, the levels of 20β-HSD and/or P450c17-II mRNA were elevated upon the stimulation of LH. Transcript level changes were also seen at the level of enzyme activity to bring about shift in steroidogenesis. Thus, 17α,20β-DP is regarded as maturation inducing hormone in several teleosts. Transcription factors such as Ad4BP/SF-1, Foxl2 and CREB are potentially contributes for the expression of these steroidogenic enzyme genes during final oocyte maturation.

Introduction

Teleosts comprise more than 21000 species, including marine and freshwater habitats, considered to be one of the largest groups of vertebrates. Most of the teleosts show oviparity, although it follows a diverse array of reproductive strategy including viviparity (Coward et al., 2002; Lubzens et al., 2010). Oogenesis is one of the chief events in female gametogenesis where gonadotropins (GTHs) and maturation inducing hormone (17α,20β-dihydroxy-4-pregnen-3-one, 17α,20β-DP in most fishes) play crucial roles. Steroid hormones regulate the steps involved in oogenesis. However, GTHs regulate these events in a coordinated manner (Nagahama, 1994; 1997; Arukwe and Goksøyr, 2003; Yaron and Levavi-Sivan, 2010). The basic structure of ovary is made up of many ovarian follicles, each of which is composed of an oocyte and a follicular envelop consists of an outer thecal and an inner granulosa cell layer, which remain unchanged all through follicular growth (Patino and Sullivan, 2002). Oocyte growth includes primary growth, formation of cortical alveoli and vitellogenesis. Following vitellogenesis, oocytes undergo final oocyte maturation (FOM) and ovulation (Wallace and Selman, 1981; Tyler and Sumpter, 1996; Breton et al., 2012). Vitellogenesis is a major event in oocyte growth which is induced by estradiol-17β (E_2) while FOM is a major event in maturation which is induced by 17α,20β-DP (Nagahama et al., 1995). This chapter reviews about oocyte growth, regulation, shift in steroidogenesis and FOM.

Oogenesis

Oogenesis has been extensively studied in numerous species. It includes the formation of primordial cells followed by oocyte growth and final steps of maturation, which have been hormonally controlled and regulated by genes/factors involved in gonadal steroidogenesis (Condeça and Canario, 2001). Oogenesis begins with meiotic arrest at diplotene stage of prophase I and the ovarian follicles were formed with perinucleolar oocytes surrounded by granulosa and thecal cells (Devlin and Nagahama, 2002; Lubzens et al., 2010). A major cytoplasmic event like intense RNA synthesis and increase in oocyte size were contributed for the development of ovarian follicle (Patino and Sullivan, 2002).

Vitellogenesis

Vitellogenesis is mediated by two major regulators, are the pituitary GTHs, luteinzing hormone (LH) and follicle-stimulating hormone (FSH) and sex steroids (Lubzens et al.,

2010). The accumulation of yolk proteins in the oocytes for its enormous growth is termed as vitellogenesis. It involves the synthesis of vitellogenin (Vtg) in the liver with the stimulation of E_2, which transverse into the oocyte through a receptor mediated endocytosis process (Patino and Sullivan, 2002). To enable this, Vtg genes encode female specific glycolipophosphoprotein varies from 300–640 kDa in several teleosts, but the functions are known to be similar (Babin et al., 2007). Vtg and its receptor mRNAs increased during this period selectively in liver and ovary, respectively (Perazzolo et al., 1999). The follicle undergoes a secondary growth phase with the sequential accumulation of endogenously synthesized glycoprotiens (the cortical alveolus stage) and lipid bodies (the lipid inclusion stage) in the ooplasm and the formation of zona radiata between the ooplasm and granulosa cell layer (Campbell et al., 2006). Lipid deposition in oocyte is normally occurring during previtellogenic growth (Selman and Wallace, 1989). It has been suggested that lipids derived from circulating very low density lipoprotein, which binds to lipoprotein receptors present in the ovary (Prat et al., 1998). The presence of multiple Vtg genes has been demonstrated in few teleosts (Buisine et al., 2002; Wang et al., 2005; Babin et al., 2007; Le Menn et al., 2007). The arrangement of Vtgs in genome shows divergence among fish species. In Salmonidae, ancestral Vtg duplications afford two paralogous gene clutster, VtgA and VtgB (Buisine et al., 2002). The two forms of Vtg A and B in rainbow trout differ from each other due to the insertion, deletion and rearrangement during evolution (Trichet et al., 2000). Intriguingly, zebrafish genome contains at least seven Vtgs (Wang et al., 2005). The final size of the egg becomes larger size (Wallace, 1985; Tyler, 1991) and the GSI increases between 50 to 100 fold at the time of vitellogenesis (Tyler and Sumpter, 1996). FSH regulates the process through E_2 secreted by the ovary. Testosterone produced in thecal cell layer is mobilized to granulosa cell layer for the production of E_2 by ovarian aromatase (cyp19a1a) for oocyte growth. In certain species like catfish and seabream LH seems to play a role in oocyte growth as well (Kirubagaran et al., 2005; Gen et al., 2000). It has been suggested that Vtg can also be regulated by several other factors such as temperature, androgens, growth hormone, prolactin, thyroid hormone and cortisol other than estrogens (Jones et al., 2000).

Hormonal Control of Oocyte Growth

In teleosts, oocyte growth is not a preprogrammed process. Conversely, it is synergistic and flexible to the existing conditions. In early stages of vitellogenesis there is dissimilarity in the oocyte size but gradually it develops uniformity at the ovulation stage (Tyler et al., 1990). Several hormones play important role in regulating the oocyte growth in fish, which include GTHs, thyroxine, triiodothyronine and growth hormone (GH) yet direct role for hormones other that GTHs have not yet determined (Nagahama, 1994). GH synergistically acts with GTH in regulating steroidogenesis, alters the rate of body growth by acting indirectly and consequently modulates the female egg production and size. In fish there are two GTHs, GTH-I and GTH-II which are homologus to FSH and LH, respectively. Hence, the terminology of GTH-I and GTH-II is also considered as FSH and LH, respectively (Suzuki et al., 1988; Kawauchi et al., 1989; Swanson et al., 1989, 1991; Swanson, 1991; Slater et al., 1994; Prat et al., 1996). The data linked with the role of LH are apparent as it plays a central role in the steps involved in FOM (Scott and Sumpter, 1983; Swanson, 1991). The tasks associated with FSH in oocyte growth are less clear, however, it has been designated that FSH favors for vitellogenesis (Tyler et al., 1991; Prat et al., 1996).

Studies on rainbow trout indicated that LH is negligible in the plasma until very close to ovulation, whereas the levels of FSH in plasma are high during vitellogenesis (Slater et al., 1994; Prat et al., 1996), it is understood that FSH controls the secretion of E_2 thereby stimulating the Vtg synthesis in liver, estrogen receptors and egg shell proteins (Hyllner et al., 1991; Flouriot et al., 1995). *In vitro* studies in salmonids have shown that physiological concentrations of FSH may increase the uptake of Vtg into developing oocytes (Tyler et al., 1991). LH seems to have very high expression than FSH during ovarian growth and therefore, LH might regulate vitellogenesis as well in red seabream (Tanaka et al., 1995; Kagawa et al., 1998; Gen et al., 2000; Kagawa et al., 2003). Thus, the regulation of E_2 production by either FSH or LH is considered important during vitellogenesis. In catfish, annual profile of LH, equivalent to GTH-II showing multiple peaks indicating possible role of LH in vitellogenesis (Kirubagaran et al., 2005) as well as maturation. Further, cyp19a1a is also targeted by FSH/LH, which is one of the rate-limiting enzymes for E_2 biosynthesis (Tanaka et al., 1992; Gen et al., 2001; Kagawa et al., 2003). It proves that the availability of precursor steroids also controls the steroidogenesis pathway for the requirement of ovarian follicular steroidogenesis (Senthilkumaran et al., 2004; Zhou et al., 2007a).

The synthesis of certain growth factors, such as Insulin like growth factor-I (IGF-I) and transforming growth factor at specific stages suggests that they have specific roles in regulating the oocyte growth and development (Duan, 1997; Maestro et al., 1997; Kohli et al., 2003; Campbell et al., 2006; Reinecke, 2010). In rainbow trout follicles, insulin stimulates the uptake of Vtg (Tyler et al., 1987; Shibata et al., 1993), an effect that may well be mediated through the IGF receptor, rather than the receptor for insulin itself. Indeed, there is considerable structural homology between the insulin receptor and IGF-I receptor (Scavo et al., 1991). Like that of mammals, growth factors also play an integral role in oocyte growth along with GTH and steroids.

Two Cell Type Model and Steroidogenic Shift

Teleosts ovarian follicles comprised of an outer thecal and an inner granulosa cell layer. Based on the two cell type model proposed by Nagahama (1997), it has been emphasized that 17α-hydroxyprogesterone produced from progesterone by thecal cell 17α-hydroxylase (P450c17) pass through the basal lamina to granulosa cells where it is being converted into 17α,20β-DP by 20β-hydroxysteroid dehydrogenase (20β-HSD), a key enzyme for FOM. The entire process is regulated by LH mediated activation of adenylate cyclase (Nagahama, 1997; Sherwood et al., 2000; Gur et al., 2002). During vitellogenesis, GTH stimulates thecal cells to synthesis testosterone through a serial cascade pathway from cholesterol which transverse into granulosa cells for the conversion of E_2 by the enzyme cyp19a1a. Thus, E_2 is produced during oocyte growth whereas 17α,20β-DP is formed during FOM.

After follicular growth, there is an immediate shift in the steroidogenic pathway from E_2 to 17α,20β-DP before the initiation of FOM. This steroidogenic shift occurs in two stages resulting in FOM. As explained earlier in two-cell type model and studies on some teleosts showed that mRNAs of enzymes regulating cholesterol biosynthesis (Kusakabe et al., 2002;

Sreenivasulu et al., 2005; 2009; Raghuveer and Senthilkumaran, 2012) were elevated during full grown immature oocyte stage to synthesize 17α-hydroxyprogesterone for the production of 17α,20β-DP (Nagahama, 1997). However, steroidogenic acute regulatory protein (StAR) and steroidogenic enzymes are essential and required immensely for the production of E_2 during ovarian growth. Then, there is a shift from E_2 to 17α,20β-DP in granulosa cells through the elevation of 20β-HSD for implementing FOM (Tanaka et al., 2002; Senthilkumaran et al., 2002; Sreenivasulu and Senthilkumaran, 2009). On the other hand, the level of cyp19a1a decreased considerably after the stimulation of 20β-HSD with transcription factor shift that involve in the regulation of these enzymes. To explain further, the steroidogenic shift by GTHs is marked through two molecular mechanisms, the first being the suppression of Ad4BP/SF-1 expression in turn aromatase, and the second, the induction of over-expression of 20β-HSD perhaps through CREB (Senthilkumaran et al., 2004). Recent study on the promoter motif of 20β-HSD gene supports this concept effectively (Sreenivasulu et al., 2012b).

FOM

After the completion of follicle growth, oocyte enters into the next stage of oogenesis through a stepwise mechanism called FOM. Oocyte maturation is extensively studied in teleosts than any other organisms as the endocrine regulation is well known. This process is regulated by three major mediators such as GTHs, 17α,20β-DP and maturation-promoting factor (MPF).

17α,20β-DP

LH stimulates the production of 17α-hydroxyprogesterone in thecal cell layer of ovary reaches the basal lamina and converted to 17α,20β-DP in granulosa cell layer by the action of 20β-HSD. The meiotic oogenesis is arrested at diplotene stage of prophase I all through development and growth of the ovarian follicle. The LH surge triggers its activity in advance during metaphase II. The involvement of 17α,20β-DP serves as the principal mediator for the secretion of LH during maturation was already discussed for several decades in fish *in vitro* (Masui and Clarke, 1979; Nagahama et al., 1985; Nagahama, 1987). Similarly *in vivo* studies on vertebrates have also proved that the preovulatory surge of LH associate with increased production of 17α,20β-DP and the onset of oocyte meiotic resumption (Kahn and Thomas, 1999). Thus, the role of 17α,20β-DP synthesis in follicles for oocyte meiotic resumption is now well-documented in lower vertebrates (Masui and Clarke, 1979; Nagahama, 1987; 1997; Senthilkumaran, 2011). Although LH is regarded for its direct involvement in meiotic maturation, it triggers the ovarian follicular cells to synthesize 17α,20β-DP through the stimulation of pre-ovulatory surge to implement the event (Nagahama, 1994; 1997). The 17α,20β-DP was first isolated from amago salmon by Nagahama and Adachi, (1985). 17α,20β-DP levels were very low in vitellogenic phase and increases sharply in mature and ovulated females. *In vitro* studies showed that 17α,20β-DP effectively induce the FOM in several fish species including amago salmon. Therefore in teleosts, 17α,20β-DP, a progesterone derived steroid have been identified as more potent hormone for FOM (Nagahama and Adachi, 1985; Tokumoto et al., 2011). In the atlantic croaker 17α,20β,21-trihydroxy-4-pregnen-3-one is considered as 17α,20β-DP (Trant and Thomas, 1989).

20β-HSD

During FOM, a steroidogenic shift from E_2 to 17α,20β-DP is demonstrated in several teleost species with the increased expression of 20β-HSD (Haider, 1997; Nakamura et al., 2005; Ohta and Matsuyama, 2002; Sreenivasulu et al., 2005; 2012a). Studies in some fish species confirmed FSH/LH promotes the 20β-HSD enzyme activity that is mimicked by forskolin and dbcAMP (Nagahama, 1997; Kazeto et al., 2001). 20β-HSD expression was shown to be up-regulated by pre-ovulatory LH surge or by induction through human chorionic gonadotropin during FOM (Senthilkumaran et al., 2002; 2004; Tanaka et al., 2002; Sreenivasulu and Senthilkumaran, 2009b). A quick increase of 20β-HSD transcripts is more than sufficient to initiate the FOM in ayu and catfish (Tanaka et al., 2002; Sreenivasulu and Senthilkumaran, 2009b). Contrastingly, a sharp elevation is needed during FOM in the Nile tilapia (Senthilkumaran et al., 2002). Conversely, in zebrafish, 17α,20β-DP production and FOM may not involve significant change of 20β-HSD like carbonyl reductase/20β-HSD expression as evidenced in the salmonids, or that there might be other isoforms of 20β-HSD whose expression is tightly controlled by endocrine and paracrine factors (Wang and Ge, 2002). Hence, it is important to investigate the role of other steroidogenic enzymes which play important role in substrate availability during FOM as there is a divergent in expression patterns of 20β-HSD in teleosts.

P450c17

The enzyme P450c17 also known as CYP17A1, that catalysis both 17α-hydroxylation of C21 steroids as well as 17,20-lyase activity to produce C19 steroids. During maturation, postvitellogenic follicles have to synthesize enormous amount of 17α,20β-DP from the immediate precursor 17α-hydroxyprogesterone. However, it is considerable that P450c17 might involve in regulating the availability of precursor steroids during the shift in steroidogenesis. Recently two forms of P450c17s (P450c17-I and P450c17-II) were identified from tilapia, medaka, fugu and tetraodon (Zhou et al., 2007a, b). Intriguingly, P450c17-I have both hydroxylase and lyase activity while P450c17-II has only hydroxylase activity. The elevated expression of P450c17-I were found during pre- and mid-vitellogenesis, which eventually declines rapidly before FOM. However, P450c17-II expression reaches peak during late to post vitellogenesis corresponds with the initiation of 17α,20β-DP production (Zhou et al., 2007a,b). The decrease of lyase activity accumulates 17α-hydroxyprogesterone in follicles favors the production of E_2 to 17α,20β- DP during the shift. Thus, P450c17-I is needed for the vitellogenesis while P450c17-II is necessary for oocyte maturation in the Nile tilapia (Nagahama and Yamashita, 2008) and in gilthead seabream (Zapater et al., 2012). The differential expression of these two forms was regulated at transcriptional level by Ad4BP/SF-1 and Foxl2. Two forms of P450c17 have not been found in other teleosts. For example, only one form has been identified in catfish, where in minor elevation of P450c17 is sufficient to increase the availability of precursor substrate, 17α-hydroxyprogesterone for the production of 17α,20β-DP. Earlier reports in tilapia and medaka relay on P450c17 expression data rather than enzyme activity and regulation of lyase activity in detail. (Zhou et al., 2007a,b). Our study in catfish seized this fact and demonstrated that increase in P450c17 need not follow elevated protein level or enzyme level during FOM. So the increase in P450c17 transcript levels might contribute for study state turnover of protein and enzyme activity. Thus comprehensive attempt to study expression and enzyme activity might provide more clues to understand precursor substrate in shift in steroidogenesis.

Cholesterol Side Chain Cleavage Enzyme (P450scc), StAR and 3β-Hydroxysteroid Dehydrogenase (3β-HSD)

High or moderate levels of either progesterone or 17α-hydroxyprogesterone in ovarian follicles might also be contributed by StAR, P450scc and 3β-HSD during oocyte maturation (Kusakabe et al., 2002; Sreenivasulu and Senthilkumaran, 2009a; Sreenivasulu et al., 2005; Raghuveer and Senthilkumaran, 2012). In thecal cell layers of amago salmon, LH triggers the formation of 17α-hydroxyprogesterone which was mimicked by forskolin, an adenylate cyclase activator and dibutyrul cAMP, a membrane-permeable cAMP analogue (Kanamori and Nagahama, 1988). Further, LH and forskolin stimulates cAMP production in thecal cell layers indicating the action of LH acts through a receptor mediated activation of adenylate cyclase for the production of cAMP (Balamurugan and Haider, 1995; Chaube and Haider, 1997; Haider, 2004). StAR a regulatory protein which transports cholesterol within the mitochondria for the production of steroid hormones where as P450scc helps in the conversion of cholesterol to pregnenalone in thecal cells (Stocco, 2001; Hauet et al., 2002; Payne and Hales, 2004; Haider, 2004). Both enzymes were elevated during FOM (Belin et al., 2000; Ings et al., 2006).

Contrastingly, 3β-HSD and P450c17-I levels were lowered during all stages of oocyte maturation (Nagahama and Yamashita, 2008). However, moderate elevation of 3β-HSD expression was evident in tilapia and catfish (Senthilkumaran et al., 2009; Raghuveer and Senthilkumaran, 2012). Thus in the expression of steroidogenic enzymes during FOM species-specific difference is always evident, through shift in steroidogenesis is seen in several teleosts (Kumar et al., 2000; Senthilkumaran et al., 2004; Mittelholzer et al., 2007).

Future Perspectives

Basic and ultrastructural studies on oocyte growth and development have afforded several insights into the complex molecular pathways and factors regulating during early stages of oogenesis, folliculogenesis and FOM. The clarification of the relationship between the early stages of oogenesis to maturation and the factors regulating these processes is still not yet completely understood.

For example, through promoter motif analysis and functional analysis of cyp19a1a was explained explicitly (Huang et al., 2009; Sridevi et al., 2011), only limited information was available for 20β-HSD promoter (Sreenivasulu et al., 2012b). It is important to understand the role of transcription factors involved in the gene regulation of cyp19a1a and 20β-HSD. In this context, role of Ad4BP/SF-1 and Foxl2 on cyp19a1a is very well known. Conversely, the regulatory role of CREB on 20β-HSD needs to be established (Senthilkumaran et al., 2004; Sreenivasulu et al., 2012b).

Over the last two decades several studies have contributed to rule out the role of endogenous and exogenous factors necessary to produce a viable oocyte, the interplay between the somatic environment and oocyte development. Recent studies were aimed to identify the environmental factors which influence the quality of egg and elucidating the endocrine pathways regulating egg formation and functional aspects that contribute the proper development of the embryo.

Another important aspect which has to be focused is the transductional pathways that are generated in the oocyte just after the activation of mullerian inhibiting substance receptor and that ultimately force the activation of the MPF and the reinitiating of meiosis.

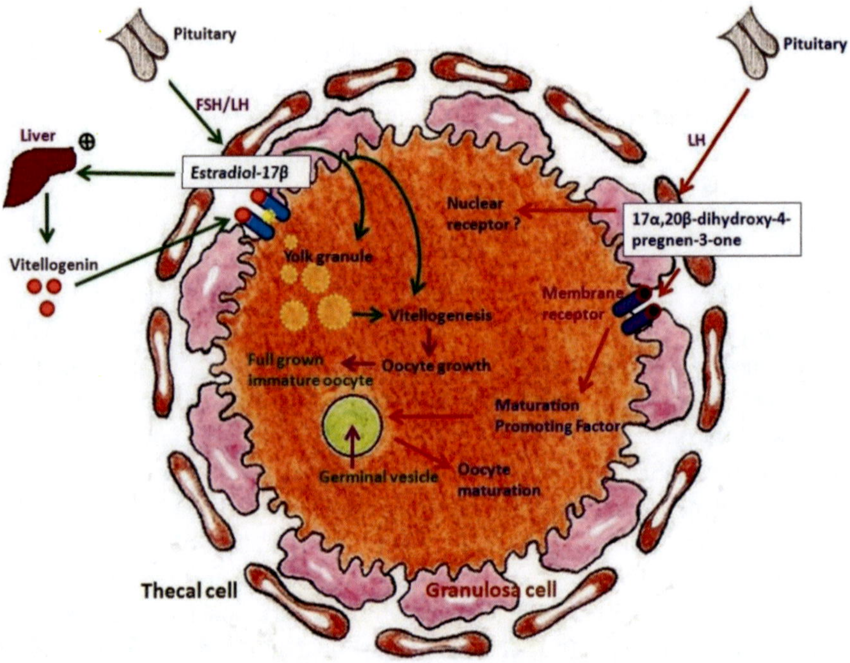

Figure 1. Schematic diagram to illustrate hormonal regulation of oocyte growth and maturation.

Conclusion

This review concludes about the concepts of oogenesis and maturation with a special note on shift in steroidogenesis among various teleosts. E_2 regulates oocyte growth which is under the regulation of FSH and LH. During this process, oocyte accumulates Vtg to reach full grown immature post vitellogenic follicle from primary oocyte. Then there is immediate shift in steroidogenic pathway leading to the production of 17α,20β-DP which ultimately leads to FOM.

This entire process has been explained using a schematic diagram (Figure 1). The process or oocyte growth involves modulation of Ad4BP/SF1, Foxl2 and cyp19a1a while FOM requires over expression of CREB and 20β-HSD.

Acknowledgments

The research work from BS laboratory mentioned in this review was supported by Grant-in-Aid (SP/SO/AS-49/2003) from Department of Science and Technology, India for which BS is thankful. RM is grateful to Council of Scientific and Industrial Research, India for Junior and Senior Research Fellowships.

References

Arukwe, A. and Goksøyr, A. (2003). Eggshell and egg yolk proteins in fish: hepatic proteins for the next generation: oogenetic, population, and evolutionary implications of endocrine disruption. *Comp. Hepatol.*, *2*, 4.

Babin, P. J., Carnevali, O., Lubzens, E. and Schneider, W. J. (2007). Molecular aspects of oocyte vitellogenesis in fish. In P. J. Babin, J. Cerdà, and E. Lubzens, (Eds.), *The fish oocyte: from basic studies to biotechnological applications*. (Springer, 39-76). Dordrecht, The Netherlands.

Balamurugan, K. and Haider, S. (1995). Effects of actinomycin D and cycloheximide on gonadotropin- or 17α,20β-dihydroxy-4-pregnen-3-one-induced *in vitro* maturation in oocytes of the catfish, *Clarias batrachus*. *J. Biosciences, 20,* 151-156.

Belin, F., Goudet, G., Duchamp, G. and Gérard, N. (2000). Intrafollicular concentrations of steroids and steroidogenic enzymes in relation to follicular development in the mare. *Biol. Reprod., 62,* 1335-1343.

Breton, T. S., Anderson, J. L., Goetz, F. W. and Berlinsky, D. L. (2012). Identification of ovarian gene expression patterns during vitellogenesis in Atlantic cod (*Gadus morhua*). *Gen. Comp. Endocrinol., 179,* 296-304.

Buisine, N., Trichet, V. and Wolff, J. (2002). Complex evolution of vitellogenin genes in salmonid fishes. *Mol. Genet. Genomics, 268,* 535-542.

Campbell, B., Dickey, J., Beckman, B., Young, G., Pierce, A., Fukada, H. and Swanson, P. (2006). Previtellogenic oocyte growth in salmon: relationships among body growth, plasma insulin-like growth factor-1, estradiol-17β, follicle-stimulating hormone and expression of ovarian genes for insulin-like growth factors, steroidogenic-acute regulatory protein and receptors for gonadotropins, growth hormone, and somatolactin. *Biol. Reprod., 75,* 34-44.

Chaube, S. K. and Haider, S. (1997). Effects of cAMP forskolin and cyanoketone on *in vitro* oocyte maturation in the catfish, *Clarias batrachus*. *J. Biosci., 22,* 255-265.

Condeça, J. B. and Canario, A. V. (2001). Gonadal steroidogenesis in response to estradiol-17β administration in the sea bream (*Sparus aurata* L.). *Gen. Comp. Endocrinol., 124,* 82-96.

Coward, K., Bromage, N. R., Hibbitt, O. and Parrington, J. (2002). Gamete physiology, fertilization and egg activation in teleost fish. *Rev. Fish Biol. Fish, 12,* 33-58.

Devlin, R. H. and Nagahama, Y. (2002). Sex determination and sex differentiation in fish: an overview of genetic, physiological, and environmental influences. *Aquaculture, 208,* 191-364.

Duan, C. (1997). The insulin-like growth factor system and its biological actions in fish. *Am Zool, 37,* 491-503.

Flouriot, G., Pakdel, F., Ducouret, B. and Valotaire, Y. (1995) Influence of xenobiotics on rainbow trout liver estrogen receptor and vitellogenin gene expression. *J. Mol. Endocrinol., 15,* 143-151.

Gen, K., Okuzawa, K., Kumakura, N., Yamaguchi, S. and Kagawa, H. (2001). Correlation between messenger RNA expression of cytochrome P450 aromatase and its enzyme activity during oocyte development in the red seabream (*Pagrus major*). *Biol. Reprod., 65,* 1186-1194.

Gen, K., Okuzawa, K., Senthilkumaran, B., Tanaka, H., Moriyama, S. and Kagawa, H. (2000). Unique expression of gonadotropin-I and -II subunit genes in male and female red seabream (*Pagrus major*) during sexual maturation. *Biol. Reprod., 63,*308-319.

Gur, G., Bonfil, D., Safarian, H., Naor, Z. and Yaron, Z. (2002). Pituitary adenylate cyclase activating polypeptide and neuropeptide Y regulation of gonadotropin subunit gene expression in tilapia: role of PKC, PKA and ERK. *Neuroendocrinol, 75,* 164-174.

Haider, S. (1997). Hormonal control of oocyte maturation in fish. In *Advances in Fish Research,* vol 2, pp 285–298. Ed BR Singh. Delhi: Narendra Publishing House.

Haider, S. G. (2004). Cell biology of Leydig cells in the testis. *Int. Rev. Cytol., 233,* 181-241

Hauet, T., Liu, J., Li, H., Gazouli, M., Culty, M. and Papadopoulos, V. (2002). PBR, StAR, and PKA: partners in cholesterol transport in steroidogenic cells. *Endocr. Res., 28,* 395-401.

Huang, W., Zhou, L., Li, Z. and Gui, J. F. (2009). Expression pattern, cellular localization and promoter activity analysis of ovarian aromatase (Cyp19a1a) in protogynous hermaphrodite red-spotted grouper. *Mol. Cell Endocrinol., 307,* 224-236.

Hyllner, S. J., Oppen-Berntsen, D. O., Helvik, J. V., Walther, B. T. and Haux, C. (1991). Oestradiol-17β induces the major vitelline envelope proteins in both sexes in teleosts. *J. Endocrinol., 131,*229-236.

Ings, J. S. and Van Der Kraak, G. J. (2006). Characterization of the mRNA expression of StAR and steroidogenic enzymes in zebrafish ovarian follicles. *Mol. Reprod. Dev., 73,* 943-954.

Jones, P. D., Tremblay, L. A., Coen, W. M. D. and Glesy, J. P. (2000). Vitellogenin as a biomarker for environmental estrogens. *Aust. J. Ecotoxicol., 6,* 45-58.

Kagawa, H., Gen, K., Okuzawa, K. and Tanaka, H. (2003). Effects of luteinizing hormone and follicle-stimulating hormone and insulin-like growth factor-I on aromatase activity and P450 aromatase gene expression in the ovarian follicles of red seabream, *Pagrus major. Biol. Reprod., 68,* 1562-1568.

Kagawa, H., Tanaka, H., Okuzawa, K. and Kobayashi, M. (1998). GTH II but not GTH I induces final oocyte maturation and the development of maturational competence of oocytes of red seabream *in vitro*. *Gen. Comp. Endocrinol., 112,* 80-88.

Kanamori, A. and Nagahama, Y. (1988). Involvement of 3′,5′-cyclic adenosine monophosphate in the control of follicular steroidogenesis of amago salmon (*Oncorhynchus rhodurus*). *Gen. Comp. Endocrinol., 72,* 39-53.

Kawauchi, H., Suzuki, K., Itoh, H., Swanson, R., Nagahama, Y., Nozaki, M., Naito, N. and Itoh, S. (1989). The duality of teleost gonadotropins. *Fish Physiol. Biochem., 7,* 29-38.

Kazeto, Y., Adachi, S. and Yamauchi, K. (2001). 20β-Hydroxysteroid dehydrogenase of the Japanese eel ovary: its cellular localization and changes in the enzymatic activity during sexual maturation. *Gen. Comp. Endocrinol., 122,* 109-115.

Khan, I. A. and Thomas, P. (1999). Ovarian Cycle, Teleost Fish. In E. Knobil, and J. D. Neill, (Eds), *Encyclopedia of Reproduction*, (Academic Press, volume 2, 193-204). San Diego.

Kirubagaran, R., Senthilkumaran, B., Sudhakumari, C. C. and Joy, K. P. (2005). Seasonal dynamics in gonadotropin secretion and E_2-binding in the catfish *Heteropneustes fossilis. Fish Physiol. Biochem., 31,* 183-188.

Kohli, G., Hu, S., Clelland, E., Di Muccio, T., Rothenstein, J. and Peng, C. (2003). Cloning of transforming growth factor-β 1 (TGF-β 1) and its type II receptor from zebrafish ovary and role of TGF-β 1 in oocyte maturation. *Endocrinology, 144,* 1931-1941.

Kumar, R. S., Ijiri, S. and Trant, J. M. (2000). Changes in the expression of genes encoding steroidogenic enzymes in the channel catfish (*Ictalurus punctatus*) ovary throughout a reproductive cycle. *Biol. Reprod., 63,* 1676-1682.

Kusakabe, M., Todo, T., McQuillan, H. J., Goetz, F. W. and Young, G. (2002). Characterization and expression of steroidogenic acute regulatory protein (StAR) and MLN64 cDNAs in trout. *Endocrinology, 143,* 2062-2070.

Le Menn, F., Cerdá J. and Babin P. J. (2007). Molecular aspects of oocyte vitellogenesis in fish. In P. J. Babin, J. Cerdá, and E. Lubzens (Eds.), *The Fish Oocyte: From Basic Studies to Biotechnological Applications*, (Springer, 39-76). Dordrecht.

Lubzens, E., Young, G., Bobe, J. and Cerdà, J. (2010). Oogenesis in teleosts: how eggs are formed. *Gen. Comp. Endocrinol., 165,* 367-389.

Maestro, M. A., Planas, J. V., Moriyama, S., Gutiérrez, J., Planas, J. and Swanson, P. (1997). Ovarian receptors for insulin and insulin-like growth factor I (IGF-I) and effects of IGF-I on steroid production by isolated follicular layers of the preovulatory coho salmon ovarian follicle. *Gen. Comp. Endocrinol., 106,* 189-201.

Masui, Y. and Clarke, H. J. (1979). Oocyte maturation. *Int. Rev. Cytol., 57,* 185-282.

Mittelholzer, C., Andersson, E., Consten, D., Hirai, T., Nagahama, Y. and Norberg, B. (2007). 20β-hydroxysteroid dehydrogenase and CYP19A1 are differentially expressed during maturation in Atlantic cod (*Gadus morhua*). *J. Mol. Endocrinol., 39,* 319-328.

Nagahama, Y. and Adachi, S. (1985). Identification of a maturation inducing steroid in a teleost, the amago salmon (*Oncorhynchus rhodurus*). *Dev. Biol., 109,* 428-435.

Nagahama, Y. and Yamashita, M. (2008). Regulation of oocyte maturation in fish. *Dev Growth Differ, 50 Suppl 1,* S195-219.

Nagahama, Y. (1987). Gonadotropin action on gametogenesis and steroidogenesis in teleost gonads. *Zool. Sci., 4,* 209-222.

Nagahama, Y. (1994). Endocrine regulation of gametogenesis in fish. *Int. J. Dev. Biol., 38,* 217-229.

Nagahama, Y. (1997). 17α,20β-Dihydroxy-4-pregnen-3-one, a maturation inducing hormone in fish oocytes: mechanisms of synthesis and action. *Steroids, 62,* 190-196.

Nagahama, Y., Kagawa, H. and Young, G. (1985). Stimulation of 17α, 20β- dihydroxy-4-pregnen-3-one production in the granulosa cells of amago salmon, *Oncorhynchus rhodurus*, by cyclic nucleotides. *J. Exp. Zool., 236,* 371-375.

Nagahama, Y., Yoshikuni, M., Yamashita, M., Tokumoto, T. and Katsu, Y. (1995). Regulation of oocyte growth and maturation in fish. *Curr. Top Dev. Biol., 30,* 103-145.

Nakamura, I., Evans, J. C., Kusakabe, M., Nagahama, Y. and Young, G. (2005). Changes in steroidogenic enzyme and steroidogenic acute regulatory protein messenger RNAs in ovarian follicles during ovarian development of rainbow trout (*Oncorhynchus mykiss*). *Gen. Comp. Endocrinol.. 144,* 224-231.

Ohta, K. and Matsuyama, M. (2002). Steroidogenic pathways to 17,20β-dihydroxy-4-pregnen-3-one and 17,20β,21-trihydroxy-4-pregnen-3-one in the ovarian follicles of the bambooleaf wrasse *Pseudolabrus sieboldi. Fisheries Sci., 68,* 41-50.

Patino, R. and Sullivan, C. V. (2002). Ovarian follicular growth, maturation, and ovulation in teleost fish. *Fish Physiol. Biochem., 26,* 57-70.

Payne, A. H. and Hales, D. B. (2004). Overview of steroidogenic enzymes in the pathway from cholesterol to active steroid hormones. *Endocr. Rev., 25,* 947-970.

Perazzolo, L. M., Coward, K., Davail, B., Normand, E., Tyler, C. R., Pakdel, F., Schneider, W. J. and Le Menn, F. (1999). Expression and localization of messenger ribonucleic acid for the vitellogenin receptor in ovarian follicles throughout oogenesis in the rainbow trout, *Oncorhynchus mykiss*. *Biol. Reprod., 60,* 1057-1068.

Prat, E., Sumpter, J. P. and Tyler, C. R. (1996). Validation of radioimmunoassays for two salmon gonadotropins (GtH I and GtH II) and their plasma concentrations throughout the reproductive cycle in male and female rainbow trout (*Oncorhynchus mykiss*). *Biol. Reprod., 54,* 1375-1382.

Raghuveer, K. and Senthilkumaran, B. (2012). Cloning and expression of 3β-hydroxysteroid dehydrogenase during gonadal recrudescence and after hCG induction in the air-breathing catfish, *Clarias gariepinus*. *Steroids, 77,* 1133-1140.

Reinecke, M. (2010). Insulin-like growth factors and fish reproduction. *Biol. Reprod., 82,* 656-661.

Scavo, L., Shuldiner, A. R., Serrano, J., Dashner, R., Roth, J. and de Pablo, F. (1991). Genes encoding receptors for insulin and insulin-like growth factor I are expressed in *Xenopus* oocytes and embryos. *Proc. Natl. Acad. Sci. USA, 88,* 6214-6218.

Scott, A. R. and Sumpter, J. E. (1983). A comparison of female reproductive cycles of autumn-spawning and winter-spawning strains of rainbow trout (*Salmo gaidneri*, Richardson). *Gen. Comp. Endocrinol., 52,* 79-85.

Selman, K. and Wallace, R. A. (1989). Cellular aspects of oocyte growth in teleosts. *Zool Sci, 6,* 211-231.

Senthilkumaran, B. (2011). Recent advances in meiotic maturation and ovulation: comparing mammals and pisces. *Front. Biosci., 16,* 1898-1914.

Senthilkumaran, B., Sudhakumari, C. C., Chang, X. T., Kobayashi, T., Oba, Y., Guan, G., Yoshiura, Y., Yoshikuni, M. and Nagahama, Y. (2002). Ovarian carbonyl reductase-like 20β-hydroxysteroid dehydrogenase shows distinct surge in messenger RNA expression during natural and gonadotropin-induced meiotic maturation in Nile tilapia. *Biol. Reprod., 67,* 1080-1086.

Senthilkumaran, B., Sudhakumari, C. C., Wang, D. S., Sreenivasulu, G., Kobayashi, T., Kobayashi, H. K., Yoshikuni, M. and Nagahama, Y. (2009). Novel 3β-hydroxysteroid dehydrogenases from gonads of the Nile tilapia: phylogenetic significance and expression during reproductive cycle. *Mol. Cell Endocrinol., 299,* 146-152.

Senthilkumaran, B., Yoshikuni, M. and Nagahama, Y. (2004). A shift in steroidogenesis occurring in ovarian follicles prior to oocyte maturation. *Mol. Cell Endocrinol., 215,* 11-18.

Sherwood, N. M., Krueckl, S. L. and McRory, J. E. (2000). The origin and function of the pituitary adenylate cyclase-activating polypeptide (PACAP)/glucagon superfamily. *Endocr. Rev., 21,* 619-670.

Shibata, N., Yoshikuni, M. and Nagahama, Y. (1993). Vitellogenin incorporation into oocytes of rainbow trout, *Oncorhynchus mykiss, in vitro*: effects of hormones on denuded oocytes. *Develop Growth Diff., 35,* 115-121.

Slater, C. H., Schreck, C. B. and Swanson, P. (1994). Plasma profiles of the sex steroids and gonadotropins in maturing female spring chinook salmon (*Oncorhynchus tshawytscha*). *Comp. Biochem. Physiol. A Comp. Physiol., 109,* 167-175.

Sreenivasulu, G. and Senthilkumaran, B. (2009a). A role for cytochrome P450 17α-hydroxylase/c17-20 lyase during shift in steroidogenesis occurring in ovarian follicles prior to oocyte maturation. *J. Steroid Biochem. Mol. Biol., 115,* 77-85.

Sreenivasulu, G. and Senthilkumaran, B. (2009b). New evidences for the involvement of 20β-hydroxysteroid dehydrogenase in final oocyte maturation of air-breathing catfish. *Gen. Comp. Endocrinol., 163,* 259-269.

Sreenivasulu, G., Senthilkumaran, B., Sridevi, P., Rajakumar, A. and Rasheeda, M. K. (2012a). Expression and immunolocalization of 20β-hydroxysteroid dehydrogenase during testicular cycle and after hCG induction, *in vivo* in the catfish, *Clarias gariepinus. Gen. Comp. Endocrinol., 175,* 48-54.

Sreenivasulu, G., Senthilkumaran, B., Sudhakumari, C. C., Guan, G., Oba, Y., Kagawa, H. and Nagahama, Y. (2012b). 20β-hydroxysteroid dehydrogenase gene promoter: Potential role for cyclic AMP and xenobiotic responsive elements. *Gene, 509,* 68-76.

Sreenivasulu, G., Sridevi, P., Sahoo, P. K., Swapna, I., Ge, W., Kirubagaran, R., Dutta-Gupta, A. and Senthilkumaran, B. (2009). Cloning and expression of StAR during gonadal cycle and hCG-induced oocyte maturation of air-breathing catfish, *Clarias gariepinus. Comp. Biochem. Physiol. B Biochem. Mol. Biol., 154,* 6-11.

Sreenivasulu, G., Swapna, I., Rasheeda, M. K., Ijiri, S., Adachi, S., Thangaraj, K. and Senthilkumaran, B. (2005). Expression of 20β-hydroxysteroid dehydrogenase and P450 17α-hydroxylase/c17-20 lyase during hCG-induced *in vitro* oocyte maturation in snake head murrel *Channa striatus. Fish Physiol. Biochem., 31,* 227-230.

Sridevi, P., Dutta-Gupta, A. and Senthilkumaran, B. (2011). Molecular cloning and expression analysis of fushi tarazu factor 1 in the brain of air-breathing catfish, *Clarias gariepinus. PLoS One, 6,* e28867.

Stocco, D. M. (2001). StAR protein and the regulation of steroid hormone biosynthesis. *Annu. Rev. Physiol., 63,* 193-213.

Suzuki, K., Kawauchi, H. and Nagahama, Y. (1988). Isolation and characterisation of two distinct gonadotropins from chum salmon pituitary glands. *Gen. Comp. Endocrinol., 71,* 292-301.

Swanson, E. (1991). Salmon gonadotropins: reconciling old and new ideas. In A. E. Scott, J. R. Sumpter, D. E. Kime, and M. S. Rolfe, (Eds.), *Proc. 4th Int. Syrup. Reproductive Physiology of Fish* (University of Sheffield Press. 2-7). Sheffield, UK.

Swanson, E., Suzuki, K., Kawauchi, H. and Dickoff, W. W. (1991). Isolation and characterisation of 2 coho salmon gonadotropins, GtH I and GtH II. *Biol. Reprod., 44,* 29-38.

Swanson, P., Bernard, M. G., Nozaki, M., Suzuki, K., Kawauchi, H. and Dickhoff, W. W. (1989). Gonadotropins I and II in juvenile coho salmon. *Fish Physiol. Biochem., 7,* 169-176.

Tanaka, H., Kagawa, H. and Hirose, K. (1995). Steroidogenic activities of two distinct gonadotropins in red seabream, *Pagrus major*. In F. W. Goetz, and P. Thomas (Eds.), *Reproductive Physiology of Fish* (Fish Symposium, 10-12), Austin, TX.

Tanaka, M., Nakajin, S., Kobayashi, D., Fukada, S., Guan, G., Todo, T., Senthilkumaran, B. and Nagahama, Y. (2002). Teleost ovarian carbonyl reductase-like 20β-hydroxysteroid dehydrogenase: potential role in the production of maturation-inducing hormone during final oocyte maturation. *Biol. Reprod., 66,* 1498-1504.

Tanaka, M., Telecky, T. M.,Fukada, S., Adachi, S., Chen S. and Nagahama,Y. (1992). Cloning and sequence analysis of the cDNA encoding P-450 aromatase (P450arom) from a rainbow trout (*Oncorhynchus mykiss*) ovary; relationship between the amount of P450arom mRNA and the production of oestradiol-17β in the ovary. *J. Mol. Endocrinol., 8,* 53-61.

Tokumoto, T., Yamaguchi, T., Ii, S. and Tokumoto, M. (2011). *In vivo* induction of oocyte maturation and ovulation in zebrafish. *PLoS One, 6,*e25206.

Trant, J. M. and Thomas, P. (1989). Isolation of a novel maturation inducing steroid produced *in vitro* by ovaries of Atlantic croaker. *Gen. Comp. Endocrinol., 75,* 397-404.

Trichet, V., Buisine, N., Mouchel, N., Morán, P., Pendás, A. M., Le Pennec, J. P. and Wolff, J. (2000). Genomic analysis of the vitellogenin locus in rainbow trout (*Oncorhynchus mykiss*) reveals a complex history of gene amplification and retroposon activity. *Mol. Gen. Genet., 263,* 828-837.

Tyler, C. R. and Sumpter, J. P. (1996). Oocyte growth and development in teleosts. *Rev. Fish Biol. Fish, 6,* 287-318.

Tyler, C. R. (1991). Vitellogenesis in Salmonids. In: A. P. Scott, J. P. Sumpter, D. E. Kime, and J. Rolfe (Eds.), *Reproductive Physiology of Fish* (Fish Symp 91, 295-299). United Kingdom: Sheffield.

Tyler, C. R., Sumpter, J. E. and Bromage, N. R. (1987). The hormonal control of vitellogenin uptake into cultured ovarian follicles of the rainbow trout, *Salmo gairdneri*. In D. Idler, L. Crim, and J. Walsh, *Proc. Third Int. Symp. Reproductive Physiology of Fish* (Memorial University Press, 220). St. John's, Newfoundland, Canada.

Tyler, C. R., Sumpter, J. P. and Witthames, P. R. (1990). The dynamics of oocyte growth during vitellogenesis in the rainbow trout, *Salmo gairdneri*. *Biol. Reprod., 43,* 202-209.

Tyler, C. R., Sumpter, J. R, Kawauchi, H. and Swanson, R. (1991). Involvement of gonadotrophin in the uptake of vitellogenin into vitellogenic follicles of the rainbow trout, *Oncorhynchus mykiss*. *Gen. Comp. Endocrinol., 84,* 291-299.

Wallace, R. A. and Selman, K. (1981). Cellular and dynamic aspects of oocyte growth in teleosts. *Am. Zool., 21,* 325-343.

Wallace, R. A. and Selman, K. (1990). Ultrastructural aspects of oogenesis and oocyte growth in fish and amphibians. *J. Electron Microsc. Tech., 16,* 175-201.

Wallace, R. A. (1985). Vitellogenesis and oocyte growth in nonmammalian vertebrates. In L. W. Browder, (Eds.), *Developmental Biology* (Vol. 1, 127–177). Plenum Press, New York.

Wang, H., Tan, J. T., Emelyanov, A., Korzh, V. and Gong, Z. (2005). Hepatic and extrahepatic expression of vitellogenin genes in the zebrafish, *Danio rerio. Gene, 356,* 91-100.

Wang, Y. and Ge, W. (2002). Cloning of zebrafish ovarian carbonyl reductase-like 20β-hydroxysteroid dehydrogenase and characterization of its spatial and temporal expression. *Gen. Comp. Endocrinol., 127,* 209-216.

Yaron, Z. and Levavi-Sivan, B. (2011). Endocrine regulation of fish reproduction. In A. P. Farrell (Ed.). *Encyclopedia of Fish Physiology: From Genome to Environment* (volume 2, 1500–1508). San Diego, Academic Press.

Zapater, C., Chauvigné, F., Scott, A. P., Gómez, A., Katsiadaki, I. and Cerdà, J. (2012). Piscine follicle-stimulating hormone triggers progestin production in gilthead seabream primary ovarian follicles. *Biol. Reprod., 87,* 111.

Zhou, L. Y., Wang, D. S., Kobayashi, T., Yano, A., Paul-Prasanth, B., Suzuki, A., Sakai, F. and Nagahama, Y. (2007a). A novel type of P450c17 lacking the lyase activity is responsible for C21-steroid biosynthesis in the fish ovary and head kidney. *Endocrinology, 148,* 4282-4291.

Zhou, L. Y., Wang, D. S., Shibata, Y., Paul-Prasanth, B., Suzuki, A. and Nagahama, Y. (2007b). Characterization, expression and transcriptional regulation of P450c17-I and -II in the medaka, *Oryzias latipes. Biochem. Biophys. Res. Comm., 362,* 619-624.

In: Sexual Plasticity and Gametogenesis in Fishes
Editor: Balasubramanian Senthilkumaran
ISBN: 978-1-62618-848-8

Chapter XVIII

Function of Germ Cells in Sex Differentiation

Toshiya Nishimura[1,2] and Minoru Tanaka[1,2,*]
[1]Laboratory of Molecular Genetics for Reproduction,
National Institute for Basic Biology, Okazaki, Japan
[2]Graduate School of Advanced Study, SOKENDAI, Okazaki, Japan

Abstract

Gonads consist of both germ cells and gonadal somatic cells. In the gonadal development, primordial germ cells (PGC) are segregated from other somatic cell lineages and migrate into genital ridges, forming sexually undifferentiated gonads. Affected by genetic factors or environmental factors, gonadal primordia develop either testes or ovaries.

It is believed that sexual differentiation of germ cells is dependent on the sex of their surrounding somatic cells, and thus germ cells have been considered as passive cells in terms of the gonadal sex differentiation. However, recent cellular analysis of germ cell-deficient gonads and gonads with hyperproliferated germ cells shows that an improper number of germ cells can lead sex reversal, suggesting germ cells have an essential role in sex differentiation.

In this chapter, we will review recent studies about involvement of germ cells in the sex differentiation using medaka. From the studies, we have proposed that sex differentiation is a process of balancing between feminization of somatic cells by germ cells and cell-autonomous masculinization of somatic cells. To validate the model to gonadal sex differentiation in other teleost fish, germ cells deficient gonads of zebrafish and loach are also discussed. The comparison of gonadal sex differentiation indicates commonality and diversification of the germ cells function in sex differentiation among teleost fishes.

[*] Corresponding Author: Minoru Tanaka, email: mtanaka@nibb.ac.jp, tel/fax: +81-564-59-5851.

Introduction

Sexual reproductive animals develop either a testis or an ovary. The decision to make an either tissue is called as sex determination. In animal kingdom, there is variation in sex determination system. In mammals and birds, their sexes are determined genetically, but they use different genetic system: XX-XY in mammals and ZZ-ZW in birds. The sex determination gene is also different: *Sry* in mammals (Koopman et al., 1990, Sinclair et al., 1990, Koopman et al., 1991) and *Dmrt1* in birds (Smith et al., 2009). In reptiles, on the other hand, their sex is determined by environmental factors such as temperature at which eggs grow up. In most animals, once the sex is determined, they follow its sex differentiation program and rarely show sex reversal in normal development. However, in some groups of teleost fishes, such as black porgy (Lee et al., 2001), anemone fish (Fricke and Fricke, 1977) and bluehead wrasse (Warner and Swearer, 1991), they do show sex reversal controlled by genetic or social system.

Thus, sex differentiation process is also varied among animals. Although there is variation in the sex determination system and sex differentiation process, these animals share common characteristics: they all form either a testis or an ovary. One important goal of research in sex differentiation is to understand the molecular machinery which ensures the dimorphism of gonadal development.

Gonadal development begins with the formation of gonadal primordia consisting of germ cells and somatic cells. Germ cells are segregated from other somatic cells lineages at a very early stage of development and migrate toward a gonadal region, followed by sex-specific gametogenesis. It has been believed that sexual different gametogenesis, an oogenesis and a spermatogenesis, depends on the sex of surrounding somatic cells, and thus germ cells have been considered as passive cells in term of the sexual differentiation of gonads. In XY mice, the presence of germ cells is not required for testis cord formation (McLaren, 1991). However, in XX mice, when very few germ cells enter a female genital ridge, granulosa cells fail to differentiate and result in a streak ovary containing only stromal cells (Mclaren, 1991, McLaren, 2000).

In most teleost fishes, germ cells proliferation and differentiation precede ovarian differentiation and are restricted during testicular development (Nakamura et al., 1998). In natural sex reversal process, as is observed in protogynous fishes, oocytes loss precedes the appearance of testicular structure. These studies suggest a role of germ cells in ovarian differentiation and/or maintenance of ovarian structure in some fishes.

We have recently obtained several pieces of evidence that suggest the importance of close interaction between germ cells and somatic cells in establishing the dimorphism of gonads using germ cell-ablated and mutant medaka (Kurokawa et al., 2007, Morinaga et al., 2007, Tanaka et al., 2008). We have proposed that the sex differentiation is a process of balancing between feminization of somatic cells by germ cells and male predisposition of somatic cells. In this chapter, we will review our recent studies about sex reversal in germ cell-ablated medaka and *hotei* mutants. To compare the sex differentiation model acquired from medaka with that of other teleost fishes, germ cell ablation studies using zebrafish and loach will be also discussed.

Gonadal Sex Differentiation in Medaka

The sex of medaka is genetically determined by an XX-XY system, and on the Y chromosome, a sex determination gene, *DMY/dmrt1Y*, has been already identified (Matsuda et al., 2002, Nanda et al., 2002). The gonadal development begins with the formation of sexually indifferent gonadal primordia at 3-4 day post fertilization (dpf) (Nakamura et al., 2006). Just after the formation, the gonadal somatic cells surrounding germ cells start to express *DMY/dmrt1Y* in XY gonads, and its expression continues throughout the entire period of gonadal development (Kobayashi et al., 2004). Germ cells in both XX and XY gonads proliferate by slow intermittent division (typeI), and XX and XY gonads are morphologically similar by 5-6dpf. Just after the stage, a subset of germ cells in XX gonads initiates two to four rounds of continuous division (typeII), forming cysts of four, eight, or sixteen cells, which subsequently enter meiosis (Saito et al., 2007). By hatching stage (7-8dpf), while germ cells in XX gonads increase the number rapidly and become differentiated, those in XY gonads persist typeI division and slightly increase the number. Therefore, one possible role of *DMY/dmrt1Y* in somatic cells is to suppress the transition from typeI to typeII division of germ cells and to keep the germ cell number low.

After hatching, female germ cells enter meiosis and start oogenesis. A female somatic cell marker, *foxl2* is expressed in supporting cells from 5-10 day after hatching (dph), which subsequently found in granulosa cells surrounding diplotene oocytes (Nakamura et al., 2008). At the same stage, another female marker, *aromatase*, which is necessary for the production of estradiol, is also expressed in the ventral epithelial cells, which is subsequently recruited as theca cells surrounding follicles (Nakamura et al., 2009).

In XY gonads, germ cells keep quiescence state by 20-30dph. A male somatic marker, *dmrt1,* is expressed from 5-10dph, but only a few or no steroidogenic enzymes such as *p450c17* and *p450c11*, which are necessary for the production of 11-keto testosterone (11-KT), are expressed at this stage. Around 20-30dph, a lobule structure is formed with the increased number of steroidogenic cells, and germ cells resume proliferation and start spermatogenesis. Thus, the production of 11-KT may be accompanied by initiation of spermatogenesis and formation of testicular structure.

Female to Male Sex Reversal in Germ Cell-Deficient Medaka

Although the sex of medaka is determined genetically, reduction of germ cells number during gonadal sex differentiation can induce sex reversal, suggesting germ cells have an essential role in sex differentiation in medaka (Kurokawa et al., 2007). When germ cells are ablated by disruption of PGC migration using *cxcr4* morpholino, the germ cell-deficient gonads form a tubule-like structure in both XX and XY adults. The tubule-like structure consists of empty lumen surrounded by a single layer of cells, which is further separated by basement membrane and surrounded by outer stromal cells. Interestingly, gene expression of the tubule-like gonads in even XX fish shows the similar pattern to testes. They fail to express female markers such as *foxl2* and *aromatase* but express a male marker, *dmrt1*, in an inner layer of cells lining the tubule, suggesting they acquire characteristics of Sertoli cells.

Steroidogenic enzymes necessary for the production of 11-KT, such as *p450c17* and *p450b11* are also expressed in outer stromal region. Furthermore, the germ cell-deficient medaka displays male secondary characteristics, accompanied by an increase level of 11-KT. Therefore, these results suggest that medaka with germ cell-deficient medaka cytologically and endocrinologically develops as a male irrespective of the genetic sex.

During the early development of germ cell-deficient gonads in XX medaka, *aromatase* is expressed in ventral epithelia as is observed in normal XX gonads. *foxl2* is weakly detected in the interstitial region. At later stage, however, *dmrt1* begins to express, and *aromatase* and *foxl2* are diminished in germ cell-deficient gonads. These developmental changes in sex specific markers suggest that female gonadal somatic cell lineages have been transdifferentiated into male cell lineages. Gonadal somatic cells in XX might have a potential to initiate a female pathway cell-autonomously, but they are not maintained without germ cells. Therefore, a possible role of germ cells during sex differentiation is to maintain feminization of gonads, and thus, germ cells are necessary for female development.

Male to Female Sex Reversal in *Hotei* Mutant

Previously, *hotei* mutants have been isolated by mutant screening in the medaka with defects of gonadal development (Morinaga et al., 2007). The *hotei* mutants develop hypertrophic gonads in both XX and XY and show large, swollen abdomens. This phenotype results from overproliferation of germ cells that initiate at around the hatching stage through adult in both XX and XY. As a result of positional cloning, the responsible gene for *hotei* mutants was identified as type II receptor of anti-Müllerian hormone (*amhrII*). The ligand of *amhrII* is anti-Müllerian hormone, also known as Müllerian-inhibiting substance. In medaka, both *amh* and *amhrII* are expressed in supporting cells surrounding germ cells, suggesting the *amh* signal indirectly control germ cells proliferation. This is further confirmed by chimeric analysis (Nakamura et al., 2012). When germ cells from XY wild-type are transplanted to XY *hotei* embryos, XY germ cells surrounded by *hotei* somatic cells proliferate rapidly and undergo cystic division. Conversely, XY *hotei* germ cells show slow intermittent division when surrounded by XY wild-type somatic cells. Therefore, the *amh* signal is required in somatic cells to control germ cell proliferation. One interesting aspect of *hotei* mutants is that a half of XY mutants display female secondary sex characteristics with a hypertrophic ovarian structure in gonads (Morinaga et al., 2007). During the early sex differentiation at the embryonic stage, *DMY/dmrt1Y* as well as a male marker, *gonadal soma derived growth facter* (*gsdf*), are expressed in gonadal somatic cells in *hotei* mutants, suggesting the early male pathway is unaffected. After hatching, uncontrolled germ cell proliferation leads to appearance of cystic and meiotic germ cells in XY *hotei* mutants, followed by oocyte formation, with expression of female markers, *foxl2* and *aromatase*, in supporting cells and stromal cells, respectively. It is important to note that gonadal somatic cells in XY mutants also express a male marker, *dmrt1*. These results suggest that gonadal somatic cells seem to undergo the male pathway, but hyperproliferation of germ cells caused by the impairment of *amh/amhrII* system leads to the feminization of gonads manifested by expression of *foxl2* and *aromatase*. Importantly, when germ cells are ablated before sex differentiation by *cxcr4* morpholino, this sex reversal is suppressed, and instead both XX and XY *hotei* mutants show the male phenotype. This result indicates that sex reversal of XY *hotei* mutants is not caused by impairment of *amh/amhrII* system itself in somatic cells but is mediated by hyperproliferated germ cells (Nakamura et al., 2012). Therefore, the studies of *hotei* mutants suggest that germ cells are sufficient for the feminization of gonads in medaka and produce secretion factors that can canalize the feminization of gonads.

A Model of Balancing Sex Mediated by Germ Cells

Until recently, studies on germ cells have been mainly focused on the process of gametogenesis and unique feature of multipotency, and less attention has been paid to the contribution of germ cells to sexual differentiation of gonads. The series of studies using medaka have suggested close association of germ cells with feminization of gonadal somatic cells. Figure 1 shows a model of balancing of sex mediated by germ cells. Germ cells-deficient gonads show testis-like characteristics, indicating gonadal somatic cells are predisposed to male without germ cell interaction. One possible role of *DMY/dmrt1Y* is to reinforce the masculinization of somatic cells. Another possible role of *DMY/dmrt1Y* is to suppress the germ cells proliferation and/or differentiation (transition from typeI to typeII division). On the other hand, the sex reversal from male to female in *hotei* mutants suggests that *amh/amhrII* system controls sex via regulating proliferation of germ cells which then canalize the feminization of gonadal somatic cells. In this model, the sex differentiation can be considered as a process of balancing between feminization of somatic cells by germ cells and cell-autonomous masculinization of somatic cells.

In XY gonads, the number of germ cells is restricted by *DMY/dmrt1Y*, so masculinization effect of somatic cells is dominant over the feminization effect by germ cells. In XX gonads, on the other hand, the increased number of germ cells reinforces the feminization of somatic cells, which overrides the cell-autonomous masculinization of somatic cells.

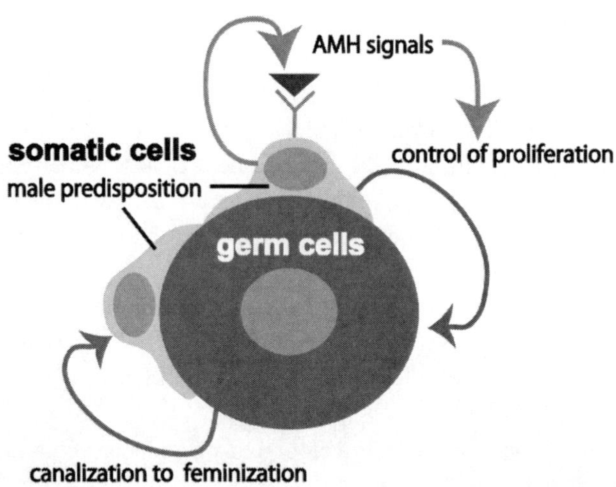

Figure 1. A model of balancing of sex mediated by germ cells. The gonadal sex differentiation can be considered as a process of balancing between canalization of somatic cells by germ cells and male predisposition of somatic cells. AMH signal functions in somatic cells and controls sex differentiation via regulating proliferation of germ cells.

Function of Germ Cells in Sex Differentiation in Other Teleost Fishes

It is intriguing whether or not the model acquired from the lines of researches using medaka is also adaptive to sex differentiation in other teleost fishes. Here, germ cell- deficient gonads in two species of teleost fishes, zebrafish and loach will be discussed.

Germ Cell-Deficient Gonads in Zebrafish

Zebrafish (*Danio rerio*) is a juvenile hermaphrodite, with all individuals having ovary like gonads during the juvenile period and the bisexual differentiation taking place at later stage (Takahashi, 1977, Maack and Segner 2003). The initial phase of gonadal development in zebrafish is similar to that of female medaka. Germ cells enter meiosis and begin oogenesis in all individuals, regardless of later gonadal sex. Meiotic germ cells are first observed around 11-14dpf (Takahashi, 1977), which is similar to that observed in medaka. Expression of *aromatase* genes is observed in somatic cells at around two weeks post fertilization (wpf) (Wang and Orban, 2007), which is also comparable with the chronological timing of *aromatase* expression in female medaka. By 4wpf, the majority of the fish possessed gonads with meiotic germ cells and diplotene oocytes, representing presumptive ovaries. Starting with 5wpf, a putative male fish shows alternation of gonadal structure from the initial ovarian structure. The alternation process first includes a decrease in the number and size of oocytes (Maack and Segner 2003). After loss of oocytes, putative spermatogonial cells are present and form a cyst-like structure, followed by spermatogenesis by 7wph. In presumptive female, developing oocytes are maintained from 4wpf and ovarian development continues. The sexually dimorphic gene expression is also observed at the period of which individual undergoes gonadal transformation. Whereas ovarian *aromatase* is downregulated and *amh* and 11β-hydroxylase (*cyp11b*) is upregulated in male gonads, *aromatase* is maintained and *amh* and *cyp11b* are not upregulated in female gonads (Wang and Orban, 2007).

In zebrafish, two reports have suggested that germ cells are necessary for female development as is observed in medaka (Slanchev et al., 2005, Siegfried and Nüsslein-Volhard, 2008). When germ cells are ablated by the biscistornic protein killer bacterial system parD or *dnd*-morpholino, zebrafish develops as a male as judged by morphological and behavioral criteria. The germ cell-deficient gonads organize tubules resembling testes, in which putative Sertoli cells lined the tubules and putative Leydig cells are observed in the interstitial space. This structure is similar to germ cell-deficient gonads in medaka. Furthermore, gene expression analysis shows that *amh, sox9a* and *cyp11b* are expressed at the comparable level to testes, suggesting Sertoli cells and Leydig cells are present in germ cell-deficient gonads.

One notable characteristic is that the development of gonadal somatic cells in the germ cell-deficient fish progresses with the similar timing to wild-type gonadal development. Ovarian *aromatase* is expressed during the period of the juvenile ovary stage in wild-type. Similar to a putative male of wild-type, the expression of *aromatse* is downregulated and *amh* and *cyp11b* are upregulated in germ cell-deficient gonads at later stage. This indicates that germ cells not required for initial expression of *aromatase* but is required for its maintenance, which is similar to germ cell-deficient gonads in female medaka. Therefore, a common role of germ cells during sex differentiation in medaka and zebrafish is to maintain feminization of gonadal somatic cells by keeping the expression of female-specific genes such as *fox2* and *aromatase*.

Germ Cell-Deficient Gonads in Loach, *Misgurnus Anguillicaudatus*

Loach, *Misrunus anguillicaudatus* (Crypriniformes: Cobitidae) is a gonochoristic species distributing across a wide area of the temperate zone of East Asia and living in rivers, lakes, ponds and paddy fields (Fujimoto et al., 2010). The sex determination of loach is concluded as male heterozygous XX-XY system because all female loaches are produced after induction of gynogenesis in which a paternal contribution to the zygotes is absence (Suzuki et al., 1985). However, it is also reported that the sex differentiation of loach is influenced by temperature, and high temperature induces female to male sex reversal (Nomura et al., 1998).

A male loach takes about 6 months to achieve sexual maturity under artificial rearing conditions whereas a female loach takes one year (Fujimoto et al., 2010). The time taken to reach maturity in loach is longer than that in medaka and zebrafish. It takes about two months from the formation of gonadal primordia to the first sign of sexual differentiation in loach, which is also much longer compared with that in medaka and zebrafish. The first sign of sexual differentiation of gonads is proliferation of germ cells, which is observed in two months after hatching. Like medaka and other teleost fish, one type of gonads begins cystic division and increases the number of germ cells, while the other type undergoes slow intermittent division and thus keeps the number of germ cells low. At later stage, dimorphic structural changes of gonadal somatic cells are observed in wild-type loach. Putative female gonads elongate to attach to a coelomic wall, which consequently form the ovarian cavity between a gonad and a coelomic wall. In putative male, a seminal lobule structure and a sperm duct begin to form. Loach with germ cell-deficient gonads is induced by *dnd*-morpholino (Fujimoto et al., 2010). Interestingly, sex-specific structural changes of somatic cells described above are also observed in germ cell-deficient gonads of loach at the similar timing. In sexually mature wild-type loach, male and female can be distinguished phenotypically by the presence or absence, respectively, of bony plates in the pectoral fins. These secondary characteristics are also observed in two-year-old *dnd*-morphant loach. The gonads from phenotypic male show a tubule-like structure. Similar to germ cell-deficient gonads of medaka and zebrafish, the tubule-like structure in loach consists of an inner layer of Sertoli-like cells lining the tubule and stromal cells surrounding outside the tubule. *dmrt1* is expressed in the tubule-like gonads at the similar level to wild-type testes, thus indicating that sexual differentiation to male normally takes place. On the other hand, the gonads from phenotypic female show a ribbon-like structure that consists of stromal cells at a ventral side and epithelia cells lining coelomic wall. One notable characteristic of ribbon-like gonads is presence of hypertrophied and ciliated epithelial cells lining coelomic wall. It is reported that hypertrophied epithelial cells lining the ovarian lumen were present in advance of ovarian maturation in medaka (Yamamoto, 1963). A study in goldfish also shows that the estrogen is capable of inducing precocious hypertrophy and ciliation of the epithelial cells lining the ovarian lumen (Takahashi and Takano, 1971). Therefore the presence of hypertrophied and ciliated epithelia suggests that the ribbon-like gonads are formed by endogenous estrogen. In support of this, *aromatase* is expressed in the gonads from phenotypic female. In addition, *foxl2* is also expressed in the ribbon-like gonads, indicating germ cell-deficient gonads in loach are normally feminized. Therefore, germ cell-deficient gonads in loach can develop into either a testicular or an ovarian structure, and thus sexual dimorphism of gonads is evident. In loach, germ cells seem not to be required for the feminization of gonads.

Comparison between Medaka, Zebrafish and Loach

What makes the difference between medaka and zebrafish, and loach in terms of gonadal sex differentiation without germ cells? It is unlikely that the pattern of gonadal sex differentiation without germ cells is correlated with close phylogenetic relationship. Whereas medaka is included in a group of Beloniformes, both zebrafish and loach are included in Cypriniformes. Therefore, the phylogenetic relationship between medaka and zebrafish is not as close as that between zebrafish and loach, but both medaka and zebrafish show all male phenotype when germ cells are ablated. In Figure 2, gonadal sex differentiations of medaka, zebrafish and loach are summarized. The female gonadal development in medaka is comparable to that in zebrafish as described previously (Saito and Tanaka, 2009). By comparing the gonadal sex differentiation of three species, two major differences are found in loach compared to medaka and zebrafish. First, the timing of ovarian morphogenesis, especially for the formation of ovarian cavity, relative to development of germ cells is different. Second, the initiation of meiosis and oogenesis is much later in loach than that in medaka and zebrafish. In teleost fishes, the timing of the formation of ovarian cavity relative to germ cell development is different from species to species (Nakamura et al., 1998). In medaka and zebrafish, germ cells enter meiosis and start oogenesis (oocyte formation) before the formation of ovarian cavity (Takahashi, 1977, Suzuki et al., 2004). In loach, on the other hand, stromal elongation of gonads for the formation of the ovarian cavity occurs coincidentally with the appearance of meiotic germ cells (Fujimoto et al., 2010). The similar pattern of ovarian formation is also observed in goldfish, *Carassius auratus* (Nakamura, 1978, Nakamura et al., 1998). Although the mechanism of the ovarian cavity formation is not fully understood, it is reported that endogenous estradiol is one of the factors required for the formation of the ovarian cavity in medaka (Suzuki et al., 2004). Therefore, it is possible that a sufficient amount of estradiol for ovarian cavity formation is already produced before meiotic entry of germ cells in loach. In Nile tilapia, *Oreochromis niloticus*, the ovarian cavity forms before the appearance of meiotic germ cells (Nakamura and Nagahama, 1985, Nakamura et al., 1998). Furthermore, although the expression levels are low, *foxl2* and *aromatase* are expressed higher in putative female than those in putative male before oogonial proliferation in tilapia (Kobayashi et al., 2000, Ijiri et al., 2008). The timing of *foxl2* and *aromatase* expression has not been reported in loach, but considering the studies in tilapia and the presence of the long period before germ cell proliferation in loach, it might be possible that *aromatase* and *foxl2* begin to be expressed and estradiol is accumulated by the time at which germ cells enter meiosis in loach. Therefore, in this case, gonadal somatic cells in loach are first feminized endocrinologically, and feminizing effects by germ cells are not required.

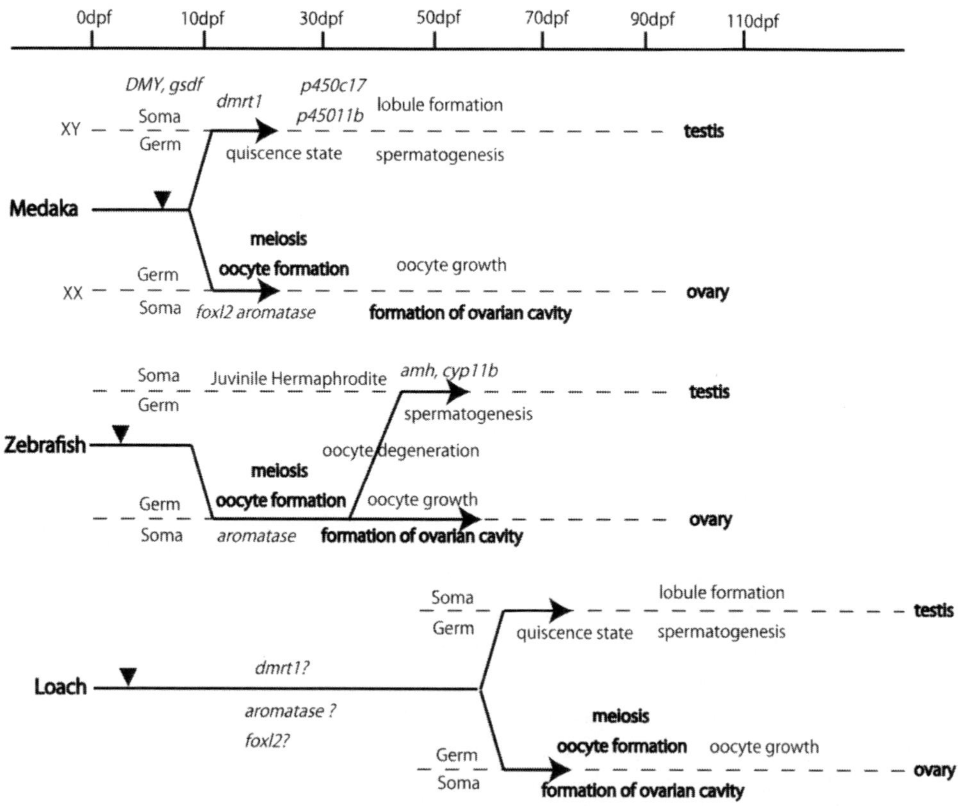

Figure 2. Comparative analysis of gonadal development in medaka, zebrafish and loach. Two major differences are found in loach compared to medaka and zebrafish. First, the timing of the formation of ovarian cavity relative to development of germ cells is different. Second, the initiation of meiosis and oogenesis is much later in loach than that in medaka and zebrafish. The type of gonadal development in goldfish and tilapia is similar to that in loach. Arrowheads indicate hatching.

However, this does not mean that feminizing effects of somatic cells by germ cells are absent in loach, but changing the timing of ovarian morphogenesis might modify the dependency of germ cells on feminization of somatic cells. From these speculations, the timing of ovarian morphogenesis relative to germ cell development might be one indicator of the dependency of germ cells on the feminization of somatic cells. The more germ cells differentiate during the ovarian morphogenesis, as is observed in medaka and zebrafish, the more the dependency of germ cells on feminization of gonadal somatic cells increases. If this model is true, germ cell-deficient goldfish and tilapia might also show dimorphic gonadal structures as is observed in loach. In the evolutionary point of view, it is interesting issues how teleost fishes modify the contribution of germ cells for the establishment of gonadal feminization depending on their diversity of life cycles and reproductive modes.

Unsolved Problems about the Feminization of Gonads by Germ Cells

There are two unsolved questions about the feminization of gonads by germ cells: 1) which stage of germ cells are critical for the feminization of somatic cells? 2) What are the secreted factors which can canalize the feminization of gonads?

For the first question, meiotic germ cells and/or oocytes may be candidate germ cells for the feminization of gonads. In teleost fishes, cystic germ cells enter meiosis and subsequently break up into separated oocytes during which supporting cells surround each oocyte and differentiate to granulose cells. In medaka, it is reported a female marker, *foxl2*, is predominantly expressed in granulose cells surrounding oocytes (Nakamoto et al., 2006, Nakamura et al., 2008), and *aromatase* expressing cells, which appear in the ventral side of epithelia, are recruited to surrounding of diplotene oocytes (Nakamura et al., 2009). The loss of *foxl2* and *aromatase* expression in germ cell- deficient gonads may result from failure of the recruitment of pre-granulose cells and *aromatase* expressing cells by meiotic germ cells and/or oocytes. In adult mammal, loss of oocytes does not induce transdifferentiation of granulose cells to Sertoli-like cells as analyzed by male markers such as *Sox9* and *Hsd17b1* (Uhlenhaut et al., 2009). However, during a fetal stage, loss of germ cells leads to failure of granulose cell differentiation and results in a streak gonad containing only stromal tissue (McLaren, 1991, McLaren, 2000). Considering that pre-granulosa cells invade inbetween the clusters of meiotic germ cells and each separated oocyte becomes surrounded by granulose cells in mammals, meiotic germ cells and/or oocytes may be required for the granulose cell differentiation during a fetal stage and thus, formation of primordial follicles. These two studies in mammals suggest that meiotic germ cells and/or oocytes are required for the establishment of feminization of gonad during a fetal stage. But once the feminization of gonads is established, oocytes may not be any longer required for the maintenance of feminized gonads. Therefore, it is possible to speculate that the requirement of meiotic germ cells and/or oocytes for the differentiation of granulose cells is conserved in both teleosts and mammals. To further elucidate this issue, a stage-specific ablation of germ cells should be necessary using, for example, a cell specific expression of Nitroreductase (Curado et al., 2008).

For the second question, candidate genes for factors for feminizing gonads originating from germ cells have not been reported so far, but their functions or roles of the genes could be addressed: upregulation of *foxl2* in supporting cell lineage and recruitment of *aromatase* expressing cells. The latter process might not be a direct effect of the feminizing factors, but might be the secondary effect of feminization of supporting cells. It is also possible that feminization factors suppress *dmrt1* in supporting cells, which is antagonized by *foxl2* in mice (Uhlenhaut et al., 2009, Matson et al., 2011). The antagonizing and overriding effects are possibly present in medaka. Loss-of-function mutants of *dmrt1* form follicles during sex differentiation and show male-to-female sex reversal in adults, suggesting induction of *foxl2* expression without *dmrt1* (Masuyama et al., 2012). In XY *hotei* mutants, the overproliferation of germ cells induces *foxl2* expression, overriding *dmrt1* function in the supporting cells and leading to complete sex reversal (Nakamura et al., 2012).

Conclusion

The gonadal sex differentiation is achieved by the close interaction between germ cells and somatic cells. Generally, a unidirectional signal is not sufficient especially for ovarian development, but a reciprocal crosstalk is required for dimorphic gonadal formation. In medaka, somatic cells control the proliferation and differentiation of germ cells using *amh/amhrII* system. On the other hand, germ cells have the ability to canalize the feminization of somatic cells. These two interactions make proper balance between feminization of somatic cells by germ cells and male predisposition of somatic cells. The impairment of the balance results in the sex reversal as observed in germ cells-deficient medaka and *hotei* mutants. Comparative analysis of germ cell-deficient gonads in medaka, zebrafish and loach reveals that germ cells are not always necessary for feminization of gonadal somatic cells among teleost fishes.

However, this result does not conclude that feminization effect of somatic cells by germ cells is absent in a group of teleost fishes, but we propose that the timing of ovarian morphogenesis modifies the dependency of germ cells on the feminization of gonadal somatic cells. To further understand the role of germ cells in sex differentiation, it is necessary to identify the feminizing factors secreted by germ cells, which provides us with more clear views on commonality and diversity in the balancing of sex differentiation.

References

Curado, S., Stainier, D. Y. R. and Anderson, R. M. (2008). Nitroreductase-mediated cell/tissue ablation in zebrafish: a spatially and temporally controlled ablation method with applications in developmental and regeneration studies. *Nature Protocol,* 3, 948-954.

Fricke, H. and Fricke, S. (1977). Monogamy and sex change by aggressive dominance in coral reef fish. *Nature,* 266, 830-832.

Fujimoto, T., Nishimura, T., Goto-Kazeto, R., Kawakami, Y., Yamaha, E. and Arai, K. (2010). Sexual dimorphism of gonadal structure and gene expression in germ cell-deficient loach, a teleost fish. *Proceedings of National Academy of Science USA,* 107, 17211-17216.

Ijiri, S., Kaneko, H., Kobayashi, T., Wang, D.S., Sakai, F., Paul-Prasanth, B., Nakamura, M. and Nagahama, Y. (2008). Sexual dimorphic expression of genes in gonads during early differentiation of a teleost fish, the nile tilapia *Oreochromis niliticus. Biology of Reproduction,* 78, 333-341.

Kobayashi, T., Kajiura-Kobayashi, H. and Nagahama, Y. (2000). Differential expression of *vasa* homologue gene in the germ cells during oogenesis and spermatogenesis in a teleost fish, tilapia, *Oreochromis niloticus. Mechanisms of Development,* 99, 139-142.

Kobayashi, T., Matsuda, M., Kajiura-Kobayashi, H., Suzuki, A., Saito, N., et al. (2004). Two DM domain genes, *DMY* and *DMRT1*, involved in testicular differentiation and development in the medaka, *Oryzias latipes. Developmental Dynamics,* 231, 518-526.

Koopman, P., Munsterberg, A., Capel, B., Vivian, N. and Lovell-Badge, R. (1990). Expression of a candidate sex determining gene during mouse testis differentiation. *Nature,* 348, 450-452.

Koopman, P., Gubbay, J., Vivian, N., Goodfellow, P. and Lovell-Badge, R. (1991). Male development of chromosomally female mice transgenic for *Sry*. *Nature,* 351, 117-121.

Kurokawa, H., Saito, D., Nakamura, S., Katoh-Fukui, Y., Ohta, K., Baba, T., Morohashi, K. and Tanaka, M. (2007). Germ cells are essential for sexual dimorphism in the medaka gonad. *Proceedings of National Academy of Science USA,* 104, 16958-16963.

Lee, Y. H., Du, J. L., Yueh, W. S., Lin, B. Y., Huang, J. D., Lee, C. Y., Lee, M. F., Lau, E. L., Lee, F. Y., Morrey, C., Nagahama, Y. and Chang, C. F. (2001). Sex change in the protandrous black porgy, Acanthopagrus schlegeli: A review in gonadal development, estradiol, estrogen receptor, aromatase activity and gonadotropin. *Journal of Experimental Zoology,* 290, 715-716.

Maack, G. and Segner, H. (2003). Morphological development of the gonads in zebrafish. *Journal of Fish Biology,* 62, 895-906.

Masuyama, H., Yamada, M., Kamei, Y., Fujiwara-Ishikawa, T., Todo, T., Nagahama, Y. and Matsuda, M. (2012). Dmrt1 mutation causes a male-to-female sex reversal after sex determination by *Dmy* in the medaka. *Chromosome Research,* 20, 163-176.

Matson, C. K., Murphy, M. W., Sarver, A. L., Griswold, M. D., Bardwell, V. J. and Zarkower, D. (2011). DMRT1 prevents female reprogramming in the postnatal mammalian testis. *Nature,* 476, 101-104.

Matsuda, M., Nagahama, Y., Shinomiya, A., Sato, T., Matsuda, C., Kobayashi, T., Morrey, C. E., Shibata, N., Asakawa, S., Shimuze, N., et al. (2002). DMY is a Y-specific DM-domain gene required for male development in the medaka fish. *Nature,* 417, 559-563.

McLaren, A. (1991). Development of the mammalian gonad: The fate of the supporting cell lineage. *BioEssays,* 3, 151-156.

McLaren, A. (2000). Germ and somatic cell lineages in the developing gonad. *Molecular and Cellular Endocrinology,* 163, 3-9.

Morinaga, C., Saito, D., Nakamura, S., Sasaki, T., Asakawa, S., Shimizu, N., Mitani, H., Furutani-Seki, M., Tanaka, M. and Kondoh, H. (2007). The *hotei* mutation of medaka in the anti-Müllerian hormone receptor causes the dysregulation of germ cell and sexual development. *Proceedings of National Academy of Science USA,* 104, 9691-9696

Nakamoto, M., Matsuda, M., Wang, D.S., Nagahama, Y. and Shibata, N. (2006). Molecular cloning and analysis of gonadal expression of *Foxl2* in the medaka, *Oryzias latipes. Biochemical and Biophysical Research Communications,* 344, 353-361.

Nakamura, M. (1978). Morphological and experimental studies on sex differentiation of the gonad in several teleost fishes. *Ph.D. thesis. Hokkaido University.*

Nakamura, M. and Nagahama, Y. (1985). Steroid producing cells during ovarian differentiation of the tilapia, *Sarotherodon niloticus. Development Growth and Differentiation,* 27, 701-708.

Nakamura, M., Kobayashi, T., Chang, X. and Nagahama, Y. (1998). Gonadal sex differentiation in teleost fish. *Journal of Experimental Zoology,* 281, 362-372.

Nakamura, S., Kobayashi, D., Aoki, Y., Yokoi, H., Ebe, Y., Wittbrodt, J. and Tanaka, M. (2006). Identification and lineage tracing of two populations of somatic gonadal precursors in medaka embryos. *Developmental Biology*, 295, 678-688.

Nakamura, S., Aoki, Y., Saito, D., Kuroki, Y., Fujiyama, A., Naruse, K. and Tanaka, M. (2008). *Sox9b/sox9a2*-EGFP transgenic medaka reveals the morphological reorganization of the gonads and a common precursor of both the female and male supporting cells. *Molecular Reproduction and Development,* 75, 472-476.

Nakamura, S., Kurokawa, H., Asakawa, S., Shimizu, N. and Tanaka, M. (2009). Two distinct types of theca cells in the medaka gonad: germ cell-dependent maintenance of *cyp19a1*-expressing theca cells. *Developmental Dynamics*, 238, 2652-2657.

Nakamura, S., Watakabe, I., Nishimura, T., Picard, J. Y., Toyoda, A., Taniguchi, Y., Clemente, N. D. and Tanaka, M. (2012). Hyperproliferation of mitotically active germ cells due to defective anti-Müllerian hormone signaling. *Development*, 139, 2283-2287.

Nanda, I., Kondo, M., Hornung, U., Asakawa, S., Winkler, C., Shimizu, A., Shan, Z., Haaf, T., Shimizu, N., Shima, A., et al. (2002). A duplicated copy of *DMRT1* in the sex-determining region of the Y chromosome of the medaka, *Oryzias latipes*. *Proceedings of National Academy of Science USA,* 99, 11778-11783.

Nomura, T., Arai, K., Hayashi, T. and Suzuki, R. (1998) Effect of temperature on sex ratios of normal and gynogenetic diploid loach. *Fisheries Science,* 64, 753-758.

Saito, D., Morinaga, C., Aoki, Y., Nakamura, S., Mitani, H., Furutani-Seiki, M., Kondoh, H. and Tanaka, M. (2007). Proliferation of germ cells during gonadal sex differentiation in medaka: Insights from germ cell-depleted mutant *zenzai*. *Developmental Biology*, 310, 280-290.

Saito, D. and Tanaka, M. (2009). Comparative aspects of gonadal sex differentiation in medaka: A conserved role of developing oocytes in the sexual canalization. *Sexual Development,* 3, 99-107.

Siegfried, K. R. and Nüsslein-Volhard, C. (2008). Germ line control of female sex determination in zebrafish. *Developmental Biology,* 324, 277-287.

Sinclair, A. H., Berta, P., Palmer, M.S., Hawkins, J. R., Griffiths, B. L., Smith, M. J., Foster, J. W., Frischauf, A. M., Lovell-Badge, R. and Goodfellow P. N. (1990). A gene from the human sex-determining region encodes a protein with homology to a conserved DNA-binding motif. *Nature,* 346, 240-244.

Slanchev, K., Stebler, J., de la Cueva-Méndez, G. and Raz, E. (2005). Development without germ cells: The role of the germ line in zebrafish sex differentiation. *Proceedings of National Academy of Science USA,* 102, 4074-4079.

Smith, C. A., Roeszler, K. N., Ohnesorg, T., Cummins, D. M., Farlie, P. G., Doran, T. J. and Sinclari, A. H. (2009). The avian Z-linked gene DMRT1 is required for male sex dtermination in the chicken. *Nature,* 461, 267-271.

Suzuki, A., Tanaka, M. and Shibata, N. (2004). Expression of *aromatase* mRNA and effects of aromatase inhibitor during ovarian development in the medaka, *Oryzias latipes*. *Journal of Experimental Zoology PartA: Comparative Experimental Biology,* 301A, 266-273.

Suzuki, R., Oshiro, T. and Nakanishi, T. (1985). Survival, growth and fertility of gynogenetic diploids induced in the cyprinid loach, *Misgurnus anguillidaudatus*. *Aquaculture,* 48, 45-55.

Takahashi, H. (1977). Juvenile hermaphroditism in the zebrafish, *Brachydanio rerio*. *Bulletin of the Faculty of Fisheries Hokkaido University*, 28, 57-65.

Takahashi, H. and Takano, K. (1971). Sex hormone-induced precocious hypertrophy and ciliation of epithelial cells in the ovarian lumen of the goldfish. *Annotationes Zoologicae Japonenses*, 44, 32-41

Tanaka, M., Saito, D., Morinaga, C. and Kurokawa, H. (2008). Cross talk between germ cells and gonadal somatic cells is critical for sex differentiation of the gonads in the teleost fish, medaka (*Oryzias latipes*). *Development, Growth and Differentiation,* 50, 273-278.

Uhlenhaut, N. H., Jakob, S., Anlag, K., Eisenberger, T., Sekido, R., Kress, J., Treier, A. C., Klugmann, C., Klasen, C., Holter, N. I. et al., (2009). Somatic sex reprogramming of adult ovaries to testes by FOXL2 ablation. *Cell,* 139, 1130-1142.

Wang, X. G. and Orban, L. (2007). Anti-Müllerian hormone and 11 β-hydroxylase show reciprocal expression to that of aromatase in the transforming gonad of zebrafish males. *Developmental Dynamics,* 236, 1329-1338.

Warner, R. R. and Swearer, S. E. (1991). Social control of sex change in the bluehead wrasse, *Thalassoma bifasciatum* (Pisces: Labridae). *Behavior,* 181, 199-204.

Yamamoto, K. (1963). Cyclical changes in the wall of ovarian lumen in the medaka, *Oryzias latipes. Annotationes Zoologicae Japonenses*, 36, 179-186.

In: Sexual Plasticity and Gametogenesis in Fishes
Editor: Balasubramanian Senthilkumaran
ISBN: 978-1-62618-848-8

Chapter XIX

Fish Gonadotropin Receptors

Yuichi Oba
Graduate School of Bioagricultural Sciences, Nagoya University,
Nagoya, Japan

Abstract

Gonadotropin hormones (GTHs) are kinds of heterodimeric glycoproteins released from pituitary, target gonadal tissues and control gonadogenesis, gametogenesis, and sex change *via* cognate receptor in vertebrates. Since 1999, the genes of GTH receptor (GTHR) have been isolated and characterized from over 20 fishes. These findings provide insights into the mechanism of GTH actions for applications in fish endocrinology, reproductive biology, and fishery science. This chapter summarize the last 13 year of literature regarding the fish follicle-stimulating hormone receptor (FSHR) and luteinizing hormone receptor (LHR) genes, particularly by focusing on the molecular cloning and evolution, ligand preference, tissue specificity, temporal and spatial distribution in gonads, which also involve in recent challenges to understand the origin of glycoprotein hormone (GpH)–receptor system during the evolution of chordates and natural/artificial sex change in fishes.

Abbreviations

LHR, luteinizing hormone receptor;
FSHR, follicle-stimulation hormone receptor;
TSHR, thyrotropin receptor;
GTHR, gonadotropin receptor;
GpHR, glycoprotein hormone receptor.

Introduction

In mammals, it has been well studied that two different gonadotropins (FSH and LH) play pivotal roles to control various reproductive aspects, especially gonadal development. FSH and LH are the heterodimeric glycoprotein hormones, consisting of common α subunit and variable (type-specific) β subunit, secreted from pituitary to the bloodstream. In the testis and ovary, there are two gonadotropin receptors FSH and LH receptor, which selectively bind to FSH and LH respectively, and transduce the pituitary signal temporarily and spatially to the gonads. In teleosts, since the duality of fish gonadotropin, FSH and LH (formerly GTH I and GTH II), was elucidated, the secretory profiles during the reproductive cycle and the responsiveness to the follicular cells, especially sex steroid production, has been studied extensively in various fish species (Nagahama, 1994). However, little was known about the characteristics of fish GTH receptors because of lacking its molecular evidences. This chapter summarizes the isolation and characterization of fish FSHR and LHR genes since two GTHRs were isolated for the first time from amago salmon (Oba et al., 1999a, b).

Molecular Cloning

The duality of GTH receptor in teleosts has long been discussed (Breton et al., 1986; Kanamori and Nagahama, 1988; Le Gac et al., 1988; Yan et al., 1992; Miwa et al., 1994) and finally identified by molecular cloning of two different receptor genes from amago salmon, *Oncorhynchus rhodurus* (Oba et al., 1999a, b), followed by channel catfish, *Ictalurus punctatus* (Kumar et al., 2001a, b), Nile tilapia, *Oreochromis niloticus* (Oba et al., 2001), African catfish, *Clarias gariepinus* (Bogerd et al., 2001; Schulz et al., 2001; Vischer and Bogerd, 2003), gilthead seabream *Sparus aurata* (Wong et al., 2004), and zebrafish *Danio rerio* (Laan et al., 2002; Kwok et al., 2005). To date, a pair of GTHR genes has been isolated from 5 major super orders of Neopterygii viz. (Elopomorpha, Ostariophysi, Eutelostei, Paracanthopterygii, and Acanthopterygii) containing 12 orders (Anguilliformes, Cypriniformes, Siluriformes, Salmoniformes, Gadiformes, Atheriniformes, Beloniformes, Cyprinodontiformes, Gasterosteiformes, Perciformes, Pleuronectiformes, and Tetraodontiformes) and over 20 species, which strongly suggests that the duality of GTH receptor are universal in fishes.

The deduced amino acid sequences showed that the fish GTHRs consist of putative signal peptides at N-terminus, long extracellular domain containing *N*-glycosylation sites, seven transmembrane domain, and intracellular C-terminal domain, as tetrapod GTHRs (Levavi-Sivan et al., 2010). Overall amino acid identities of FSHRs between teleosts and tetrapods are approximately 50%, and similar identity values are found with those of LHRs. The identities are relatively high in seven transmembrane domains.

Molecular Evolution

It had been unclear whether two different types of GTRHs in teleosts are evolutionary homologous to FSHR and LHR in tetrapod lineages when sequence information of fish

GTHR was very liminted (Oba et al., 1999a, b). After accumulating cloning reports from wide range of teleosts species, it became clear that two GTHRs are the orthologue of FSHR and LHR (see the comprehensive maximum likelihood tree in Chauvigné et al., 2010). This finding is also supported by the presence of extra-intron unique to tetrapod and fish LHRs (Kumar et al., 2001a, b; Kumar and Trant, 2001; Oba et al., 2001; Vischer et al., 2003) and by the syntenic analysis (Chauvigné et al., 2010). Therefore, two types of GTHRs in teleosts are now appropriate to be named as the FSHR and LHR, as those in tetrapods, and we can easily judge whether the cloned fish GTHR is FSHR or LHR only by sequence analysis.

It is of interest that fish LHRs comprise two distinct clades, and those of channel catfish and African catfish are in each clade, as well as those of tongue sole/ Senegalese sole, and gilthead seabream/ goldlined seabream. To address this enigma, Chauvigné et al. (2010) speculated by their extensive phylogenetic and syntenic analyses that interallelic gene conversion or tandem gene duplication might have occurred during the evolution of LHR in teleosts lineage.

Whole genome shotgun sequence of the elephant shark, *Callorhinchus milii* (Elasmoblanchii), revealed that the two putative GTHR genes are tandemly aligned on the chromosome, as the case with higher tetrapods FSHR and LHR (Chauvigné et al., 2010). This suggests that the origin of FSHR and LHR might trace back even before the divergence of elasmobranches. Two GTH β subunits, corresponding to FSHβ and LHβ, were isolated from the dogfish shark, *Scyliorhinus canicula* (Quérat et al., 2001), suggesting that the date of origin for FSH and LH might also be before the divergence of elasmobranches.

Two putative genes of GpHRs were cloned from the testis and thyroid of sea lamprey, *Petromyzon marinus* (Freamat et al., 2006, Freamat and Sower, 2008a). Molecular phylogenetic analyses showed that these two putative genes are placed outside of the clade containing FSHR, LHR and TSHR in teleosts and tetrapods (Chauvigné et al., 2010). Sower et al. (2006) suggested that sea lamprey possesses single GpH, and Freamat and Sower (2008b, 2010) hypothesized that the two GpHR share single GpH ligand. In brown hagfish, *Paramyxine atami,* single GpH β subunit was isolated, and purified native GpH induced the release of gonadal sex steroids *in vitro* (Uchida et al., 2010), which implicates the presence of GpHR. In amphioxus, *Branchiostoma belcheri*, glycoprotein subunit genes similar to the thyrostimulin were isolated (Tando and Kubokawa, 2009a, b; Kubokawa et al., 2010), but no evidence for the presence of a GpHR in amphioxus (Freamat and Sower, 2010). These recent findings suggest that GpHR might have originated prior to the evolutionary split between agnathans and gnathostomes.

Functional Characterization

In order to characterize GTH receptor, it is necessary to determine the properties of ligand binding and/or ligand response for the gene product. The gene products of fish GTH receptors have been expressed in mammalian COS (Oba et al., 1999a, b; Kumar et al., 2001a, b; Laan et al., 2002; Ko et al., 2007; Sambroni et al., 2007), HEK (Bogerd et al., 2001; Vischer et al., 2003; Rocha et al., 2007; Andersson et al., 2009), and CHO (So et al., 2005) cells or *Xenopus* oocyte (Chauvigné et al., 2010), and the functional responses have been determined by measuring cAMP accumulation (Oba et al., 1999a, b; Laan et al., 2002; Ko et al., 2007; Chauvigné et al., 2010) or reporter gene assay, such as firefly luciferase, driven by cAMP responsive promoter (Kumar et al., 2001a, b; Bogerd et al., 2001; Vischer et al., 2003; So et al., 2005; Rocha et al., 2007; Ko et al., 2007; Sambroni et al., 2007; Andersson et al., 2009). Ko et al. (2007) performed both cAMP measurement and reporter gene assay, showing that these independent results were in good agreement. Although they did not provide EC_{50} concentrations, as pointed out by Andersson et al., 2009, the facts may ensure compatibility of these two assay methods.

The results for the ligand selectivity of fish GTHR are varied, but mostly FSHR and LHR exhibit binding preference to FSH and LH respectively, regardless of its homologous/heterologous or purified/ recombinant origins. These findings are consistent with the facts that heterologous (including mammalian origins') or recombinant GTHs can induce the gametogenesis, sex change and steroidogenesis of fish *in vivo* (Devlin and Nagahama, 2002). Furthermore, the results suggest that the binding preference of FSHR and LH to the corresponding hormones (FSH and LH, respectively) were established prior to the split between teleost and tetrapod lineages. In this viewpoint, studying the ligand specificity for elasmobranch GTHRs will be of particular interest. However, more detailed interpretation of ligand selectivity should also be analyzed carefully because of the following reasons: (i) most of the reports used heterologous ligands; (ii) it is uncertain that recombinant GTH possess same property as that of naturally purified GTH from fish pituitary; (iii) it is also uncertain whether fish GTHR expressed in mammalian cells is functionally equivalent to those in fish gonad (Levavi-Sivan et al., 2010).

Studies on the ligand specificity of FSHR and LHR using homologous FSH and LH are limited. Vischer et al. (2003) studied ligand specificities of African catfish FSHR and LHR expressed in HEK cell using recombinant African catfish FSH and LH, showing that FSHR was activated by homologous FSH and LH and LHR was activated selectively by LH at ng/ml order. García-López et al. (2009) studied African catfish FSHR and LHR using recombinant channel catfish FSH and LH, which showed similar responsiveness as those of homologous FSH and LH. Sambroni et al. (2007) showed analyzed ligand specificities of rainbow trout FSHR and LHR expressed in COS cells using purified FSH and LH from rainbow trout pituitary, which had revealed that FSHR and LHR were activated selectively by homologous FSH and LH at ng/ml order, respectively. They also showed ligand specificity of rainbow trout FSHR and LHR for heterologous FSH and LH purified from Chinook salmon pituitary, and the results were similar to those of using homologous FSH and LH.

Tissue Distribution

Most of fish GTHR genes have been cloned from either testis or ovary. Tissue distribution has been studied by RT-PCR, showing that fish FSHR and LHR are commonly expressed in testis and ovary.

Extragonadal expression was also observed, for example: in kidney (channel catfish LHR, Kumar et al., 2001a; zebrafish FSHR, Kwok et al., 2005; Atlantic halibut FSHR and LHR, Kobayashi et al., 2008), in liver (zebrafish LHR, Kwok et al., 2005), in brain (amago salmon LHR, Oba et al., 1999a; African catfish LHR, Vischer and Bogerd, 2003; Atlantic salmon LHR, Maugars and Schmitz, 2006; Atlantic halibut LHR, Kobayashi et al., 2008; Atlantic salmon FSHR, Andersson et al., 2009; Atlantic cod FSHR and LHR, Mittelholzer et al., 2009; orange-spotted grouper FSHR and LHR, Hu et al., 2011; Japanese eel FSHR, Kazeto et al., 2012), in gill (channel catfish LHR; Kumar et al., 2001a; Atlantic salmon LHR, Maugars and Schmitz, 2006; Atlantic halibut LHR, Kobayashi et al., 2008; Senegalese sole FSHR, Chauvigné et al., 2010; eel FSHR and LHR, Kazeto et al., 2012), in muscle (channel catfish LHR, Kumar et al., 2001a; African catfish LHR, Vischer and Bogerd, 2003; Atlantic salmon LHR, Andersson et al., 2009), in spleen (channel catfish FSHR, Kumar et al., 2001b; Okinawan rubble goby LHR, Kobayashi et al., 2009), heart (Atlantic halibut LHR, Kobayashi et al., 2008), in head-kidney (African catfish LHR, Vischer and Bogerd, 2003), in gut (zebrafish LHR, Wong and van Eenennaam, 2004), and in pituitary (zebrafish LHR, Wong and van Eenennaam, 2004; Atlantic halibut FSHR and LHR, Kobayashi et al., 2008; Atlantic salmon FSHR, Andersson et al., 2009). However, some of these extragonadal expression profiles are not congruent. For example, expression of LHR in zebrafish pituitary (Wong and van Eenennaam, 2004) was not detected in other report (Kwok et al., 2005). The expression of LHR in kidney was not detected in amago salmon (Oba et al., 1999a), pejerrey (Shinoda et al., 2010), and Senegalese sole (Chauvigné et al., 2010). Further studies are necessary to explain these discrepancies with careful consideration of other possible factors such as the sensitivity and selectivity of RT-PCR and condition of the specimen including sex and life stage.

Cellular Localization in Gonad

In testis, the cell type specificity of the fish GTHR gene expression has been studied by immunohistochemistry, *in situ* hybridization or quantitative PCR (qPCR). In general, fish FSHR gene was expressed in Sertoli and/or Leydig cells, fish LHR was in Leydig but not in Sertoli cells (amago salmon and tilapia, Figures 1 and 2, Oba et al., 2001; Japanese eel, Ohta et al., 2007; African catfish, García-López et al., 2009; honeycomb grouper, Alam et al., 2010). These results are in agreement with the previous report by *in vitro* ligand autoradiographies using coho salmon testis (Miwa et al., 1994). In zebrafish testis, by *in situ* hybridization, FSHR transcript was detected in Leydig, but not in Sertoli and germ cells.
On the other hand, the expression of LHR was not detected in any testicular cells. However, significant expression of FSHR and LHR in both Leydig and Sertoli cells of zebrafish testis was shown by qPCR (García-López et al., 2010). The authors suggested that the observation of LHR in Sertoli cells is unusual and this might be related to the undifferentiated gonochoristic mode of gonadal sex differentiation.

In ovary, the studies on the cellular localization of fish GTHRs were very limited. *In situ* hybridization studies showed that the expression of FSHR gene was detected in granulosa cells of vitellogenic follicles in tilapia, and LHR was in granulosa cells of full-grown follicles in amago salmon (Figures 1 and 2, Oba et al., 2001). In Atlantic salmon, the expression of LHR gene was detected in the granulosa and weakly in the theca cells of full-grown follicles, but the expression of FSHR was not detected by *in situ* hybridization (Andersson et al., 2009). In Atlantic halibut, the expression of FSHR was detected in the granulosa cells of late vitellogenic follicles (Kobayashi et al., 2008). These results are largely congruent with those by ligand binding analyses in coho salmon, which suggested that FSHR is present in both theca and granulosa cells whereas LHR is only in granulosa (Yan et al., 1992; Miwa et al., 1994).

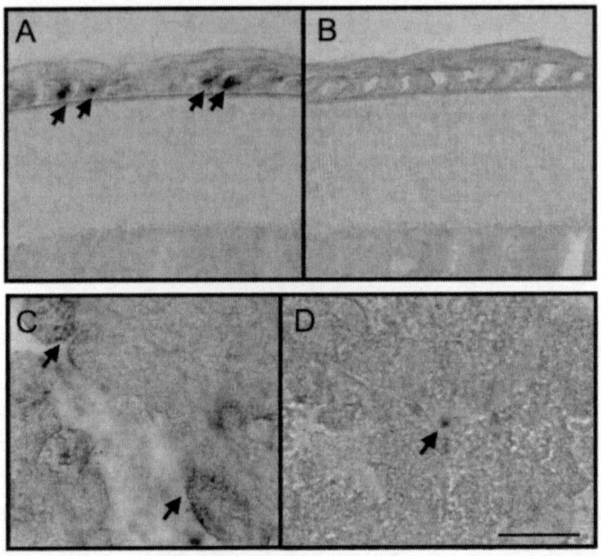

Figure 1. *In situ* hybridization of amago salmon full-grown ovary (A, B) and matured testis (C, D), using DIG-labeled antisense LHR (A, D), sense LHR (B), and antisense FSHR (C). Arrows in panel A represent the expression of LHR in granulosa cells. Arrows in panel C represent the expression of FSHR in Sertoli cells. Arrow in panel D represents the expression of LHR in Leydig cell. Bar shows 50 μm.

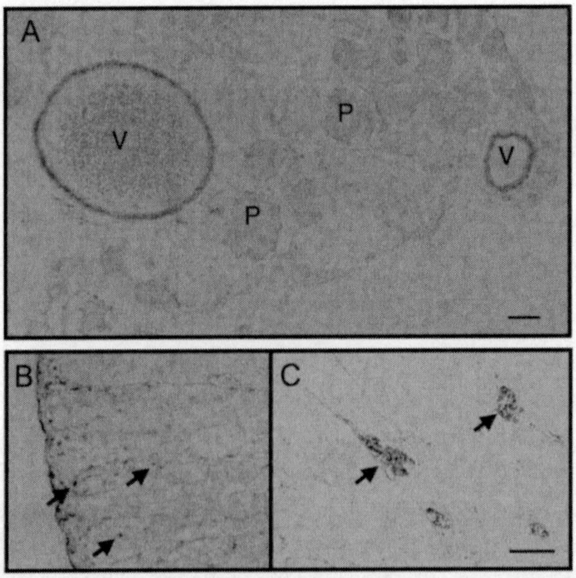

Figure 2. *In situ* hybridization of Nile tilapia ovary (A) and matured testis (B, C), using DIG-labeled antisense FSHR (A, B) and antisense LHR (C). In panel A, P and V represent pre-vitellogenic and vitellogenic oocytes, respectively. FSHR is expression in granulosa cells in vitellogenic stage but not in pre-vitellogenic stage. Arrows in panels B represent the expression of FSHR in Sertoli cells. Arrows in panel C represent the expression of LHR in Leydig cells. Bars show 100 μm.

Stage Specificity in Gonad

The temporal profiles of the gene expression for fish FSHR and LHR in ovary and testis have been analysed for studying gonadogenesis, gametogenesis, sex differentiation, and sex change. Most of these studies have performed using qPCR, but care must be taken for interpreting these data, because alternative splicing forms have found in some fish FSHRs (Laan et al., 2002; Rocha et al., 2007; Kobayashi and Andersen, 2008) though their roles are uncertain.

In female ovary, fish FSHR and LHR transcripts are generally expressed in vitellogenic and full-grown follicles, respectively (Figures 3 and 4) in amago salmin and tilapia as well as other teleosts (Oba et al., 2001; Levavi-Sivan et al., 2010; Kwok et al., 2006; Kitano et al., 2011). Eel ovarian growth and gametogenesis can be induced by gonadotropic treatments using fish pituitary extracts, and Jeng et al. (2007) showed that Japanese eel ovary injected by salmon pituitary extract exhibit the elevation of both FSHR and LHR transcripts. These results support the involvement of FSHR in the recruitment of new oocyte clutches and vitellogenesis, and LHR in the shift of the stages from oocyte growth to maturation. In addition, increase in the expression of salmon FSHR (Figure 3) was observed at oocyte maturation and ovulation stages (Oba et al., 2001; Sambroni et al., 2007) though the physiological significance of this increase in salmonids remains unclear. Bobe et al. (2003) demonstrated that rainbow trout follicles possessing maturational competence for pituitary GTHs express significantly higher FSHR than those of less competent follicles.

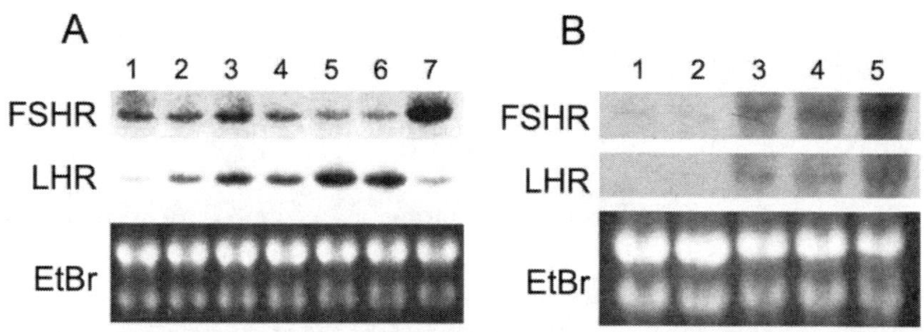

Figure 3. Northern blot hybridization of FSHR and LHR in amago salmon. Panel A: ovarian follicles. lane 1, yolk vesicle stage (GSI= 0.56); lanes 2-5, yolk globule stages (GSI= 0.84, 2.67, 6.67, 18.72, respectively); lane 6, full-grown but not ovulated stage (GSI= 22.95); lane 7, after ovulation (GSI= 1.28). Panel B: testis. lanes 1-4, spermatogenesis (GSI= 0.12, 0.23, 1.65, 5.40, respectively); lane 5, spermiation (GSI= 5.17). EtBr represents the ethidium bromide staining of the gel.

In male, both fish FSHR and LHR transcripts significantly increase during spermatogenesis, suggesting the functions of FSHR and LH in the male gametogenesis and spermiation in fish (Figures 3 and 4), but the temporal profiles vary in certain fish species (Oba et al., 2001; Rahman et al., 2003; Kusakabe et al., 2006; Sambroni et al., 2007; Maugars and Schmitz, 2008; García-López et al., 2009; Rocha et al., 2009; Chauvigné et al., 2010; de Almeida et al., 2011; Mechaly et al., 2012). When injecting recombinant zebrafish LH, the expression of FSHR gene is increased in zebrafish testis (García-López et al., 2010).

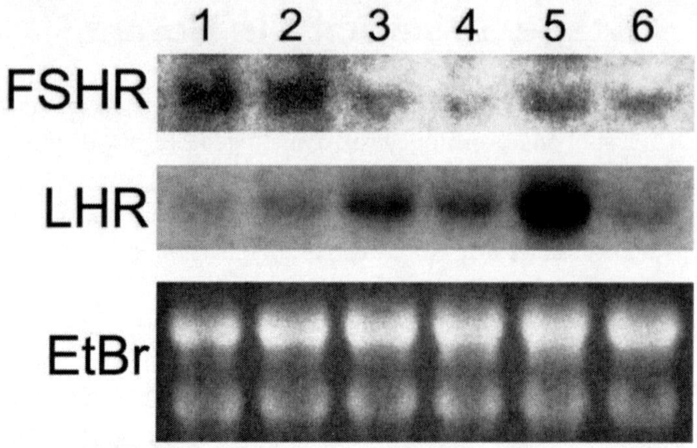

Figure 4. Northern blot hybridization of FSHR and LHR in the first clutch intact follicles of Nile tilapia during the ovarian development. lanes 1-2, mid-vitellogenic stages; lanes 3-4, late-vitellogenic stages; lane 5, full-grown stage; lane 6, after germinal vesicle braked down. EtBr represents the ethidium bromide staining of the gel.

The transcript levels of GTHRs in gilthead seabream were measured during the first 36 days after fertilization by qPCR, showing the continuous expression of both FSHR and LHR with concomitant fluctuation (Wong et al., 2004). This result suggests that FSHR and LHR may be involved in the gonadal development and germ cell proliferation. The expression profiles of fish FSHR and LHR during gonadogenesis was studied by qPCR using juvenile of pejerrey, a species of temperature-dependent sex determination (Shinoda et al., 2010). The results showed that the expression level of LHR is significantly higher at female producing temperature than those in male producing temperature during sex differentiation period. On the other hand, the relationships between the expression of FSHR and sex differentiation was not clear (Shinoda et al., 2010).

The involvement of fish FSHR and LHR during sex change have recently studied by qPCR. An et al. (2009, 2010) showed in protandrous hermaphrodite fishes, black porgy *Acanthopagrus schlegeli* and cinnamon clownfish *Amphiprion melanopus*, that FSHR and LHR are significantly expressed during sex change in gonads, but the transcript levels are lower than those in mature female. Kobayashi et al. (2009) studied the expression manners of FSHR and LHR transcripts during sex change using *Trimma okinawae*, a gobiid fish changing their sex in both directions successively. This fish possesses both ovarian and testicular tissues, and the result by qPCR showed that both FSHR and LHR increased in ovary and decreased in testis when the fish changes from male to female. When the fish changes from female to male, both FSHR and LHR increased in testis and decreased in ovary (Kobayashi et al., 2009). In honeycomb grouper, FSHR transcript in the gonadal tissue significantly elevated at early stages of sex change from male to female (Alam et al., 2010). Hu et al. (2011) demonstrated by qPCR that sex change in the orange-spotted grouper from female to male by injecting 17α-methyltestosterone induces elevation of LHR and reduction of FSHR. The authors discussed that the contradictory results between two groupers imply the different regulatory mechanisms of FSHR signaling in natural and artificial sex change. Involvements of GTHRs in sex differentiation and sex change are ongoing topics of interest.

Conclusion

The last 13 years of research into the fish FSHR and LHR genes has revealed the crucial actions of GTHs in fish reproduction. Basic characteristics of fish FSHR and LHR, such as protein structure, tissue distribution, and cell locality, are similar to those in mammals and avians. On the other hand, ligand affinities of fish FSHR and LHR to the respective ligands (fish FSH and LH) seem to be less selective than those in mammalian FSHR and LHR. However, it should be pointed out that this fact does not mean primitive or unsophisticated characteristics of GTH/GTHR systems in fish. Fish exhibits a wide variety of reproduction strategies, iteroparous/ semelparous, environmental/ genetic sex determination, protogynous/ protandrous sex change, and GTH/GTHR systems are largely involved in these mechanisms (Devlin and Nagahama, 2002). Knowledge from the molecular analyses of GTHR in fish has also provided new insights into the evolution of hormonal reproductive systems on hypothalamic-pituitary-gonadal axis. Recent studies in amphioxus, hagfish, lamprey, and elasmobranch assisted by whole genome sequence projects are exploring its deep origins.

Acknowledgment

The author thanks Prof. Yoshitaka Nagahama for allowing to use experimental figures in this chapter.

References

Alam, M. A., Kobayashi, Y., Hirai, T. and Nakamura, M. (2010). Isolation, characterization and expression analyses of FSH receptor in protogynous grouper. *Comp. Biochem. Physiol. A,* 156, 364-371.

An, K. W., Lee, K. -Y., Yun, S. G. and Choi, C. Y. (2009). Molecular characterization of gonadotropin subunits and gonadotropin receptors in black porgy, *Acanthopagrus schlegeli*: Effects of estradiol-17β on mRNA expression profiles. *Comp. Biochem. Physiol. B,*152, 177-188.

An, K. W., Lee, J. and Choi, C. Y. (2010). Expression of three gonadotropin subunits and gonadotropin receptor mRNA during male-to-female sex change in the cinnamon clownfish, Amphiprion melanopus. *Comp. Biochem. Physiol. A,* 156, 407-415.

Andersson, E., Nijenhuis, W. Male, R., Swanson, P., Bogerd, J., Taranger, G. L. and Schulz R. W. (2009). Pharmacological characterization, localization and quantification of expression of gonadotropin receptors in Atlantic salmon (*Salmo salar* L.) ovaries. *Gen. Comp. Endocrinol.*, 163, 329-339.

Bobe, J., Maugars, G., Nguyen, T., Rime, H. and Jalabert, B. (2003). Rainbow trout follicular maturational competence acquisition is associated with an increased expression of follicle stimulating hormone receptor and insulin-like growth factor 2 messenger RNAs. *Mol. Reprod. Dev.,* 66, 46-53.

Bogerd, J., Andersson, B. E., van der Putten, H. H. A. G. M, Tensen, C. P., Vischer, H. F., Granneman, J. C. M., Janssen-Dommerholt, C., Goos, H. J. Th. and Schulz, R. W.

(2001). Discrepancy between molecular structure and ligand selectivity of a testicular follicle-stimulating hormone receptor of the African catfish (*Clarias gariepinus*). *Biol. Reprod.*, 64, 1633-1643.

Breton, B., Le Gac, F. and Sambroni, E. (1986). Gonadotropin hormone (GtH) receptors in the ovary of the brown trout *Salmo trutta* L. *in vitro* studies. *Gen. Comp. Endocrinol.*, 64, 163-171.

Chauvigné, F. Tingaud-Sequeira, A., Agulleiro, M. J., Calusinska, M., Gómez, A., Finn, R. N. and Cerdà, J. (2010). Functional and evolutionary analysis of flatfish gonadotropin receptors reveals cladal and lineage-level divergence of the teleost glycoprotein receptor family. *Biol. Reprod.*, 82, 1088-1102.

de Almeida, F. F. L., Andersson, E., Mittelholzer, C., Karlsen, Ø., Taranger, G. L. and Schulz, R. W. (2011). Pituitary gonadotropin and testicular gonadotropin receptor expression in Atlantic cod (*Gadus morhua* L.) during the first reproductive season: Effects of photoperiod modulation. *Gen. Comp. Endocrinol.*, 173, 111-119.

Devlin, R. H. and Nagahama, Y. (2002). Sex determination and sex differentiation in fish: an overview of genetic, physiological, and environmental influences. *Aquaculture*, 208, 191-364.

Freamat, M., Kawauchi, H., Nozaki, M. and Sower, S. A. (2006). Identification and cloning of a glycoprotein hormone receptor from sea lamprey, *Petromyzon marinus*. *J. Mol. Endocrinol.*, 37, 135-146.

Freamat, M. and Sower, S. A. (2008a). A sea lamprey glycoprotein hormone receptor similar with gnathostome thyrotropin hormone receptor. *J. Mol. Endocrinol.*, 41, 219-228.

Freamat, M. and Sower, S. A. (2008b). Glycoprotein hormone receptors in the sea lamprey Petromyzon marinus. *Zool. Sci.*, 25, 1037-1044.

Freamat, M., and Sower, S. A. (2010). Functional divergence of glycoprotein hormone receptors. *Integ. Comp. Biol.*, 50, 110-123.

García-López, Á., Bogerd, J., Granneman, J. C. M., van Dijk, W., Trant, J. M., Taranger, G. L. and Schulz, R. W. (2009). Leydig cells express follicle-stimulating hormone receptors in African catfish. *Endocrinology*, 150, 357-365.

García-López, Á., de Jonge, H., Nóbrega, R. H., de Waal, P. P., van Dijk W., Hemrika, W., Taranger, G. L., Bogerd, J. and Schulz, R. W. (2010). Studies in zebrafish reveal unusual cellular expression patterns of gonadotropin receptor messenger ribonucleic acids in the testis and unexpected functional differentiation of the gonadotropins. *Endocrinology*, 151, 2349-2360.

Hu, X., Liu, X., Zhang, H., Zhang, Y., Li, S., Sang, Q., Wang, Q., Luo, W., Liu, Q., Lu, D., Meng, Z. and Lin, H. (2011). Expression profiles of gonadotropins and their receptors during 17α-methyltestosterone implantation-induced sex change in the orange-spotted grouper (*Epinephelus coioides*). *Mol. Reprod. Dev.*, 78, 376-390.

Jeng, S. -R., Yueh, W.-S., Chen, G. -R., Lee, Y.-H., Dufour, S. and Chang, C.-F. (2007). Differential expression and regulation of gonadotropins and their receptors in the Japanese eel, Anguilla japonica. *Gen. Comp. Endocrinol.*, 154, 161-173.

Kanamori, A. and Nagahama, Y. (1988). Developmental changes in the properties of gonadotropin receptors in the ovarian follicles of amago salmon (*Oncorhynchus rhodurus*) during oogenesis. *Gen. Comp. Endocrinol.*, 72, 25-38.

Kazeto, Y., Kohara, M., Tosaka, R., Gen, K., Yokoyama, M., Miura, C., Miura, T., Adachi, S. and Yamauchi, K. (2012) Molecular characterization and gene expression of Japanese eel (*Anguilla japonica*) gonadotropin receptors. *Zool. Sci.*, 29, 204-211.

Kitano, H., Irie, S., Ohta, K., Hirai, T., Yamaguchi, A. and Matsuyama, M. (2011). Molecular cloning of two gonadotropin receptors and their distinct mRNA expression profiles in daily oogenesis of the wrasse Pseudolabrus sieboldi. *Gen. Comp. Endocrinol.*, 172, 268-276.

Ko, H., Park, W., Kim, D. -J., Kobayashi, M. and Sohn Y. C. (2007). Biological activities of recombinant Manchurian trout FSH and LH: their receptor specificity, steroidogenic and vitellogenic potencies. *J. Mol. Endocrinol.*, *38*, 99-111.

Kobayashi, T. and Andersen, Ø. (2008). The gonadotropin receptors FSH-R and LH-R of Atlantic halibut (*Hippoglossus hippoglossus*), 1: Isolation of multiple transcripts encoding full-length and truncated variants of FSH-R. *Gen. Comp. Endocrinol.*, 156, 584-594.

Kobayashi, T., Pakarinen, P., Torgersen, J., Huhtaniemi, I., and Andersen, Ø. (2008). The gonadotropin receptors FSH-R and LH-R of Atlantic halibut (*Hippoglossus hippoglossus*)—2. Differential follicle expression and asynchronous oogenesis. *Gen. Comp. Endocrinol.,* 156, 595-602.

Kobayashi, Y., Nakamura, M., Sunobe, T., Usami, T., Kobayashi, T., Manabe, H., Paul-Prasanth, B., Suzuki, N. and Nagahama, Y. (2009). Sex change in the gobiid fish is mediate through rapid switching of gonadotropin receptors from ovarian to testicular portion or vice versa. *Endocrinology,* 150, 1503-1511.

Kubokawa, K., Tando, Y. and Roy, S. (2010). Evolution of the reproductive endocrine system in chordates. *Integ. Comp. Biol.*, 50, 53-62.

Kumar, R. S., Ijiri, S. and Trant, J. M. (2001a). Molecular biology of channel catfish gonadotropin receptors: 1. Cloning and functional luteinizing hormone receptor and preovulatory induction of gene expression. *Biol. Reprod.*, 64, 1010-1018.

Kumar, R. S., Ijiri, S. and Trant, J. M. (2001b). Molecular biology of the channel catfish gonadotropin receptors: 2. Complementary DNA cloning, functional expression, and seasonal gene expression of the follicle-stimulating hormone receptor. *Biol. Reprod.*, 65, 710-717.

Kumar, R. S. and Trant, J. M. (2001). Piscine glycoprotein hormone (gonadotropin and thyrotropin) receptors: a review of recent development. *Comp. Biochem. Physiol.*, *B129*, 347-355.

Kusakabe, M., Nakamura, I., Evans, J., Swanson, P. and Young, G. (2006). Changes in mRNAs encoding steroidogenic acute regulatory protein, steroidogenic enzymes and receptors for gonadotropins during spermatogenesis in rainbow trout testes. *J. Endocrinol.*, 189, 541-554.

Kwok, H. -F., So, W. -K., Wang, Y. and Ge, W. (2005). Zebrafish gonadotropin and their receptors: I. Cloning and characterization of zebrafish follicle-stimulating hormone and luteinizing hormone receptors— evidence for their distinct functions in follicle development. *Biol. Reprod.*, 72, 1370-1381.

Laan, M., Richmond, H., He, C. and Campbell, R. K. (2002). Zebrafish as a model for vertebrate reproduction: characterization of the first functional zebrafish (*Danio rerio*) gonadotropin receptor. *Gen. Comp. Endocrinol.*, 125, 349-364.

Le Gac, F., Breton, B. and Bougoussa, M. (1988) Gonadotropic hormone (GtH) receptors in the testis of the trout *Salmo gairdneri*: *in vitro* studies. *Fish Physiol. Biochem., 5*, 209-217.

Levavi-Sivan, B., Bogerd, J., Mañanós, E. L., Gómez, A. and Lareyre, J. J. (2010) Perspectives on fish gonadotropins and their receptors. *Gen. Comp. Endocrinol.*, 165, 412-437.

Maugars, G. and Schmitz, M. (2006) Molecular cloning and characterization of FSH and LH receptors in Atlantic salmon (*Salmo salar* L.). *Gen. Comp. Endocrinol.*, 149, 108-117.

Mechaly, A. S., Viñas, J. and Piferrer, F. (2012) Sex-specific changes in the expression of kisspeptin, kisspeptin receptor, gonadotropins and gonadotropin receptors in the Senegalese sole (*Solea senegalensis*) during a full reproductive cycle. *Comp. Biochem. Physiol. A*, 162, 364-371.

Mittelholzer, C., Andersson, E., Taranger, G. L., Consten, D., Hirai, T., Senthilkumaran, B., Nagahama, Y. and Norberg, B. (2009) Molecular characterization and quantification of the gonadotropin receptors FSH-R and LH-R from Atlantic cod (*Gadus morhua*). *Gen. Comp. Endocrinol.*, 160, 47-58.

Miwa, S., Yan, L. and Swanson, P. (1994). Localization of two gonadotropin receptors in the salmon gonad by in vitro ligand autoradiography. *Biol. Reprod.*, 50, 629-642.

Nagahama, Y. (1994). Endocrine regulation of gametogenesis in fish. *Int. J. Dev. Biol.,* 38, 217-229.

Oba, Y., Hirai, T., Yoshiura, Y., Yoshikuni, M., Kawauchi, H. and Nagahama, Y. (1999a). Cloning, functional characterization, and expression of a gonadotropin receptor cDNA in the ovary and testis of amago salmon (*Oncorhynchus rhodurus*). *Biochem. Biophys. Res. Commun.*, 263, 584-590.

Oba, Y., Hirai, T., Yoshiura, Y., Yoshikuni, M., Kawauchi, H. and Nagahama, Y. (1999b). The duality of fish gonadotropin receptors: cloning and functional characterization of a second gonadotropin receptor cDNA expressed in the ovary and testis of amago salmon (*Oncorhynchus rhodurus*). *Biochem. Biophys. Res. Commun., 265*, 366-371.

Oba, Y., Hirai, T., Yoshiura, Y., Kobayashi, T. and Nagahama, Y. (2001). Fish gonadotropin and thyrotropin receptors: the evolution of glycoprotein hormone receptors in vertebrates. *Comp. Biochem. Physiol., B129*, 441-448.

Ohta, T., Miyake, H., Miura, C., Kamei, H., Aida, K. and Miura, T. (2007). Follicle-stimulating hormone induces spermatogenesis mediated by androgen production in Japanese eel, *Anguilla japonica Biol. Reprod.*, 77, 970-977.

Quérat, B., Tonnerre-Doncarli, C., Géniès, F. and Salmon, C. (2001). Duality of gonadotropins in gnathostomes. *Gen. Comp. Endocrinol., 124*, 308-314.

Rahman, M. A., Ohta, K., Yamaguchi, A., Chuda, H., Hirai, T. and Matsuyama, M. (2003). Gonadotropins, gonadotropin receptors and their expressions during sexual maturation in yellowtail, a carangid fish. *Fish Phys. Biochem.*, *28*, 81-83.

Rocha, A., Gómez, A., Zanuy, S., Cerdá-Reverter, J. M. and Carrillo, M. (2007). Molecular characterization of two sea bass gonadotropin receptors: cDNA cloning, expression analysis, and functional activity. *Mol. Cell. Endocrinol., 272*, 63-76.

Rocha, A., Zanuy, S., Garrillo, M. and Gómez, A. (2009). Seasonal changes in gonadal expression of gonadotropin receptors, steroidogenic acute regulatory protein and steroidogenic enzymes in the European sea bass. *Gen. Comp. Endocrinol.*, *162*, 265-275.

Sambroni, E., Le Gac, F., Breton, B. and Lareyre, J. -J. (2007). Functional specificity of the rainbow trout (*Oncorhynchus mykiss*) gonadotropin receptors as assayed in a mammalian cell line. *J. Endocrinol.*, 195, 213-228.

Schulz, R. W., Vischer, H. F., Cavaco, J. E. B., Santos, E. M., Tyler, C. R., Goos, H. J. Th. and Bogerd, J. (2001). Gonadotropin, their receptors, and the regulation of testicular functions in fish. *Comp. Biochem. Physiol.*, *B129*, 407-417.

Shinoda, T., Miranda, L. A., Okuma, K., Hattori, R. S., Fernandino, J. I., Yoshizaki, G., Somoza, G. M. and Strüssmann, C. A. (2010). Molecular cloning and expression analysis of *Fshr* and *Lhr* in relation to *Fshb* and *Lhb* subunits during the period of temperature-dependent sex determination in pejerrey *Odontesthes bonariensis*. *Mol. Reprod. Dev.*, 77, 521-532.

So, W. -K., Kwok, H. -F. and Ge, W. (2005). Zebrafish gonadotropins and their receptors: II. Cloning and characterization of zebrafish follicle-stimulating hormone and luteinizing hormone subunits— their spatial-temporal expression patterns and receptor specificity. *Biol. Reprod.*, 72, 1382-1396.

Sower, S. A., Moriyama, S., Kasahara, M., Takahashi, A., Nozaki, M., Uchida, K., Dahlstrom, J. M. and Kawauchi, H. (2006). Identification of sea lamprey GTHβ-like cDNA and its evolutionary implications. *Gen. Comp. Endocrinol.*, 148, 22-32.

Tando, Y. and Kubokawa, K. (2009a). A homolog of the vertebrate thyrostimulin glycoprotein hormone α subunit (GPA2) is expressed in amphioxus neuron. *Zool. Sci.*, 26, 409-414.

Tando, Y. and Kubokawa, K. (2009b). Expression of the gene for ancestral glycoprotein hormone β subunit in the nerve cord of amphioxus. *Gen. Comp. Endocrinol.*, 162, 329-339.

Uchida, K., Moriyama, S., Chiba, H., Shimotani, T., Honda, K., Miki, M., Takahashi, A., Sower, S. A. and Nozaki, M. (2010). Evolutionary origin of a functional gonadotropin in the pituitary of the most primitive vertebrate, hagfish. *Proc. Natl. Acad. Sci. USA,* 107, 15832-15837.

Vischer, H. F. and Bogerd, J. (2003). Cloning and functional characterization of a gonadal luteinizing hormone receptor complementary DNA from the African catfish (*Clarias gariepinus*). *Biol. Reprod.*, 68, 262-271.

Vischer, H. F., Granneman, J. C. M., Linskens, M. H. K., Schulz, R. W. and Bogerd, J. (2003). Both recombinant African catfish LH and FSH are able to activate the African catfish FSH receptor. *J. Mol. Endocrinol.*, 31, 133-140.

Wong, A. C. and van Eenennaam, A. L. (2004). Gonadotropin hormone and receptor sequences from model teleost species. *Zebrafish,* 1, 203-221.

Wong, T. -T., Gothilf, Y., Zmora, N., Kight, K. E., Meiri, I., Elizur, A. and Zohar, Y. (2004). Developmental expression of three forms of gonadotropin-releasing hormone and ontogeny of the hypothalamic-pituitary-gonadal axis in gilthead seabream (*Sparus aurata*). *Biol. Reprod.*, 71, 1026-1035.

Yan, L., Swanson, P. and Dickhoff, W. W. (1992). A two-receptor model for salmon gonadotropins (GTH I and GTH II). *Biol. Reprod.,* 47, 418-427.

In: Sexual Plasticity and Gametogenesis in Fishes
Editor: Balasubramanian Senthilkumaran
ISBN: 978-1-62618-848-8

Chapter XX

Regulation of Teleost Reproduction by Brain-Pituitary Axis

P. D. Prasada Rao
National Environmental Engineering Research Institute,
Nehru Marg, Nagpur, India

Abstract

The brain-pituitary (BP) complex regulates the teleost reproduction through various hormones/factors that act at different levels of the BP-gonadal axis, in conjunction with intrinsic physiological processes, environmental cues and social interactions. Hypophysial gonadotropins (GtHs), FSH and LH are directly regulated by hypophysiotropic neurons, especially by gonadotropin-releasing hormone (GnRH) neurons which in turn are modulated by brain factors and sensory inputs. GnRH neurons exist in 3 different locations. The preoptic area GnRH-I neurons are involved in direct GtH stimulation, but dopamine inhibits GtH release; blockage of dopamine actions by domperidone has been crucial in formulation of the fish-spawning agent, Ovaprim. Midbrain tegmental GnRH-II and rostral forebrain GnRH-III perikarya partake in neuromodulatory and auxiliary reproductive functions. Recent knowledge on Kiss1 and Kiss2 neurons, including their receptors, association with GnRH neurons and functional significance has been summarized; Kiss2 neurons may stimulate or inhibit GtH depending upon the reproductive state. Kisspeptins serve as gate-keepers allowing steroid feedback, photoperiod, temperature and energy balance signals. The information about the role of many other GtH regulators such as GnIH, serotonin, melatonin, vasotocin, monoamines, GABA, nitric oxide, noradrenalin, Neuropeptide Y and Ghrelin has also been included. Participation of other factors such as cholecystokinin, pituitary adenylate cyclase-activating polypeptide, inhibin, activin, insulin-like growth factor-1, glucocorticoids, leptin, gonadal steroids in teleost reproduction has also been indicated. Having traced the significance of environmental factors in tropical fish breeding, the usefulness of these factors in evolving novel, non-invasive fish breeding techniques has been proposed.

1. Introduction

Teleost reproduction encompasses processes such as gonad differentiation, puberty, male and female gametogenesis, cyclic histocytological, hormonal and neurotrasmitter changes in the brain-pituitary-gonadal (BPG) axis, social and behavioral interaction, status change, and the timing of actual spawning These physiological processes are regulated by several neuroendocrine, endocrine, paracrine and autocrine factors synthesized and secreted by different components of the BPG axis (Reviews: Trudeau, 1997; Yaron et al., 2003; Abe and Oka, 2010; Zohar et al., 2010; Kanda and Oka, 2012; Tena-Sempere et al., 2012). Furthermore, the reproductive functions also interact with other important intrinsic physiological processes such as feeding, growth, nutrition, osmoregulation, and response to stress factors, pheromones and extrinsic factors including photoperiod, biological and physicochemical properties of water (see Suresh et al., 2008; Zohar et al., 2010) etc. at different stages depending on the species. Added to this complexity of reproduction is the occurrence of about 30,000 extant teleosts that occupy a wide variety of environments with a staggering diversity in reproductive strategies for successful breeding, survival of the eggs, young ones and frequency of spawning. The rainbow trout breeds only once in lifetime, a few others once in a few years, most tropical carps and catfishes breed once in a year, some breed once in a month or even every few weeks; interestingly, the black bream, and greenback flounder are multiple-repeat spawners probably on a daily basis (see Suresh et al., 2008). However, most tropical cyprinids do not breed in ordinary perennial tanks, but breed only in rivers, inundated fields and tanks that become flooded during monsoon period (Bhattacharya et al., 2007; Moniruzzaman and Maitra, 2012).

Primarily, the reproductive activity is controlled by the pituitary gonadotropins (GtHs) which in turn are regulated by specific hormones manifested through the brain-pituitary (BP) axis. Although early studies ascribed the primarily role in GtH regulation to the hypothalamic nuclei, subsequent research unfolded the role played by the forebrain and other areas of the brain. There is also a growing body of evidence demonstrating the influence of various environmental cues and social interactions in modulation of brain hormones and regulation of reproduction.

2. Brain-Pituitary Complex

The pituitary gland (PG), the source of GtHs is composed of a glandular part, the adenohypophysis and a neural part, the neurohypophysis. The adenohypophysis has two distinct lobes: the pars distalis (PD) and pars intermedia (PI). The PD consists of an anterior rostral pars distalis (RPD), a middle proximal pars distalis (PPD) and caudal pars intermedia (PI). Immunocytochemical studies revealed that the RPD contains prolactin-, adrenocorticotropin (ACTH)-, occasionally thyrotropin (TSH)-secreting cells, while the PPD has GtH, TSH and growth hormone (GH)-secreting cells. Earlier attempts to delineate two types of GtH cells based upon tinctorial properties required indisputable confirmation. However, experimental and imunocychemical approaches have helped in unequivocal functional identification of FSH and LH cells. The close proximity of the pituitary with hypothalamus, and their intricate association both in developmental and adult stages lend a clue to the brain regulation of the PG and this aspect has been reviewed (Trudeau, 1997; Yaron et al., 2003; Zohar et al., 2010).

2.1. Functional Link between the Brain and Adenohypophysis

In tetrapods, the median eminence (ME) acts as a functional link between the brain and hypophysis; different brain hormones are released into the capillary plexus of the ME, reach their target hormone-secreting cells of the adenohypophysis *via* the portal system and regulate the hypophysial cell activity. However, teleosts do not have an ME with established homology to that of teleosts. Different peptidergic nerve terminals originating in the brain may establish direct contacts - sometimes like synaptic contacts - with various hormone-secreting cells or may terminate on basement membrane and regulate the adenohypophysial cell functions (Zohar et al., 2010). In view of the unique features exhibited by the pituitary cells and their relationship with different neurohormone-releasing nerve terminals, localization of hypophysiotropic neurons has become important.

2.2. Hypophysiotropic Neurons

Brain neurons that project to the PG have been identified in the hypothalamic and extrahypothalamic areas in several teleosts, and different aspects of the hypopysiotropic neurons have been been reviewed by different investigators (Prasada Rao, 1999; Zohar et al., 2010). Retrogradely labeled perikarya were demonstrated in the olfactory bulb (OB), medial olfactory tract (MOT), the nucleus entopeduncularis (NE) and ventral telencephalon (VT). In the hypothalamus, they were identified in the preoptic area (POA), suprachiasmatic nucleus (SCN), nucleus preopticus periventricularis (NPP), nucleus recessus lateralis (NRL) and nucleus recessus posterioris (NRP) using different neuronal tract tracing techniques in goldfish (Anglade et al., 1993), electric fish (Johnston and Maler, 1992) and catfish (Prasada Rao et al., 1993).

2.3. Gonadotropins and Co-Gonadotropins

The first isolation of two types of GtHs, that showed predominant vitellogenic or maturational activities, was achieved in the plaice (Ng and Idler, 1978). Later, two distinct salmonid GtHs *viz.,* GtH-I and GtH-II were isolated and characterized by their respective steroidogenic potency *in vitro* (Swanson et al., 1989). Subsequent studies have identified two distinct GtHs in many teleost species, using classical purification methods and through sequencing and homology cDNA cloning (Yaron et al., 2003). GTH-I stimulates gonadal steroidogenesis, whereas GTH-II stimulates steroidogenesis, gametogenesis and ovulation (or sperm release) and is thus an important regulator of fertility (see Trudeau, 1997). In goldfish, growth hormone (GH) secreted by some PPD cells, plays a role in reproduction, and has therefore been termed co-gonadotropin (Hull and Harvey, 2000). Serum GH concentrations were also high when gonadal size underwent an increase in the winter indicating that GH may participate in seasonal reproduction in goldfish (Trudeau, 1997).

3. Brain Factors Regulating Gonadotropins

Several brain factors affect GtH and reproduction in fish, but GnRH plays a pivotal role in this important phenomenon through its action on pituitary GnRH receptors (GnRH-Rs) to stimulate the synthesis and release of GtHs, which in turn regulate the gonadal activity (review: Trudeau, 1997; Zohar et al., 2010). However, certain factors such as dopamine inhibit the release of GtHs in some species (Dufour Peter et al., 1988; 2010), while others stimulate and/or inhibit depending upon the reproductive state of the fish. Other brain factors including many neuropeptides, neurotransmitters, amines and even a gaseous molecule, nitric oxide seem to regulate fish reproduction.

3.1. Gonadotropin-Releasing Hormone (GnRH)

Fish GnRH was first isolated from chum salmon (Sherwood et al., 1983). Following the first discovery of sGnRH structure, three different forms of GnRH *viz.*, sGnRH, cGnRH-II and sbGnRH were reported in the brains of gilthead seabream, the pacu (Powell et al., 1994) and other teleosts. cDNAs for the three forms were found in the brains of some fish species (White et al., 1995); HPLC and/or RIA studies also revealed the presence of the three forms in several fishes (Senthilkumaran et al., 1999; Chen and Fernald, 2008; Zohar et al., 2010; Kim et al., 2012). The sea bass exhibits five GnRHs with high affinity for ligand GnRH-2 (Lethimonier et al., 2004; Servili et al., 2012).

3.1.1. Categorization of GnRH Isoforms and Functional Significance

Earlier investigators have named each molecular form of GnRH variants after the specific name of the animal species in which it was first discovered, e.g., mammalian GnRH (mGnRH), chicken GnRH-I (cGnRH-I), chicken GnRH-II (cGnRH-II), seabream GnRH (sbGnRH), catfish GnRH (cfGnRH) and so on (see Okuzawa and Kobayashi, 1999). These names were assigned despite the occurrence of a specified isoform in other species. In order to prevent misunderstanding, a novel classification of the GnRH variants on the basis of phylogenetic analysis of known sequences and their respective sites of expression was proposed (Chen and Fernald, 2008). Basically, three main GnRH branches were delineated: the first, termed GnRH-1, contains hypophysiotropic variants, expressed mainly in the hypothalamus in a number of fish; the second, GnRH-2 consists of GnRH forms expressed in the mesencephalic tegmentum (MT), and the third, termed GnRH-3 comprises GnRH isoforms, mainly expressed in the OB, OT, olfactory nerve (OLN) and VT.

The neuronal assemblage in each of the 3 groups expresses different molecular forms of GnRH peptide and each group is produced by one of the three paralogous GnRH genes *viz.*, gnrh1, gnrh2 and gnrh3; these are considered to have originated from gene duplications (see Oka, 2010). In most teleosts, the neurons located in the POA-GnRH1 project to the PG (Anglade et al., 1993; Prasada Rao et al., 1993; Yamamoto et al., 1998) and facilitate the release of GtHs from the pituitary. GnRH-II is expressed by midbrain tegmentum (MT) neurons (Baby et al., 1993) and GnRH-III by neurons located at the terminal nerve ganglion (TNG) and adjacent areas (Okuzawa and Kobayashi, 1999; Oka, 2010). The GnRH-II and GnRH-III systems operate as modulators of excitability of other neurons over wide brain areas (Abe and Oka, 2007; Oka, 2010), and GnRH-II modulates the pineal functions (Servili et al., 2010).

3.1.2. GnRH Receptors in the Brain

Subsequent to the first description of molecular characterization of a full-length fish GnRH receptor (GnRHR) in African catfish (Tensen et al., 1997), cDNA encoding GnRHR has been obtained in several teleosts; the number of receptors in the brain ranged from two to four (Lethimonier et al. 2004; Guilgur et al., 2009) in many fishes. As many as five GnRHR genes and four GnRHRs were identified in the brain of Atlantic cod (Hildahl et al., 2011). In the brain of masu salmon expression of five different types of putative GnRH receptor genes was identified (Jodo et al., 2003). In goldfish, there exist two GnRHRs widely distributed in several brain areas including thalamus, epithalamus, optic tectum, motor zone of the vagal lobes and motor neurons of the fifth, seventh and tenth cranial nerves, besides forebrain areas (Peter et al., 2003). It was hypothesized that GnRH receptors have evolved separately and are independent from their ligands; they have possible GnRH effects on sensory, motor, cognitive, energy control and other physiological roles across vertebrates (Chen and Fernald, 2008).

3.1.3. GnRH Receptors in the Pituitary

In European sea bass, GnRH receptors were reported only in LH-secreting cells (González-Martínez et al., 2004). In goldfish, GnRHR-1A and -1B showed an overlapping expression and present mainly in GtH cells and to a lesser extent in GH cells (Illing et al., 1999). Using heterologous antisera in Nile tilapia, immunoreactive GnRHR-1A was discerned in the FSH and LH cells, GnRH-1B in FSH, LH cells and in a cell cluster in the PI of the pituitary, and GnRHR-2A and -2C receptors in GH cells (Parhar et al., 2002). In pejerrey pituitary two forms of GnRH receptors (pjGnRH-RI and pjGnRH-RII) were found (Guilgur et al., 2009).

3.1.4. GnRH effects on GtH

Correlative changes occur in the GnRH neuronal system, FSH and LH levels, gonadal state or levels of gonadal steroids in different fishes (Parhar et al., 1998; Senthilkumaran et al., 1999). Similar correlation was also found in mRNA expression of GnRH variants and receptors in the brain, pituitary and in pejerrey (Guilgur et al., 2009). Stimulatory effect of different forms of GnRH on release of LH and FSH was observed after *in vivo* and *in vitro* studies using dispersed pituitary cells or pituitary fragments (Murthy et al., 1994; Levavi-Sivan et al., 2010). Some GnRH forms are more effective than others and co-administration with testosterone (T) potentiates GnRH-induced GtH release (see Trudeau, 1997). Such varying effects on FSH and LH release might be caused by multiple factors operating in the teleostean BP compex. GtH release is a valuable tool to induce the fish to breed. Primarily, it is the failure of GtH release that prevents fish from breeding; this key factor has prompted investigators to accomplish fish breeding, as desired, through augmenting GtH release by adopting different approaches.

3.2. Dopamine

In goldfish, GtH inhibition is caused by dopamine, and blocking dopaminergic inhibition plays a crucial role to confront the problem of induced breeding. Dopamine was shown to act through D2 receptors to inhibit LH and FSH release and on GnRH pituitary terminals, and through D1 receptors on the GnRH neurons (Yu and Peter, 1990; Yu et al., 1991; Dufour et al., 2010). Thus, dopamine in goldfish has a dual effect, inhibiting GtH release by direct action on the GtH cells and by reducing GnRH secretion (Trudeau, 1997). However, dopamine inhibition of GtH release was not observed in some teleosts, particularly in marine species (see for details: Zohar et al., 2010). Domperidone and pimozide are most effective dopamine blockers and blocking dopaminergic inhibition on hypophysial GtH is crucial for fish spawning (Chang and Peter, 1983; Dufour et al., 2010). A combined treatment with a GnRH agonist and DA-D2 receptor antagonist (pimozide/domperidone) proved to be highly effective in inducing spawning in several species of mature female fishes (Lin et al., 1993). This finding by Professor H.R. Lin and Professor R.E. Peter was termed the "Linpe" method after the names of these two investigators. This formulation, "Ovaprim" has become established globally as an excellent fish-spawning agent, also for tropical carps (Yu and Peter, 1990).

3.3. Kisspeptins

Kisspeptins have been intensely studied in fishes during the past decade and more than a dozen elegant reviews appeared on vertebrate Kisspeptins during the past two years. Kisspeptins are the peptide products of the Kiss1 gene, which were initially identified in mammals as ligands of the G protein-coupled receptor 54 (GPR54; also termed Kiss1R) with ability to suppress tumor metastasis. Kisspeptins play a pivotal role in sexual differentiation of the brain, maturation and function of the reproductive axis, including timing of puberty, adult regulation of GtHs by gonadal hormones, and the control of fertility by metabolic and environmental cues such as photoperiod (Servili et al., 2011; Tena-Sempere et al., 2012; Zmora et al., 2012).

3.3.1. Localization of Kisspeptin Neurons

Until recently, visualization of kisspeptins was hampered by non-availability of appropriate antibodies or the cross-reactivity produced by the available ones. Consequently, most information derived on the kisspeptins in fish brain has been based on studies conducted only at the mRNA level (see Servili et al., 2011). Having developed specific antibodies, these authors successfully distinguished two independent systems of Kisspeptin neurons in zebrafish. The Kiss1 neurons occur in the ventromedial habenula and Kiss2 somata occur in the mediobasal hypothalamus, POA, ventral hypothalamus and in the posterior tuberal nucleus. Another Kiss2 neuronal assemblage was found at the level of the periventricular nucleus of the posterior tuberculum. In zebrafish, the prokiss1 neuronal processes of the ventromedial habenula extend caudally into the interpeduncular nucleus (IPN) and into the raphe nuclei, permitting the speculation that Kiss1 neurons regulate serotonin (Ogawa et al., 2012). In red seabream neurons that express Kiss2 mRNA were distributed in the hypothalamic NRL, POA and NVT. Some prokiss2 fibres also revealed contacts with preoptic

GnRH neurons (Servili et al., 2011). However, in puffer fish, Senegalese sole and three-spined stickleback, only Kiss2 neurons were present. In striped bass, Kiss1 and Kiss2 neurons were co-expressed in the NRL and both GRP54-1 and GRP54-2 neurons are associated with GnRH1 neurons (Zmora et al., 2012). The PG of zebrafish does not receive any proKiss1 or proKiss2 fibers, but showed GnRH3 fibers and positive cells in the PI (Servili et al., 2011).

3.3.2. Kisspeptin Receptors

In tilapia, potential anatomical association of KiSS1R with the GnRH system was found. Using laser capture technology and RT-PCR in single cells, co-expression of GnRH and KiSS1R was demonstrated (Parhar et al., 2004). KiSS1R expression undergoes correlative changes with the gonadal cycle in several fish species. KiSS1R, GnRH2 and GnRH3 activity was highest during early stages of gonadal development in female gray mullet, (Nocillado et al., 2007), cobia (Mohamed et al., 2007) and fathead minnow (Filby et al., 2008), indicating that KiSS1R and GnRH play a role in the puberty and reproduction in many teleosts, although there exists some gender-related pattern.

3.3.3. Kisspeptin in Relation to Age and Gender

In female zebrafish, kiss1 mRNA levels gradually increased during the first few weeks of life and attained maximum levels in fish with mature oocytes in ovaries, while in males kiss1 mRNA levels peaked as the testes exhibited initial stages of spermatogenesis and decreased after puberty (Biran et al., 2008). A significant increase in Kiss1, Kiss2, GnRH2, and GnRH3 mRNA levels was evident at the onset of pubertal phase, which remained high in adult stage (Kitahashi et al., 2009). In medaka, the Kiss1 neurons in the NVT show sexual dimorphism, males having significantly more neurons than in females (Kanda et al., 2008).

3.3.4. Functional Significance of Kisspeptin

Both kisspeptins, Kiss1 and Kiss2, and/or their receptors have been shown to participate in a variety of reproductive functions of sea bass (Felip et al., 2008), zebrafish (Biran et al., 2008), goldfish (Li et al., 2009; Kanda et al., 2012), red sea bream (Shimizu et al., 2012), fathead minnow (Tena-Sempre et al., 2012), striped- and hybrid-bass (Zmora et al., 2012) and white- and striped-bass (Beck et al., 2012).

3.3.4.1. Kisspeptin in Relation to Reproductive State

Generally the teleostean Kiss1 and Kiss2 neuronal populations function differently depending upon the gonadal state. In medaka, NVT Kiss1 expression was strongly correlated with breeding condition. In goldfish of both sexes, the number of kiss2 expressing neurons in the hypothalamic dorsal NRL was higher during the spawning season than in the post-spawning period. The number of GnRH1 neurons in the POA also increased in the dorsal NRL in both sexes (Kanda et al., 2012). In male red seabream, the number of kiss2 neurons in the dorsal and ventral NRL was higher in maturing fish than in the post-spawned ones; however, significant changes were not detected in females (Shimizu et al., 2012).

3.3.4.2. Kisspeptin Effects on BP Axis and Gonads

In many fishes, primary regulator of reproductive BPG axis is Kisspeptin2, whereas Kisspeptin1 acts as a weak modulator. Kisspeptin also influences extrahypothalamic GnRH neuronal assemblages (Kitahashi et al., 2009; Li et al., 2009; Zhao and Wayne, 2012). In teleosts, Kisspeptin acts directly on GnRH neurons through activation of Kiss1R and Kiss2R or indirectly by regulating synaptic inputs via kiss1R on interneurons (Ten-Sempere, 2010; Zhao and Wayne, 2012). However, the actions of Kiss1 and Kiss2 vary with the reproductive status of the fish (Zmora et al., 2012). Kiss2 administration to sexually mature female zebrafish caused a significant increase in FSHβ, and LHβ mRNA levels in the pituitary, but not after Kiss1 administration (Kitahashi et al., 2009). In European sea bass, Kiss1 peptide was less potent than Kiss2 in elevating plasma LH levels, and in up-regulating GnRH1 and GPR54-2 expression in prepubertal hybrid bass *in vivo* (Felip et al., 2009). In contrast, during recrudescence, Kiss1 was more potent than Kiss2 in inducing LH release, and Kiss2 down-regulated GnRH1 and GPR54-2 expression, indicating two different modes of GnRH1 regulation. In early-mid pubertal fathead minnow, Kisspeptin injection resulted in an increase in GPR54-2b and GnRH3 expression in the brain, but not of GnRH2 (Tena-Sempre et al., 2012). These data indicate that kiss2 neurons are involved in pubertal processes via regulatory influences on GnRH1 in goldfish (Li et al., 2009), zebrafish (Kitahashi et al., 2009) and red seabream (Shimizu et al., 2012). In sexually mature female zebrafish, Kiss2 peptide administration stimulated LHβ-subunit and FSHβ-subunit mRNA levels in the pituitary (Kitahashi et al., 2009). *In vivo* study in prepubertal hybrid bass revealed that Kiss1 peptide was less potent than Kiss2 in elevating plasma LH levels, and in up-regulating GnRH1 and GRP54-2 expression. In contrast, during recrudescence, Kiss1 was more potent than Kiss2 in inducing LH release, and Kiss2 down-regulated GnRH1 and GPR54-2 *e*xpression. Intraperitoneal administration of gfKiss1–10 to sexually mature female goldfish increased serum LH levels, but there was no significant influence on LH release from goldfish pituitary cells in primary culture, indicating that the peptide does not exert its actions at the pituitary level. On the other hand *in vitro* and *in vivo* studies revealed that gfKiss2–10 was much less potent (Li et al., 2009). Intramuscular injections of Kisspeptin in sexually mature white- and striped-bass increased the gonad weight and gonadosomatic index, and treatment of mature white bass with Kisspeptins resulted in a potential improvement of fecundity (Beck et al., 2012).

3.3.5. Steroid Sensitivity of Kisspeptins

Kisspeptin plays a crucial role in stimulating GnRH through acting as a gatekeeper to the onset of puberty through providing steroid hormone feedback signals to GnRH neurons. In medaka, ovariectomy reduces significantly the Kiss1 expression in neurons of the nucleus ventralis tuberis, which was restored after estrogen replacement (Kanda et al., 2012). In this fish, Kiss1 neurons express estrogen receptor-alpha, respond to estrogen and mediate feedback actions of this steroid (Mitani et al., 2010). Thus, the feedback actions of estrogen and regulation of reproduction in medaka are mediated by Kiss1 neurons that in turn stimulate the GnRH neurons (Mitani et al., 2010; Servili et al., 2011), suggesting that these neurons are involved in a positive feedback control of the BPG axis. Treatment of juvenile zebrafish with E2 also increases the activity and number of kiss2 neurons in the hypothalamus (Servili et al., 2011). Unlike the results in zebrafish, Kanda et al. (2012) found in goldfish that the POA kiss2 neurons show clear steroid sensitivity. In the adult goldfish during the breeding condition, the kiss1 neurons are localized in the habenula, whereas the kiss2 neurons are located in the NLT, NRL, and POA. Among these neurons, the POA Kiss2 neurons decreased in number after ovariectomy, and the reduction was recovered by estrogen implant. These differences in responses among different species may be due to species specificity or other physiological conditions that need to be investigated further in teleosts.

3.4. Gonadotropin-Inhibitory Hormone (GnIH)

Just over a decade ago, another hypothalamic neuropeptide, gonadotropin-inhibitory hormone (GnIH) that belongs to the LPXRF-amide family of peptides was discovered in quail and found to inhibit the gonadotropin release (Tsutsi et al., 2000). Existence of GnIH in fishes was also demonstrated recently (Sawada et al., 2002; Tsutsui et al., 2007). The zebrafish GnIH precursor contains three putative LPXRFamide peptides (Zhang et al., 2010). These authors have shown that intraperitoneal administration of mature zebrafish GnIH peptide (LPXRFa peptide-3) caused a significant reduction of the basal serum LH level in goldfish. However, the knowledge on fish GnIH remains fragmentary todate. Three orthologous GnIH genes were identified in zebrafish, stickleback, medaka and Takifugu (Zhang et al., 2010). In goldfish, GnIH neurons occur in the posterior periventricular nucleus and in the TN area, and fibers reach several BP areas (Sawada et al., 2002). In goldfish, GnIH orthologs (gfLPXRFa-1, -2 and -3) stimulated the release of GtHs *in vitro* (Amano et al., 2006). Intraperitoneal injections of goldfish GnIH (gfGnIH) resulted in an increase in pituitary LHβ and FSHβ mRNA levels at early to late gonadal recrudescence, but a reduction in serum LH and hypophysial gfGnIH-R mRNA levels occurred at other reproductive phases (Moussavi et al. 2012). These authors further showed that static incubation with gfGnIH augmented LH secretion from dispersed pituitary cell cultures from prespawning fish, but not from hypophysial cells at other reproductive stages. This treatment has suppressed LHβ mRNA levels at early recrudescence and prespawning, but elevated LHβ at mid-recrudescence and consistently attenuated FSHβ mRNA in a dose-specific manner. Moussavi et al. (2012) therefore suggested that in goldfish, regulation of LH secretion and GtH subunit mRNA levels depend on maturational status and administration route. For a better understanding of the role of GnIH in fish reproduction, further studies on more teleosts are warranted.

3.5. 5-Hydroxytryptamine (5-HT)

As reported in Senegalese sole, 5-HT neurons occur among the cerebrospinal fluid-contacting 5-HT neurons in the NRL, paraventriculat organ (PVO), and NRP and in the rhombencephalic reticular formation and inferior olivary region. In the brainstem, 5-HT somata were found in the superior and inferior raphe, the nucleus reticularis superioris, NE and the inferior olive. Although positive fibers were not found in the neurohypophysis, a few 5-HT-ir cells were identified in the adenohypophysis (Rodríguez-Gómez et al., 2000). In catfish, tryptophan hydroxylase (TH) is the key regulator in serotonin (5-HT) and stimulates the release of GnRH and GtHs by acting at the level of BP axis (Raghuveer et al., 2011). Administration of 5HT has been shown to facilitate GnRH and/or GTH II release in catfish (Senthilkumaran and Joy, 1996). 5-HT stimulates GTH-II release via a 5HT2-like receptor in goldfish (Somoza and Peter, 1991) and in Atlantic croaker (Khan and Thomas, 1994).

3.6. Melatonin (MLT)

MLT receptors are widely expressed in the brain, PG, and gonads of carp (Moniruzzaman and Maitra, 2012). These authors have shown that the content of both MLT receptor proteins varied significantly in relation to time of the day (peak at midnight and fall at midday), and reproductive phase in the annual cycle (highest in the spawning phase and lowest in the post-spawning phase). The available data on the role of pineal hormone, MLT in fish reproduction are rather conflicting, being either pro- or anti-gonadal. A pro-gonadal effect of MLT was found in turbot (Alvariño et al., 2002) and in catfish (Ghosh and Nath, 2005). MLT acts on GnRH, dopaminergic system, hypophysial LH/FSH and on gonads (Falcón et al., 2010). In croaker, sea bass, and in cultured carp pituitary, MLT induced an increase in LHβ, secretion, while in eel, implant of MLT induced a decrease in LHβ and FSHβ; in zebrafish, MLT induces a significant increase in fecundity (Carnevali et al. 2011).

Some investigators reported antigonadal effects. A previous study on catfish revealed an inhibitory influence of MLT on the ovarian functions (Senthilkumaran and Joy, 1995). In murrel, MLT exhibits pro- or anti-GtH effects depending upon the photoperiodicity/gonadal state of the fish (*Renuka* and *Joshi,* 2010). Having observed a direct link between the MT GnRH-2 cells and the pineal, Servile et al. (2010) suggested that GnRH-2 a has functional role in modulating the activity of the pineal organ in sea bass.

3.7. Arginine Vasotocin (AVT)

AVT is the basic neurohypophysial nonapeptide in teleosts synthesized in the nucleus preopticus (NPO) of the hypothalamus, transported and stored in the neurohypophysis for further release. In catfish and other teleosts, AVT neurons in the POA show gender-related differences. AVT mediates the physiology of reproduction besides regulating other functions (Singh and Joy, 2008). They have demonstrated changes in AVT in the catfish brain and ovary by immunocytochemical, HPLC, and EIA through different phases of the reproductive cycle. Administration of AVT affects reproductive and behavioral activities such as spawning, courtship, egg laying, clasping and song production in different lower vertebrates. Changes in GnRH and AVT expression in the brain influence the sexual status or the expression of morph specific display behaviors which are important in reproduction (Bass and Grober, 2001; Forlano and Bass, 2011).

3.8. Gamma-Aminobutyric Acid (GABA)

In teleosts GABA perikarya occur in the telencephalon and hypothalamus, and innervate the PD (Kah et al., 1987; Kim et al., 2004). GtH and GH occur colocalized in the goldfish pituitary, and both are innervated by GABA neurons; involvement of GABA in release of GtH-II and LH was shown in goldfish (see Trudeau, 1997). Stimulation of LH release by GABA appears to result from combined effects involving GnRH release, potentiation of GnRH action, decreased dopaminergic activity and in some cases directly at the LH cell. Further, these actions appear to depend upon such factors as sex or sex steroid levels, besides species differences. Muscimol (a $GABA_A$ receptor agonist) stimulates GTH-II release, and bicuculline (a $GABA_A$ receptor antagonist) blocks GABA-stimulated GTH-II release *in vivo* in goldfish, indicating the involvement of a $GABA_A$-like receptor (Trudeau et al., 1997).

3.9. Noradrenalin (NA)

Noradrenalin (NA) has a stimulatory effect on GnRH in goldfish (Peter et al., 191) and on GtH in catfish (Senthilkumaran and Joy, 1996). Stimulatory effect of NA on GTH-II release is mediated by a-1 receptor mechanism and sexually regressed goldfish was more responsive than mature fish (Peter et al., 1991). In European eel, NA fibers connect the POA and the PD (Fremberg et al., 1977). Peripherally released NA probably acts on GnRH nerve terminals and directly on the GtH cells to stimulate GTH-II release, while centrally released NA may activate the GnRH neurons (Yu et al., 1991).

3.10. Nitric Oxide (NO)

NO, a gaseous molecule has been known to stimulate the release of GnRH in mammals and play a role in reproduction (Brann et al., 1997). However, the information on the role of NO in fish reproduction is fragementary. Several nicotinamide adenine dinucleotide phosphate-diaphorase (NADPHd)-positive granules that represent the fiber terminals or receptors were found on the GnRH3 neurons of the OB, olfactory nerve (OLN), MOT and VT in bitterling (Baby et al., 2000). A close association exists between GNRH2 neurons and NO neurons in the midbrain tegmentum (MT). The presence of fiber terminals may represent potential contacts between the nitrinergic and GnRH neurons in the olfactory system. In bitterling, the MT GnRH neurons showed soma-soma, soma-process or process-process contacts with NADPHd somata (Baby et al., 2000), indicating that there exists some functional influence of NADPHd neurons on the GnRH neurons through NO. The presence of nitrinergic neurons in proximity of GnRH neurons led to the proposition of a direct action of NOS on GnRH neurons (Bhat et al., 1995). Previous reports suggest that these contacts between the MT GnRH neurons and NO somata might be responsible for the neuromodulatory function, biosynthesis, and/or secretion of GnRH. MT GnRH neurons seem to belong to the nucleus of the medial longitudinal fasciculus (NMLF) that projects to the spinal cord in salmon (Oka et al., 1986) and goldfish (Prasada Rao et al., 1987), and might play a role in control of gonadal duct motility, and gamete release (Carolsfeld et al., 1997; Forlano and Bass, 2011).

4. Metabolic Modulators

Metabolic hormones such as leptin, ghrelin, neuropeptide Y and melanin-concentrating hormone affect the reproductive physiology. The metabolic state and body energy reserves play an important role in the regulation of puberty onset and fertility. Further, reproductive disorders may also result from metabolic stress caused by energy insufficiency or obesity (Castellano et al., 2010). However, the information available on the targets of leptin, glucocorticoids, IGF-1 and other growth factors in the BP complex is fragmentary, and how these factors modulate the neuroendocrine systems in fishes is inadequate (Zohar et al., 2010).

4.1. Leptin

The knowledge about the adipose-hormone, leptin, and its implication in reproduction stems mainly from the mammalian system. Leptin provides a crucial signal for transmitting metabolic information to the brain neuronal groups governing puberty and reproduction. The mode of action of leptin on GnRH neurons in mammals has remained controversial for several years. More recently, Kiss neurons have been shown to express leptin receptors and provide a link between metabolic and reproductive processes through actions on GnRH neurons (Castellano et al., 2010). Our knowledge about its role in fish is still in its infancy.

4.2. Neuropeptide Y (NPY)

NPY plays an important role the regulation of feeding in fish and mammals. Immunocytochemical and *in situ* hybridization studies revealed the presence of NPY in forebrain regions, particularly in the NE of the telencephalon, POA, OB and also in various thalamic regions in goldfish; this peptide was shown to stimulate GTH release (Kah et al., 1989). It was further shown that in goldfish NPY stimulates the release of LH at the level of the pituitary, besides promoting GnRH release from GnRH terminals in the pituitary or preoptic region slices (Peng et al., 1993a).

The effect of NPY in stimulating GTH-II release from the pituitary is high in sexually mature female goldfish as compared to fish at other stages of gonadal state.

The stimulatory effect of NPY on GtH-II release was enhanced when combined with testosterone (T) and estradiol (E2) treatment, particularly in sexually regressed fish when endogenous steroid concentrations are low. Similarly, NPY-induced GnRH release from the POA-anterior hypothalamic slices *in vitro* is also potentiated by implantation of T and E2 *in vivo* (Peng et al., 1993b).

4.3. Ghrelin (GRLN)

GRLN is a multifunctional hormone, expressed in a wide array of tissues including the brain and pituitary having a main role in regulation of GH secretion, but it is also implicated in stimulation of LH release in goldfish (Unniyappan and Peter, 2004; Grey and Chang, 2011).

The structure of the cDNA encoding GRLN was reported in goldfish (Unniyappan et al., 2002), and GRLN has also been identified from other fishes, such as eel, tilapia, and rainbow trout (see Peter and Unniyappan, 2004). GRLN and its receptor was identified and characterized in goldfish brain and the pituitary (Chang and Grey, 2011). Similarly, challenging male or female pituitary cell incubations of common carp, either with combination of GRLN and sGnRH-A or with sGnRH alone, revealed that GRLN functions as

LH-stimulating hormone and that it acts directly on GtH cells, potentiating the action of sGnRH (Sokołowska-Mikołajczyk et al., 2009).

5. Other Factors

Besides the above, there are other brain factors that play a role in teleost reproduction; these include cholecystokinin, pituitary adenylate cyclase-activating polypeptide (PACAP), insulin-like growth factor-1(IGF-1), other growth factors in the brain, glucocorticoids, gonadal steroids, activin and inhibin.

Interestingly various brain factors affect GtH release/inhibition through interaction with several neuroendocrine circuits within the brain interacting with environmental signals received through different sensory systems.

6. Environmental Regulation of BPG Axis

Kisspeptin has been shown to stimulate GnRH through functioning as a relay for environmental cues such as photoperiodicity and odorant information aided by visual and olfactory organs respectivley.

In teleosts, the terminal nerve (TN)-GnRH neuronal processes project to both the mitral and granule cell layers in the olfactory bulb and widely in the brain (Abe and Oka, 2000; Kawai et al., 2010). GnRH neurons in the olfactory bulb show nitric oxide (NO) fiber terminals on their surface (Baby et al., 2000) and NO have been known to stimulate the release of GnRH in mammals (Bran et al., 1997).

The OB efferents extend into the forebrain and show terminal fields in the telencephalon and in the posterior tuberal region in winter flounder (Prasada Rao and Finger, 1984). It has been known that the tropical carps breed only when exposed to flooded water caused by monsoon rains with altered physicochemical properties that trigger spawning activity, whereas the same species in perennial tanks do not breed. OT manipulations have been reported to affect the gonadal function with reduced levels of GtHs in goldfish (Dulka et al., 1987); in tilapia, the olfactory rosette was shown to be crucial for modulating nest-building behavior through second order olfactory pathways interacting with GnRH1 and GnRH2 neuronal systems (Uchida et al., 2005). While the role of photoperiod and temperature has been investigated to some extent, there is a paucity of information on the effect of different altered physicochemical properties such as pH, water flow, vegetation etc. on Kisspeptin and GnRH neurons. So is also the case with information pertaining to the effect of inputs from electroreceptors, mechanoreceptors, thermoreceptors, olfactory, gustatory, pheromonal, auditory systems and social cues on the BP axis. Attempts to unravel the significance of signals from the above-mentioned systems might yield fruitful information on the role of environmental cues in fish reproduction.

6.1. Photoperiod

Photic cues affect the BPG axis primarily via the retina and pineal organ in fishes (Grens et al., 2005: Moniruzzaman and Maitra, 2012). The type-I and type-II GnRH receptors occur in the retinal amacrine and ganglion cells respectively. Amacrine cells are involved in processing visual information and ganglion cells relay the processed information to the brain in order to influence of GnRH in this process (Grens et al., 2005). In catfishes, retinal efferents project to the suprachiasmatic nucleus (SCN) besides other brain areas (Prasada Rao and Sharma, 1982) and the SCN neurons project to the pituitary (Prasada Rao et al., 1993b). Exposure of fish to prolonged photoperiod results in rapid gonadal maturation (Bhattacharya et al., 2007). In European sea bass, the pineal organ receives GnRH2 fibers from the MT GnRH neurons and the pinealofugal fibers terminate in the tegmental GnRH neuronal area suggesting bidirectional connectivity.

The pineal cells also express GnRH receptors (Servili et al., 2010). The pineal hormone, melatonin, affects GnRH, dopamineric system, FSH, LH and gonads (Falcón et al., 2010).

6.2. Temperature

In blue gourami maintained at 27°C, GnRH3, PACAP, insulin-like growth factor-1(IGF-1), βLH, and prolactin were significantly higher than in fish at 23°C or 31°C. βFSH mRNA levels were significantly lower when maintained at 31°C than at the other temperatures tested (David and Degani, 2011). These authors have suggested that temperature has a more marked effect on transcription of genes associated with reproduction, than on those pertaining to growth.

Conclusion

Laudable advances have been made in fish reproductive physiology through investigations on GnRH, followed by exhaustive work on Kisspeptin by several investigators within the past decade. However, wide adaptive radiation and manifold reproductive strategies exhibited by teleosts signify the need for extended research. GnIH and melatonin have attracted limited attention and the intricacy of melatonin needs to be deciphered. Tropical carps do not breed unless they receive stimulus generated by flooded monsoon water.

For induced spawning of carps, required for improved aquaculture practices, no noteworthy method has been evolved after the discovery of injectible “Ovaprim”. It would be magnificent if a non-invasive method could be evolved through suitable alteration of physicochemical properties of water in order to facilitate triggering spawning as it happens in nature; it is known that many neuropeptides including GnRH exist in the olfactory epithelium, pineal organ and visual system. These sensory systems might help in transmission of environmental stimuli and stimulate the release of GtH to induce profuse breeding after interaction with neuroendocrine circuits within the brain. A thorough understanding of the involved environmental parameters and mechanisms that influence the neuropeptides to induce GtH release might help in developing an easy and non-invasive procedure for induced fish breeding.

Acknowledgments

Support received from Indian National Science Academy (INSA), New Delhi, India, through "INSA Honorary Scientist" award is acknowledged. Thanks are due to Director, National Environmental Engineering Reseach Institute, Nagpur, for facilities.

References

Abe, H. and Oka, Y. (2000). Modulation of pacemaker activity by salmon gonadotropin-releasing hormone (sGnRH) in terminal nerve (TN)-GnRH neurons. *J Neurophysiol,* 83, 3196-4200.

Abe, H. and Oka, Y. (2007). Neuromodulatory functions of terminal nerve-GnRH neurons. In T.J. Hara, and B. Zielenski (Eds.) Fish Physiology (pp. 453 - 501). San Diego, CA: Elsevier.

Alvariño, J. M. R., Rebollar, P. G., Olmedo, M., Alvarez-Blázquez, B., Ubilla, E. and Peleteiro, J. B. (2002). Effects of melatonin implants on reproduction and growth of turbot broodstock. *Aquacult Internat,* 9, 477- 487.

Amano, M., Moriyama, S., Iigo, M., Kitamura, S., Amiya, N., Yamamori, K., Ukena, K. and Tsutsui, K. (2006). Novel fish hypothalamic neuropeptides stimulate the release of gonadotrophins and growth hormone from the pituitary of sockeye salmon. *J Endocrinol,* 188, 417–423.

Anglade, I., Zandbergen, T. and Kah, O. (1993). Origin of pituitary innervations in the goldfish. *Cell Tiss Res,* 27, 345–355.

Baby, S. M., Ueck, M. and Prasada Rao, P. D. (2000). Gonadotropin-releasing hormone-immunoreactive neurons and associated nicotinamide adenine dinucleotide phosphate-diaphorase-positive neurons in the brain of a teleost, *Rhodeus amarus. Gen Comp Endocrinol,* 120, 44-54.

Bass, A. H. and Grober, M. S. (2001). Social and neural modulation of sexual plasticity in teleost fish. *Brain Behav Evol,* 57, 293-300.

Beck, B. H., Fuller, S. A., Peatman, E., McEntire, M. E., Darwish, A. and Freeman, D. W. (2012). Chronic exogenous kisspeptin administration accelerates gonadal development in basses of the genus Morone. *Comp Biochem Physiol A Mol Integr Physiol,* 162, 265-273.

Bhat, G. K., Mahesh, V. B., Lamar, C. A., Ping, L., Aguan, K. and Brann, D. W. (1995). Histochemical localization of nitric oxide neurons in the hypothalamus: association with gonadotropin-releasing hormone neurons and co-localization with N-methyl-D-aspartate receptors. *Neuroendocrinol,* 62, 187-197.

Bhattacharya, S., Chattoraj, A. and Maitra. S. K. (2007). Melatonin in the regulation of annual testicular events in carp Catla catla: evidence from the studies on the effects of exogenous melatonin, continuous light, and continuous darkness. *Chronobiol Int,* 24, 629-650.

Biran, J., Ben-Dor, S. and Levavi-Sivan, B. (2008). Molecular identification and functional characterization of the kisspeptin/kisspeptin receptor system in lower vertebrates. *Biol Reprod,* 79, 776-786.

Brann, D. W., Bhat, G. K., Lamar, C. A. and Mahesh, V. B. (1997). Gaseous transmitters and neuroendocrine regulation. *Neuroendocrinol,* 65, 385–395.

Carnevali, O., Gioacchini, G., Maradonna, F., Olivotto, I. and Migliarini, B. (2011). Melatonin induces follicle maturation in *Danio rerio*. *PLoS One,* 6, e19978.

Carolsfeld, J., Tester, M., Kreiberg, H. and Sherwood, N. M. (1997). Pheromone-induced spawning of Pacific herring. I. Behavioral characterization. *Horm Behav,* 31, 256-268.

Castellano, J. M., Bentsen, A. H., Mikkelsen, J. D. and Tena-Sempere, M. (2010). Kisspeptins: bridging energy homeostasis and reproduction. *Brain Res.,* 1364, 129-138.

Chen, C. C. and Fernald, R. D. (2008). GnRH and GnRH receptors: distribution, function and evolution. *J. Fish Biol.,* 73, 1099–1120.

David, D. and Degani, G. (2011). Temperature affects brain and pituitary gene expression related to reproduction and growth in the male blue gouramis, *Trichogaster trichopterus*. *J Exp Zool A Ecol Genet Physiol,* 315, 203-214.

Dufour, S., Sebert, M. E., Weltzien, F. A., Rousseau, K. and Pasqualini, C. (2010). Neuroendocrine control by dopamine of teleost reproduction. *J. Fish Biol.,* 76, 129-160.

Dulka, J. G., Stacey, N. E., Sorensen, P. W., Van der Kraak, G. J. and Marchant, T. A. (1987). A sex pheromone system in goldfish: is the nervus terminalis involved? *Ann N Y Acad Sci.,* 519, 411–420.

Falcón, J., Migaud, H., Muñoz-Cueto, J. A. and Carrillo, M. (2010). Current knowledge on the melatonin system in teleost fish. *Gen. Comp. Endocrinol.,* 165, 469-482.

Felip, A., Zanuy, S., Pineda, R., Pinilla, L., Carrillo, M., Tena-Sempere, M. and Gómez A. (2009). Evidence for two distinct KiSS genes in non-placental vertebrates that encode kisspeptins with different gonadotropin-releasing activities in fish and mammals. *Mol. Cell Endocrinol.,* 312, 61-71.

Filby, A. L., van Aerle, R., Duitman, J. and Tyler, C. R. (2008). The kisspeptin/gonadotropin-releasing hormone pathway and molecular signaling of puberty in fish. *Biol. Reprod.,* 78, 278-289.

Forlano, P. M. and Bass, A. H. (2011). Neural and hormonal mechanisms of reproductive-related arousal in fishes. *Horm. Behav.,* 59, 616–629.

Fremberg, M., van Veen., Th. and Hartwig, H. G. (1977). Formaldehyde-induced fluorescence in the telencephalon and diencephalon of the eel *(Anguilla anguilla). A* fluorescence-microscopic and microspectro-fluoremetric investigation with special reference to the innervation of the pituitary. *Cell Tiss Res.,* 176, 1-22.

Ghosh, J. and Nath P. (2005). Seasonal effects of melatonin on ovary and plasma gonadotropins and vitellogenin levels in intact and pinealectomized catfish, *Clarias batrachus (*Linn*). Indian J. Exp. Biol.,* 43, 224–232.

González-Martínez, D., Madigou, T., Mañanos, E., Cerdá-Reverter, J.M., Zanuy, S., Kah, O. and Muñoz-Cueto, J. A. (2004). Cloning and expression of gonadotropin-releasing hormone receptor in the brain and pituitary of the European sea bass: an *in situ* hybridization study. *Biol. Reprod.,* 70, 1380-1391.

Grens, K. E., Greenwood, A. K. and Fernald, R. D. (2005). Two visual processing pathways are targeted by gonadotropin-releasing hormone in the retina. *Brain Behav. Evol.,* 66, 1-9.

Grey, C. L. and Chang, J. P. (2011). Differential involvement of protein kinase C and protein kinase A in ghrelin-induced growth hormone and gonadotrophin release from goldfish (Carassius auratus) *pituitary* cells. *J. Neuroendocrinol.,* 23, 1273-1287.

Guilgur, L. G.,. Strüssmann, C. A. and Somoza, G. M. (2009). mRNA expression of GnRH variants and receptors in the brain, pituitary and ovaries of *pejerrey (Odontesthes bonariensis) in* relation to the reproductive status. *Fish Physiol. Biochem.,* 35, 157-166.

Hildahl, J., Sandvik, G. K., Edvardsen, R. B,. Norberg, B., Haug, T. M. and Weltzien, F. A. (2011). Four gonadotropin releasing hormone receptor genes in Atlantic cod are differentially expressed in the brain and pituitary during puberty. *Gen. Comp. Endocrinol.,* 173, 333-345.

Hull, K. L. and Harvey, S. (2002). GH as a co-gonadotropin: the relevance of correlative changes in GH secretion and reproductive state. *J. Endocrinol.,* 172, 1–19.

Illing, N., Troskie, B. E., Nahorniak, C. S., Hapgood, J. P., Peter, R. E. and Millar R. P. (1999). Two gonadotropin-releasing hormone receptor subtypes with distinct ligand selectivity and differential distribution in brain and pituitary in the *goldfish (Carassius auratus). Proc. Natl. Acad. Sci. USA,* 96, 2526-2531.

Jodo, A., Ando, H. and Urano, A. (2003). Five different types of putative GnRH receptor gene are expressed in the brain of masu salmon *(Oncorhynchus masou). Zool Sci.,* 20, 1117–1125.

Johnston, S. A. and Maler, L. (1992). Anatomical organization of the hypophysiotrophic systems in the electric fish, *Apteronotus leptorhynchus. J. Comp. Neurol.,* 317, 421-437.

Kah, O., Danger, J. M., Dubourg, P., Pelletier, G., Vaudry, H. and Calas, A. (1989). Characterization, cerebral distribution and gonadotrophin-release activity of neuropeptide Y (NPY) in the goldfish. *Fish Physiol. Biochem.,* 7, 69–76.

Kah, O., Dubourg, P., Martinoli, M. G., Geffard, M. and Calas, A. (1987). Morphological evidence for a direct neuroendocrine GABAergic control of the anterior pituitary in teleosts. *Experientia,* 43, 300-302.

Kanda, S., Akazome, Y., Matsunaga, T., Yamamoto, N., Yamada, S., Tsukamura, H., Maeda, K. and Oka, Y. (2008). Identification of KiSS-1 product kisspeptin and steroidsensitive sexually dimorphic kisspeptin neurons in medaka (*Oryzias latipes*). *Endocrinol,* 149, 2467–2476.

Kanda, S., Karigo, T. and Oka, Y. (2012). Steroid Sensitive kiss2 Neurones in the Goldfish: Evolutionary Insights into the Duplicate Kisspeptin Gene-Expressing Neurones. *J Neuroendocrinol,* 24, 897-906.

Kawai, T., Abe, H., Akazome, Y. and Oka, Y. (2010). Neuromodulatory effect of GnRH on the synaptic transmission of the olfactory bulbar neural circuit in goldfish, *Carassius auratus. J. Neurophysiol.,* 104, 3540-3550.

Khan, I. A. and Thomas, P. (1994). Seasonal and daily variations in the plasma gonadotropin-II response to an LHRH analog and serotonin in Altantic croaker (*Micropogonias undulatus*): evidence for mediation by 5-HT2 receptors. *J. Exp. Zool.,* 269, 531-*537.*

Kim, Y. J., Nam, R. H., Yoo, Y. M. and Lee, C. J. (2004). Identification and functional evidence of GABAergic neurons in parts of the brain of adult zebrafish (*Danio rerio*). *Neurosci. Lett.,* 355, 29-32.

Kim, N. N., Shin, H. S., Habibi, H. R., Lee, J. and Choi, C. Y. (2012). Expression profiles of three types of GnRH during sex-change in the protandrous cinnamon clownfish,

Amphiprion melanopus: Effects of exogenous GnRHs. *Comp. Biochem. Physiol. B Biochem. Mol. Biol.,* 161, 124-133.

Kitahashi, T., Ogawa, S. and Parhar, I. S. (2009). Cloning and expression of kiss2 in the zebrafish and medaka. *Endocrinol.,*150, 821-831.

Lethimonier, C., Madigou, T., Muñoz-Cueto, J. A., Lareyre, J. J. and Kah, O. (2004). Evolutionary aspects of GnRHs, GnRH neuronal systems and GnRH receptors in teleost fish. *Gen. Comp. Endocrinol.,* 135, 1-16.

Levavi-Sivan, B., Bogerd, J., Mañanós, E. L., Gómez, A. and Lareyre, J. J. (2010). Perspectives on fish gonadotropins and their receptors. *Gen. Comp. Endocrinol.,* 165, 412-437.

Li, S., Zhang, Y., Liu, Y., Huang, X., Huang, W., Lu, D., Zhu, P., Shi, Y., Cheng, C. H. K., Liu, X. and Lin, H. (2009). Structural and functional multiplicity of the kisspeptin/GPR54 system in goldfish (*Carassius auratus*). *J. Endocrinol.,* 201,407-418.

Lin, X. W., Lin, H. R. and Peter, R. E. (1993). Growth hormone and gonadotropin secretion in the common carp (*Cyprinus carpio* L.): *in vitro* interactions of gonadotropin-releasing hormone, somatostatin, and the dopamine agonist apomorphine. *Gen. Comp. Endocrinol.,* 89, 62-71.

Mitani, Y., Kanda, S., Akazome, Y., Zempo, B. and Oka, Y. (2010). Hypothalamic Kiss1 but not Kiss2 neurons are involved in estrogen feedback in medaka (*Oryzias latipes*). *Endocrinol.,* 151, 1751-1759.

Mohamed, J. S., Benninghoff, A. D., Holt, G. J. and Khan, I. A. (2007). Developmental expression of the G protein-coupled receptor 54 and three GnRH mRNAs in the teleost fish cobia. *J. Mol. Endocrinol.,* 38, 235–244.

Moniruzzaman, M. and Maitra, S. K. (2012). Influence of altered photoperiods on serum melatonin and its receptors (mt1 and mt2) in the brain, retina, and ovary in carp, *Catla catla. Chronobiol. Int.,* 29, 175-188.

Moussavi, M., Wlasichuk, M., Chang, J. P. and Habibi, H. R. (2012). Seasonal effect of GnIH on gonadotrope functions in the pituitary of goldfish. *Mol. Cell Endocrinol.,* 350, 53-60.

Murthy, C. K., Turner, R. J., Wong, A. O. L., Prasada Rao, P. D., Rivier, J. E. and Peter, R. E. (1994). Differential actions of a mammalian gonadotropin-releasing hormone antagonist on gonadotropin-II and growth hormone release in goldfish, *Carassius auratus. Neuroendocrinol.,* 59: 561-571.

Ng, T. B. and Idler, D. R. (1978). A vitellogenic hormone with a large and a small form from salmon pituitaries. *Gen. Comp. Endocrinol.,* 35, 189-195.

Nocillado, J. N., Levavi-Sivan, B., Carrick, F. and Elizur, A. (2007). Temporal expression of G-protein-coupled receptor 54 (GPR54), gonadotropin-releasing hormones (GnRH), and dopamine receptor D2 (drd2) in pubertal female grey mullet, *Mugil cephalus. Gen. Comp. Endocrinol.,* 150, 278-87.

Ogawa, S., Ng, K. W., Ramadasan, P. N., Nathan, F. M. and Parhar, I. S. (2012). Habenular kiss1 neurons modulate the serotonergic system in the brain of zebrafish. *Endocrinol.,* 153, 2398–2407.

Oka, Y. (2010). Electrophysiological characteristics of gonadotrophin-releasing hormone 1–3 neurones: insights from a study of fish brains. *J. Neuroendocrinol.,* 22, 659–663.

Oka, Y., Satou, M. and Ueda, K. (1986). Descending pathways to the spinal cord in the himé salmon (landlocked red salmon, *Oncorhynchus nerka). J. Comp. Neurol.,* 254, 91-103.

Okuzawa, K. and Kobayashi, M. (1999). Gonadotropin-releasing hormone neuronal systems in the teleostean brain and functional significance. In P. D. Prasada Rao, R. E. Peter, (Eds.), Neural regulation in the Vertebrate Endocrine System (pp. 85–100). Kluwer Academic/Plenum Publishers, New York.

Parhar, I. S., Soga, T., Ishikawa, Y., Nagahama, Y. and Sakuma, Y. (1998). Neurons synthesizing gonadotropin-releasing hormone mRNA subtypes have multiple developmental origins in the medaka. *J. Comp. Neurol.,* 401, 217–226.

Parhar, I. S., Soga, T., Sakuma, Y. and Millar, R. P. (2002). Spatio-temporal expression of gonadotropin-releasing hormone receptor subtypes in gonadotropes, somatotropes and lactotropes in the cichlid fish. *J. Neuroendocrinol.,* 14, 657–665.

Parhar, I. S., Ogawa, S. and Sakuma, Y. (2004). Laser-captured single digoxigenin-labeled neurons of gonadotropin releasing hormone types reveal a novel G protein-coupled receptor (Gpr54) during maturation in cichlid fish. *Endocrinol.,* 145, 3613-3618.

Peng, C., Chang, J. P., Yu, K. L., Wong, A. O. L., Van Goor, F., Peter, R. E. and Rivier, J. E. (1993a). Neuropeptide-Y stimulates growth hormone and gonadotropin-II secretion in the goldfish pituitary: involvement of both presynaptic and pituitary cell actions. *Endocrinol.,* 132, 1820–1829.

Peng, C., Trudeau, V. L. and Peter, R. E. (1993b). Seasonal variation of neuropeptide Y actions on growth hormone and gonadotropin-II secretion in the goldfish: effects of sex steroids. *J. Neuroendocrinol.,* 5, 273–280.

Peter, R. E., Prasada Rao, P. D., Baby, S. M., Illing, N. and Millar, R. P. (2003). Differential brain distribution of gonadotropin-releasing hormone receptors in the goldfish. *Gen. Comp. Endocrinol.,* 132, 399-408.

Peter, R..E., Trudeau, V. L. and Sloley, B. D. (1991) Brain regulation of reproduction in teleosts. Bulletin of the Institute of Zoology (Taipei). *Academica Sinica Monograph,* 16, 89–118.

Peter, R. E., Lin, H. R. and Van Der Kraak, G. (1988). Induced ovulation and spawning of cultured freshwater fish in China: advances in application of GnRH analogues and dopamine antagonists. *Aquaculture,* 74, 1–10.

Powell, J. F., Zohar, Y., Elizur, A., Park, M., Fischer, W. H., Craig, A. G., Rivier, J. E., Lovejoy, D. A. and Sherwood, N. M. (1994). Three forms of gonadotropin-releasing hormone from brains of one species. *Proc. Natl .Acad. Sci., USA,* 91, 12081–12085.

Prasada Rao, P. D. (1999). Hypophysiotropic neurons in the brain of teleosts. In P. D. Prasada Rao and R. E. Peter (Eds.), Neural Regulation in the Vertebrate Endocrine System (pp. 23-40). New York, Kluwer Academic/Plenum Publishers.

Prasada Rao, P. D. and Finger, T. E. (1984). Asymmetry of the olfactory system in the brain of the winter flounder, *Pseudopleuronectes americanus. J, Comp, Neurol,,* 225, 492-510.

Prasada Rao, P. D., Jadhao, A. G. and Sharma, S. C. (1987). Descending projection neurons to the spinal cord of the goldfish, *Carassius auratus. J, Comp, Neurol,,* 265, 96-108.

Prasada Rao, P. D., Job, T. C. and Schreibman, M. P. (*1993).* Hypophysiotropic Neurons in the hypothalamus of the catfish Clarias batrachus: A cobaltous lysine and HRP study. *Brain Behav, Evol,, 42, 24–38.*

Prasada Rao, P. D. and Sharma, S. C. (1982). Retinofugal pathways in juvenile and adult channel catfish, Ictalurus (Ameiurus) punctatus: An HRP and autoradiographic study. *J, Comp, Neurol,,* 210, 37-48.

Raghuveer, K., Sudhakumari, C. C., Senthilkumaran, B., Kagawa, H., Dutta-Gupta, A. and Nagahama, Y. (2011). Gender differences in tryptophan hydroxylase-2 mRNA, serotonin, and 5-hydroxytryptophan levels in the brain of catfish, Clarias gariepinus, during sex differentiation. *Gen, Comp, Endocrinol,,* 171, 94-104.

Renuka, K. and Joshi, B. N. (2010). Melatonin-induced changes in ovarian function in the freshwater fish Channa punctatus (Bloch) held in long days and continuous light. *Gen Comp Endocrinol,* 165, 42-46.

Rodríguez-Gómez, F. J., Rendón-Unceta, M. C., Sarasquete, C., and Muñoz-Cueto, J. A. (2000). Distribution of serotonin in the brain of the Senegalese sole, Solea senegalensis: an immunohistochemical study. *J. Chem. Neuroanat.,* 18, 103–115.

Sawada, K., Ukena, K., Satake, H., Iwakoshi, E., Minakata, H. and Tsutsui K. (2002). Novel fish hypothalamic neuropeptide: Cloning of a cDNA encoding the precursor polypeptide and identification and localization of the mature peptide. *Eur. J. Biochem.,* 269, 6000–6008.

Senthilkumaran, B. and Joy, K. P. (1995). Effects of melatonin, p-chlorophenylalanine, and a-methylparatyrosine on plasma gonadotropin level and ovarian activity in the catfish, Heteropneustes fossilis:A study correlating changes in hypothalamic monoamines. *Fish Physiol. Biochem.,* 14, 471–480.

Senthilkumaran, B. and Joy, K. P. (1996). Effects of administration of some monoamine-synthesis blockers and precursors on ovariectomy-induced rise in plasma gonadotropin II in the catfish *Heteropneustes fossilis. Gen Comp Endocrinol, 101*, 220-226.

Senthilkumaran, B., Okuzawa, K., Gen, K., Ookura, T. and Kagawa, H. (1999). Distribution and seasonal variations in levels of three native GnRHs in the brain and pituitary of perciform fish. *J. Neuroendocrinol.,* 11, 181-186.

Servili, A., Le Page, Y., Leprince, J., Caraty, A., Escobar, S., Parhar, I. S., Seong, J. Y., Vaudry, H. and Kah, O. (2011). Organization of two independent kisspeptin systems derived from evolutionary-ancient kiss genes in the brain of zebrafish, *Endocrinol.,* 152, 1527–1540.

Servili, A., Lethimonier, C., Lareyre, J. J., López-Olmeda, J. F., Sánchez-Vázquez, F. J., Kah, O. and Munóz-Cueto, J. A. (2010). The Highly conserved gonadotropin-releasing hormone-2 form acts as a melatonin-releasing factor in the pineal of a teleost fish, the European sea bass *Dicentrarchus labrax. Endocrinol.,* 151, 2265–2275.

Servili, A., Herrera-Pérez, P., Kah, O. and Muñoz-Cueto, J. A. (2012). The retina is a target for GnRH-3 system in the European sea bass, *Dicentrarchus labrax. Gen. Comp. Endocrinol.,* 175, 398-406.

Sherwood, N., Eiden, L., Brownstein, M., Spiess, J., Rivier, J. and Vale, W. (1983). Characterization of a teleost gonadotropin-releasing hormone. *Proc. Natl. Acad. Sci. USA,* 80, 2794–2798.

Shimizu, Y., Tomikawa, J., Hirano, K., Nanikawa, Y., Akazome, Y., Kanda, S., Kazeto, Y., Okuzawa, K., Uenoyama, Y., Ohkura, S., Tsukamura, H., Maeda, K., Gen, K., Oka, Y. and Yamamoto, N. (2012). Central distribution of kiss2 neurons and peri-pubertal changes in their expression in the brain of male and female red seabream *Pagrus major. Gen. Comp. Endocrinol.,* 175, 432-42.

Singh, V. and Joy, K. P. (2008). Immunocytochemical localization, HPLC characterization, and seasonal dynamics of vasotocin in the brain, blood plasma and gonads of the catfish *Heteropneustes fossilis. Gen. Comp. Endocrinol.,* 159, 215-225.

Sokołowska-Mikołajczy, K. M., Socha, M., Szczerbik, P. and Epler, P. (2009). The effects of ghrelin on the in vitro spontaneous and sGnRH-A stimulated luteinizing hormone (LH) release from the pituitary cells of common carp (*Cyprinus carpio* L.). *Comp. Biochem. Physiol. A Mol. Integr. Physiol.,* 153, 386-90.

Somoza, G. M. and Peter, R. E. (1991). Effects of serotonin on gonadotropin and growth hormone release from *in vitro* perifused goldfish pituitary fragments. *Gen. Comp. Endocrinol.,* 82, 103-110.

Suresh, D. V. N. S., Baile, V. V. and Prasada Rao, P. D. (2008). Annual reproductive phase-related profile of sex steroids and their carrier, SHBG, in the Indian major carp, *Labeo rohita. Gen. Comp. Endocrinol.,* 159, 143–149.

Swanson, P., Bernard, M., Nozaki, M., Suzuki, K., Kawauchi, H. and Dickhoff, W. W. (1989). Gonadotropins I and II in juvenile coho salmon. *Fish Physiol. Biochem.,* 7, 169–176.

Tena-Sempere, M. (2010). Kisspeptin signaling in the brain: recent developments and future challenges. *Mol. Cell Endocrinol.,* 314, 164–169.

Tena-Sempere, M., Felip, A., Gómez, A., Zanuy, S. and Carrillo, M. (2012). Comparative insights of the kisspeptin/kisspeptin receptor system: lessons from non-mammalian vertebrates. *Gen. Comp. Endocrinol.,* 175, 234-243.

Tensen, C., Okuzawa, K., Blomenröhr, M., Rebers, F., Leurs, R., Bogerd, J., Schulz, R. and Goos, H. (1997). Distinct efficacies for two endogenous ligands on a single cognate gonadoliberin receptor. *Eur. J. Biochem.,* 243, 134–40.

Trudeau, V. L. (1997). Neuroendocrine regulation of gonadotrophin II release and gonadal growth in the goldfish, *Carassius auratus. Rev. Reproduction.,* 2, 55-68.

Tsutsui, K., Saigoh, E., Ukena, K., Teranishi, H., Fujisawa, Y., Kikuchi, M., Ishii, S. and Sharp, P. J. (2000). A novel avian hypothalamic peptide inhibiting gonadotropin release. *Biochem. Biophys. Res. Commun.,* 275, 661–667.

Uchida, H., Ogawa, S., Harada, M., Matushita, M., Iwata, M., Sakuma, Y. and Parhar, I. S. (2005). The Olfactory organ modulates gonadotropin-releasing hormone types and nest-building behavior in the tilapia *Oreochromis niloticus. J. Neurobiol.,* 65, 1–11.

Unniappan, S., Lin, X., Cervini, L., Rivier, J., Kaiya, H., Kangawa, K. and Peter, R. E. (2002). Goldfish ghrelin: molecular characterization of the complementary deoxyribonucleic acid, partial gene structure, and evidence for its stimulatory role in food intake. *Endocrinol.,* 143, 4143–4146.

Unniappan, S. and Peter, R. E. (2004). In vitro and in vivo effects of ghrelin on luteinizing hormone and growth hormone release in goldfish. *Am. J. Physiol. Regul. Integr. Comp. Physiol.,* 286, 1093-1101.

White, S. A., Kasten, T. L., Bond, C. T., Adelman, J. P. and Fernald, R. D. (1995). Three gonadotropin-releasing hormone genes in one organism suggest novel role for an ancient peptide. *PNAS USA,* 92, 8363–8367.

Yamamoto, N., Parhar, I. S., Sawai, N., Oka, Y. and Ito, H. (1998). Preoptic gonadotropin releasing hormone (GnRH) neurones innervate the pituitary in teleosts. *Neurosci. Res.,* 31, 31–38.

Yaron, Z., Gur, G., Melamed, P., Rosenfeld, H., Elizur, A. and Levavi-Sivan, B. (2003). regulation of fish gonadotropins. *Int. Rev. Cytol.,* 225, 131–185.

Yu, K. L. and Peter, R. E. (1990). Dopaminergic regulation of brain gonadotropin-releasing hormone in male goldfish during spawning behavior. *Neuroendocrinol.,* 52, 276-83.

Yu, K. L., Rosenblum, P. M. and Peter, R. E. (1991). In vitro release of gonadotropin-releasing hormone from the brain preoptic-anterior hypothalamic region and pituitary of female goldfish. *Gen. Comp. Endocrinol.,* 81, 256-267.

Zhang, Y., Li S., Liu, Y., Lu, D., Chen, H., Huang, X., Liu, X., Meng, Z., Lin, H. and Cheng, C. H. (2010). Structural diversity of the GnIH/GnIH receptor system in teleost: its involvement in early development and the negative control of LH release. *Peptides,* 31, 1034-1043.

Zhao, Y. and Wayne, N. L. (2012). Effects of Kisspeptin1 on electrical activity of an extrahypothalamic population of gonadotropin-releasing hormone neurons in Medaka (Oryzias latipes). *PLoS ONE,* 7, e37909. doi:10.1371/journal.pone.0037909.

Zmora, N., Stubblefield, J., Zulperi, Z., Biran, J., Levavi-Sivan, B., Muñoz-Cueto, J. A. and Zohar, Y. (2012). Differential and gonad stage-dependent roles of kisspeptin1 and kisspeptin2 in reproduction in the modern teleosts, morone species. *Biol. Reprod.,* 86, 177.

Zohar, Y., Muñoz-Cueto, J. A., Eizur, A. and Kah, O. (2010). Neuroendocrinology of reproduction in teleost fish. *Gen. Comp. Endocrinol.,* 165: 438-455.

In: Sexual Plasticity and Gametogenesis in Fishes
Editor: Balasubramanian Senthilkumaran
ISBN: 978-1-62618-848-8

Chapter XXI

Perspectives on the Gonadotropin-Releasing Hormone Gonadotropins-Monoamines: Relevance to Gonadal Differentiation and Maturation in Teleosts

***Yarikipati Prathibha*[†], *Chawngte Laldinsangi*[†] and *Balasubramanian Senthilkumaran*[*]**
Department of Animal Sciences, School of Life Sciences-Centre for Advanced Studies, University of Hyderabad, P. O. Central University, Hyderabad, Andhra Pradesh, India

Abstract

Reproduction in fish is essentially regulated by factors originating from hypothalamo-hypophyseal axis. The neuropeptide, gonadotropin-releasing hormone regulates the synthesis and release of the gonadotropins (GTHs) from the pituitary through monoaminergic systems, which in turn regulate gonadal development. There is an extensive documentation on the evolution and functional roles of GnRH family members and their variants. Monoamines influence the GnRH–GTH release through feedback regulation by steroids. The release and modulation of GnRH and GTH levels play an essential role in puberty and spawning. The role of GnRH-GTH and monoamines with reference to sexual development is an emerging topic with a lot of interesting findings in the past decade. In this review, we have discussed several reports explaining the possibility of the interplay of GnRH-GTH and monoamine axis with reference to sex differentiation.

[*] Corresponding author: bsksl@uohyd.ernet.in/senthilkumaranb@yahoo.com.
[†] These authors contributed equally.

Introduction

Gonadotropin-releasing hormone (GnRH), previously known as luteinizing-hormone-releasing hormone or luliberin, is a trophic decaneuropeptide that plays a vital role in entraining hypothalamo–hypophyseal–gonadal (H-H-G) axis to regulate reproduction and development (Vadakkadath and Atwood, 2005; Somoza et al., 2006; Maruska and Fernald, 2011). It also mediates several other processes in the central nervous system (CNS) such as reproductive behaviour and neuromodulation through its receptors (Hildahl et al., 2011b). Impaired GnRH affects gonadal development and reproductive cycle as it is a key regulator of gonadotropins (GTHs). It provides a more balanced integration of the events of reproduction with its pulsatile secretion differing in females when compared to males, wherein a preovulatory surge could be observed just before ovulation (Brodie and Crowley, 1985; Navarro et al., 2012). The pulsatile secretion of GnRH is responsible for synthesis and release of follicle stimulating hormone (FSH) and luteinizing hormone (LH) from the anterior pituitary, and is known to be synthesized and released from neurons within the hypothalamus (Brodie and Crowley, 1985; Wierman et al., 1995). The pituitary gonadotropins, LH and FSH are known to orchestrate gonadal function during reproductive cycle. It is well known that GTHs influence puberty, recrudescence and gamete maturation. However, their role in initiating gonadal sex development is at large, vague. It is also conceivable to expect that GnRH *per se* can dictate GTHs to entrain sexual development as they together control gonadal steroidogenesis. In spite of these notions, neither GnRH nor GTH have been considered as pivotal molecules for sex determination/differentiation. However, the feedback regulation from gonads is required as an over power to initiate the GnRH-GTH axis. In this context, monoamines mediate steroidal feedback to execute action on the GnRH-GTH axis. It is also plausible to imply that the initiation of monoaminergic system might be a pre-requisite for the origin of GnRH-GTH system vis-à-vis gonadal development. Among monoamines (MAs), serotonin (5-HT), dopamine (DA), norepinephrine (NE) and GABA (γ-aminobutyric acid) play important roles in GnRH-GTH release in teleosts. In this review, we highlighted certain important findings that propose the relevance of GnRH to gonadal differentiation and maturation by creating sub-headings on GnRH, GTH and MAs. We also emphasized the influence of GnRH and GTH on puberty, maturation and other related factors that have a control over reproduction.

GnRH

GnRH was first isolated from porcine and ovine hypothalamic extracts and has long been considered to be a unique molecular form, but subsequent studies have reported the occurrence of variants of GnRH in numerous species and currently include 25 isoforms representing 14 and 11 in vertebrate and invertebrate species, respectively (Kavanaugh et al., 2008). The identity of GnRH was clarified in 1977 by Nobel Laureates, Roger Guillemin and Andrew V. Schally as -pyroGlu-His-Trp-Ser-Tyr-Gly-Leu-Arg-Pro-Gly-NH_2 (Tan and Rousseau, 1982). When this peptide was first identified in a porcine species, it was named as LHRH (Matsuo et al., 1971). The naming of LHRH was replaced by GnRH that became more appropriate as it regulates FSH as well. However, later discoveries using several fish species emphasized the presence of multiple forms of GnRH and are distributed throughout brain including hypothalamus (Sherwood et al., 1986; Bogerd et al., 1992; Powell et al., 1994; Somoza et al., 2002; Parhar, 2002). Gilthead seabream, *Sparus aurata* was first reported (Powell et al., 1994) to have three forms of GnRH, in which the first form was named as Seabream GnRH (sbGnRH), the second was identical to chicken GnRH-II (cGnRH-II) and the third was found to be salmon GnRH (sGnRH). Interestingly, the studies so far in relation to the evolution of GnRH genes using various teleost models have demonstrated that there are three paralogous forms of these genes viz. sGnRH, sbGnRH and cGnRH and are considered to have originated due to genome duplication (Levy and Degani, 2012). Later reports indicated that most vertebrate species possess two GnRH genes, GnRH-1 and GnRH-2 with the exception of modern mammals (Servili et al., 2010). Presence of three GnRH systems is not always seen in all teleosts as some of them seem to have lost the GnRH1 or GnRH3 gene (Okubo and Nagahama, 2008). These authors further suggested that GnRH 1, GnRH2 and GnRH3 genes might have resulted from two rounds of tetraploidization in early vertebrate evolution. In this viewpoint, in mammals also, the GnRH2 gene has been deleted or silenced in certain species. They further indicated that the GnRH3 gene which is currently known only in teleosts has most likely been lost in the tetrapod lineage which resulted in the identification of fourth GnRH gene, named GnRH4 (Tostivint, 2011). Thus, teleosts serve as fundamental models in various aspects of GnRH research. Furthermore, teleosts possess many similar vertebrate functions as well as diverse features of growth and reproduction. The divergence of GnRH variants came into limelight from the studies using different teleost species, which revealed that more than half of the total 16 variants identified till date were from teleosts, which are mostly species-specific (Parhar et al., 1996; Somoza et al., 2002). Various teleost models have been used to elucidate the functions of GnRH during developmental and reproductive stages. Salmon was the first model in which a variant of GnRH, named as sGnRH, was isolated (Sherwood et al., 1983) which comprised of a two amino acid residue difference from mammalian GnRH (mGnRH). Two distinct forms of GnRH were isolated in lamprey, an ancient class of jawless vertebrates called Agnatha, and the first peptide named lGnRH-I have amino acid substitutions at positions 3 and 6 as compared to mGnRH. The second form, lamprey II GnRH showed variation at 3 amino acid residues with respect to type I (Sherwood et al., 1986) and a third molecular form, lGnRH-III was also identified (Sower et al., 1993).

Different species of catfish have also served in the identification of new GnRH variants. Two forms have been characterized from the African catfish, *Clarias gariepinus*, (Bogerd et al., 1992) and the Thai catfish, *Clarias macrocephalus* (Ngamvongchon *et al.*, 1992). The first form of catfish GnRH (cfGnRH) was found to be identical in both catfish species, and the second form of cfGnRH was identical to cGnRH-II. A novel form of GnRH was identified from the teleost herring, subsequently named as hGnRH, along with two other forms which were identified as cGnRH-II and sGnRH (Carolsfeld et al., 2000). The teleost, pejerry is another candidate in which a novel form of the peptide named as pjGnRH has been isolated, and is found to be distinct from the sequences of other vertebrate GnRHs (Somoza et al., 2002). Adams et al., (2002) examined a basal salmonid, the lake whitefish, in which they identified another novel form of GnRH, the wfGnRH, and another recent variant has been isolated from medaka named mdGnRH, which has been characterized as a homolog of sbGnRH and mGnRH types according to phylogenetic study. Interestingly, exceptional to these teleosts, eel has a single form of GnRH, comparable to mGnRH (Dufour et al., 1982).

All these discoveries indicate that the species-specific form of GnRH is relevant to reproduction based on its distribution in hypothalamus, preoptic area-hypothalamus (POA-H) and innervations to pituitary. Species-specific form of GnRH showed correlative changes to gonado-somatic index and annual reproductive cycle (Senthilkumaran et al., 1999). Interestingly, after having identified multiple forms of GnRH, multiple receptor subtypes have also been identified in several fish species (Illing et al., 1999; Okubo et al., 2000; Bogerd et al., 2002; Jodo et al., 2003; Moncaut et al., 2005; Parhar et al., 2005; Flanagan et al., 2007; Joseph et al., 2012). These reports together indicated that GnRH, especially extra-hypothalamal GnRH, for e.g. the GnRH localized in telencephalon, thalamus and other brain regions play essential roles in spawning, migration and neuromodulation (Kim et al., 1995; Kitani et al., 2003; Kawai et al., 2010; Onuma et al., 2010).

FSH and LH

The pituitary GTHs, FSH and LH, are heterodimeric glycoprotein-hormones consisting of a non-covalently linked common glycoprotein-hormone α-subunit (GSUα; also shared with thyroid-stimulating hormone) and a specific β-subunit (FSHβ or LHβ) for conferring their biological activity through G-protein coupled receptors (Pierce and Parsons, 1981; Aizen et al., 2012). In recent years, the availability of molecular cloning techniques allowed the isolation of the genes coding for the GTH subunits in 56 fish species representing at least 14 teleost orders (Levavi-Sivan et al., 2010). Two types of GTHs, different for structure and chemical roles, were identified in the teleosts: GTH-I (or FSH), involved in the initial stages of gametogenesis (vitellogenesis), and GTH-II (or LH), which implement final oocyte maturation, ovulation, spermiogenesis and spermiation (Mylonas et al., 2010). Interestingly, in catfish only one form of GTH could be purified homologous to LH (Schulz et al., 1995) and which shows multiple peaks corresponding to both GTH I and II (Kirubagaran et al., 2005). However, later reports could identify two GTH genes in catfish, yet purification of FSH was not feasible by large. Nevertheless, siluriformes tend to evolutionarily diverge by having a single gonadotropin molecule, LH to entrain complete gonadal function. Many studies are still underway to debate on the presence of FSH in this species. When GTH expression was examined in goldfish, both GTH I ß and II ß mRNA levels in the pituitary gland increased in accordance with gonadal development, unlike the sequential expression of GTH subunits in salmonids. FSH mRNA levels increased in association with gonadal development in males, whereas the levels remain low throughout sexual maturation in females. On the contrary, LH mRNA levels of both sexes were maintained at high levels from the beginning of gametogenesis to spawning phase, more specifically in females (Gen et al., 2000). Thus, the role of FSH and LH showed distinct differences across species but at times LH is capable of initiation and maintenance of gonadal function.

Role of Monoamines in Entraining GnRH and GTH

In teleosts, pulsatile release of GnRH vis-à-vis secretion of GTHs from pituitary is under the control of stimulatory and inhibitory central monoaminergic systems at the level of POA-

H and pituitary (Goos et al., 1999). Among the central MAs, DA is inhibitory while 5-HT, NE and GABA are stimulatory to LH secretion (Peter et al., 1991; Senthilkumaran and Joy, 1996; Joy et al., 1998; Goos et al., 1999; Senthilkumaran et al., 2001). A number of studies indicated the involvement of MA-related enzymes such as tryptophan hydroxylase (tph), monoamine oxidase (MAO) and catecholamine-o-methyl transferase (COMT) in the steroid-mediated negative feedback regulation of gonadal steroids on GnRH-GTH axis (Saligaut et al., 1992a; Larson et al., 2003). More reports on this line demonstrated the existence of differential negative feedback exerted by estradiol-17ß (E_2) on GnRH-GTH axis and hypothalamic MA metabolism (Saligaut et al., 1992b; Levavi-Sivan et al., 2006). In light of this, it is conceivable that MAs play a prominent role to orchestrate GnRH-GTH axis. Let us now review different MAs in the perspective of GnRH-GTH regulation.

5-HT is well known for modulating the release of GnRH and GTH in teleosts and appears to be an integral part of the pathway mediating neuroendocrine events (Peter et al., 1991; Hery et al., 1997; Goos et al., 1999; Senthilkumaran et al., 2001; Wong et al., 2004; Nakamura and Hasegawa, 2007). Hypothalamic 5-HT is well known to induce GnRH release through 5-HT receptor subtype-2 and GTH release from pituitary (Somoza and Peter, 1991; Khan and Thomas 1994; Trudeau, 1997). Number of studies in teleosts indicates the involvement of 5-HT and MAO in the exertion of negative feedback by E_2 to modulate GTH release (Khan and Joy, 1988; Senthilkumaran and Joy, 1995b; Yaron et al., 2003). 5-HT is also considered as an important MA as it transduces environmental cues such as photoperiod and temperature to modulate GnRH-GTH axis in response to reproductive cycle (Khan and Thomas, 1994; Tsai et al., 2002). Another stimulatory MA, NE is known for its involvement in GnRH-GTH release through specific NE receptors (Chang et al., 1984; Chang et al., 1991; Peter et al., 1991; Yu and Peter, 1992). Hypothalamic NE and DA also shows changes like that of 5-HT during negative feedback effects by E_2 and also after artificial modulation of photothermal conditions (Senthilkumaran and Joy, 1995a). Thus, both 5-HT and NE play significant roles in the regulation of GnRH-GTH axis. Intriguingly, COMT also showed changes correlative to NE like that of MAO (Saligaut et al., 1992a; Sebert et al., 2008). Likewise 5-HT, NE is also capable to initiate and complete gonadal sex reversal as well as color change perhaps through H-H-G axis. Unlike 5-HT or NE, GABA showed both stimulatory and inhibitory tone on GTH release in the Atlantic croaker (Khan and Thomas 1999). Interestingly, no inhibitory DA tone influence on GnRH-GTH release is evident in this species (Copeland and Thomas, 1989). On the other hand, GABA was stimulatory via receptors in inducing sbGnRH release *in vitro* in the POA-H slices and pituitary fragments of red seabream (Senthilkumaran et al., 2001). In goldfish also, GABA is known for its stimulatory influence on GnRH-GTH axis (Kah et al., 1992; Trudeau et al., 1993; Trudeau, 1997). Despite these studies, no attempt was made to stimulate GnRH release using stimulatory MAs, NE or GABA, yet successful spawning trials were seen using 5-HT in combination with GnRH analogue (GnRHa) in teleosts (Goos et al., 1987; Joy et al., 1998; Hassin et al., 2000). While GnRH is considered as the major hypothalamic factor controlling pituitary GTHs through excitatory MAs in teleosts and most other vertebrates, its stimulatory actions may be opposed by the potent inhibitory actions of DA in teleosts (Peter et al., 1991; Goos et al., 1999; Dufour et al., 2010). Dufour et al. (2010) also explained that this dual neuroendocrine control of reproduction by GnRH and DA has been demonstrated in various, but not all, adult teleosts, where DA exerts an inhibitory tone in the neuroendocrine regulation of gametogenesis especially during final oocyte maturation and ovulation in females, and spermiation in males. In addition, DA may also play an inhibitory role during the early stages of gametogenesis in some teleost species, and thus interact with GnRH in the control of puberty (Dufour et al., 2010). This concept has major implications for inducing spawning in aquaculture. In this context, the important discovery was made by understanding the neuroendocrine role of GnRH and inhibitory tone of DA on GnRH-GTH release from several teleosts (De Leeuw et al., 1986; Peter et al., 1991; Yu and Peter, 1992; Goos et al., 1999; Vacher et al., 2000). Exploiting this phenomenon, a new technique known as 'LinPe' has been developed for several teleosts to induce spawning in aquaculture wherein GnRHa and DA receptor subtype-2 blocker, domperidone or pimozide were essentially used and formulated as ovaprim (Billard et al., 1984; Peter et al., 1988; Halder et al., 1991; Peter et al., 1991; Alok et al., 1993; Sharaf, 2012). Further studies on this line clearly indicated that changes in preovulatory GTH surge in response to GnRH or ovaprim implantation is brought about by the modulation of MAs and MA related enzymes during pre/periovulation (Peter et al., 1991; Joy et al., 1998; Goos et al., 1999; Sharaf, 2012). Thus, both GnRH-GTH and MA coordination is essential to implement maturation and spawning in teleosts (Peter et al., 1991; Goos et al., 1999). Surprisingly, though DA inhibitory tone is evident in eel (Dufour et al., 1988), favourable spawning response could be induced by implantation of a single osmotic pump loaded with human chorionic gonadotropin (Kagawa et al., 2009).

Induction of Puberty: Pivotal Role of GnRH-GTH Axis

The process of gonadal maturation and reproductive competence involve a number of internal as well as external factors which provide crucial cues for reproductive success for the organism, and in most vertebrates, GnRH seems to play a prominent role in puberty through differential expression of gonadotropin receptors (Kumakura et al., 2004; Hildahl et al., 2011b). Development of GnRH cells followed by increased GnRH levels consequently elevates GTHs, thereby setting up gonadal physiology for puberty and reproduction (Dubois et al., 2000; Whitlock et al., 2006). In teleosts as well as other vertebrates, sexual maturation is marked by the transition of an inactive H-H-G axis to a functional state, and studies have indicated variation in the regulation profiles of GTHs which may also be species-specific (Goos et al., 1999; Chang et al., 2000). Such a phenomenon could also be evident during gonadal recrudescence of annual breeders (Peter et al., 1991; Goos et al., 1999; Kumar and Trant, 2004). Studies on sexually immature female red seabream to identify the pubertal development of the brain pituitary gonadal axis showed increased levels of LH-β subunit in pituitary and serum, in response to implanted cholesterol pellets containing GnRHa (Kumakura et al., 2004). Levels of both brain and pituitary sbGnRH increased from immature phase and reached a peak in spawning phase in red seabream, revealing the rise in gonado somatic index and vitellogenesis (Senthilkumaran et al., 1999). In black porgy, *Acanthopagrus schlegeli*, the role of GnRH was investigated in immature fish treated with GnRHa by examining the mRNA levels of GTH subunits (GTH-α, FSH-β, and LH-β) in the pituitary as well as plasma E_2 level, which showed an increase in of both correlates (An et al., 2008). The comparison of GnRH and GTH levels in the pituitary at various developmental stages also implied their role in gonadal development and maturation. Three forms of GnRH expressed in the pituitary of male European sea bass, *Dicentrarchus labrax* showed significant changes during the stages of sex differentiation and also in early puberty, i.e. the first seasonal spawning (Rodriguez et al., 2000). Gonadal development leading to puberty in male catfish, *Clarias gariepinus* involves the increase in basal plasma levels of 11-ketotestosterone, testosteone and pituitary GnRH sensitivity (Schulz et al., 1995) which suggested the presence of a positive signal that enhances pituitary function during puberty, and might be linked to the action of steroids in GTH-II production (Blazquez et al., 1998). The development of GnRH neuronal systems in the brain of pubertal catfish was found to be more advanced than the pre-pubertal or juvenile stages (Dubois et al., 1998). The GnRH levels of Atlantic cod, *Gadus morhua*, (gmGnRH) in the brain showed a significant increase during spawning, and was found to be expressed in the pituitary at pubertal stages (Hildahl et al., 2011a). Similarly, in red seabream, rise in sbGnRH is considered an important physiological event for the onset of puberty (Kumakura et al., 2003). In a step further to these research reports, a superior strategy for successful breeding of sexually immature marine teleost during non-breeding phase by modulating environmental variables with GnRH implantation by osmotic pump has been demonstrated (Kanemaru et al., 2012).

The importance on the gonadotropins FSH and LH varies with respect to species, and several studies have shown their different roles during sexual maturation. Several studies have shown the more prominent role of FSH in pubertal stages of some salmonids, where the plasma levels of FSH in blood and FSH mRNA were found to be much higher than LH during puberty in coho salmon, masu salmon and rainbow trout (Dickey and Swanson, 2000; Furukuma et al., 2008; Sambroni et al., 2012; Okuzawa et al., 2002). In female striped bass, the FSHβ mRNA level in pituitary was significantly higher at the pre-pubertal stages, followed by an increase of LHβ expression only at late puberty (Hassin et al., 1999). However, ontogenic studies of FSH and LH expression in female zebrafish brain and pituitary showed the abundant expression of FSHβ before puberty while LH was found to be much higher at the onset, and expression of both was found to increase during pubertal stages which suggests a potential role of LH in the onset of puberty in this species (Chen and Ge, 2012). Plasma levels of LH also showed an increase during early puberty in the European seabass which also suggests its significant role in the onset of puberty (Rodriguez et al., 2000). These findings concur that both FSH and LH under the regulation of GnRH, have relative and different roles during the onset of puberty, but may vary within species, and may even show sexual dimorphisms in their expression profiles within the same species (Tanaka et al., 1993; Gen et al., 2000). Recent evidences indicate that Kisspeptin through GPR54 expression play a prominent role by regulating GnRH in neurons vis-à-vis gonadotropin release, thereby Kisspeptin is also known as gatekeeper of puberty and reproduction (Filby et al., 2008; Zohar et al., 2010).

Relevance of GnRH-GTH Axis during Sexual Development/Sex Differentiation

In the previous section, we had explained lot of interesting findings that describe the precise role of GnRH and GTH in relation to puberty and gamete maturation. In this section, we are highlighting some important findings indicating changes in GnRH and GTH in response to gonadal sex differentiation. In this perspective, it is important to analyze the events of gonadal differentiation with the onset of GnRH-GTH axis. Occurrence of GnRH neurons in POA-H coincides with gonadal sex differentiation (Soga et al., 2005). Miranda et al. (2003) also showed changes in POA-H in response to temperature during the sensitive period of temperature sex determination in the pejerrey, *Odontesthes bonariensis*, suggesting a plausible role of the CNS and GnRH in the process of sex determination. Corroborating to these studies, there are reports in sex changing fish and hermophrodites quoting changes in the levels of POA-H GnRH during crucial stages of sex reversal (Lee et al., 2001; Du et al., 2005). Further, minor variations in GnRH levels and localization may also influence differences in sexual behaviour. In this perspective, Swapna et al. (2008) showed changes in GnRH immunolocalization in the XX and XY Nile tilapia. This study revealed the presence of sbGnRH immunoreactive neurons as early as 5 and 15 days post hatch (dph) in the brains of XY and XX tilapia, respectively. Higher immunoreactivity was detected in the brain and pituitary of XY tilapia, however, qualitative drop was evident between 20 and 30 dph. On the other hand in XX fish, authors observed a steady localization of sbGnRH in the brain and pituitary from 15 dph onwards. The apparent reason for the differential localisation of GnRH between males and females might have been due to factors that regulate GnRH, for e.g., MAs or it is a response to gonadal differentiation, as explained by the authors and also from our viewpoint. This needs to be analyzed extensively with additional experiments with different animal models. Earlier, Pandolfi et al., (2002) showed differential distribution and ontogeny of three distinct GnRH during the period of sex differentiation in the South American cichlid fish. The authors could even identify GnRH neurons three days after fertilization, yet, they could not perform the analysis sex-wise. Corroborative to these reports, expression of FSHβ and LHβ could be observed from 5 dph in the Nile tilapia (Fan et al., 2003). In addition, immunolocalization of both FSH and LH was analyzed in tilapia using heterologous antisera (Sakai et al., 2005). This report revealed that LH immunoreactivity was found to be lower in the pituitary of XX fish when compared to XY tilapia. GTH I and II in pejerry also showed differential localization and the cells appeared even before histological gonadal differentiation where temperature sex differentiation is operative (Miranda et al., 2001). Thus, GnRH and GTH have some association to gonadal ontogeny, but the significance of these correlates in gonadal sex differentiation is yet to be established. In addition to GnRH and GTH, certain molecular level changes are also seen in MA systems. This information will be reviewed in the subsequent sections to establish their relative networks.

Monoamines Involvement in Sexual Development

MAs play an important role in reproduction and sexual behaviour in most vertebrates by being the first endogenous chemical signals in the regulation of the H-H-G axis (Larson et al., 2003). Although MAs were considered to be key molecules in sex reversal of teleosts as they play an important role in the regulation of the H-H-G axis, their direct involvement in sexual development is still not understood. Studies on this line revealed a certain role for tryphtophan hydroxylase (Tph) vis-à-vis 5-HT in relation to male and female brain development (Tsai et al., 2000; Sudhakumari et al., 2010; Raghuveer et al., 2011). In tilapia, the blockade of Tph by *para-chlorophenylalanine* (p-CPA) during critical periods of sexual differentiation resulted in the reduction of brain 5-HT levels that caused an increased female sex population similar to that of ethinyl estradiol (EE_2) treatment (Tsai et al., 2000). This implied a significant role of Tph in the process of male brain sex differentiation which was established by Sudhakumari et al. (2010) using semi-quantitative RT-PCR and *in situ* hybridisation for Tph in brain. Studies on air-breathing catfish, *Clarias gariepinus* Tph mRNA level in male and female brains during early developmental stages also showed gender specific differences where male brains showed a significantly higher level of Tph expression as compared to the females (Raghuveer et al., 2011). Ontogenic brain studies also showed higher transcript level in male brains at 50 and 75 dph which are critical periods for gonadal differentiation and development for catfish (Raghuveer et al., 2011). Treatment of the juvenile catfish with methyltestosterone (MT) and p-CPA gave rise to skewed female:male population in which MT produced a higher number of males, and p-CPA gave rise to more females which was verified by real time PCR analysis of Tph transcripts (Raghuveer et al., 2011). These findings might implicate that the process of gonadal sex differentiation acts in concert with brain sex differentiation, with a possible involvement of MA systems. Earlier studies on this line revealed changes in 5-HT and brain aromatase activity up on administration of sex steroids during the critical period of sexual differentiation in tilapia, *Oreochromis mossambicus* (Tsai et al., 2000). Further, the effects of E_2 on the secretion and turnover of serotonin in the hypothalamic fragments of male tilapia demonstrate that the moderate dose of E_2 increases the serotonin activity by increasing the serotonin concentration, whereas the high dose of E_2 increases the serotonin activity by increasing the ratio of 5-hydroxyindoleacetic acid and 5-HT (Tsai and Wang, 1997).

Involvement of Aromatase

The relevance of brain aromatase and estrogens as possible cues for gonadal development is raised in several fish species having thermo-labile sex determination (Blázquez and Somoza, 2010). Although the variations in the levels of sex steroids, especially aromatases (cyp19a1b, brain form), during developmental stages of both sexes in teleosts have not been fully established, studies have shown a correlation between their levels and the process of sexual differentiation and development (Rasheeda et al., 2010). This can even be correlated with differential localization patterns of GnRH in XX and XY tilapia (Swapna et al., 2008). In tilapia, the role of aromatase along with estrogen receptors was studied in the ontogenic expression influenced by temperature factors, which implicated the role of aromatase in the process of brain sex differentiation (Tsai et al., 2003). Studies on the critical timeline of the expression of brain aromatase in zebrafish, under the influence of pharmacological steroids like EE_2 and 17α-methyltestosterone, have shown a significant up regulation of aromatase gene expression which may be associated with sexual differentiation during developmental stages (Trant et al., 2001). These findings correlate with that of the studies done on the roach, *Rutilus rutilus* in which the expression levels of brain aromatase in response to their exposure to EE_2 during early life showed marked effects which also suggests their association with sex differentiation in the roach (Lange et al., 2008). In rainbow trout, a distinct and significant dimorphic pattern of brain aromatase expression was shown, with higher levels found in male brain during the period of gonadal morphology development (Vizziano-Cantonnet et al.,

2011). Expression analysis of brain aromatase in Nile tilapia at the time of gonadal differentiation showed a steady level of expression in testis, in contrast to females where it reappeared only at later stages of ovarian differentiation (Chang et al., 2005; Sudhakumari et al., 2005). It was reported that, in teleosts cyp19a1b is regulated by FTZ-F1 and FOXL2 based on their synchronous expression pattern in the brain (Rasheeda et al., 2010; Sridevi et al., 2012). Interestingly, FTZ-F1 expression is also found to be higher in females. This indicates that brain aromatase and its transcriptional regulator FTZ-F1 may play a certain role in ovarian differentiation (Sridevi et al., 2011). Nevertheless, cyp19a1a and its transcriptional regulator Ad4BP/SF1 (Yoshiura et al., 2003) and co-regulator Foxl2 (Wang et al., 2007) might play decisive roles for ovarian differentiation considering their sexual dimorphism. From these findings, it might be a plausible assumption that brain aromatase expression might either be a cause or participant in the process of sex differentiation, perhaps through MAergic system as well, and changes in brain aromatase and FTZ-F1 may indicate a cause or consequence of gonadal differentiation which remains to be elucidated.

Conclusion

The role of GnRH, GTH and MA are very well established to entrain reproductive axis by regulating the events of gametogenesis (Peter et al., 1991; Goos et al., 1999; Yaron *et al.*, 2003, Zohar et al., 2010). Regardless of this, the role of GnRH and GTH in initiating gonadal development is always debatable mainly due to late occurrence of GnRH-GTH systems (Sakai et al., 2005; Swapna et al., 2008). Further, the appearance gonadal specific marker genes usually precede GnRH and GTH (Nagahama, 2005). This profoundly raises a question whether gonadal or brain sex differentiation is the beginning of sexual development. Nevertheless, in serial sex changing species, the sexual behavioural changes precede gonadal sex differentiation (Grober and Sunobe, 1996; Kobayashi et al., 2009). Similarly, in species that follow temperature dependent sex determination, GnRH and GTH have been shown to play very prominent roles. In addition, MAs also play very important roles during the early differentiation and even blockade of 5-HT mimic estradiol and causes sex reversal (Tsai et al., 2000; Raghuveer et al., 2011). To explain this further, based on our studies, we provided sex-wise schematic representation of GnRH, GTHs, 5-HT and Tph occurrence during the early development of Tilapia and catfish (figure 1) in figures (Sakai et al., 2005; Swapna et al., 2008; Sudhakumari et al., 2010; Raghuveer et al., 2011). Taken together, this area of neuroendocrine research is definitely an emerging and prominent area. Studies on this line warrant better understanding towards gonadal and brain sex differentiation. The knowledge gained in these studies can be correlated to higher mammals as the concept of MA and GnRH-GTH regulation prevails in those systems as well. Future research can be focussed on the utilization of specific blockers or modern molecular biology strategies that target GnRH, GTH or MA or even MA related enzymes. Since MAs are also neurotransmitters, careful planning of experiments is essential considering their neuromodulatory role. Our review may seed new thoughts to understand gonadal sex differentiation by looking at factors which are non-gonadal such as brain or pituitary. Such an approach might solve the discrepancies that still exist on this contentious matter, and will contribute to our ever growing knowledge on gonadal factors that comprehend the complex events during gonadal sex determination/differentiation and maturation including puberty.

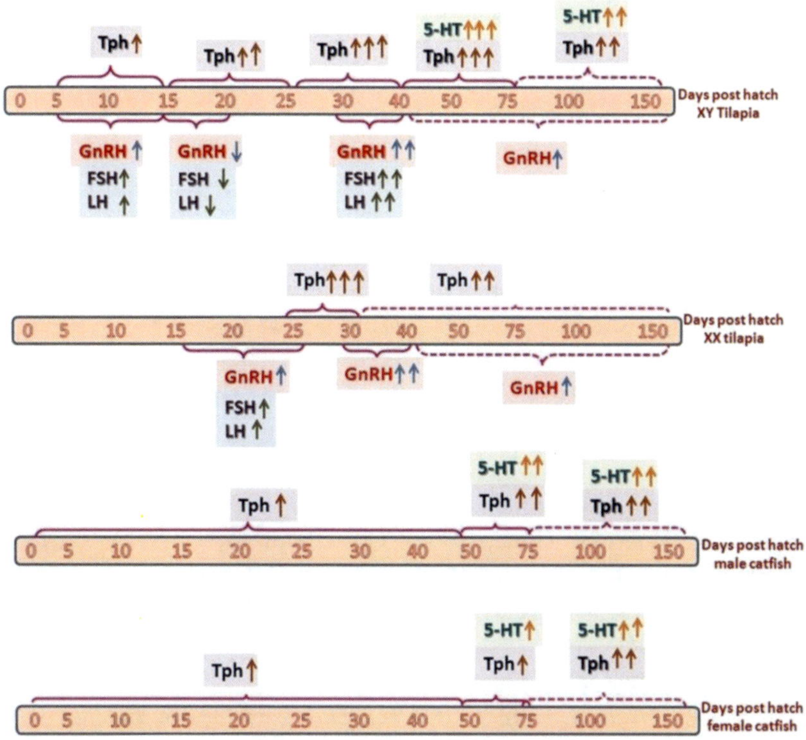

Figure 1: Schematic representation of sex-wise abundance of gonadotropin –releasing hormone, gonadotropins, tryptophan hydroxylase and serotonin during early sexual development in Tilapia and Catfish (Note: Sex-wise ontogeny studies were not performed till 50dph in catfish).

Acknowledgments

We acknowledge the Grant-in-Aids to B.S. from University Grants Commission (F.30-195/2004 [SR]), India and Women Scientist Scheme (WOS-A) from Department of Science and Technology, India to Dr. C. C. Sudhakumari. YP is grateful to Council of Scientific and Industrial Research, India for Junior and senior research fellowship. CL is grateful to Department of Biotechnology, India for Junior Research fellowship.

References

Adams, B. A., Vickers, E. D., Warby, C., Park, M., Fischer, W. H., Grey Craig, A., Rivier, J. E. and Sherwood, N. M. (2002). Three forms of gonadotropin-releasing hormone, including a novel form, in a basal salmonid, *Coregonus clupeaformis*. *Biol. Reprod.,* 67, 232–239.

Aizen, J., Kobayashi, M., Selicharova, I., Sohn, Y. C., Yoshizaki, G. and Levavi-Sivan, B. (2012). Steroidogenic response of carp ovaries to piscine FSH and LH depends on the reproductive phase. *Gen. Comp. Endocrinol.,* 178, 28-36.

Alok, D., Krishnan, T., Talwar, G. P. and Grag, L. C. (1993). Induced spawning of catfish, *Heteropneustes fossilis* (Bloch), using D-Lys6 salmon gonadotropin releasing hormone analogue. *Aquaculture,* 115, 159–167.

An, K. W., Nelson, E. R., Habibi, H. R. and Choi, C. Y. (2008). Molecular characterization and expression of three GnRH forms mRNA during gonad sex-change process, and effect of GnRHa on GTH subunits mRNA in the protandrous black porgy (*Acanthopagrus schlegeli*). *Gen. Comp. Endocrinol.*, 159, 38-45.

Billard, R., Reinaud, P., Hollebecq, M. G. and Breton, B. (1984). Advancement and synchronization of spawning in Salmo gairdneri and *S. trutta* following administration of LRH-A combined or not with pimozide. *Aquaculture,* 43, 57– 66.

Blázquez, M., Bosma, P. T., Chang, J. P., Docherty, K. and Trudeau, V. L. (1998). Gamma-aminobutyric acid up-regulates the expression of a novel secretogranin-II messenger ribonucleic acid in the goldfish pituitary. *Endocrinology*, 139, 4870-4880.

Blázquez, M. and Somoza, G. M. (2010). Fish with thermolabile sex determination (TSD) as models to study brain sex differentiation. *Gen Comp Endocrinol,* 166, 470-477.

Bogerd, J., Diepenbroek, W. B., Hund, E., van Oosterhout, F., Teves, A. C., Leurs, R. and Blomenröhr, M. (2002). Two gonadotropin-releasing hormone receptors in the African catfish: no differences in ligand selectivity, but differences in tissue distribution. *Endocrinology,* 143, 4673-4682.

Bogerd, J., Li, K. W., Janssen-Dommerholt, C. and Goos, H. (1992). Two gonadotropin-releasing hormones from African catfish (*Clarias gariepinus*). *Biochem Biophys Res Commun,* 187, 127-134.

Brodie, T._D. and Crowley, W._F. Jr. (1985). Neuroendocrine control of reproduction and its therapeutic manipulation with GnRH and its analogs. *Int. J. Fertil.,* 30, 66-70, 74-75.

Carolsfeld, J., Powell, J. F., Park, M., Fischer, W. H., Craig, A. G., Chang, J. P., Rivier, J. E. and Sherwood, N. M. (2000). Primary structure and function of three gonadotropin-releasing hormones, including a novel form, from an ancient teleost, herring. *Endocrinology,* 141, 505-512.

Chang, J. P., Johnson, J. D., Van Goor, F., Wong, C. J. H., Yunker, W. K., Uretsky, A. D., Taylor, D., Jobin, R. M., Wong, A. O. L., and Goldberg, J. I. (2000). Signal transduction mechanisms mediating secretion in goldfish gonadotropes and somatotropes. *Biochem. Cell Biol.,* 78, 139–153.

Chang, J. P., MacKenzie, D. S., Gould, D. R. and Peter, R. E. (1984). Effects of dopamine and norepinephrine on the in vitro spontaneous and gonadotropin releasing hormone-induced gonadotropin release by dispersed cells or fragments of the goldfish pituitary. *Life Sci.,* 35, 2027–2033.

Chang, J. P., Van Goof, F. and Acharya, S. (1991). Influence of norepinephrine, and adrenergic agonists and antagonist on gonadotropin secretion from dispersed pituitary cells of goldfish, *Carassius auratus*. *Neuroendocrinology,* 54, 202–210.

Chang, X., Kobayashi, T., Senthilkumaran, B., Kobayashi-Kajura, H., Sudhakumari, C. C. and Nagahama, Y. (2005). Two types of aromatase with different encoding genes, tissue distribution and developmental expression in Nile tilapia (*Oreochromis niloticus*). *Gen. Comp. Endocrinol.,* 141, 101-115.

Chen, W. and Ge, W. (2012). Ontogenic expression profiles of gonadotropins (fshβ and lhβ) and growth hormone (gh) during sexual differentiation and puberty onset in female zebrafish. *Biol. Reprod.,* 86, 73.

Copeland, P. A. and Thomas, P. (1989). Control of gonadotropin release in the Atlantic croaker (*Micropogonias undulatus*): evidence for lack of dopaminergic inhibition. *Gen. Comp. Endocrinol.,* 74, 474-483.

De Leeuw, R., Goos, H. J. Th. and Van Oordt, P. G. W. J. (1986). The dopaminergic inhibition of the gonadotropin-releasing hormone induced gonadotropin release: an in vitro study with fragments and cell suspensions from pituitaries of African catfish, *Clarias gariepinus* (Burchell). *Gen. Comp. Endocrinol.,* 63, 171–177.

Dickey, J. T. and Swanson, P. (2000). Effects of salmon gonadotropin-releasing hormone on follicle stimulating hormone secretion and subunit gene expression in coho salmon (*Oncorhynchus kisutch*). *Gen. Comp. Endocrinol.,* 118, 436-449.

Du, J. L., Lee, Y. H., Yueh, W. S. and Chang, C. F. (2005). Seasonal profiles of brain and pituitary gonadotropin-releasing hormone and plasma luteinizing hormone in relation to sex change of protandrous black porgy, *Acanthopagrus schlegeli. Biol. Reprod.,* 72, 922-931.

Dubois, E. A., Florijn, M. A., Zandbergen, M. A., Peute, J. and Goos, H._J. (1998). Testosterone accelerates the development of the catfish GnRH system in the brain of immature African catfish (*Clarias gariepinus*). *Gen. Comp. Endocrinol.,* 112, 383-393.

Dubois, E. A., Zandbergen, M. A., Peute, J., Hassing, I., van Dijk, W., Schulz, R. W. and Goos H. J. (2000). Gonadotropin-releasing hormone fibers innervate the pituitary of the male African catfish (*Clarias gariepinus*) during puberty. *Neuroendocrinology,* 72, 252-262.

Dufour, S., Lopez, E., Le Menn, F., Le Belle, N., Baloche, S. and Fontaine, Y. A. (1988). Stimulation of gonadotropin release and ovarian development by administration of a gonadoliberin agonist and of dopamine antagonists in female silver eel pretreated with estradiol. *Gen. Comp. Endocrinol.,* 70, 20–30.

Dufour, S., Pasqualini, C., Kerdelhué, B. and Fontaine, Y. A. (1982). Presence and distribution of radioimmunoassayable LHRH in the European eel, *Anguilla anguilla. Neuropeptides,* 3, 159-171.

Dufour, S., Sebert, M. -E., Weltzien, F. -A., Rousseau, K. and Pasqualini, C. (2010). Neuroendocrine control by dopamine of teleost Reproduction. *J Fish Biol, 76*, 129–160.

Fan, H. G., Wang, D. S., Kobayashi, T., Senthilkumaran, B., Sudhakumari, C .C., Yoshikuni, M. and Nagahama, Y. (2003). Molecular cloning of the three gonadotropin subunits and early expression of FSHβ during sex differentiation in the Nile tilapia (*Oreochromis niloticus*). *Fish. Physiol. Biochem.,* 28, 143–144.

Filby, A. L., van Aerle, R., Duitman, J. and Tyler, C. R. (2008). The kisspeptin/gonadotropin-releasing hormone pathway and molecular signaling of puberty in fish. *Biol. Reprod.,* 78, 278-289.

Flanagan, C._A., Chen, C._C., Coetsee, M., Mamputha, S., Whitlock, K._E., Bredenkamp, N., Grosenick, L., Fernald, R. D. and Illing, N. (2007). Expression, structure, function, and evolution of gonadotropin-releasing hormone (GnRH) receptors GnRH-R1SHS and GnRH-R2PEY in the teleost, *Astatotilapia burtoni. Endocrinology,* 148, 5060-5071.

Furukuma, S., Onuma, T., Swanson, P., Luo, Q., Koide, N., Okada, H., Urano, A. and Ando, H. (2008). Stimulatory effects of insulin-like growth factor 1 on expression of gonadotropin subunit genes and release of follicle-stimulating hormone and luteinizing hormone in masu salmon pituitary cells early in gametogenesis. *Zoolog. Sci.,* 25, 88-98.

Gen, K., Okuzawa, K., Senthilkumaran, B., Tanaka H., Moriyama, S. and Kagawa H. (2000). Unique Expression of Gonadotropin-I and -II Subunit Genes in Male and Female Red Seabream (*Pagrus major*) During Sexual Maturation. *Biol. Reprod.,* 63, 308–319.

Goos, H. J. Th., Joy, K. P., de Leeuw, R., Van Oordt, P. G. W. J., Van Delft, A. M. L. and Gielen, J. Th. (1987). The effect of luteinizing hormone releasing hormone analogue (LHRHa) in combination with different drugs with anti-dopamine and anti-serotinin properties on gonadotropin release and ovulation in the African catfish, *Clarias gariepinus*. *Aquaculture,* 63, 143–156.

Goos, H. J., Senthilkumaran, B. and Joy, K. P. (1999). Neuroendocrine integrative mechanisms in the control of gonadotropin secretion in teleosts. In K. P. Joy, A. Krishna and C. Haldar, (Eds.), Comparative Endocrinology and Reproduction. Narousa Publishing House. (pp. 113–136). New Delhi, India.

Grober, M._S. and Sunobe, T. (1996). Serial adult sex change involves rapid and reversible changes in forebrain neurochemistry. *Neuroreport,* 7, 2945-2949.

Halder, S., Sen, S., Bhattacharaya, S., Ray, A. K., Ghosh, A. and Jhingran, A.G. (1991). Induced spawning of Indian major carps and maturation of a perch and a catfish by murrel gonadotropin releasing hormone, pimozide and calcium. *Aquaculture,* 97, 373–382.

Hassin, S., Claire, M., Holland, H. and Zohar, Y. (1999). Ontogeny of follicle-stimulating hormone and luteinizing hormone gene expression during pubertal development in the female striped bass, *Morone saxatilis* (Teleostei). *Biol. Reprod.,* 61, 1608-1615.

Hassin, S., Claire, M., Holland, H., and Zohar, Y. (2000). Early maturity in the male striped bass, Morone saxatilis: follicle-stimulating hormone and luteinizing hormone gene expression and their regulation by gonadotropin-releasing hormone analogue and testosterone. *Biol. Reprod.,* 63, 1691-1697.

Hery, M., Francois-Bellan, A. M., Hery, F., Deprez, P. and Becquet, D. (1997). Serotonin directly stimulates luteinizing hormone-releasing hormone release from GT1 cells via 5-HT7 receptors. *Endocrine,* 7, 261–265.

Hildahl, J., Sandvik, G. K., Edvardsen, R. B., Fagernes, C., Norberg, B., Haug, T._M. and Weltzien, F. A. (2011a). Identification and gene expression analysis of three GnRH genes in female Atlantic cod during puberty provides insight into GnRH variant gene loss in fish. *Gen. Comp. Endocrinol.,* 172, 458-467.

Hildahl, J., Sandvik, G. K., Edvardsen, R. B., Norberg, B., Haug, T. M., and Weltzien, F. A. (2011b). Four gonadotropin releasing hormone receptor genes in Atlantic cod are differentially expressed in the brain and pituitary during puberty. *Gen. Comp. Endocrinol.,* 173, 333-345.

Illing, N., Troskie, B._E., Nahorniak, C. S., Hapgood, J. P., Peter, R. E. and Millar, R. P. (1999). Two gonadotropin-releasing hormone receptor subtypes with distinct ligand selectivity and differential distribution in brain and pituitary in the goldfish (*Carassius auratus*). *Proc. Natl. Acad. Sci. U S A,* 96, 2526-2531.

Jodo, A., Ando, H. and Urano, A. (2003). Five different types of putative GnRH receptor gene are expressed in the brain of masu salmon (*Oncorhynchus masou*). *Zoolog. Sci.,* 20, 1117-1125.

Joseph, N._T., Aquilina-Beck, A., MacDonald, C., Decatur, W. A., Hall, J. A., Kavanaugh, S. I. and Sower S. A. (2012). Molecular cloning and pharmacological characterization of two novel GnRH receptors in the lamprey (*Petromyzon marinus*). *Endocrinology,* 153, 3345-3356.

Joy, K. P., Senthilkumaran, B. and Sudhakumari, C._C. (1998). Periovulatory changes in hypothalamic and pituitary monoamines following GnRH analogue treatment in the catfish *Heteropneustes fossilis*: a study correlating changes in plasma hormone profiles. *J. Endocrinol.,* 156, 365-372.

Kagawa, H., Kasuga, Y., Adachi, J., Nishi, A., Hashimoto, H., Imaizumi, H. and Kaji, S., (2009). Effects of continuous administration of human chorionic gonadotropin, salmon pituitary extract, and gonadotropin-releasing hormone using osmotic pumps on induction of sexual maturation in male Japanese eel, *Anguilla japonica. Aquaculture,* 296, 117–122.

Kah, O., Trudeau, V. L., Sloley, B._D., Chang, J. P., Dubourg, P., Yu, K._L. and Peter, R._E. (1992). Influence of GABA on gonadotrophin release in the goldfish. *Neuroendocrinology,* 55, 396-404.

Kanemaru,T., Nakamura, M., Murata, R., Kuroki, K., Horie, H., Uchida, K., Senthilkumaran, B., and Kagawa, H. (2012). Induction of sexual maturation of the female honeycomb grouper, *Epinephelus merra,* in the non-breeding season by modulating environmental factors with GnRH analogue implantation. *Aquaculture,* 358–359, 85–91

Kavanaugh, S. I., Nozaki, M. and Sower, S. A. (2008). Origins of Gonadotropin-Releasing Hormone (GnRH) in Vertebrates: Identification of a Novel GnRH in a Basal Vertebrate, the Sea Lamprey. *Endocrinology,* 149, 3860–3869.

Kawai, T., Abe, H., Akazome, Y., and Oka Y. (2010). Neuromodulatory Effect of GnRH on the Synaptic Transmission of the Olfactory Bulbar Neural Circuit in Goldfish, *Carassius auratus. J. Neurophysiol.,* 104, 3540–3550.

Khan, I. A. and Joy, K. P. (1988). Seasonal and daily variations in hypothalamic monoamine levels and monoamine oxidase activity in the teleost *Channa punctatus* (Bloch). *Chronobiol. Int., 5*, 311-316.

Khan, I. A. and Thomas, P. (1994). Seasonal and daily variations in the plasma gonadotropin-II response to an LHRH analog and serotonin in Altantic croaker (*Micropogonias undulatus*): evidence for mediation by 5-HT2 receptors. *J. Exp. Zool.,* 269, 531–537.

Khan, I. A. and Thomas, P. (1999). GABA exerts stimulatory and inhibitory influences on gonadotropin II secretion in the Atlantic croaker (*Micropogonias undulatus*). *Neuroendocrinology,* 69, 261-268.

Kim, M. H., Oka, Y., Amano, M., Kobayashi, M., Okuzawa, K., Hasegawa, Y., Kawashima, S., Suzuki, Y. and Aida, K. (1995). Immunocytochemical localization of sGnRH and cGnRH-II in the brain of goldfish, *Carassius auratus. J. Comp. Neurol.,* 356, 72-82.

Kirubagaran, R., Senthilkumaran, B., Sudhakumari, C. C. and Joy, K. P. (2005). Seasonal dynamics in gonadotropin secretion and E(2)-binding in the catfish *Heteropneustes fossilis. Fish Physiol. Biochem.,* 31, 183-188.

Kitani, T., Matsumoto, S., Yamada H., Amano, M., Iwata, M. and Ueda, H. (2003). Changes in GnRH levels in the brain and pituitary gland during migrations of sockeye salmon and chum salmon. *Fish Physiol. Biochem.,* 28, 269–270.

Kobayashi, Y., Nakamura, M., Sunobe, T., Usami, T., Kobayashi, T., Manabe, H., Paul-Prasanth, B., Suzuki, N. and Nagahama, Y. (2009). Sex change in the Gobiid fish is mediated through rapid switching of gonadotropin receptors from ovarian to testicular portion or vice versa. *Endocrinology,* 150, 1503-1511.

Kumakura, N., Okuzawa, K., Gen, K. and Kagawa, H. (2003). Effects of gonadotropin-releasing hormone agonist and dopamine antagonist on hypothalamus-pituitary-gonadal axis of pre-pubertal female red seabream (*Pagrus major*). *Gen. Comp. Endocrinol.,* 131, 264-273.

Kumakura, N., Okuzawa, K., Gen, K., Yamaguchi, S., Lim, B. S. and Kagawa H. (2004). Effects of gonadotropin-releasing hormone on pituitary-ovarian axis of one-year old pre-pubertal red seabream. *Gen. Comp. Endocrinol.,* 138, 105-112.

Kumar, R. S and Trant, J. M. (2004). Hypophyseal gene expression profiles of FSH-beta, LH-beta, and glycoprotein hormone-alpha subunits in Ictalurus punctatus throughout a reproductive cycle. *Gen. Comp. Endocrinol.,* 136, 82-89.

Lange, A., Katsu, Y., Ichikawa, R., Paull, G. C., Chidgey, L. L., Coe, T. S., Iguchi, T. and Tyler, C. R. (2008). Altered sexual development in roach (Rutilus rutilus) exposed to environmental concentrations of the pharmaceutical 17alpha-ethinylestradiol and associated expression dynamics of aromatases and estrogen receptors. *Toxicol. Sci.,* 106, 113-123.

Larson, E. T., Norris, D. O., Gordon, Grau, E. and Summers, C. H. (2003). Monoamines stimulate sex reversal in the saddleback wrasse. *Gen. Comp. Endocrinol.,* 130, 289-298.

Lee, Y. H., Du, J. L., Yueh, W. S., Lin, B. Y., Huang, J. D., Lee, C. Y., Lee, M. F., Lau, E. L., Lee, F. Y., Morrey, C., Nagahama, Y. and Chang C. F. (2001). Sex change in the protandrous black porgy, *Acanthopagrus schlegeli*: a review in gonadal development, estradiol, estrogen receptor, aromatase activity and gonadotropin. *J. Exp. Zool.*, 290, 715-726.

Levavi-Sivan, B., Biran, J. and Fireman, E., (2006). Sex steroids are involved in the regulation of gonadotropin-releasing hormone and dopamine D2 receptors in female tilapia pituitary. *Biol. Reprod.*, 75, 642-650.

Levavi-Sivan, B., Bogerd, J., Mañanós, E. L., Gómez, A. and Lareyre, J. J. (2010). Perspectives on fish gonadotropins and their receptors. *Gen. Comp. Endocrinol.,* 165, 412-437.

Levy, G. and Degani, G. (2012). Involvement of GnRH, PACAP and PRP in the Reproduction of Blue Gourami Females (*Trichogaster trichopterus*). *Mol Neurosci*, 48, 603-616.

Maruska, K. P. and Fernald, R. D. (2011). Social regulation of gene expression in the hypothalamic-pituitary-gonadal axis. *Physiology (Bethesda),* 26, 412-423.

Matsuo, H., Baba, Y., Nair, R. M., Arimura, A. and Schally, A. V. (1971). Structure of the porcine LH- and FSH-releasing hormone. I. The proposed amino acid sequence. *Biochem. Biophys. Res. Commun.,* 43, 1334–1339.

Miranda, L. A., Strobl-Mazzulla, P. H., Strüssmann, C. A., Parhar, I. and Somoza, G. M. (2003). Gonadotropin-releasing hormone neuronal development during the sensitive period of temperature sex determination in the pejerrey fish, *Odontesthes bonariensis. Gen. Comp. Endocrinol.,* 132, 444-453.

Miranda, L. A., Strüssmann, C. A. and Somoza, G. M. (2001). Immunocytochemical identification of GtH1 and GtH2 cells during the temperature-sensitive period for sex determination in pejerrey, *Odontesthes bonariensis. Gen. Comp. Endocrinol.,* 124, 45-52.

Moncaut, N., Somoza, G., Power, D. M. and Canário, A. V. (2005). Five gonadotrophin-releasing hormone receptors in a teleost fish: isolation, tissue distribution and phylogenetic relationships. *J. Mol. Endocrinol.,* 34, 767-779.

Mylonas, C. C., Fostier, A. and Zanuy, S. (2010). Broodstock management and hormonal manipulations of fish reproduction. *Gen. Comp. Endocrinol.,* 165, 516-534.

Nagahama Y. (2005). Molecular mechanisms of sex determination and gonadal sex differentiation in fish. *Fish Physiol. Biochem.,* 31, 105–109.

Nakamura, K. and Hasegawa, H. (2007). Developmental role of tryptophan hydroxylase in the nervous system. *Mol. Neurobiol.,* 35, 45–54.

Navarro, V. M. (2012). New insights into the control of pulsatile GnRH release: the role of Kiss1/neurokinin B neurons. *Front Endocrinol (Lausanne),* 3, 48.

Ngamvongchon, S., Sherwood, N. M., Warby, C. M. and Rivier, J. E. (1992). Gonadotropin-releasing hormone from thai catfish: chromatographic and physiological studies. *Gen. Comp. Endocrinol.,* 87, 266-274.

Okubo, K., and Nagahama, Y. (2008). Structural and functional evolution of gonadotropin-releasing hormone in vertebrates. *Acta. Physiol. (Oxf),* 193, 3-15.

Okubo, K., Suetake, H., Usami, T. and Aida, K. (2000). Molecular cloning and tissue-specific expression of a gonadotropin-releasing hormone receptor in the Japanese eel. *Gen. Comp. Endocrinol.,* 119, 181-192.

Okuzawa, K., Kumakura, N., Mori, A., Gen, K., Yamaguchi, S. and Kagawa, H. (2002). Regulation of GnRH and its receptor in a teleost, red seabream. *Prog. Brain. Res.,* 141, 95-110.

Onuma, T. A., Makino, K., Ando, H., Ban, M., Fukuwaka, M. A., Azumaya, T. and Urano, A. (2010). Expression of GnRH genes is elevated in discrete brain loci of chum salmon before initiation of homing behavior and during spawning migration. *Gen. Comp. Endocrinol.,* 168, 356-368.

Pandolfi, M., Parhar, I. S., Ravaglia, M. A., Meijide, F. J., Maggese, M. C. and Paz, D. A. (2002). Ontogeny and distribution of gonadotropin-releasing hormone (GnRH) neuronal systems in the brain of the cichlid fish *Cichlasoma dimerus. Anat. Embryol. (Berl),* 205, 271-281.

Parhar, I.S. (2002). Cell migration and evolutionary significance of GnRH subtypes. *Prog. Brain Res.,* 141, 3-17.

Parhar, I. S., Ogawa, S. and Sakuma Y. (2005). Three GnRH receptor types in laser-captured single cells of the cichlid pituitary display cellular and functional heterogeneity. *Proc. Natl. Acad. Sci. U S A,* 102, 2204-2209.

Parhar, I. S., Pfaff, D. W., and Schwanzel-Fukuda, M. (1996). Gonadotropin-releasing hormone gene expression in teleosts. *Brain Res. Mol. Brain Res.,* 41, 216-227.

Peter, R. E., Lin, H. R. and Van Der Kraak, G. (1988). Induced ovulation and spawning in cultured fresh water fish in China: advanced in application of GnRH analogue and dopamine antagonists. *Aquaculture,* 74, 1–10.

Peter, R. E., Trudeau, V. L. and Sloley, B. D. (1991). Brain regulation of reproduction in teleosts. *Bull. Inst. Zool. Acad .Sin. Monogr.,* 16, 89–118.

Pierce, J. G. and Parsons, T. F. (1981). Glycoprotein hormones: structure and function. *Annu. Rev. Biochem.,* 50, 465–495.

Powell, J. F., Zohar, Y., Elizur, A., Park, M., Fischer, W. H., Craig, A. G., Rivier, J. E., Lovejoy, D. A. and Sherwood, N. M. (1994). Three forms of gonadotropin-releasing hormone characterized from brains of one species. *Proc. Natl. Acad. Sci.* U S A, 91, 12081-12085.

Raghuveer, K., Sudhakumari, C. C., Senthilkumaran, B., Kagawa. H., Dutta-Gupta, A. and Nagahama, Y. (2011). Gender differences in tryptophan hydroxylase-2 mRNA, serotonin, and 5-hydroxytryptophan levels in the brain of catfish, *Clarias gariepinus*, during sex differentiation. *Gen. Comp. Endocrinol.,* 171, 94-104.

Rasheeda, M. K., Sridevi, P. and Senthilkumaran, B. (2010). Cytochrome P450 aromatases: Impact on gonadal development, recrudescence and effect of hCG in the catfish, *Clarias gariepinus*. *Gen. Comp. Endocrinol.,* 167, 234-245.

Rodríguez, L., Carrillo, M., Sorbera, L. A., Soubrier, M. A., Mañanós, E., Holland, M. C., Zohar, Y. and Zanuy, S. (2000). Pituitary levels of three forms of GnRH in the male European sea bass (*Dicentrarchus labrax*, L.) during sex differentiation and first spawning season. *Gen. Comp. Endocrinol.,* 120, 67-74.

Sakai, F., Swapna, I., Sudhakumari, C. C., Ganesh, M. V., Kagawa, H., Kobayashi, T., Fan, H., Nagahama, Y. and Senthilkumaran B. (2005). Immunocytochemical localization of gonadotropins during the development of XX and XY Nile tilapia. *Fish Physiol. Biochem.,* 31, 177-181.

Saligaut, C., Garnier, D. H., Bennani, S., Salbert, G., Bailhache, T. and Jego, P. (1992b). Effects of estradiol on brain aminergic turnover of the female rainbow trout (*Oncorhynchus mykiss*) at the beginning of vitellogenesis. *Gen. Comp. Endocrinol.,* 88, 209-216.

Saligaut, C., Salbert, G., Bailhache, T., Bennani, S. and Jego, P. (1992a). Serotonin and dopamine turnover in the female rainbow trout (*Oncorhynchus mykiss*) brain and pituitary: changes during the annual reproductive cycle. *Gen. Comp. Endocrinol.,* 85, 261-268.

Sambroni, E., Rolland, A. D., Lareyre, J. J. and Le Gac, F. (2012). Fsh and Lh have common and distinct effects on gene expression in rainbow trout testis. *J. Mol. Endocrinol.,* 50, 1-18.

Schulz, R. W., Renes, I. B., Zandbergen, M. A., van Dijk, W., Peute, J. and Goos, H. J. (1995). Pubertal development of male African catfish (*Clarias gariepinus*). Pituitary ultrastructure and responsiveness to gonadotropin-releasing hormone. *Biol. Reprod.,* 53, 940-950.

Sébert, M. E., Weltzien, F. A., Moisan, C., Pasqualini, C. *and* Dufour, S. *(*2008*)*. Dopaminergic systems in the European eel: characterization, brain distribution and potential role in migration and reproduction. *Hydrobiologia,* 602, 27–46.

Senthilkumaran, B. and Joy, K. P. (1995a). Changes in hypothalamic catecholamines, dopamine-b hydroxylase, and phenylethanolamine-*N*-methyltransferase in the catfish Heteropneustes fossilis in relation to season, raised photoperiod and temperature, ovariectomy, and estradiol 17-b replacement. *Gen. Comp. Endocrinol.,* 97, 121–134.

Senthilkumaran, B. and Joy, K. P. (1995b). A turnover study of hypothalamic monoamine oxidase (MAO) and effects of MAO inhibition on gonadotropin secretion in the female catfish, *Heteropneustes fossilis*. *Gen. Comp. Endocrinol.,* 97, 1–12.

Senthilkumaran, B. and Joy, K. P. (1996). Effects of administration of some monoamine-synthesis blockers and precursors on ovariectomy-induced rise in plasma gonadotropin II in the catfish *Heteropneustes fossilis*. *Gen. Comp. Endocrinol.,* 101, 220–226.

Senthilkumaran, B., Okuzawa, K., Gen, K. and Kagawa, H. (2001). Effects of serotonin, GABA and neuropeptide Y on seabream gonadotropin releasing hormone release in vitro from preoptic-anterior hypothalamus and pituitary of red seabream, *Pagrus major. J. Neuroendocrinol.,* 13, 395-400.

Senthilkumaran, B., Okuzawa, K., Gen, K., Ookura, T. and Kagawa, H. (1999). Distribution and seasonal variations in levels of three native GnRHs in the brain and pituitary of perciform fish. *J. Neuroendocrinol.,* 11, 181-186.

Servili, A., Lethimonier, C., Lareyre, J. J., López-Olmeda, J. F., Sánchez-Vázquez, F. J., Kah, O. and Muñoz-Cueto, J. A. (2010). The Highly conserved gonadotropin-releasing hormone-2 form acts as a melatonin-releasing factor in the pineal of a teleost fish, the european sea bass *Dicentrarchus labrax. Endocrinology,* 151, 2265-2275.

Sharaf, S.M. (2012). Effect of GnRHa, pimozide and Ovaprim on ovulation and plasma sex steroid hormones in African catfish *Clarias gariepinus. Theriogenology*, 77, 1709-1716.

Sherwood, N., Eiden, L., Brownstein, M., Spiess, J., Rivier, J. and Vale, W. (1983). Characterization of a teleost gonadotropin-releasing hormone. *Proc. Natl. Acad. Sci. U S A,* 80, 2794-2798.

Sherwood, N. M., Sower, S. A., Marshak, D. R., Fraser, B. A. and Brownstein, M. J. (1986). Primary structure of gonadotropin-releasing hormone from lamprey brain. *J. Biol. Chem.*, 261, 4812-4819.

Soga, T., Ogawa, S., Millar, R. P., Sakuma, Y. and Parhar, I. S. (2005). Localization of the three GnRH types and GnRH receptors in the brain of a cichlid fish: Insights into their neuroendocrine and neuromodulator functions. *J. Comp. Neurol.,* 487, 28-41.

Somoza, G. M, Miranda, L. A., Strobl-Mazzulla, P. and Guilgur, L. G. (2002). Gonadotropin-releasing hormone (GnRH): from fish to mammalian brains. *Cell Mol. Neurobiol.,* 22, 589-609.

Somoza, G. M., Miranda, L. A., Guilgur, L. G. and Strobl-Mazzulla P. H. (2006). Characterization of the Brain-pituitary axis in pejerrey *Odontesthes bonariensis* Biocell v.30 n.1

Somoza, G. M., and Peter, R. E. (1991). Effects of serotonin on gonadotropin and growth hormone release from *in vitro* perifused goldfish pituitary fragments. *Gen. Comp. Endocrinol.,* 82, 103–110.

Sower, S. A., Chiang, Y. C., Lovas, S. and Conlon, J. M. (1993). Primary structure and biological activity of a third gonadotropin-releasing hormone from lamprey brain. *Endocrinology,* 132, 1125-1131.

Sridevi, P., Chaitanya, R. K., Dutta-Gupta, A. and Senthilkumaran, B. (2012). FTZ-F1 and FOXL2 up-regulate catfish brain aromatase gene transcription by specific binding to the promoter motifs. *Biochim. Biophys. Acta.,* 1819, 57-66.

Sridevi, P., Dutta-Gupta, A. and Senthilkumaran, B. (2011). Molecular cloning and expression analysis of fushi tarazu factor 1 in the brain of air-breathing catfish, *Clarias gariepinus. PLoS One,* 6, e28867.

Sudhakumari, C. C., Senthilkumaran, B., Kobayashi, T., Kajiura-Kobayashi, H., Wang, D. S., Yoshikuni, M. and Nagahama, Y. (2005). Ontogenic expression patterns of several nuclear receptors and cytochrome P450 aromatases in brain and gonads of the Nile tilapia *Oreochromis niloticus* suggests their involvement in sex differentiation. *Fish Physiol. Biochem.,* 31, 129-135.

Sudhakumari, C. C., Senthilkumaran, B., Raghuveer, K., Wang, D. S., Kobayashi, T., Kagawa, H., Krishnaiah, Ch., Dutta-Gupta, A., and Nagahama, Y. (2010). Dimorphic expression of tryptophan hydroxylase in the brain of XX and XY Nile tilapia during early development. *Gen. Comp. Endocrinol.,* 166, 320-329.

Swapna, I., Sudhakumari, C. C., Sakai, F., Sreenivasulu, G., Kobayashi, T., Kagawa, H., Nagahama, Y. and Senthilkumaran, B. (2008). Seabream GnRH immunoreactivity in brain and pituitary of XX and XY Nile tilapia, *Oreochromis niloticus* during early development. *J. Exp. Zool. A Ecol. Genet. Physiol.,* 309, 419-426.

Tan, L. and Rousseau, P. (1982). The chemical identity of the immunoreactive LHRH-like peptide biosynthesized in the human placenta. *Biochem. Biophys. Res. Commun.,* 109, 1061–1071.

Tanaka, H., Kagawa, H., Okuzawa, K. and Hirose, K. (1993). Purification of gonadotropins (PMGtH-I and PMGtHII) from red seabream (*Pargus major*) and development of a homologous radioimmunoassay for PMGtH-II. *Fish Physiol. Biochem.,* 10, 409–418.

Tostivint, H. (2011). Evolution of the gonadotropin-releasing hormone (GnRH) gene family in relation to vertebrate tetraploidizations. *Gen. Comp. Endocrinol.,* 170, 575-581.

Trant, J. M., Gavasso, S., Ackers, J., Chung, B. C. and Place, A. R. (2001). Developmental expression of cytochrome P450 aromatase genes (CYP19a and CYP19b) in zebrafish fry (*Danio rerio*). *J. Exp. Zool.,* 290, 475-483.

Trudeau, V.L. (1997). Neuroendocrine regulation of gonadotrophin II release and gonadal growth in the goldfish, *Carassius auratus*. *Rev. Reprod.,* 2, 55-68.

Trudeau, V. L., Sloley, B. D. and Peter, R. E. (1993). GABA stimulation of gonadotropin-II release in goldfish: involvement of GABAA receptors, dopamine, and sex steroids. *Am. J. Physiol.,* 265, R348-355.

Tsai, C. L., Wang, L. H. and Tsai, C. C. (2002). Role of serotonin, gamma-aminobutyric acid, and glutamate in the behavioral thermoregulation of female tilapia during the prespawning phase. *J. Exp. Zool.*, 293, 443-449.

Tsai, C. L., Chang, S. L., Wang, L. H. and Chao, T. Y. (2003). Temperature influences the ontogenetic expression of aromatase and oestrogen receptor mRNA in the developing tilapia (*Oreochromis mossambicus*) brain. *J. Neuroendocrinol., 15*, 97-102.

Tsai, C. L. and Wang, L. H. (1997). Effects of estradiol on the serotonin secretion and turnover in the hypothalamus of male tilapia, *Oreochromis mossambicus, in vitro*. *Gen. Comp. Endocrinol.,* 106, 175-180.

Tsai, C. L., Wang, L. H., Chang, C. F. and Kao, C. C. (2000). Effects of gonadal steroids on brain serotonergic and aromatase activity during the critical period of sexual differentiation in tilapia, *Oreochromis mossambicus*. *J. Neuroendocrinol.,* 12, 894-898.

Vacher, C., Mañanos, E. L., Breton, B., Marmignon, M. H. and Saligaut, C. (2000). Modulation of pituitary dopamine D1 or D2 receptors and secretion of follicle stimulating hormone and luteinizing hormone during the annual reproductive cycle of female rainbow trout. *J. Neuroendocrinol.,* 12, 1219-1226.

Vadakkadath, M. S. and Atwood, C. S. (2005). The role of hypothalamic-pituitary-gonadal hormones in the normal structure and functioning of the brain. *Cell Mol. Life Sci.,* 62, 257-270.

Vizziano-Cantonnet, D., Anglade, I., Pellegrini, E., Gueguen, M. M., Fostier, A., Guiguen, Y. and Kah, O. (2011). Sexual dimorphism in the brain aromatase expression and activity, and in the central expression of other steroidogenic enzymes during the period of sex differentiation in monosex rainbow trout populations. *Gen. Comp. Endocrinol.,* 70, 346-355.

Wang, D. S., Kobayashi, T., Zhou, L. Y., Paul-Prasanth, B., Ijiri, S., Sakai, F., Okubo, K., Morohashi, K. and Nagahama, Y. (2007). Foxl2 up-regulates aromatase gene transcription in a female specific manner by binding to the promoter as well as interacting with ad4 binding protein/steroidogenic factor 1. *Mol. Endocrinol.,* 21, 712–725.

Whitlock, K. E., Illing, N., Brideau, N. J., Smith, K. M. and Twomey, S. (2006). Development of GnRH cells: Setting the stage for puberty. *Mol. Cell Endocrinol.,* 25, 254-255.

Wierman, M. E., Bruder, J. M. and Kepa, J. K. (1995). Regulation of gonadotropin-releasing hormone (GnRH) gene expression in hypothalamic neuronal cells. *Cell Mol. Neurobiol., 15*, 79-88.

Wong, A. O. L., Murphy, C., Chang, J. P., Neumann, C. M., Lo, A. and Peter, R. E. (2004). Direct actions of serotonin on gonadotropin-II and growth hormone release from goldfish pituitary cells: interactions with gonadotropin-releasing hormone and dopamine and further evaluation of serotonin receptor specificity. *Fish Physiol. Biochem.,* 19, 23–34.

Yaron, Z., Gur, G., Melamed, P., Rosenfeld, H., Elizur, A. and Levavi-Sivan, B. (2003). Regulation of fish gonadotropins. *Int. Rev. Cytol.*, 225, 131-185.

Yoshiura, Y., Senthilkumaran, B., Watanabe, M., Oba, Y., Kobayashi, T. and Nagahama, Y. (2003). Synergistic expression of Ad4BP/SF-1 and cytochrome P-450 aromatase (ovarian type) in the ovary of Nile tilapia, *Oreochromis niloticus*, during vitellogenesis suggests transcriptional interaction. *Biol. Reprod.,* 68, 1545-1553.

Yu, K. L. and Peter, R. E., (1992). Adrenergic and dopaminergic regulation of gonadotropin releasing hormone release from goldfish preoptic-anterior hypothalamus and pituitary *in vitro*. *Gen. Comp. Endocrinol.*, 85, 138–146.

Zohar, Y., Muñoz-Cueto, J. A., Elizur, A. and Kah, O. (2010). Neuroendocrinology of reproduction in teleost fish. *Gen. Comp. Endocrinol.*, 165, 438-55.

In: Sexual Plasticity and Gametogenesis in Fishes
Editor: Balasubramanian Senthilkumaran

ISBN: 978-1-62618-848-8

Chapter XXII

Zebrafish as a Vertebrate Model System for Developmental Biology and Biomedical Research

K. Raghuveer *

Section on Molecular Endocrinology,
Eunice Kennedy Shriver National Institute of Child Health and Human Development,
National Institutes of Health, Bethesda, Maryland, US

Abstract

Over the past two decades zebrafish has been extensively used in biomedical research for studying vertebrate development and molecular pathogenesis of human diseases. Zebrafish has potential as a genetic model due to its external development, transparency of the embryos, relatively short generation time and with the application of transgenic reporter lines which allows direct visualization of tissue morphogenesis. Large scale forward genetic screens have been used to identify novel genes in the pathways affected by cancers. Several zebrafish mutant phenotypes have been characterized carrying mutations in genes essential for numerous processes including morphogenesis, pattern formation, organogenesis, and differentiation. Major advances in reverse genetic methods allowed rapid development of gene-based zebrafish disease models. The transgenic zebrafish models can be monitored frequently to study the behavior of cell populations carrying the reporter gene of interest *in vivo* and in real time. Recent improvements in molecular imaging techniques and development of transparent zebrafish strains offer the unique opportunity to follow normal and aberrant developmental and physiological processes at subcellular level. In addition, zebrafish is an effective *in vivo* model system to perform large-scale high throughput screening of chemicals/drugs that suppress genetic diseases.

* Email: kavarthapur@mail.nih.gov.

Introduction

Zebrafish (*Danio rerio*), is a tropical freshwater teleost fish originally found in slow streams and rice paddies in several parts of the world including the Ganges river in East India and Burma. The females are often able to spawn on daily basis and sometimes at intervals of two to three days, laying hundreds of eggs. The breeding is easy and the embryos grow quickly. Zebrafish reach sexual maturity by three months of age. In early 1930's zebrafish was initially used as a classical model to study vertebrate embryonic development. Later on the development of genetic techniques such as mutagenesis, transgenesis and positional cloning approaches has made zebrafish an ideal model to quickly assess gene function *in vivo (see review Udvadia & Linney, 2003)*. Unlike mouse, zebrafish embryos are transparent and develop externally therefore it is easy to visualize cells under microscope during development and in later stages it also enables to monitor organ development. Furthermore, it is possible to perform a variety of technical manipulations such as microinjections, cell transplantations, microsurgery, and cell ablations. These benefits allow rapid phenotypic analysis of early embryonic defects. Generating transgenic lines is a popular approach to study the *in vivo* function of genes (functional genomics) in zebrafish wherein the gene of interest is tag with a fluorescent protein which is driven by a tissue-specific promoter. It allows us direct visualization of tissue morphogenesis in developing embryo. The genome of zebrafish is completely sequenced and almost all human genes can also be found in the zebrafish. Both shotgun and minimum tiling path sequencing are being combined to generate a high-quality genome resource for zebrafish research which has immensely led to the identification disease causing mutations. It was found that approximately 20% of the human genes have two orthologs in zebrafish (Postlethwait et al., 1998). In this chapter, I focused on how zebrafish has emerged as a wonderful *in vivo* model system to study vertebrate development, genetics of human diseases and cancer research. I also reviewed the advances in the transgenic techniques that are essential for generating gene based disease models in zebrafish.

Zebrafish Model for Developmental Biology Research

Zebrafish is a fast escalating and an ideal model for studying various developmental processes in vertebrates. As an experimental model organism several new molecular techniques have been developed for imaging and manipulating early developmental processes in zebrafish which led to many significant discoveries in vertebrate development. There is a high genomic and molecular similarity between zebrafish and other vertebrates including higher mammals. Zebrafish models have been well established to study heart, kidney, liver, muscle and eye development and now it is being extensively used to understand human physiological processes like hematopoiesis, angiogenesis and neurogenesis.

Heart Development

In recent years, the zebrafish model has emerged as a powerful genetic system to study cardiac development and function. Transcription factors *hand*, *nkx*2.5, *t-box*, and *gata* essential for cardiac specification and differentiation in zebrafish are also observed during mammalian cardiac development (Tu & Chi, 2012). Zebrafish homologues of genes like *myosin light chain*, *titin*, *cardiac troponin T* and *β myosin heavy chain* can be corroborated to their counterparts in human cardiac development. Any of these genes defective in zebrafish mutants showed reduced contractility, myofibril disarray and dysmorphic cardiac chambers (Tu & Chi, 2012). For example, zebrafish *silent heart* mutant exhibit non-contractile heart and disorganized sarcomeres due to the defect in *cardiac troponin T*. Zebrafish *ilk* mutant heart has severe myocardial dysfunction resembling human dilated cardiomyopathy and knockdown of *laminin α4* using morpholinos also results in cardiac dysfunction similar to that observed in *ilk* mutants (Nguyen et al., 2008). *Nexilin* is a highly enriched protein in cardiac muscles and knock down of this gene in zebrafish causes severely reduced cardiac contractile function which later on led to the identification of *NEXILIN* mutations in human dilated cardiomyopathy patients (Tu & Chi, 2012).

Kidney Development

During development zebrafish forms a simple pronephros comprised of a pair of nephrons and will last for over several weeks of larval life. The zebrafish juvenile eventually develops a mesonephros, which contains several hundred nephrons. The renal progenitor cells that reside throughout the stroma of the mesonephros can produce entirely new nephrons after injury. Transcription factors like *pax2a* and *cdx4/1* are essential for axial positioning and early patterning of pronephros field. The renal progenitor cells express several transcription factors, including *pax2a*, *pax8*, *wt1b*, and *lhx1a which drives the formation of new nephrons* (Gerlach & Wingert, 2012). Defects in cilia formation or function lead to kidney cyst formation. Mutations in polycystins, PC1 and PC2 which localize to the cilium, lead to polycystic kidney disease (PKD) in humans. In zebrafish also mutation in these two genes results in renal cysts and pathogenesis of the disease is similar to PKD (Sun et al., 2004).

Liver Development

In zebrafish at early gastrula stage the endoderm cells located in the ventral part give rise to hepatoblast which differentiate to liver bud. Based on studies using the zebrafish transgenic *Tg(gutGFP)S584* it has been revealed that liver morphogenesis occurs in two phases: budding and growth. Signaling molecules *wnt2bb*, *bmp2b*, *fgf* and *RA* play key roles in specification of hepatoblasts. *Hhex* controls the initiation and budding of the liver and *prox1* function to promote outgrowth of liver bud and migration of hepatoblasts. Genes like *foxA1-A3*, *gata4-6*, *sox17*, *hhex*, *prox1* and *hnf4* are found to be expressed in liver primodium during the budding stage. Loss-of-function of *gata5* in the '*faust*' mutant leads to a liverless or small liver phenotype and the double knockdown of *gata4* and *gata6* in zebrafish completely abolish liver development and leads to a liverless phenotype. The γ-hexachlorocyclohexane, thioacetamide and alcohol can induce hepatic steatosis also know as fatty liver disease in zebrafish similar to that observed in humans which makes it as good model to study liver diseases (Tao & Peng, 2009).

Eye Development

Zebrafish cornea and lens show similar kind of morphology as other vertebrates including humans. The cornea develops very rapidly in embryogenesis making it handy to loss-of-function genetic analysis. Further several corneal proteins initially characterized in higher vertebrates are conserved in the zebrafish. For these reasons, zebrafish is a suitable model for examining corneal development and function. The corneal protein, BIGH3, keratin 3, K3, and corneal keratan sulfate proteoglycan, CKS in zebrafish showed conserved expression patterns as that in higher vertebrates. *BIGH3* gene is responsible for several types of human corneal dystrophies. K3 is an excellent marker of corneal epithelium while CKS is expressed specifically in the stroma. Zebrafish cornea contains the Bowman's layer which plays a crucial role in corneal wound healing and it is affected by several human disorders, including Reis-Bücklers' corneal dystrophy and keratoconus (Zhoa et al., 2006).

Hematopoiesis

Hematopoiesis in the vertebrate is characterized by the initiation of ventral mesoderm to form hematopoietic stem cells (HSC) and the differentiation of these progenitors finally gives rise to peripheral blood lineages. Several genes are involved in the differentiation and development of hematopoietic progenitor cells. In the last decade, zebrafish has become a useful genetic model to study hematopoietic development as it enables direct visualization of embryonic process of HSCs development in mesoderm and migration of blood cells. Most of the critical hematopoietic transcription factors (*scl, gata1/2, lmo2, fli1a and runx*) identified in mammals has orthologues in zebrafish. The genetic programming governing the process of hematopoiesis was found to be highly conserved. *Stem cell leukemia* (*scl*) gene which is crucial for HSC formation has been extensively characterized in both mammals and zebrafish. Mutation in this gene leads to acute lymphoblastic leukemia. Several mutant phenotypes lacking hematopoetic cells and other vasculature defects in zebrafish have been identified as result of mutations in these hematopoietic transcription factors (Davidson & Zon et al., 2004).

Angiogenesis

Angiogenesis is the process of formation of new blood vessels and its controlled propagation is very important for wound healing and tissue growth. Several methods have been developed to visualize blood vessels in zebrafish like *in situ* hybridization, microangiography and transgenics. Researchers have developed transgenic fish lines with fluorescent blood vessels wherein GFP expression is driven by a promoter for the vascular endothelial growth factor receptor-2 gene which is specifically expressed in blood vessels. Several studies in zebrafish have identified new mechanism involved in angiogenesis.

Characterization of zebrafish mutant phenotypes with defects in artery development showed mutations in Vegf receptor, *kdr-like* (*kdrl*), and *phospholipase c γ 1 which are essential for vascular morphogenesis (Covassin* et al., *2009).*

Neurogenesis

The zebrafish is an excellent model organism for understanding the process of neural development because it has simple brain architecture homologous to those of other vertebrates. It provides the opportunity to visualize the initial development and coalescence of neurons, as well as the functional organization of complete neural networks. Several genes have been identified that play an important role during early neural development using the mutant screening and transgenic methods in zebrafish. One specific gene *hmx4* in zebrafish which is strongly expressed in early brain development and its knock down resulted in phenotypes with failure of the neural tube to close, small ears, and loss of the pectoral fins. In humans, mutations in *HMX* genes have been linked to neural mis-patterning and neural tube closure defects (Gongal et al., 2011). In mammals, neural crest gives rise to the pigment cells of skin and neurons and glia of the peripheral nervous system and dorsal root ganglia (DRG). This process is regulated through the activity of two *neurogenin* genes, *neurog1 and neurog2. However, in zebrafish the d*ifferentiation of neurons in the DRG is dependent on the activity of *neurog1 which is very crucial for primary neurogenesis and there is no neurog2*. Mutation in *neurog1* resulted in embryos showing severe reduction in spinal sensory neurons and failed to form DRG neurons. This study clearly indicates that *neurog1* acts as key factor in this lineage to direct the formation of sensory neurons (McGraw et al., 2008). Several zebrafish disease models have been generated to understand the genetics of neurological diseases commonly noticed in humans.

Table 1.

Disease	Gene
Muscular dystrophy	Dystrophin, dystroglycan and *Dp71*
Diabetes	*Insulin, IA-2 autoantigen* and *IA-2b autoantigen*
Alzheimer's disease	*presenilin-1, presenilin-2, Presenilin Enhancer 2*, acetylcholinesterase, amyloid precursor protein and *apoE*
Cardiomyopathy	cardiac *troponin T* and *titin*
Huntington's disease	*Huntingtin*
Leukemia	*runx1, cbfb*
Amyotrophic lateral sclerosis	*sod-1*
Thrombosis	Factor VII, *cox-1* and *cox-2*
Parkinson's disease	*parkin, pink1, dj-1*
Joubert syndrome	*CEP290 (NPHP6)*
Congenital sideroblastic anemia	*ALAS 2*
Nephrotic syndrome	*plce1*
Erythropoietic protoporphyria	*Ferrochelatase*
Spinal muscular atrophy	*SMN*

Zebrafish Model of Human Genetic Diseases

In the last decade, zebrafish has been successfully employed as an organism to elucidate the etiology of human disease. Zebrafish models of human disease are widely used in many different areas of biomedical research to understand the genetic mechanisms of pathogenesis. During 1990s, two large-scale genetic screens were conducted parallelely by two independent groups in Boston and in Tubingen (Driever et al., 1996; Mullins et al., 1996). It resulted in the identification of several embryonic mutant phenotypes that exhibit various developmental and physiological disorders and many of which resembled human genetic diseases. The combination of easy mutagenesis and phenotypic screens of the earliest developmental stages of zebrafish resulted in the undertaking of large-scale screens. Further recent advances in transgenic techniques has made easier for the scientist to develop human disease models in zebrafish that are specific to a particular gene. Several human disease genes as orthologs have been identified in zebrafish (Table 1; see review Rubinstein, 2003; Kari et al., 2007). Some of these genes are already cloned from zebrafish and diseases model have been developed. The Table 1 below summarizes examples of human disease related genes found in zebrafish.

Zebrafish Models of Genetic Disease Can Be Developed by Two Different Approaches, Forward and Reverse Genetics Approaches

Forward Genetics Approach

Zebrafish is well known for its effectiveness as a forward genetic tool. Scientists have extensively used this model for large-scale forward genetic screens in which phenotypic defects are identified before the identification of the gene causing these defects. In this approach, the mutations are introduced into the genome of adult fish either through chemical mutagenesis or γ irradiation. The offsprings from the mutagenized fish are screened for abnormal phenotypes. Genes that are carrying the genetic mutations are identified through genetic mapping, sequencing analysis, and phenotype validation. N-ethyl-N-nitrosourea (ENU) is the most common and effective chemical mutagen used to generate large number of screens (Haffter et al., 1996; Driever et al., 1996; Amsterdam & Hopkins, 2006). Male adult fish are exposed to ENU which induces point mutations throughout the genome in germ cells. Mutagenized fish are crossed with wildtype females. F1 progeny which are heterozygous for individual mutations are selected and crossed again with wildtype females to get F2 progeny. The F2 generation is intercrossed randomly to produce F3 families in which homozygous mutations occur. Most ENU induced mutations are recessive. Finally, the F3 embryos are selected based on the abnormalities during development using microscopic examination. Positional cloning can then be used to identify the affected gene in these mutant phenotypes. Large numbers of mutations that disrupt embryonic development have now been isolated in the zebrafish, many of which may serve as models for human congenital diseases and result in phenotypic similarities (Haffter et al., 1996; Driever et al., 1996). For example, a zebrafish mutant *sapje* has a defect that impairs the muscle protein dystrophin which is similar to the human gene that causes Duchenne muscular dystrophy (Bassett et al., 2003). Cardiac contractile defects identified in zebrafish using forward genetic screen mutants, such as *ilk* and *titin*, have also now been detected in human cardiomyopathies. Zebrafish anaemic mutants *yquem* and *dracula* are models of hepatoerythropoietic porphyria and erythropoietic protoporphyria, respectively (Wang et al., 1998). Phenotypic similarities between *Tbx1* mutant mice and the zebrafish *van gogh* mutant also implicated *TBX1* in the human DiGeorge syndrome (Childs et al., 2000). Major drawback of forward genetics approach is that it is slow and laborious due to positional cloning methods.

Another approach to induce mutations disrupting developmental processes in zebrafish is radiation, mainly γ-rays. Mutagenesis using γ produces a very high locus mutation rate in contrast to chemical mutagenesis. γ-rays induce translocations and large deletions at high frequency in the zebrafish genome (Fritz et al., 1996).

Reverse Genetics Approach

Reverse genetics involves studying the phenotypic consequences by disturbing the function of a gene of interest. There are several different methods which are currently employed in reverse genetic approach to develop zebrafish mutants. These include TILLING, morpholinos, retroviral mediated mutagenesis and transgenesis.

Tilling

TILLING (Targeting Induced Local Lesions IN Genomes) was the first reverse genetic approach used in zebrafish to induce germline mutations in a desired target gene. It allows rare mutagenic events to be recovered for subsequent phenotypic analysis. This method involves random induced mutations by ENU and subsequent screening for mutations in target genes. Lesions within the library are identified through the analysis of PCR amplicons spanning genomic regions of interest. This screen is an enzyme-mediated (CELI endonuclease) mismatch recognition procedure to detect heterozygous germ line mutations in the F1 generation. This method leads to generation of more number of Null mutants in zebrafish (Wienholds et al., 2003; Sood et al., 2006). The first zebrafish mutant to be generated using this method has mutation in gene *rag1* (*recombination activating gene 1*) associated with human severe combined immunodeficiency (Wienholds et al., 2002). Zebrafish mutants containing mutation in the tumor suppressor gene *tp53* were identified using target-selected mutagenesis which developed malignant peripheral nerve sheath tumors (Berghmans et al., 2005).

Morpholinos

Morpholinos have proven to be a valuable tool for assessing gene function in developmental biology. Morpholinos (MOs) are synthetic oligonucleotides of 25 bases, which hybridize specifically to complementary sequences of mRNA, disrupting translation initiation or pre-mRNA splicing. The backbone of MOs is similar to the backbone of DNA or RNA, with minor changes (Summerton & Weller, 1997). Ekker and colleagues were the first to demonstrate the use of MOs in zebrafish by targeting the translation start site for a number of development related genes. This technology is effectively used to confirm new mutant gene identification to explore the functions of genes where no mutant exists (Nasevicius & Ekker, 2000). *Pipetail* (*wnt5*) and *silberblick* (*wnt11*) zebrafish mutants developed using morpholinos undergo abnormal gastrulation with an undulated notochord and cyclopic embryos, respectively (Rauch et al., 1997; Heisenberg et al., 2000). Zebrafish diseases models similar to human diseases were created to evaluate the function of specific genes more precisely. Hypertrophic cardiomyopathy model was developed using morpholino targeting zebrafish *cardiac troponin T* (*tnnt2*) gene which is characterized by myocardial hypertrophy, myocyte and myofibrillar disarray (Becker, 2011). Mutation of the human *TBX5* gene causes a disorder known as Holt-Oram syndrome, characterized by deformities in the heart and upper limbs (Li et al., 1997). Zebrafish with this mutation have analogous defects both in the heart and pectoral fins. *Npc1* morphants display numerous abnormalities, including cholesterol mislocalisation, cell-movement defects that is seen in Niemann-Pick disease, type C1. Zebrafish *vegf-A* morphant embryos develop with an enlarged pericardium and with major blood vessel deficiencies. Zebrafish *pdx1* morphant displays defects in pancreas development and digestive organ chirality, and potentially identifies a multipotent pancreas progenitor cell (Yee et al., 2001). Knockdown of the Zebrafish ortholog of the Retinitis Pigmentosa 2 (*RP2*) gene results in retinal degeneration (Shu et al., 2011).

Retroviral

Retroviral based mutagenesis systems can also be employed for reverse genetic strategies to identify modified alleles of a target gene. It has the potential to provide collections of disrupted alleles for the majority of genes within the zebrafish genome. The inserted DNA can be tailored to interfere with gene expression, while also providing readout of the expression pattern of the gene through the incorporation of an expressed tag such as GFP. A pseudo-typed retrovirus derived from MLV with its envelope protein can able to infect zebrafish embryos and transmit efficiently through germline following insertion into the genome (Amsterdam & Becker, 2005). It gets integrated into the genomic regions close to transcription start sites. By utilizing the retroviral sequence as a molecular tag for mapping, the genomic sequence flanking the insertion site can be easily identified through inverse PCR or linker-mediated PCR. Using this approach, 315 genes essential for zebrafish development has been successfully identified (Amsterdam & Varshney, 2011).

Transgenic Approach

Transgenic techniques in zebrafish are widely used to characterize the function of many newly identified genes. The first transgenic zebrafish was generated by injecting the plasmid DNA or bacterial artificial chromosomes into the cytoplasm of fertilized eggs which integrated into the genome and stably transmitted to subsequent generations (Stuart et al., 1988). DNA integration is best achieved when foreign DNA molecules are injected at one-cell or two-cell stages. In this approach, a large number of eggs need to be injected and screened to compensate for the low germ-line transmission efficiency to the F1 generation. In addition, the inserted foreign DNA tends to integrate as concatemers with multiple tandem copies. Later on other several techniques have been developed to improve germ-line transmission efficiency which includes transposon, Gal4-UAS, tetracycline (Tet-On), and the Cre/loxP based systems.

Transposon *Tol2*-based transgenesis has emerged as a reliable and effective system for transient and stable expression of transgenes in zebrafish. The *Tol2* elements usually inserted into the genome as a single copy and therefore do not dramatically rearrange the genomic sequences flanking the insertion site, for this reason, they are less susceptible to methylation. Transposon mediated transgenesis by simple injection of constructs into fertilized eggs can result in a 50–80% efficiency of germ line trangenesis (Suster et al., 2009).

The Gal4-UAS, tetracycline (Tet-On) and the Cre/loxP systems have been used to achieve tissue-specific expression of transgenes in zebrafish transgenic models. Each of these methods can be used to activate gene expression in a regulated manner. It allows regulation of the transgene expression either temporally or spatially. This kind of tight regulation of transgene is essential because many of the cancer genes involved in tumorigenesis play crucial roles during development (see review Skromne & Prince, 2008). The gal4-UAS system uses the yeast transcriptional activator GAL4 to drive expression of a gene of interest via its specific *cis*-binding sites known as upstream activating sequences (UAS). When driver lines expressing GAL4 are crossed to UAS regulated reporter lines, the activation of the reporter genes can be tightly controlled by the presence of Gal4 in a promoter-specific manner. The transgenesis efficiency has improved by using the more potent GAL4-VP16 fusion as an activator (Koster & Fraser, 2001). The Tet-On system again depends on a driver and an effector gene where in temporal control is also dependent upon drug treatment. This system is activated when the stable Tet derivative doxycycline is added to the medium. Transient GFP expression was achieved either by using a single plasmid carrying both rtTA driver and GFP responder (Huang et al., 2005).

Another well-established genetic method to regulate transgene expression is the Cre/LoxP system. The Cre/loxP system is widely used in mouse, in particular to allow conditional knockouts. First time this system was used in zebrafish to demonstrate that Cre recombinase was introduced into fish embryos through microinjection has successfully excised the *dsRED* gene from a *CMV-loxP-dsRED2-loxP-EGFP* transgene and green fluorescent protein was expressed in the transgenic fish (Langenau et al., 2005). Using this system lymphoblastic leukemia model has been developed in zebrafish. The *rag2-loxP-dsRED2-loxP-EGFP-mMyc* transgene which is introduced in to the embryos induces Myc activity in T cells causing leukemia (Langenau et al., 2005).

Targeted Gene Inactivation Using Zinc-Finger Nucleases

More recently engineered Zinc-Finger Nucleases (ZFNs) are efficiently used for targeted gene inactivation in zebrafish (Meng et al., 2008). ZFNs are artificial restriction enzymes generated by fusing a zinc finger DNA-binding domain to a DNA-cleavage domain which is a restriction endonuclease Fok I derived from bacteria. Zinc finger domains can be engineered to target unique DNA sequences within the genome. The α helix domain of each zinc finger recognizes and binds to a specific 3 bp DNA sequence. Each ZFN typically contains three to six individual zinc finger motifs, which can recognize a total of 9–18 bp target DNA sequence. Since *Fok* I has to dimerize to become fully functional, ZFNs usually function in pairs, where they bind to DNA with appropriate orientation and spacing to induce double strand breaks (DSBs) at its target locus. The DSB could be repaired through error-prone non-homologous end joining or homology-directed repair, leading to either indel mutations or DNA replacement. Multiple pairs of ZFNs can also be used to completely remove entire large segments of genomic sequence (Lee et al., 2009). To monitor the editing activity, a PCR of the target area will amplify both alleles and if one contains an insertion, deletion, or mutation, it will result in a heteroduplex single stranded bubble which cleavage assays can easily detect. ZFN technology has been successfully applied in *in vitro* cultured cells as well as many model organisms including zebrafish. Recently, researchers have successfully used the ZFN technology to generate transgenic fish embryos with no tail (ntl) that lack notochord and tail and *kdrl* mutant fishes have specific vasculature defects (Amacher, 2008). The ZFN encoding mRNA was injected into one-cell embryos and a high percentage of animals carried the desired mutations and phenotypes. ZFNs can specifically and efficiently create heritable mutant alleles at loci of interest in the germ line, and ZFN-induced alleles can be propagated in subsequent generations.

Transgenesis also enables the over expression of dominantly acting disease gene within a specific tissue using tissue specific promoters and this trangenesis approach has extensively been used to generate several cancer based models in zebrafish. Most of the cancer related genes that are identified in humans were introduced into zebrafish embryos which are driven by different promoters to induce cancers in specific tissues (Table 2; Bourque & Houvras, 2011).

The Table 2 below summarizes transgenic zebrafish models for cancer and other genetically based human diseases:

Table 2.

Cancer model	Oncogene or tumor suppressor
T-cell *acute lymphoblastic leukemia*	*c-Myc* , *Notch1*
B-cell *acute lymphoblastic leukemia*	*elfa*
Acute myeloid leukemia (AML)	*AML-ETO, MOZ-TIF2*
Melanoma	*BRAF, HRAS, NRAS*
Pancreatic carcinoma	*KRAS*
Neuroendocrine carcinoma	MYCN
Embryonal rhabdomyosarcoma	*KRASG12*
Malignant Peripheral Nerve Sheath Tumor or Neurofibrosarcoma	*tp53, rp, emi, msh2, msh6*
Neuroblastoma	*fbxw4*
Ocular tumor	*ptenb*
Intestinal, pancreatic, and hepatic adenomas	*apc*
Testicular germ cell tumor	*Lamc1*

Conclusion

Zebrafish is an emerging and effective vertebrate animal model not only for developmental biology but also for biomedical research. Morphology based approach of gene knockdown deciphered the role of several genes that play crucial role in the early embryogenesis. The power of forward genetics and improvements in reverse genetics techniques have greatly assisted in the development of zebrafish disease models for a variety of human genetic diseases and cancers which exhibit comparable features both histopathologically and genetically. One of the important technical challenges in zebrafish transgenics is developing conditional knockouts of specific genes. This has been achieved using conditional Cre/lox system and Zinc finger nuclease technology. With ongoing development in zebrafish research, zebrafish models of disease offer valuable insights into disease pathogenesis and provide a tool to develop new and effective therapies for a various human diseases. Scientists are working on zebrafish disease models for large-scale screening of chemicals/drugs that can suppress genetically caused diseases and in near future it will aid in the development of new therapeutic drugs.

References

Amacher, S. L. (2008). Emerging gene knockout technology in zebrafish: zinc-finger nucleases. *Brief Funct Genomic Proteomic, 7,* 460-464.

Amsterdam, A. & Becker, T. S. (2005). Transgenes as screening tools to probe and manipulate the zebrafish genome. *Dev Dyn, 234*, 255-268.

Amsterdam, A. & Hopkins, N. (2006). Mutagenesis strategies in zebrafish for identifying genes involved in development and disease. *Trends Genet, 2,* 473-478.

Amsterdam, A. & Varshney, G. K. (2011). Burgess SM. Retroviral-mediated Insertional Mutagenesis in Zebrafish. *Methods Cell Biol, 104*, 59-82.

Bassett, D.I., Bryson-Richardson, R. J., Daggett, D. F., Gautier, P., Keenan, D. G. & Currie, P. D. (2003). Dystrophin is required for the formation of stable muscle attachments in the zebrafish embryo. *Development, 130*, 5851-5860.

Becker, J. R., Deo, R. C., Werdich, A. A., Panàkovà, D., Coy, S. & MacRae, C. A. (2011). Human cardiomyopathy mutations induce myocyte hyperplasia and activate hypertrophic pathways during cardiogenesis in zebrafish. *Dis Model Mech, 4*, 400–410.

Berghmans, S., Murphey, R. D., Wienholds, E., Neuberg, D., Kutok, J. L., Fletcher, C. D., Morris. J. P., Liu, T. X., Schulte-Merker, S., Kanki, J. P., Plasterk, R., Zon, L. I. & Look, A. T. (2005). tp53 mutant zebrafish develop malignant peripheral nerve sheath tumors. *Proc Natl Acad Sci USA, 102,* 407-412.

Bourque, C. & Houvras, Y. (2011). Hooked on zebrafish: insights into development and cancer of endocrine tissues. *Endocr Relat Cancer, 18*, 149-164.

Childs, S., Weinstein, B. M., Mohideen, M. A., Donohue, S., Bonkovsky, H. & Fishman, M. C. (2000). Zebrafish dracula encodesferrochelatase and its mutation provides a model for erythropoietic protoporphyria. *Curr Biol, 10*, 1001-1004.

Covassin, L. D., Villefranc, J. A., Kacergis, M. C., Weinstein, B. M. & Lawson, N. D. (2006). Distinct genetic interactions between multiple Vegf receptors are required for development of different blood vessel types in zebrafish. *Proc Natl Acad Sci USA, 103*, 6554-6559.

Driever, W., Solnica-Krezel, L., Schier, A. F., Neuhauss, S. C., Malicki, J., Stemple D. L., Stainier, D. Y., Zwartkruis, F., Abdelilah, S., Rangini, Z., Belak, J. & Boggs, C. (1996). A genetic screen for mutations affecting embryogenesis in zebrafish. *Development, 123*, 37-46.

Fritz, A., Rozowski, M., Walker, C. & Westerfield, M. (1996). Identification of selected γ-ray induced deficiencies in zebrafish using multiplex polymerase chain reaction. *Genetics, 144*, 1735-1745.

Gerlach, G. F. & Wingert, R. A. (2012). Kidney organogenesis in the zebrafish: insights into vertebrate nephrogenesis and regeneration. *Dev Biol,* doi: 10.1002/wdev.92.

Gongal, P. A., March, L. D., Holly, V. L., Pillay, L. M., Berry-Wynne, K. M., Kagechika, H. & Waskiewicz, A. J. (2011). Hmx4 regulates Sonic hedgehog signaling through control of retinoic acid synthesis during forebrain patterning. *Dev Biol, 355*, 55-64

Gore, A. V., Monzo K., Cha, Y. T., Pan, W. & Weinstein, B. M. (2012). Vascular Development in the Zebrafish *Cold Spring Harb Perspect Med, 2,* a006684.

Davidson, A. J. & Zon, L. I. (2004). The 'definitive' (and 'primitive') guide to zebrafish hematopoiesis. *Oncogene*, *23*, 7233-7246.

Haffter, P. Granato, M., Brand, M., Mullins, M. C., Hammerschmidt, M., Kane, D. A., Odenthal, J., van Eeden, F. J., Jiang, Y. J., Heisenberg, C. P., Kelsh, R. N., Furutani-Seiki, M., Vogelsang, E., Beuchle, D., Schach, U., Fabian, C. & Nüsslein-Volhard C. (1996). The identification of genes with unique and essential functions in the development of the zebrafish, *Danio rerio*. *Development, 123*, 1-36.

Heisenberg C. P., Tada, M., Rauch, G. J., Saude, L., Concha, M. L., Geisler, R., Stemple, D. L., Smith, J. C. & Wilson, S. W. (2000). Silberblick/Wnt11 mediates convergent extension movements during zebrafish gastrulation. *Nature, 405*, 76–81.

Huang, C. J., Jou, T. S., Ho, Y. L., Lee, W. H., Jeng, Y. T., Hsieh, F. J. & Tsai, H. J. (2005). Conditional expression of a myocardium-specific transgene in zebrafish transgenic lines. *Dev Dyn, 233,* 1294-1303.

Kari, G., Rodeck, U. & Dicke, A. P. (2007). Zebrafish: an emerging model system for human disease and drug discovery. *Clin Pharmacol Ther, 82*, 70-80.

Koster, R. W. & Fraser, S. E. (2001). Tracing transgene expression in living zebrafish embryos. *Dev Biol, 233,* 329–346.

Langenau, D. M., Feng, H., Berghmans, S., Kanki, J. P., Kutok, J. L. & Look, A. T. (2005). Cre/lox-regulated transgenic zebrafish model with conditional myc-induced T cell acute lymphoblastic leukemia. *Proc Natl Acad Sci USA,102,* 6068-6073.

Lee, H. J., Kim, E. & Kim, J. S. (2009). Targeted chromosomal deletions in human cells using zinc finger nucleases. *Genome Res, 20,* 81-89.

Li, Q. Y., Newbury-Ecob, R. A., Terrett, J. A., Wilson, D. I., Curtis, A. R., Yi, C. H., Gebuhr, T., Bullen, P. J., Robson, S. C., Strachan, T., Bonnet, D., Lyonnet, S., Young, I. D., Raeburn, J. A., Buckler, A. J., Law, D. J. & Brook, J. D. (1997). Holt-Oram syndrome is caused by mutations in TBX5, a member of the Brachyury (T) gene family. *Nat Genet, 15*, 21-29.

McGraw, H. F., Nechiporuk, A. & Raible, D. W. (2008). Zebrafish dorsal root ganglia neural precursor cells adopt a glial fate in the absence of neurogenin1. *J Neurosci, 28*, 12558-69.

Meng, X., Noyes, M. B., Zhu, L. J., Lawson, N. D. & Wolfe, S. A. (2008). Targeted gene inactivation in zebrafish using engineered zinc-finger nucleases. *Nat Biotechnol, 26*, 695-701.

Mullins, M. C., Hammerschmidt, M., Kane, D. A., Odenthal, J., Brand, M., van Eeden, F. J., Furutani-Seiki, M., Granato, M., Haffter, P., Heisenberg, C. P., Jiang, Y. J., Kelsh, R. N. & Nüsslein-Volhard, C. (1996). Genes establishing dorsoventral pattern formation in the zebrafish embryo: the ventral specifying genes. *Development, 123*, 81-93.

Nasevicius, A., & Ekker, S. C. (2000). Effective targeted gene 'knockdown' in zebrafish. *Nat Genet, 26*, 216-220.

Nguyen, C. T., Lu, Q., Wang, Y., Chen & J. N. (2008). Zebrafish as a model for cardiovascular development and disease. *Drug Discov Today Dis Models*, 5, 135-140.

Postlethwait, J. H., Yan, Y. L., Gates, M. A., Horne, S., Amores, A., Brownlie, A., Donovan, A., Egan, E. S., Force, A., Gong, Z., Goutel, C., Fritz, A., Kelsh, R., Knapik, E., Liao, E., Paw, B., Ransom, D., Singer, A., Thomson, M., Abduljabbar, T. S., Yelick, P., Beier, D., Joly, J. S., Larhammar, D., Rosa, F., Westerfield, M., Zon, L.I., Johnson, S.L. & Talbot WS. (1998). Vertebrate genome evolution and the zebrafish gene map. *Nat Genet, 18*, 345-349

Rauch G. J., *Hammerschmidt, M., Blader, P., Schauerte, H. E., Strahle, U., Ingham, P. W., McMahon, A. P., & Haffter, P. (*1997*). Wnt5 is required for tail formation in the zebrafish embryo. Cold Spring Harb Symp Quant Biol, 62, 227-234.*

Rubinstein, A. L. (2003) Zebrafish: From disease modeling to drug discovery. *Curr Opin Drug Discov Develop, 6*, 218-223.

Schulte-Merker, S., van Eeden, F. J. M., Halpern, M. E., Kimmel, C. B. & Nüsslein-Volhard, C. (1994). no tail (ntl) is the zebrafish homologue of the mouse T (Brachyury) gene. *Development, 120*, 1009–1015.

Shu, X., Zeng, Z., Gautier, P., Lennon, A., Gakovic, M., Cheetham, M. E., Patton, E. E. & Wright, A. F. (2011). Knockdown of the zebrafish ortholog of the retinitis pigmentosa 2 (RP2) gene results in retinal degeneration. *Invest Ophthalmol Vis Sci, 52*, 2960-2966.

Skromne, I. & Prince V. E. (2008). Current perspectives in zebrafish reverse genetics: moving forward. *Dev Dyn, 237*, 861-882.

Sood, R., English, M. A., Jones, M., Mullikin, J., Wang, D. M., Anderson, M., Wu, D., Chandrasekharappa, S. C., Yu, J., Zhang, J. & Paul Liu, P. (2006). Methods for reverse genetic screening in zebrafish by resequencing and TILLING. *Methods, 39*, 220-227.

Stuart, G. W., McMurray, J. V. & Westerfield, M. (1988). Replication, integration and stable germ-line transmission of foreign sequences injected into early zebrafish embryos. *Development, 103*, 403-12.

Summerton, J. & Weller, D. (1997). Morpholino antisense oligomers: design, preparation, and properties. *Antisense Nucleic Acid Drug Dev, 7*, 187-195.

Sun, Z., Amsterdam, A., Pazour, G. J., Cole. D. G., Miller, M. S. & Hopkins, N. (2004). Genetic screen in zebrafish identifies cilia genes as a principal cause of cystic kidney. Development, 131, 4085-4093.

Suster, M. L., Kikuta, H., Urasaki, A., Asakawa, K. & Kawakami, K. (2009). Transgenesis in zebrafish with the tol2 transposon system. *Methods Mol Biol, 561*, 41-63.

Tao, T., & Peng, J. (2009) Liver development in zebrafish (*Danio rerio*). *J Genet Genomics, 36*, 325-334.

Tu, S. & Chi, N. C. (2012). Zebrafish models in cardiac development and congenital heart birth defects. *Differentiation, 84*, 4-16.

Udvadia, A. J. & Linney, E. (2003). Windows into development: historic, current, and future perspectives on transgenic zebrafish. *Dev Biol, 256*, 1-17.

Wang, H., Long, Q., Marty, S. D., Sassa, S. & Lin, S. A zebrafish model for hepatoerythropoietic porphyria. *Nature Genet, 20*, 239-243 (1998).

Wienholds, E., Schulte-Merker, S., Walderich, B. & Plasterk, R. H. (2002). Target-selected inactivation of the zebrafish rag1 gene. *Science, 297*, 99-102.

Wienholds, E., van Eeden, F., Kosters, M., Mudde, J., Plasterk, R. H. & Cuppen, E. (2003). Efficient target-selected mutagenesis in zebrafish. *Genome Res, 13*, 2700-2707.

Yee, N. S., Yusuff, S. & Pack, M. (2001). Zebrafish pdx1 morphant displays defects in pancreas development and digestive organ chirality, and potentially identifies a multipotent pancreas progenitor cell. *Genesis, 30*, 37-40.

Zhao, X. C., Yee, R.W., Norcom, E., Burgess, H., Avanesov, A. S., Barrish, J. P. & Malicki, J. (2006). The zebrafish cornea: structure and development. *Invest Ophthalmol Vis Sci, 47*, 4341-4348.

In: Sexual Plasticity and Gametogenesis in Fishes
Editor: Balasubramanian Senthilkumaran
ISBN: 978-1-62618-848-8

Chapter XXIII

Sperm Maturation in Teleosts: Role of Androgens and Progestins in Our Present Understanding to Emerging New Concepts

Anbazhagan Rajakumar
and Balasubramanian Senthilkumaran*
Department of Animal Sciences, School of Life Sciences-Centre for Advanced Studies,
University of Hyderabad, P. O. Central University,
Hyderabad, Andhra Pradesh, India

Abstract

Reproductive success is vital for the sustainability of a species and it is tightly regulated by sex steroids which exert variable effects based on the sex and stage of gonadal development in addition to gametogenesis to time reproduction. Teleosts are good model for understanding spermatogenesis as it is synchronous and this process is conserved among many vertebrate species and is strongly regulated by gonadotropins and sex steroids. The role of sex steroids in the regulation of spermatogenesis and sperm maturation has been hypothesized and proved in many instances. In teleosts, 11-ketotestosterone is a potent androgen, which along with progestin, 17,20β-dihydroxypregn-4-en-3-one, mediates the progression of meiosis during spermatogenesis vis-à-vis sperm maturation. In this chapter, we comprehend the role of androgens and progestins in teleostean sperm maturation by emphasizing the hormone production, action and how gonadotropins or other related factors regulate this process. In addition, we also discussed about the regulation of hypothalamo-hypophyseal-gonadal axis and clarified how modulation of any of these processes assists sperm maturation and spermiation, which are critical for better seed production in pisciculture.

* Corresponding author: bsksl@uohyd.ernet.in/senthilkumaranb@yahoo.com.

Introduction

For any animal, reproductive function is very critical to sustain its own generation in environment and for better reproductive function, tightly regulated gametogenesis and fertilization is essential. Gametogenesis i.e. formation of gametes, sperm and oocyte, is under the control of gonadotropins and sex steroids like progestins, androgens and estrogens. Hypothalamo-hypophyseal-gonadal axis (H-H-G) maintains hormonal homoeostasis vis-à-vis gametogenesis and these regulations are well conserved among the class of vertebrates. Androgens, the C19 class of steroids regulate gonadal development/function and spermatogenesis and thereby promoting fertility of males. In teleosts, 11-ketotestosterone (11-KT) and testosterone (T) are the principle androgens (Borg, 1994), and in addition, they are the important stimulators for secondary sexual characters. The levels of 11-KT is higher than T in all stages of testicular development (Borg, 1994) and *in vivo/in vitro* studies showed that 11-KT is a direct stimulator of spermatogenesis (Miura *et al.*, 1991a; Borg, 1994; Cavaco et al., 1998; Amer et al., 2001; Miura and Miura, 2003), while T is a stimulator of H-H-G regulation of gonadal function (Goos et al., 1986; Amano et al., 1994; Montero et al., 1995). Progestins are C21-class of steroids with wider role which can serve as a neuro- and reproductive steroids. In teleosts, progestins include pregnenalone, progesterone (P), 17,20β-dihydroxypregn-4-en-3-one (17α,20β-DP) and 17,20β,21-trihydroxy-4-pregnen-3-one (20β-S). All these progestins can exert rapid non-genomic action through membrane progestin receptors (mPRs) which are well characterized in lower vertebrates like fishes (Thomas et al., 2006; 2009) and in turn observed in higher vertebrates including primates.

Spermatogenesis is the process by which diploid spermatogonia undergo meiosis to become primary and secondary spermatocytes, and in turn haploid spermatids. These spermatids elongate and differentiate into flagellated sperm to undergo sperm maturation and acquire motility. Many studies are available on the importance of androgens and progestins in fish gamete maturation. However, species specific differences exist along with seasonal variation in several fishes like the Japanese eel, catfish and goldfish (Ueda et al., 1985; Miura et al., 1991a; Cavaco et al., 2001).

In this chapter, we have comprehended and highlighted the importance of sex steroids, androgens and progestins, and gonadotropins in the regulation of sperm maturation of teleosts. The in-depth analysis and understanding of this process will revolutionize the fisheries industry immensely.

Spermatogenesis and Sperm Maturation in Teleosts

Spermatogenesis is conserved among many species with very few minor changes. It can be cystic (fishes and amphibians) or non-cystic type of spermatogenesis (reptiles, birds and mammals). In cystic type (fishes), spermatogenesis occurs in seminiferous tubules within cysts, which are formed of cytoplasmic extensions of Sertoli cells (Matta et al., 2002; Schulz and Miura, 2002; Schulz et al., 2005). In non-cystic type, spermatogenesis occurs in seminiferous tubules as a whole. In cystic type, the cyst forming Sertoli cells can proliferate even in adult animals, unlike non-cystic type, wherein their proliferation stops at puberty. In addition, the cysts contain synchronously developing germ cell clones and hence cystic Sertoli cells can support spermatogenesis better by supplying specific growth factors than non-cystic, where, they support different germ cell clones, which are in different stages of development (Schulz et al., 2010). In general, spermatogenesis of teleosts comprises of five processes:

1. Spermatogonial renewal
2. Spermatogonial proliferation
3. Meiosis
4. Spermiogenesis
5. Sperm maturation

During spermatogonial phases, spermatogonia undergo mitosis and form spermatogonial stem cells, and differentiating spermatogonia. The differentiated spermatogonia undergo meiosis and produce short lived primary and secondary spermatocytes, which further undergo spermiogenic phase to produce haploid spermatids. These sperrmatids undergoes spermiogenesis, wherein, a series of morphological changes like nuclear condensation, elimination of organelles and cytoplasm, flagellum formation, and the rearrangement of cellular organelles occur, which lead to the differentiation of spermatids into sperm (Jamieson, 1991; Shimizu et al., 2000). The final stage of spermatogenesis or sperm maturation, involves activation of non-functional gamete into mature/motile sperm. This process involves physiological rather than morphological changes. The role of sex steroid hormones is critical for entire spermatogenesis. Spermatogonial renewal is regulated by estradiol-17β (E_2), while spermatogonial proliferation, meiosis and spermiogenesis are regulated by 11-KT (Miura et al., 1991a; Borg, 1994; Amer et al., 2001). Progestin (17,20β-DP) regulates the sperm maturation and acquisition of motility (Miura et al., 1991b; Miura et al., 1992; Miura and Miura, 2003).

Role of Steroidogenic Enzymes in Sperm Maturation: Does, Shift in Steroidogenesis Occur in Male Teleosts?

Steroid hormones, 17α,20β-DP, 11-KT and E_2 can induce DNA synthesis, spermatogonial renewal, and/or spermatogonial proliferation *in vitro,* which impart crucial role in spermatogenesis (Amer et al., 2001). In general, androgens and progestins control mutual biosynthesis (Chen et al., 2010), further, the production of progestins and androgens are critically depends on substrate availability which in turn depends on the action of steroidogenic enzymes at different levels of steroidogenesis. The main substrate for the synthesis of T, 11-KT and 17α,20β-DP is 17α(OH)P4 or 17-P. The enzymes, 20β-hydroxysteroid dehydrogenase (17α,20β-DP production) and P450c17 (T and 11-KT production) are critical for the maintenance of androgen and progestin levels. During sperm maturation and spermiation, a clear increase in the production of 17α,20β-DP was observed in many fish species than that of T and 11-KT (Ueda et al., 1983; Sakai et al., 1989; Asturiano et al., 2002; Onuma et al., 2003; Scott et al., 2010), which is under the critical control of gonadotropins (Chen et al., 2010). In salmonids and eel, it is suggested that gonadotropins stimulate the testicular somatic cells to produce 17-P, which is then converted to 17α,20β-DP by 20β-HSD activity of spermatozoa (Miura et al., 1992; Nagahama, 1994). However, few other studies using immature testis (devoid of sperm) also showed 20β-HSD activity and thus the production of 17α,20β-DP can occur at different locations in different classes of fishes, which remain to be elucidated in detail (Vizziano and LeGac, 1998). The enzyme 20β-HSD was localized in Leydig cells and spermatogonia /spermatocytes for the first time using different stages of catfish testis and further demonstrated the stage-dependent expression of 20β-HSD during testicular cycle and ontogeny (Sreenivasulu et al., 2012). In zebrafish, 17α,20β-DP contributes to the regulation of Leydig cell steroidogenesis while in the Japanese eel, it specifically enhances 11-KT production to indirectly regulate spermatogenesis (Ozaki et al., 2006; Chen et al., 2010). Potential gonadotropic stimulation results in progestin (17α,20β-DP) production, whereas moderate gonadotropic stimulation would mainly result in androgen (T and11-KT) production (Chen et al., 2010). Whether such potential gonadotropic stimulation alone can activate steroidogenic enzyme genes critical for 17-P and 17α,20β-DP production needs to be elucidated. In the present scenario, further studies are necessary to clearly delineate the shift in steroidogenesis in male teleosts.

Role of T and 11-KT in Sperm Maturation and Regulation of Spermatogenesis

The androgens are critical for various cell processes including spermatogenesis. In teleosts, the potent androgens, T and 11-KT can bind to androgen receptors (ARs) and play vital role during testis formation and differentiation, (Borg, 1994; Sperry and Thomas, 1999). The ARs are shown to be predominantly expressed in Sertoli and interstitial cells but not in germ cells. The plasma levels of T and 11-KT increase gradually during initial stages of spermatogenesis and declined at sperm maturation. This leads to the speculation that these can play important roles during spermatogenesis, which was subsequently proven using *in vitro* studies (Miura et al., 1991a; Ozaki et al., 2006). Further, increased plasma levels of T and 11-KT were associated with increasing testicular mass throughout the reproductive cycle. During the spermatogenesis, plasma levels of 11-KT was higher than T (around four-fold) during all stages of spermatogenesis and thus 11-KT might have an important role in spermatogenic processes (Borg, 1994; Weltzien et al., 2002). During initial phases of spermatogenesis, 11-KT can induce the switch from stem cell renewal to proliferation towards meiosis. In many teleosts, 11-KT is proved to be an initiator of spermatogenesis, and sperm production while T is most effective as a stimulator of H-H-G and thereby regulates levels of gonadotropins through feedback (Goos et al., 1986; Miura et al., 1991a; Kobayashi et al., 1991; Borg, 1994). In catfish, at the testicular level, 11-KT stimulated spermatogenesis, while T inhibited 11-KT-induced spermatogenesis (Cavaco et al., 2001). Further, 11-KT is essential for regulation of spermiation and expression of secondary sexual characters (Borg, 1994; Amer et al., 2001). Based on these, we can propose that both androgens regulate spermatogenesis and sperm maturation however, they are less effective than progestins (Ueda et al., 1985).

In juvenile eel testis, 17α,20β-DP increases enzyme activity of 11β-hsd, an enzyme involved in final step of 11-KT production (Ozaki et al., 2006). Thus, 17α,20β-DP can regulate initial stages of spermatogenesis and sperm maturation indirectly through 11-KT. On the other hand, androgens can stimulate 17α,20β-DP production by down regulating P450c17 activity in the Japanese eel and catfish (Cavaco et al., 1999; Miura et al., 2006). Taken together, both androgens and progestins (17α,20β-DP) interact to control the process of spermatogenesis and spermiation.

Role of Progestins in Sperm Maturation and Spermiation

In teleosts, progestins mainly includes 17α,20β-DP, and 20β-S. In several fishes, 17α,20β-DP increase drastically during reproductive cycle and sperm maturation. The importance of progestins (17α,20β-DP and 20β-S) in the induction of final gamete maturation has been proved in many fish species, yet, molecular mechanisms underlying these processes are not understood clearly (Ueda et al., 1985; Miura et al., 1992; Nagahama, 1994; Asturiano et al., 2002; Miura and Miura, 2003).

The plasma levels of 17α,20β-DP increased during the final stages of sperm maturation, and was shown to play critical role in increasing sperm volume, stimulating sperm hydration and acquisition of sperm motility (Ueda et al., 1985; Miura et al., 1991b; 1992; Scott et al., 2010; Tubbs et al., 2011). In sea bass, successive elevations of plasma T and 11-KT levels were observed prior to 17α,20β-DP elevations resulted from the shift in gonadal steroidogenesis which coincided with the increase in sperm production (Asturiano et al., 2002). Further, the plasma levels of 17α,20β-DP in male fishes correlated well with sperm maturation (Scott et al., 2010). In salmonids 17α,20β-DP is predominantly produced by the testis, which is under the control of luteinizing hormone (LH). In testis, Leydig cells and sperm are involved in its synthesis and the factors involved in the regulation of the switch from androgen to 17α,20β-DP production, seems to occur in many species just at the time of sperm maturation and spermiation (Asturiano et al., 2002; Scott et al., 2010). We now, shift our focus to elaborate the role of 17α,20β-DP in sperm maturation and spermiation to compare with the influence of androgens on these events.

The levels of 17α,20β-DP were mainly regulated by androgens and critical steroidogenic enzymes like P450c17. However, there is a clear evidence for females wherein, the enzyme P450c17-II directs the pathway towards 17α,20β-DP at the time of final oocyte maturation (Zhou et al., 2007), yet, existence of similar scenario during sperm maturation is not clear at present. Hence, a better understanding of regulation of the enzymatic activity of P450c17 is important to delineate the mechanisms of testicular 17α,20β-DP production in teleosts. Recent studies showed that 17α,20β-DP induces male germ cells to enter meiosis I (Miura, 2006; Scott et al., 2010). In salmoniods and few other class of fishes, plasma levels of 17α,20β-DP correlated well with milt production. Spermiation can be induced early ahead of normal breeding season using 17α,20β-DP injection and implants (Ueda et al., 1985; Milla et al., 2008).

Role of Androgens and Progestins in the Regulation of Sperm Motility in Teleosts

In fishes, motility of the sperm is very critical for fertilization which is dependent on the composition of seminal fluid like ion, sialic acid, protein and lactate concentrations (Senthilkumaran and Joy, 1993; Lahnsteiner et al., 1996; 1999; Alavi and Cosson, 2005; 2006; Chowdhury and Joy, 2007; Morita et al., 2010).

In salmonids, rainbow trout, Japanese eel, and many other species the sperm do not acquire motility until they are activated by the elevated pH of the sperm duct which results in the increase of cAMP levels (Morisawa and Morisawa, 1988; Miura et al., 1992; Ohta et al., 1997a; Schulz and Miura, 2002; Alavi and Cosson, 2005). The process is also regulated by various factors, from seminal ion concentration to steroid hormones (Morisawa and Morisawa, 1988; Ohta et al., 1997a; Alavi and Cosson, 2006). The progestin, 17α,20β-DP increase seminal plasma pH and thus plays main role in acquisition of sperm motility (Miura and Miura, 2001) but the mechanism involved in this process is not clearly understood. *In vitro* treatment of milt with 20β-S improved sperm motility in the Atlantic croaker (Thomas et al., 2004). Miura and Miura (2003) showed that 17α,20β-DP binds to mPRs present on the sperm and stimulate eSRS22, a homologue of a carbonic anhydrase (CA) which results in the increase of seminal plasma pH and thus sperm acquire motility. Recently, 11-KT was shown to play an important role along with 20β-S in sperm motility and density (Hajirezaee et al., 2011; Jafaryan et al., 2012).

Other Factors/Regulators Involved in Spermatogenesis and Sperm Maturation

Sertoli Cell - A Master Regulator

Sertoli cells are the master regulators of spermatogenesis which produces various factors in response to sex steroids which are necessary to support the entire process of spermatogenesis (Ikeuchi et al., 2001; Yaron and Levavi-Sivan, 2011).

The main factors produced by Sertoli cells includes insulin-like growth factor-I (IGF-I), activins, c-kit ligand (stem cell factor), inhibin and glial cell line-derived neurotrophic factor (Yaron and Levavi-Sivan, 2011). Out of these, IGF-I and activins play critical role in spermatogenesis either directly or indirectly. IGF-I stimulates DNA synthesis in spermatogonia and support the action of 11-KT in eel spermatogenesis and hence it is critical for continuation of the process (Loir, 1994). Activins play an important role in the differential regulation of gonadotropins, follicle stimulating hormone (FSH) and LH biosynthesis (Ge, 2000; Yuen and Ge, 2004). Activins are composed of two β subunits that can form either homo-dimers (activin A (βa + βa), activin B (ββ + ββ) and/or hetero dimmers activin BA (bβa + ββ). Activin B and 17α,20β-DP indirectly regulate 11-KT mediated spermatogonial proliferation towards meiosis (Miura and Miura, 2003). Follistatin, an activin-binding protein, is anatogonistic to activin function (Yuen and Ge, 2004). Owing to the importance of Sertoli cells, number of functional Sertoli cells is directly proportional to the quantity and quality of spermatogenesis/milt production (Matta et al., 2002).

Aquaporins

Aquaporins are integral membrane protein water channels, which facilitate the rapid and highly selective flux of water and other small solutes across biological membranes (Cerda and Finn, 2010; Chauvigne et al., 2011). There are several aquaporins, which are found in testicular germ and somatic cells in mammals (Yeung, 2010). Interestingly in fishes, many aquaporin types have been cloned and were even shown to contribute to the rapid water influx during oocyte maturation (Fabra et al., 2006; Cerda 2009; Kagawa et al., 2009; 2011). A detailed nomenclature of aquaporins in fishes, permeability properties and expression profiles, are provided by Cerda and Finn (2010) and Finn and Cerda (2011). Aquaporins may also be involved in critical processes like the hydration of the seminal fluid as well as sperm motility. The aquaporins, Aqp1aa and Aqp10b are present in the head and flagellum of gilt head seabream (*Sparus aurata*) spermatozoa and are involved in sperm activation (Zilli et al., 2009).

The acquisition of sperm motility can be triggered by the sequence of events in marine fishes, the change in intracellular concentration of ions triggers cAMP production by adenylyl cyclase, further, cAMP mediated signaling pathway triggers the phosphorylation of the flagellar proteins which causes the initiation of sperm motility (Cosson et al., 2008; Zilli et al., 2009). Whether the expression of aquaporins during teleost gametogenesis is hormonally controlled needs to be clarified (Chauvigne et al., 2011) which may address several molecular regulatory mechanisms in sperm maturation. Further, it is very much essential to understand the role of aquaporin in gamete maturation, activation, and viability, which will be critical for successful cryopreservation techniques.

Trypsin

A low dose of trypsin induces DNA synthesis and the expression of Spo11 (a molecular marker of meiosis), in the testicular germ cells of Japanese eel. Higher dose of trypsin partially induced spermiogenesis and thus trypsin and/or a trypsin-like protease can play an essential role and to be regarded as a multifunctional factor in spermatogenesis (Miura et al., 2009).

Zinc (Zn)

Zn is essential for the maintenance of germ cells, progression spermatogenesis, and regulation of sperm motility (Yamaguchi et al., 2009). In the Japanese eel, germ cell proliferation, induced by 11-KT and sperm motility, regulated by $17\alpha,20\beta$-DP is maintained by Zn. Once Zn is chelated, the germ cells undergo apoptosis and motility rate of sperm decreased considerably.

Double sex and Mab-3 Related Transcription Factor 1 (Dmrt1)

Dmrt1 was shown to play an important role in testicular differentiation, development and maintenance of testicular architecture in wide variety of species including mammals (Raymond et al. 1998; Guan et al., 2000; Raghuveer and Senthilkumaran, 2009).

The sex determining gene, *DMY* is a duplicated form of *dmrt1* present in Y chromosome of the Japanese medaka, *Oryzias latipes* (Matsuda et al. 2002). Recently dmrt1 has been proved to be a gate keeper for meiosis entry during spermatogenesis in mammals (Jorgensen et al., 2012; Matson et al., 2010) and loss/mutation in/of *dmrt1* leads to the transdiffrentiation of testis into ovary in both mammals and pisces (Matson et al., 2011; Masuyama, 2012).

Regulation of Gonadotropin Synthesis and Action at the Level of H-H-G

Hypothalamus regulates synthesis and release of gonadotropins through multiple neurohormones. Gonadotropin-releasing hormone (GnRH) is a main member of hypothalamus regulation which along with several other regulators like serotonin (Somoza and Peter, 1991; Senthilkumaran et al., 2001), dopamine (Dufour et al., 2010), norepinephrine, γ-aminobutyric acid (Senthilkumaran et al., 2001), neuropeptide Y (Senthilkumaran et al., 2001), kisspeptins/GPR54 system (Biran et al., 2008; Li et al., 2009), pituitary adenylate cyclase-activating peptide, IGF-I (Wood et al., 2005; Reinecke, 2010), ghrelin (Unniappan and Peter, 2005) and leptin (Peyon et al., 2001; Trombley and Schmitz, 2012) is critical for the regulation of H-H-G (Yaron and Levavi-Sivan, 2011).

Regulation and Action of Gonadotropins at the Level of Gonads

Gonadotropins, FSH and LH regulate gametogenesis, synthesis and release of sex steroids (Chen et al., 2010) by acting via specific receptors (FSHr and LHr) in Leydig and Sertoli cells (Yoshiura et al., 1999; Schulz et al., 2010).

The control of LH and FSH synthesis is complex which involves interplay between the hypothalamus, pituitary and gonads. In teleosts, FSH and LH show different expression patterns based on sex and at different phases of reproductive cycle and gametogenesis (Planas and Swanson, 1995; Mylonas et al., 1997; Gen et al., 2000). In salmonids, plasma levels of FSH levels are elevated during the mid and late stages of spermatogenesis and it is decreased during spermiation stage. Whereas plasma levels of LH are low through most stages of spermatogenesis and it is increased during spermiation (Swanson, 1991; Mylonas et al., 1997; Gomez et al., 1999). In general, FSH and LH are equipotent in stimulating 11-KT and T production in males (Planas and Swanson, 1995). FSH and LH together stimulate 11-KT production during early spermatogenesis, while LH alone can better stimulate 17α,20β-DP production at sperm maturation and spawning in males (Planas et al., 1993; Planas and Swanson, 1995).

In teleosts, the process is delineated in many species but, the full molecular mechanisms underlying in it and the support of hormones/factors needs to be explored, which will be critical for controlled spawning. Sertoli cells are critical for regulation of spermatogenesis in all classes of animals including fishes. In mammals, FSH is an important stimulator of Sertoli cell proliferation, but in fishes, there are no specific studies available to support the notion. However, in few fishes, FSH levels increased (with low LH levels) during testicular growth and Sertoli cell proliferation stages (Gomez et al., 1999).

LH and FSH receptors are rhodopsin-like G protein-coupled receptors. In teleosts, LH receptors (LHrs) are found in both Leydig and Sertoli cells, whereas, FSH receptors (FSHrs) are expressed in Leydig cells, Sertoli cells, and possibly in early germ cells (Garcia-Lopez et al., 2009; Levavi-Sivan et al., 2010). Both LH and FSH can equally stimulate T, 11-KT and 17α,20β-DP during early and mid-stages of spermatogenesis by binding to its specific receptors (more specifically FSHr). In fish, there is some degree of cross-activation of the FSHr by LH which results in the potent steroidogenic activity at early stages of spermatogenesis (Levavi-Sivan et al., 2010). But during sperm maturation and spermiation, levels of FSH decreased and Leydig cells start expressing LHr. Hence steroidogenic potency of LH is increased drastically, which results in more 17α,20β-DP production and thus LH can likely regulate sperm maturation and motility (Swanson, 1991; Planas et al., 1993; Planas and Swanson, 1995).

Regulation of Gonadal Maturation and Reproduction by Sex Steroid Receptors

Gonadal development and gametogenesis can be controlled by steroid hormones, whose action/sensitivity is directly related to expression level of its steroid receptors. T and 11-KT acts through ARs with different specificities, while E_2 acts through estrogen receptors (ERs). In anguilliformes and perciformes, two AR were found (Ikeuchi et al., 1999; 2001), in other fishes like cypriniformes, including zebrafish only one AR is found (Ogino et al., 2009). While, in most fishes, one ERα and two ERβ forms were identified (Hawkins et al., 2000; Bardet et al., 2002; Menuet et al., 2004). Besides the genomic action, sex steroids exhibits rapid, non-genomic signaling, which is initiated at the cell-membrane. Receptors involved in estrogen signaling includes G-protein-coupled receptor 30 (GPR30) along with membrane-localized forms of ERα and ERβ (Thomas et al., 2005). Along with estrogens, androgens do exhibit non-genomic action (Braun and Thomas, 2004; Thomas et al., 2006). Progestins acts through both G-protein-coupled gestagen receptors like mPRα, mPRβ, and mPRγ along with nuclear progestin receptor (nPR; Thomas et al., 2009; Hanna et al., 2010). Progestin receptors are also found in plasma membrane of sperm (Thomas et al., 1997; 2009; Tubbs and Thomas, 2008).

In teleosts, only one isoform of nPR has been identified. In zebrafish, nuclear progesterone receptor (nPGR) regulates germ cell differentiation and steroidogenesis, whose expression is stimulated by high levels of gonadotropins. It is expressed by Leydig and Sertoli cells and is best activated by its natural ligand 17α,20β-DP (Chen et al., 2010; 2011). The levels of testicular nPGR mRNA and plasma 17α,20β-DP increased initially during the late stages of type B spermatogonia, which further increased during the entry of meiosis (Chen et al., 2012), indicating that the actions of progestins (17α,20β-DP) are also mediated through nPGRs.

Spawning in Male Fish: Challenges and Disturbances in Natural and Artificial Environment

In teleosts, spawning in natural environment is influenced by photoperiod, temperature and food availability and thus brood stock maintenance and natural spawning in laboratory/reared fish is very challenging. Any drastic changes in any of these factors create stress, which is a major limiting factor in the reproductive success of fish. Stress can significantly decrease the plasma levels of progestins (P and 17α-P) and androgens (T and 11-KT; Pickering et al., 1987; Campbell et al., 1994; Castranova, 2005) and the stressed fish produce gametes in lower number and poor quality, further, the progeny of the stressed fish had significantly lower survival rate (Campbell et al., 1992; 1994).

The other challenging factor is the manner of spermiation, in teleosts few fishes like tilapia can be made to spermiate by gentle pressure of the abdomen whereas, others like catfish, male fish needs to be sacrificed for breeding purpose in laboratory conditions (Viveiros et al., 2002; Mylonas et al., 2010), further the sperm in milt should be viable. As the sperm has increased levels of unsaturated fatty acid, it makes them extremely susceptible to damage by reactive oxygen species and thus diet rich in antioxidants along with proteins are critical for increased reproductive performance of the fishes (Izquierdo et al., 2001). Food should be free of endocrine disruptors like phyto-estrogens to avoid any possible disruption in native hormonal homeostatis.

In natural environments, fertilizer and pesticidal runoff along with sewage contamination increases biological oxidation demand and thus hypoxia, which can skew the sex of the fish population towards male along with induction of inter-sex fishes (Wu et al., 2003; Thomas and Rahman, 2011). Similarly in laboratory conditions, treatment of hormone analogues and/or pesticides caused differential distribution of spermatogenic cells (Rajakumar et al., 2012) and in few cases it lead to the development of inter-sex fish (Figure 1) wherein both spermatogenic and ovarian cell types were present (Raghuveer and Senthilkumaran, 2009).

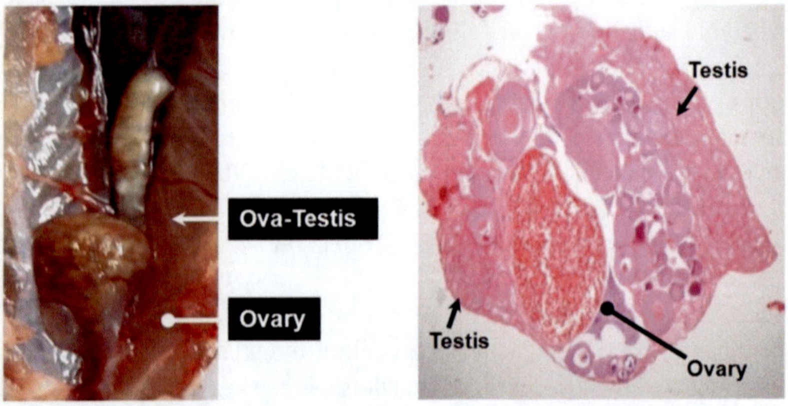

Figure 1. Representative photographs showing morphology and histology of ova-testis and ovary of catfish after methyltestosterone treatment. Detailed results regarding these has been published in Journal of Molecular Endocrinology (2009) 42, 437-448.

Common Strategies Used from Research Bench to Field Studies

Various strategies, natural methods to hormonal manipulations are employed to enhance the gamete maturation and fecundity. Hormonal manipulations are considered important in controlling puberty, as it may target H-H-G both directly and indirectly. Exogenous pituitary extracts, synthetic gonadotropins, LH preparations, synthetic agonists of GnRH (GnRHa) and human chorionic gonadotropin (hCG) are used regularly by fish farmers to increase seed production. The LH preparations, pituitary extracts and hCG can act at the level of gonads directly and even exert potent and prompt response, when the native LH stores are low. GnRHa have longer half life than native GnRH, as it is resistant to endopeptidases degradations and hence it is 30-100 times more potent than the native GnRHs in inducing LH release (Peter et al., 1988; Zohar et al., 1989; Zohar et al., 1990). Resistance to endopeptidases degradation is due to the substitutions at position 6 and 10 of GnRH decapeptide with a dextrorotatory amino acid and ethylamide group, respectively (Goren et al., 1990; Ulloa-Aguirre and Timossi, 2000). Because of this change, it remains in circulation for much longer time than the native GnRH and also exhibits increased binding affinity to the pituitary GnRH receptors (Pagelson and Zohar, 1992).

In salmonoids, catfishes and little other class of fishes, dopamine acts as a GnRH antagonist and thus it prevents the release of FSH and LH from the pituitary gonadotrophs by acting through D1 and D2 receptor (Yu and Peter, 1990; 1992; Peter et al., 1991; Silverstein et al., 1999; Yaron et al., 2003; Yaron and Levavi-Sivan, 2011). Hence, GnRHa in combination with D2 blocker like domperidone, pimozide is effective in inducing spawning (Tharakan and Joy, 1996; Joy and Tharakan, 1999) which has been officially approved for use in hormonal therapies in aquaculture fish and thus it provides immense help to fish farmers and aquaculture industry.

GnRHa acts at the level of the pituitary and can induce release of the endogenous LH/FSH, which in turn acts at the level of gonad to induce steroidogenesis, sperm/oocyte maturation and spermiation/ovulation. The main advantage of GnRHa is that it is synthesized in pure form so there are no chances for disease transmission. Further, it acts at higher levels in H-H-G and hence there is a more integrated response than hCG or gonadotropin preparations, where regulation of GH, IGFs and thyroid hormone are also maintained (Le Gac et al., 1993; Negatu et al., 1998). Synthetic GnRH/gonadotrophins were shown to be effective in inducing sexual maturation/spawning of several commercially important fishes (Matsuyama et al., 1995; Joy et al., 1998; Kumakura et al., 2003).

Recent progress in the understanding of mechanisms of gamete maturation helped the aquaculture sector to produce economically important fishes by hormonal induction and maturation. One of the best examples is maturation of glass eel in the controlled/laboratory conditions (Tanaka, 2003; Kagawa et al., 2005). The commercial hCG, can easily induce maturation and spermiation (Khan et al., 1987; Miura et al., 1991c; Ohta et al., 1996; Ohta et al., 1997b; Ohta and Tanaka, 1997; Schiavone et al., 2006) in fishes. Usually hCG is given in a single dose (vary from 100-4000 international units (IU)/Kg) which has longer half-life in circulation, but the efficacy depends on its rate of clearance, which vary among different fish species (Ohta and Tanaka, 1997). Hence, modulation/supplementation of other factors are needed that can best suit hCG treatments to induce spermiation.

Off Season Spawning in Seasonal Breeders and Spermiation Enhancement

Injection of 17α,20β-DP or delivery through silastic implants to rainbow trout also induced spermiation in males ahead of the normal season (Marshall et al., 1989; Milla et al., 2008; Scott et al., 2010). Further, artificial induction of sexual maturation and spawning of immature fishes using GnRHa in non-spawning season were also carried out using various modes of delivery systems including osmotic pump (Kumakura et al., 2003; Kanemaru et al., 2012). In addition, 17α,20β-DP and 20β-S injection in already spermiating males increased the milt volume several folds within a short time (Pankhurst, 1994 ; Yueh and Chang, 1997). The injections of 17-P (precursor for 17α,20β-DP) and hCG are also effective in inducing spermiation. But, studies regarding the role of 17α,20β-DP in increasing milt volume are quite contradictory. In, rainbow trout, 17α,20β-DP helps to spermiate the fish early but there was no increase in milt volume (Milla et al., 2008). Further, few studies showed that the main role of 17α,20β-DP is to increase the percentage of motile sperm by increasing seminal plasma pH and cAMP levels (Morisawa and Morisawa, 1988; Miura et al., 1992; Ohta et al., 1997a; Schulz and Miura, 2002; Alavi and Cosson, 2005). Based on these, we can conclude that 17α,20β-DP advances the onset of spermiation in non-spermiating males and increases milt production in already spermiating fishes.

Future Perspectives

Improved aquaculture practices had revolutionized the brood stock management and seed production, however, further understanding of molecular mechanisms underlying hormone synthesis and action at native physiological conditions, will assist to develop better reproductive strategies.

Effective New Analogue Preparations with Combination

Most methodologies and strategies are used commonly for most of the fishes in general, but each and every class of fishes has slightly varied regulatory responses at H-H-G during gametogenesis and gamete maturation. Hence, a targeted approach for a given class of fishes will be more appropriate. Further, a combination of more than one analogue can evoke a more effective and specific response, which will enable fish farmers to attain better fertilization rate and seed production.

Osmotic Pump Mediated Hormone or Hormone Supplement Delivery

Mature fishes are supplemented with hormones/hormone supplements in order to provoke gamete maturation and spawning responses. Most of the time these administration has been done via injections, which leads to repeated animal handling and increase in stress to the animal. Hence sustained release of hormones in the form of implants and mini-osmotic pumps will be effective. Cholesterol and polymer pellets were commonly used for hormone/hormone analogue release however, osmotic pump mediated delivery is a better option for long term delivery (Kagawa et al., 2009). Osmotic pump is a simple delivery device with no electronic parts, which releases loaded hormones or drugs constantly into the targeted tissue just merely through osmotic process. Osmotic pump mediated hormone delivery in fishes was shown to be very effective in inducing spawning response with less stress to the animal. Sustained release of hCG, salmon pituitary extract and GnRHa using osmotic pump was shown to induce sexual maturation and spermatogenesis in the Japanese eel (Kagawa et al., 2009). Further this report is the first of its kind to show, sustained release of hCG using single osmotic pump stimulating spermatogenesis and spermiation in sexually immature Japanese eels. This will be the best option for sustained hormone delivery in bigger fishes, where handling and monitoring the animals is a cumbersome job. Further, details regarding osmotic pump mediated hormone delivery and sexual maturation are discussed in detail in the chapter written by Prof. Hirohiko Kagawa in this book.

Improving Sperm and Sperm Motility Parameters

Active motile sperm is vital for successfull fertilization. Sperm motility varies with composition of seminal plasma which itself is regulated by nutritional diet of the animal. Seminal plasma provide optimal environment for sperm and can increase its viability/ motility, which are critical for the success of fertilization (Alavi and Cosson, 2006; Bozkurt et al., 2009). Composition of the seminal plasma is species-specific and hence artificial seminal plasma (ASP) solutions can be prepared, specifically for a given species and used for milt dilution, short-term storage and cryopreservation.

Modifying the composition of seminal plasma can improve sperm motility and other parameters significantly and the importance of ASP in regulating the acquisition of motility in sperm is long known (Morisawa and Morisawa, 1988; Ohta and Unuma, 2003). Further, the rate of motility of the spermatozoa can be controlled by adjusting the ion and sugar concentrations of ASP. This can produce "Super sperm" with higher motility, which can increase fertilization rates with less quantity of milt and can minimize the number of males required for the *in vitro* fertilization.

A targeted approach with efficient novel hormone analogues and improved delivery systems will pave way for successful controlled fish reproduction which will revolutionize the brood stock management and aquaculture practices.

Conclusion

This review presents the current understanding of the regulation of sperm maturation by androgens and progestins with a special emphasis on gonadotropins and identified non-endocrinal regulators, involved in this process.

Further, identification of novel factors and molecular mechanisms underlying this, assist to, "turn on" and "turn off" the expression of these specific factors at specific times during various stages of spermatogenesis which may provide control over the timing of annual reproductive cycle and sperm maturation. Combined strategies targeting different controls at the level of brain can produce a cohesive response similar to native hormonal conditions, which will enable fish farmers to maintain quality brood stocks to increase seed production profoundly for sustainable aquaculture.

Acknowledgments

Research work mentioned in this review was supported by the grant-in-aids from Department of Biotechnology (BT/PR11263/AAQ/03/419/2008) and University Grants Commission (34-412/2008, 30-195/2004[SR]) of the Government of India for which the authors are thankful. AR is thankful to the Council of Scientific and Industrial Research, India for junior and senior research fellowships.

References

Alavi, S. M. and Cosson, J. (2005). Sperm motility in fishes. I. Effects of temperature and pH: a review. *Cell Biol. Int.*, 29, 101-110.

Alavi, S. M. and Cosson, J. (2006). Sperm motility in fishes. II. Effects of ions and osmolality: a review. *Cell Biol. Int., 30,* 1-14.

Amer, M. A, Miura, T., Miura, C. and Yamauchi, K. (2001). Involvement of sex steroid hormones in the early stages of spermatogenesis in Japanese huchen (*Hucho perryi*). *Biol. Reprod.,* 65, 1057-1066.

Asturiano, J. F., Sorbera, L. A., Ramos, J., Kime, D.E., Carrillo, M. and Zanuy, S. (2002). Group-synchronous ovarian development, ovulation and spermiation in the European sea bass (*Dicentrarchus labrax* L.) could be regulated by shifts in gonadal steroidogenesis. *Sci. Mar.,* 66, 273-282

Bardet, P-L., Horard, B., Robinson-Rechavi, M., Laudet, V. and Vanacker, J-M. (2002). Characterization of oestrogen receptors in zebrafish (*Danio rerio*). *J. Mol. Endocrinol.,* 28, 153-163.

Biran, J., Ben-Dor, S. and Levavi-Sivan, B. (2008). Molecular identification and functional characterization of the kisspeptin/kisspeptin receptor system in lower vertebrates. *Biol. Reprod.,* 79, 776-786.

Borg, B. (1994). Androgens in teleost fishes. *Comp. Biochem. Physiol.,* 109C, 219-245.

Bozkurt, Y., Ogretmen, F., Secer, F. S. and Ercin, U. (2009). Relationship between seminal plasma composition and spermatological parameters in scaly carp (*Cyprinus carpio*). *J. Anim. Vet. Adv.,* 8, 2745-2749.

Braun, A. and Thomas, P. (2004). Biochemical characterization of a membrane androgen receptor in the ovary of the Atlantic croaker, (*Micropogonias undulatus*) *Biol. Reprod.,* 71, 146-155.

Campbell, P. M., Pottinger, T. G. and Sumpter, J. P. (1992). Stress reduces the quality of gametes produced by rainbow trout. *Biol. Reprod.,* 47, 1140-1150.

Campbell, P. M., Pottinger, T. G. and Sumpter, J. P. (1994). Preliminary evidence that chronic confinement stress reduced the quality of gametes produced by brown and rainbow trout. *Aquaculture,* 120, 151-169.

Castranova, D. A., King, V. W. and Woods III, L. C. (2005). The effects of stress on androgen production, spermiation response and sperm quality in high and low cortisol responsive domesticated male striped bass. *Aquaculture,* 246, 413-422.

Cavaco, J. E. B., Blijswijk, Van B., Leatherland, J. F., Goos, H. J. T. and Schulz, R. W. (1999). Androgen-induced changes in Leydig cell ultrastructure and steroidogenesis in juvenile African catfish, *Clarias gariepinus. Cell Tissue Res.,* 297, 291-299.

Cavaco, J. E. B., Bogerd, J., Goos, H. J. Th. and Schulz, R. W. (2001). Testosterone inhibits 11-Ketotestosterone induced spermatogenesis in African catfish (*Clarias gariepinus*). *Biol. Reprod.,* 65, 1807-1812.

Cavaco, J. E. B., Vilrokx, C., Trudeau, V. L., Schulz, R. W. and Goos, H. J. T. (1998). Sex steroids and the initiation of puberty in male African catfish, *Clarias gariepinus. Am. J. Physiol.,* 275, 1793-1802.

Cerda, J. (2009). Molecular pathways during marine fish egg hydration: the role of aquaporins. *J. Fish Biol.,* 75, 2175-2196.

Cerda, J. and Finn, R. N. (2010). Piscine aquaporins: an overview of recent advances. *J. Exp. Zool.,* 313A, 623-650.

Charlton, H. (2008). Hypothalamic control of anterior pituitary function: a history. *J. Neuroendocrinol.,* 20, 641-646.

Chauvigne, F., Zapater, C. and Cerda, J. (2011). Role of aquaporins during teleost gametogenesis and early embryogenesis. *Front. Physio.,* 2, 1-6.

Chen, S. X., Almeida, F. F. L., Andersson, E., Taranger, G. L., Schmidt, R., Schulz, R. W. and Bogerd, J. (2012). Cloning, pharmacological characterization and expression analysis of Atlantic cod (*Gadus morhua,* L.) nuclear progesterone receptor. *Gen. Comp. Endocrinol.,* 179, 71-77.

Chen, S. X., Bogerd, J., Andersson, E., Almeida, F. F. L., Taranger, G. L. and Schulz, R. W. (2011). Cloning, pharmacological characterization, and expression analysis of Atlantic salmon (*Salmo salar* L.) nuclear progesterone receptor. *Reproduction,* 141, 491-500.

Chen, S. X., Bogerd, J., Garcia-Lopez, A., de Jonge, H., de Waal, P. P., Hong, W. S. and Schulz, R. W. (2010). Molecular cloning and functional characterization of a zebrafish nuclear progesterone receptor. *Biol. Reprod.,* 82, 171-181.

Chowdhury, I. and Joy, K. P. (2007). Seminal vesicle and its role in the reproduction of teleosts. *Fish Physiol. Biochem.,* 33, 383-398.

Cosson, J., Dreanno, C., Fauvel, C., Groison, A. L., Suquet, M. and Billard, R. (2008). Marine fish spermatozoa: racing ephemeral swimmers. *Reproduction,* 136, 277-294.

De Leeuw, R., Goos, H. J. and Van Oordt, P. G. (1986). The Dopaminergic Inhibition of the Gonadotropin-Releasing Hormone-Induced Gonadotropine Release: an *in vitro* Study with Fragments and Cell Suspensions from Pituitaries of the African Catfish, *Clarias gariepinus* (Burchell), *Gen. Comp. Endocrinol.,* 63, 171-177.

Dufour, S., Sebert, M. E., Weltzien, F. A., Rousseau, K. and Pasqualini, C. (2010). Neuroendocrine control by dopamine of teleost reproduction. *J. Fish Biol.,* 76, 129-160.

Fabra, M., Raldua, D., Bozzo, M. G., Deen, P. M. T., Lubzens, E. and Cerda, J. (2006). Yolk proteolysis and aquaporin-1o play essential roles to regulate fish oocyte hydration during meiosis resumption. *Dev. Biol.,* 295, 250-262.

Filby, A. L., van Aerle, R., Duitman, J. and Tyler, C. R. (2008). The kisspeptin/gonadotropin-releasing hormone pathway and molecular signaling of puberty in fish. *Biol. Reprod.,* 78, 278-289.

Finn, R. N. and Cerda, J. (2011). Aquaporin evolution in fishes. *Front. Physiol.,* 2, 44.

Garcia-Lopez, A., Bogerd, J., Granneman, J. C. M., van Dijk, W., Trant, J. M., Taranger, G. L. and Schulz, R. W. (2009). Leydig cells express follicle-stimulating hormone receptors in African catfish. *Endocrinology,* 150, 357–365.

Ge, W., (2000). Roles of the activin regulatory system in fish reproduction. *Can. J. Physiol. Pharmacol.,* 78, 1077-1085.

Gen, K., Okuzawa, K., Senthilkumaran, B., Tanaka, H., Moriyama, S. and Kagawa, H. (2000). Unique expression of gonadotropin-I and –II subunit genes in male and female red seabream (*Pagrus major*) during sexual maturation. *Biol. Reprod.,* 63, 308-319.

Gomez, J. M., Weil, C., Ollitrault, M., Le Bail, P.-Y., Breton, B. and Le Gac, F. L. (1999). Growth hormone (GH) and gonadotropin subunit gene expression and pituitary and plasma changes during spermatogenesis and oogenesis in rainbow trout (*Oncorhynchus mykiss*). *Gen. Comp. Endocrinol.,* 113, 413-428.

Goos, H. J. Th., de Leeuw, R., Cook, H. and van Oordt, P. G. W. J. (1986). Gonadotropin-releasing hormone (GRH) bioactivity in the brain of the immature rainbow trout, *Salmo gairdneri*: the effect of testosterone. *Gen. Comp. Endocrinol.,* 64, 80-84.

Goren, A., Zohar, Y., Fridkin, M., Elhanati, E. and Koch, Y. (1990). Degradation of gonadotropin-releasing hormones in the gilthead seabream, Sparus aurata I. Cleavage of native salmon GnRH and mammalian LHRH in the pituitary. *Gen. Comp. Endocrinol.,* 79, 291-305.

Guan, G., Kobayashi, T. and Nagahama, Y. (2000). Sexually dimorphic expression of two types of DM (Doublesex/Mab-3)-domain genes in a teleost fish, the tilapia (*Oreochromis niloticus*). *Biochem. Biophys. Res. Commun.,* 272, 662-666.

Hajirezaee, S., Rafiee, Gh. R. and Hushangi, R. (2011). Comparative analysis of milt quality and steroid levels in blood and seminal fluid of Persian sturgeon males, *Acipenser persicus* during final maturation induced by hormonal treatment. *Biologia,* 66,160-169.

Hanna, R. N., Daly, S. C., Pang, Y., Anglade, I., Kah, O., Thomas, P. and Zhu, Y. (2010). Characterization and expression of the nuclear progestin receptor in zebrafish gonads and brain. *Biol. Reprod.,* 82, 112-122.

Hawkins, M. B, Thornton, J. W, Crews, D., Skipper, J. K., Dotte, A. and Thomas, P. (2000). Identification of a third distinct estrogen receptor and reclassification of estrogen receptors in teleosts. *Proc. Natl. Acad. Sci. USA,* 97, 10751-10756.

Ikeuchi, T., Todo, T., Kobayashi, T. and Nagahama, Y. (1999). cDNA cloning of a novel androgen receptor subtype. *J. Biol. Chem.,* 274, 25205-25209.

Ikeuchi, T., Todo, T., Kobayashi, T. and Nagahama, Y. (2001). Two subtypes of androgen and progestogen receptors in fish testes. *Comp. Biochem. Physiol.,* 129B, 449-455.

Izquierdo, M. S., Fernandez-Palacios, H. and Tacon, A. G. J. (2001). Effect of broodstock nutrition on reproductive performance of fish. *Aquaculture,* 197, 25-42.

Jafaryan, H., Akhoundian, M., Mehrpoosh, M., Hajirezaee, S. and Golpour, A. (2012). Relationships between sex steroids of seminal fluid and milt quality parameters of Persian sturgeon, *Acipenser persicus*. *J. Appl. Biol. Sci.,* 6, 41-44.

Jorgensen, A., Nielsen, J. E., Jensen, M. B., Graem, N. and Meyts, E. R. (2012). Analysis of meiosis regulators in human gonads: a sexually dimorphic spatio-temporal expression pattern suggests involvement of DMRT1 in meiotic entry. *Mol. Hum. Reprod.,* 18, 523-534.

Joy, K. P., Senthilkumaran, B. and Sudhakumari, C. C. (1998). Periovulatory changes in hypothalamic and pituitary monoamines following GnRH analogue treatment in the catfish *Heteropneustes fossilis*: a study correlating changes in plasma hormone profiles. *J. Endocrinol.,* 156, 365-372.

Joy, K. P. and Tharakan, B. (2008). Induced spawning of the Indian catfish, *Heteropneustes fossilis*, by GnRH analogue alone or in combination with dopamine-affecting drugs. *J. Appl. Aquacult.,* 9, 23-32

Kagawa, H., Kasuga, Y., Adachi, J., Nishi, A., Hashimoto, H., Imaizumi, H. and Kaji, S. (2009). Effects of continuous administration of human chorionic gonadotropin, salmon pituitary extract, and gonadotropin-releasing hormone using osmotic pumps on induction of sexual maturation in male Japanese eel, *Anguilla japonica*. *Aquaculture,* 296,117-122.

Kagawa, H., Kishi, T., Gen, K., Kazeto, Y., Tosaka, R., Matsubara, H., Matsubara, T. and Sawaguchi, S. (2011). Expression and localization of aquaporin1b during oocyte development in the Japanese eel (*Anguilla japonica*). *Reprod. Biol. Endocrinol.,* 9, 71.

Kagawa, H., Tanaka, H., Ohta, H., Unuma, T. and Nomura, K., (2005). The first success of glass eel production in the world: basic biology on fish reproduction advances new applied technology in aquaculture. *Fish Physiol. Biochem.,* 31, 193-199.

Kanemaru, T., Nakamura, M., Murata, R., Kuroki, K., Horie, H., Uchida, K., Senthilkumaran, B. and Kagawa, H. (2012). Induction of sexual maturation of the female honeycomb grouper, *Epinephelus merra*, in the non-breeding season by modulating environmental factors with GnRH analogue implantation. *Aquaculture,* 358-359, 85-91.

Khan, I. A., Lopez, E. and Leloup-Hatey, J. (1987). Induction of spermatogenesis and spermiation by a single injection of human chorionic gonadotropin in intact and hypophysectomized immature European eel (*Anguilla anguilla*). *Gen. Comp. Endocrinol.,* 68, 91-103.

Kobayashi, M., Aida, K. and Stacey, N. E. (1991). Induction of testis development by implantation of 11-ketotestosterone in female goldfish. *Zool. Sci.,* 8, 389-393.

Kumakura, N., Okuzawa, K., Gen, K. and Kagawa, H. (2003). Effects of gonadotropin-releasing hormone agonist and dopamine antagonist on hypothalamus-pituitary-gonadal axis of pre-pubertal female red seabream (*Pagrus major*). *Gen. Comp. Endocrinol.,* 131, 264-273.

Lahnsteiner, F., Berger, B. and Weismann, T. (1999). Sperm metabolism of the teleost fishes *Chalcalburnus chalcoides* and *Oncorhynchus mykiss* and its relation to motility and viability. *J. Exp. Zool.,* 284, 454-465.

Lahnsteiner, F., Berger, B., Weismann, T. and Patzner, R. (1996). Motility of spermatozoa of *Alburnus alburnus* (Cyprinidae) and its relationship to seminal plasma composition and sperm metabolism. *Fish Physiol. Biochem.,* 15, 167-179.

Le Gac, F., Blaise, O., Fostier, A., Le Bail, P. Y., Loir, M., Mourot, B. and Weil, C. (1993). Growth hormone (GH) and reproduction: a review. *Fish Physiol. Biochem.,* 11, 219-232.

Levavi-Sivan, B., Bogerd, J., Mananos, E. L., Gomez, A. and Lareyre, J. J. (2010). Perspectives on fish gonadotropins and their receptors. *Gen. Comp. Endocrinol.,* 1653, 412-437.

Li, S., Zhang, Y., Liu, Y., Huang, X., Huang, W., Lu, D., Zhu, P., Shi, Y., Cheng, C. H. K., Liu, X. and Lin, H. (2009). Structural and functional multiplicity of the kisspeptin/GPR54 system in goldfish (*Carassius auratus*). *J. Endocrinol.,* 201, 407-418.

Marshall, W. S., Bryson, S. E. and Idler, D. E. (1989). Control of ion transport by the sperm duct epithelium of brook trout (*Salvelinus fontinalis*). *Fish Physiol. Biochem.,* 7, 331-336.

Masuyama, H., Yamada, M., Kamei, Y., Fujiwara-Ishikawa, T., Todo, T., Nagahama, Y. and Matsuda, M. (2012). Dmrt1 mutation causes a male-to-female sex reversal after the sex determination by Dmy in the medaka. *Chromosome Res.,* 20, 163-176.

Matson, C. K., Murphy, M. W., Griswold, M. D., Yoshida, S., Bardwell, V. J. and Zarkower, D. (2010). The mammalian doublesex homolog DMRT1 is a transcriptional gatekeeper that controls the mitosis versus meiosis decision in male germ cells. *Dev. Cell,* 19, 612-624.

Matson, C. K., Murphy, M. W., Sarver, A. L., Griswold, M. D., Bardwell, V. J. and Zarkower, D. (2011). DMRT1 prevents female reprogramming in the postnatal mammalian testis. *Nature,* 476, 101-104.

Matsuda, M., Nagahama, Y., Shinomiya, A., Sato, T., Matsuda, C., Kobayashi, T., Morrey, C. E., Shibata, N., Asakawa, S., Shimizu, N., Hori, H., Hamaguchi, S. and Sakaizumi, M. (2002). DMY is a Y-specific DM-domain gene required for male development in the medaka fish. *Nature,* 417, 559-63.

Matsuyama, M., Takeuchi, H., Kashiwagi, M., Hirose, K. and Kagawa, H. (1995). Induced gonadal development and spawning of immature red sea bream Pagrus major with LHRHa administration in different ways during winter season. *Fish. Sci.,* 61, 472-477.

Matta, S. L., Vilela, D. A., Godinho, H. P. and França, L. R. (2002). The goitrogen 6-n-propyl-2-thiouracil (PTU) given during testis development increases Sertoli and germ cell numbers per cyst in fish: the tilapia (*Oreochromis niloticus*) model. *Endocrinology,* 143, 970-978.

Menuet, A., Le Page, Y., Torres, O., Kern, L., Kah, O. and Pakdel, F. (2004). Analysis of the estrogen regulation of the zebrafish estrogen receptor (ER) reveals distinct effects of ERα, ERβ1 and ERβ2. *J. Mol. Endocrinol.,* 32, 975-986.

Milla, S., Terrien, X., Sturm, A., Ibrahim, F., Giton, F., Fiet, J., Prunet, P. and Le Gac, F. (2008). Plasma 11-deoxycorticosterone (DOC) and mineralocorticoid receptor testicular expression during rainbow trout *Oncorhynchus mykiss* spermiation: implication with 17α,20β-dihydroxy-progesterone on milt fluidity. *Reprod. Biol. Endocrinol.,* 6, 19.

Miura, C., Ohta, T., Ozaki, Y., Tanaka, H. and Miura T. (2009). Trypsin is a multifunctional factor in spermatogenesis. *Proc. Natl. Acad. Sci. USA,* 106, 20972-20977.

Miura, T., Higuchi, M., Ozaki, Y., Ohta, T. and Miura, C. (2006). Progestin is an essential factor for the initiation of the meiosis in spermatogenetic cells of the eel. *Proc. Natl. Acad. Sci. USA,* 103, 7333-7338.

Miura, T., Yamauchi, K., Nagahama, Y. and Takahashi, H. (1991c). Induction of spermatogenesis in male Japanese eel, *Anguilla japonica*, by a single injection of human chorionic gonadotropin. *Zool. Sci.,* 8, 63-73.

Miura, T., Yamauchi, K., Takahashi, H. and Nagahama, Y. (1991b). Involvement of steroid hormones in gonadotropin-induced testicular maturation in male Japanese eel (*Anguilla japonica*). *Biomed. Res.,* 12, 241-248.

Miura, T., Yamauchi, K., Takahashi, H. and Nagahama, Y. (1992). The role of hormones in the acquisition of sperm motility in salmonid fish. *J. Exp. Zool.,* 261, 359-363.

Miura, T. and Miura, C. I. (2001). Japanese eel: a model for analysis of spermatogenesis. *Zool. Sci.,* 18, 1055-1063.

Miura, T. and Miura, C. I. (2003). Molecular control mechanisms of fish spermatogenesis. *Fish Physiol. Biochem.,* 28, 181-186.

Miura, T., Yamauchi, K., Takahashi, H. and Nagahama, Y. (1991a). Hormonal induction of all stages of spermatogenesis *in vitro* in the male Japanese eel (*Anguilla japonica*). *Proc. Nat.Acad. Sci. USA,* 88, 5774-5778.

Montero, M., Le Belle, N., King, J. A., Millar, R. P. and Dufour, S. (1995). Differential regulation of the two forms of gonadotropin- releasing hormone (mGnRH and cGnRH-II) by sex steroids in the European female silver eel (*Anguilla anguilla*). *Neuroendocrinology,* 61, 525-535.

Morisawa, S. and Morisawa, M. (1988). Induction of potential for sperm motility by bicarbonate and pH in rainbow trout and chum salmon. *J. Exp. Biol.,* 136, 13-22.

Morita, M., Awata, S., Takahashi, T., Takemura, A. and Kohda, M. (2010). Sperm motility adaptation to ion-differing aquatic environments in the Tanganyikan cichlid, *Astatotilapia burtoni. J. Exp. Zool.,* 313A, 169-177.

Mylonas, C. C., Scott, A. P. and Zohar, Y. (1997). Plasma gonadotropin II, sex steroids, and thyroid hormones in wild striped bass (*Morone saxatilis*) during spermiation and final oocyte maturation. *Gen. Comp. Endocrinol.,* 108, 223-236.

Mylonas, C. C., Fostier, A. and Zanuy, S. (2010). Broodstock management and hormonal manipulations of fish reproduction. *Gen. Comp. Endocrinol.,* 165, 516-534.

Nagahama, Y., 1994. Endocrine regulation of gametogenesis in fish. *Int. J. Dev. Biol.,* 38, 217-229.

Negatu, Z., Hsiao, S. M. and Wallace, R. A. (1998). Effects of insulin-like growth factor-I on final oocyte maturation and steroid production in *Fundulus heteroclitus. Fish Physiol. Biochem.,* 19, 13-21.

Ogino, Y., Katoh, H., Kuraku, S. and Yamada, G. (2009). Evolutionary history and functional characterization of androgen receptor genes in jawed vertebrates. *Endocrinology,* 150, 5415-5427.

Ohta, H., Kagawa, H., Tanaka, H., Okuzawa, K. and Hirose, K. (1996). Milt production in the Japanese eel *Anguilla japonica* induced by repeated injections of human chorionic gonadotropin. *Fish. Sci.,* 62, 44-49.

Ohta, H., Ikeda, K. and Izawa, T. (1997a). Increase in concentrations of potassium and bicarbonate ions promote acquisition of motility *in vitro* by Japanese eel spermatozoa. *J. Exp. Zool.,* 277, 171-180.

Ohta, H., Kagawa, H., Tanaka, H., Okuzawa, K., Iimura, N. and Hirose, K. (1997b). Artificial induction of maturation and fertilization in the Japanese eel, *Anguilla japonica. Fish Physiol. Biochem.,* 17, 163-169.

Ohta, H. and Tanaka, H. (1997). Relationship between serum levels of human chorionic gonadotropin (hCG) and 11-ketotestosterone after a single injection of hCG and induced maturity in the male Japanese eel, *Anguilla japonica. Aquaculture,* 153, 123-134.

Ohta, H. and Unuma, T. (2003). Induction of sperm maturation. In: K. Aida, K. Tsukamoto, K. Yamauchi (Eds.), Eel biology (pp. 415-423). Melbourne: Blackwell Publishing Asia.

Onuma, T., Higashi, Y., Ando, H., Ban, M., Ueda, H. and Urano, A. (2003). Year-to-year differences in plasma levels of steroid hormones in pre-spawning chum salmon. *Gen. Comp. Endocrinol.,* 133, 199-215.

Ozaki, Y., Higuchi, M., Miura, C., Yamaguchi, S., Tozawa, Y. and Miura T. (2006). Roles of 11β-hydroxysteroid dehydrogenase in fish spermatogenesis. *Endocrinology,* 147, 5139-5146.

Pagelson, G. and Zohar, Y. (1992). Characterization of gonadotropin-releasing hormone binding to pituitary receptors in the gilthead seabream (*Sparus aurata*). *Biol. Reprod.,* 47, 1004-1008.

Pankhurst, N. W. (1994). Effects of gonadotropin releasing hormone analogue, human chorionic gonadotropin and gonadal steroids on milt volume in the New Zealand snapper, *Pagrus auratus* (Sparidae). *Aquaculture,* 125, 185-197.

Peter, R. E., Lin, H. R. and Van Der Kraak, G. (1988). Induced ovulation and spawning of culture freshwater fish in China: advances in application of GnRH analogues and dopamine antagonists. *Aquaculture,* 74, 1-10.

Peter, R.E., Trudeau, V. L. and Sloley, B. D. (1991). Brain regulation of reproduction in teleosts. *Bull. Inst. Zool. Acad. Sin. Monogr.,* 16, 89-118.

Peyon, P., Zanuy, S. and Carrillo, M. (2001). Action of leptin on *in vitro* luteinizing hormone release in the European sea bass (*Dicentrarchus labrax), Biol. Reprod.,* 65, 1573-1578.

Pickering, A. D., Pottinger, T. G., Carragher, J. and Sumpter, J. P. (1987). The effects of acute and chronic stress on the levels of reproductive hormones in plasma of mature male brown trout, *Salmo trutta* L. *Gen. Comp. Endocrinol.,* 68, 249-259.

Planas, J. V. and Swanson, P. (1995). Maturation-associated changes in the response of the salmon testis to the steroidogenic actions of gonadotropins (GTH I and GTH II) *in vitro. Biol. Reprod.,* 52, 697-704.

Planas, J. Y., Swanson, P. and Dickhoff, W. W. (1993). Regulation of testicular steroid production in vitro by gonadotropins (GTHI and GTH I) and cyclic AMP in coho salmon (*Oncorhynchus kisutch*). *Gen. Comp. Endocrinol.,* 91, 8-24.

Raghuveer, K. and Senthilkumaran, B. (2009). Identification of multiple *dmrt1*s in catfish: localization, dimorphic expression pattern, changes during testicular cycle and after methyltestosterone treatment. *J. Mol. Endocrinol.,* 42, 437-448.

Rajakumar, A., Singh, R., Chakrabarty, S., Murugananthkumar, R., Laldinsangi, C., Prathibha, Y., Sudhakumari, C. C., Dutta-Gupta, A. and Senthilkumaran, B. (2012). Endosulfan and flutamide impair testicular development in the juvenile Asian catfish, *Clarias batrachus, Aquat. Toxicol.,* 110-111, 123-132.

Raymond, C. S., Shamu, C. E., Shen, M. M., Seifert, K. J., Hirsch, B., Hodgkin, J. and Zarkower, D. (1998). Evidence for evolutionary conservation of sexdetermining genes. *Nature,* 391, 691-695.

Reinecke, M. (2010). Insulin-like growth factors and fish reproduction. *Biol. Reprod., 82,* 656-661.

Sakai, N., Ueda, H., Suzuki, N. and Nagahama, Y. (1989). Steroid production by amago salmon (*Oncorhynchus rhodurus*) testes at different developmental stages. *Gen. Comp. Endocrinol.,* 75, 231-240.

Schiavone, R., Zilli, L., Vilella, S. and Fauvel, C. (2006). Human chorionic gonadotropin induces spermatogenesis and spermiation in 1-year-old European sea bass (*Dicentrarchus labrax*): assessment of sperm quality. *Aquaculture,* 255, 522–531.

Schulz, R. W., de Francc, L. R, Lareyre, J. J, LeGac, F., Chiarini-Garciac, H., Nobrega, R. H. and Miura, T. (2010). Spermatogenesis in fish. *Gen. Comp. Endocrinol.*, 165, 390-411.

Schulz, R. W., Menting, S., Bogerd, J., Franca, L. R., Vilela, D. A. R. and Godinho, H. P. (2005). Sertoli cell proliferation in the adult testis: evidence from two fish species belonging to different orders. *Biol. Reprod.,* 73, 891-898.

Schulz, R. W. and Miura, T. (2002). Spermatogenesis and its endocrine regulation. *Fish Physiol. Biochem.,* 26, 43-56.

Scott, A. P., Sumpter, J. P. and Stacey, N. (2010). The role of the maturation-inducing steroid, 17,20β-dihydroxypregn-4-en-3- one, in male fishes: a review. *J. Fish Biol., 76, 183*-224.

Senthilkumaran, B. and Joy, K. P. (1993). Annual cyclic, and castration and cyproterone acetate-induced, changes in sialic acid content of the seminal vesicle of the catfish, *Heteropneustes fossilis* (Bloch). *Fish Physiol. Biochem.,* 10, 425-430.

Senthilkumaran, B., Okuzawa, K., Gen, K. and Kagawa, H. (2001). Effects of serotonin, GABA and neuropeptide Y on seabream gonadotropin releasing hormone *in vitro* from preoptic-anterior hypothalamus and pituitary of red seabream, *Pagrus major. J. Neuroendocrinol.,* 13, 395-400.

Shimizu, Y., Mita, K., Tamura, M., Onitake, K. and Yamashita, M. (2000). Requirement of protamine for maintaining nuclear condensation of medaka (*Oryzias latipes*) spermatozoa shed into water but not for promoting nuclear condensation during spermatogenesis. *Int. J. Dev. Biol.,* 44, 195-199.

Silverstein, J. T., Bosworth, B. G. and Wolters, W. R. (1999). Evaluation of dual injection of LHRHa and the dopamine receptor antagonist pimozide in cage spawning of channel catfish *Ictalurus punctatus*. *J. World Aquacult. Soc.,* 30, 263-268.

Somoza, G. M. and Peter, R. E. (1991). Effects of serotonin on gonadotropin and growth hormone release from *in vitro* perifused goldfish pituitary fragments. *Gen. Comp. Endocrinol.,* 82, 103-110.

Sperry, T. S. and Thomas, P. (1999). Characterization of two nuclear androgen receptors in Atlantic croaker: comparison of their biochemical properties and binding specificities. *Endocrinology,* 140, 1602-1611.

Sreenivasulu, G., Senthilkumaran, B., Sridevi, P., Rajakumar, A. and Rasheeda, M. K. (2012). Expression and immunolocalization of 20□-hydroxysteroid dehydrogenase during testicular cycle and after hCG induction, *in vivo* in the catfish, *Clarias gariepinus. Gen. Comp. Endocrinol.,* 175, 48-54.

Swanson, P. (1991). Salmon gonadotropins: reconciling old and new ideas. *In*: A. P. Scott, J. P. Sumpter, D. E. Kime, M. S. Rolfe (Eds.), Reproductive Physiology of Fish (pp. 2-7). Sheffield: Fish Symp.

Tanaka, H. (2003). Techniques for larval rearing. *In*: K. Aida, K. Tsukamoto, K. Yamauchi (Eds), Eel biology (pp. 427-434) Tokyo: Springer.

Tharakan, B. and Joy, K. P. (1996). Effects of mammalian gonadotropin-releasing hormone analogue, pimozide and the combination on plasma gonadotropin levels in different seasons, and induction of ovulation in female catfish. *J. Fish Biol.*, 48, 623-632.

Thomas, P., Breckenridge-Miller, D. and Detweiler, C. (1997). Binding characteristics and regulation of the 17,20β,21-trihydroxy-4-pregnen-3-one (20β-S) receptor on testicular and sperm plasma membranes of spotted seatrout (*Cynoscion nebulosus*). *Fish Physiol. Biochem.,* 17, 109-116.

Thomas, P., Dressing, G., Pang, Y., Berg, H., Tubbs, C., Benninghoff, A. and Doughty, K. (2006). Progestin, estrogen and androgen G-protein coupled receptors in fish gonads. *Steroids,* 71, 310-316.

Thomas, P., Pang, Y., Filardo, E.J. and Dong, J. (2005). Identity of an estrogen membrane receptor coupled to a G-protein in human breast cancer cells. *Endocrinology,* 146, 624-632.

Thomas, P., Pang, Y., Zhu, Y., Detweiler, C. and Doughty, K. (2004). Multiple rapid progestin actions and progestin membrane receptor subtypes in fish. *Steroids,* 69, 567-573.

Thomas, P. and Rahman, M. S. (2009). Biomarkers of hypoxia exposure and reproductive function in Atlantic croaker: a review with some preliminary findings from the northern Gulf of Mexico hypoxic zone. *J. Exp. Mar. Biol. Ecol.,* 381, 38-50.

Thomas, P. and Rahman, M. S. (2011). Extensive reproductive disruption, ovarian masculinization and aromatase suppression in Atlantic croaker in the northern Gulf of Mexico hypoxic zone. *Proc. R. Soc.,* 279B, 28-38.

Thomas, P., Tubbs, C. and Garry, V.F. (2009). Progestin functions in vertebrate gametes mediated by membrane progestin receptors (mPRs): identification of mPRα on human sperm and its association with sperm motility. *Steroids,* 74, 614-621.

Trombley, S. and Schmitz, M. (2013). Leptin in fish: possible role in sexual maturation in male Atlantic salmon. Fish Physiol. Biochem., 39, 103-106.

Tubbs, C., Tan, W., Shi, B. and Thomas, P. (2011). Identification of 17,20β,21-trihydroxy-4-pregnen- 3-one (20β-S) receptor binding and membrane progestin receptor α on southern flounder sperm (*Paralichthys lethostigma*) and their likely role in 20β-S stimulation of sperm hypermotility. *Gen. Comp. Endocrinol.,* 170, 629-639.

Tubbs, C. and Thomas, P. (2008). Functional characteristics of membrane progestin receptor alpha (mPRα) subtypes: a review with new data showing mPRα expression in seatrout sperm and its association with sperm motility. *Steroids,* 73, 935-941.

Ueda, H., Kanbegawa, A. and Nagahama, Y. (1985). Involvement of gonadotrophin and steroid hormones in spermiation in the amago salmon, *Oncorhynchus rhodurus*, and goldfish, *Carassius auratus*. *Gen. Comp. Endocrinol., 59,* 24-30.

Ueda, H., Young, G., Crim, L. W., Kambegawa, A. and Nagahama, Y. (1983). 17*α*,20*β*-dihydroxy-4-pregnen-3-one: plasma levels during sexual maturation and *in vitro* production by the testes of amago salmon (*Oncorhynchus rhodurus*) and rainbow trout (*Salmo gairdneri*). *Gen. Comp. Endocrinol.,* 51, 106-112.

Ulloa-Aguirre, A. and Timossi, C. (2000). Biochemical and functional aspects of gonadotrophin-releasing hormone and gonadotrpophins. *Reprod. Biomed. Online, 1,* 48-62.

Unniappan, S. and Peter, R. E. 2005. Structure, distribution and physiological functions of ghrelin in fish. *Comp. Biochem. Physiol.,* 140A, 396-408.

Vilela, D. A. R., Silva, S. B. B., Peixoto, M. T. D., Godinho, H. P. and Franca, L. R. (2003). Spermatogenesis in teleost: insights from the Nile tilapia (*Oreochromis niloticus*) model. *Fish Physiol. Biochem.,* 28, 187-190.

Viveiros, A. T. M., Fessehaye, Y., ter Veld, M., Schulz, R. W. and Komen, J. (2002). Handstripping of semen and semen quality after maturational hormone treatments, in African catfish *Clarias gariepinus*. *Aquaculture,* 213, 373-386.

Vizziano, D. and LeGac, F. (1998). Effect of gonadotropin type II and 17-hydroxy-4-pregnene-3,20-dione on 17,20β-dihydroxy-4-pregnen-3-one production by rainbow trout testes at different developmental stages. *Fish Physiol. Biochem.,* 19, 269-277.

Watanabe, T. and Kiron, V. (1995). Broodstock management and nutritional approaches for quality offsprings in the red sea bream. *In*: N. R. Bromage, R. J. Roberts (Eds.), Broodstock Management and Egg and Larval Quality (pp. 424). Cambridge: Cambridge University Press.

Weltzien, F.-A., Taranger, G. L., Karlsen, O. and Norberg, B. (2002). Spermatogenesis and related plasma androgen levels in Atlantic halibut (*Hippoglossus hippoglossus* L.). *Comp. Biochem. Physiol.,* 132A, 567–575.

Wood, A. W., Duan, C. and Bern, H. A. (2005). Insulin-like growth factor signaling in fish. *Int. Rev. Cytol.,* 243, 215-285.

Wu, R. S. S., Zhou, B. S., Randall, D. J., Woo, N. Y. S. and Lam, P. K. S. (2003). Aquatic hypoxia is an endocrine disruptor and impairs fish reproduction. *Environ. Sci. Technol.,* 37, 1137-1141.

Yamaguchi, S., Miura, C., Kikuchi, K., Celino, F. T., Agusa, T., Tanabe, S. and Miura, T. (2009). Zinc is an essential trace element for spermatogenesis. *Proc. Natl. Acad. Sci. USA,* 106, 10859-10864.

Yaron, Z. (1995). Endocrine control of gametogenesis and spawning induction in the carp. *Aquaculture,* 129, 49-73.

Yaron, Z. and Levavi-Sivan, B. (2011). Endocrine Regulation of Fish Reproduction. *In*: A.P. Farrell (Ed.), Encyclopedia of Fish Physiology: From Genome to Environment (volume 2, pp. 1500-1508). San Diego: Academic Press.

Yaron, Z., Gur, G., Melamed, P., Rosenfeld, H., Elizur, A. and Levavi-Sivan, B. (2003). Regulation of fish gonadotropins. *Intl. Rev. Cytol.,* 225, 131-185.

Yeung, C. H. (2010). Aquaporins in spermatozoa and testicular germ cells: identification and potential role. *Asian J. Androl.,* 12, 490-499.

Yoshiura, Y., Suetake, H. and Aida, K. (1999). Duality of gonadotropin in a primitive teleost, Japanese eel (*Anguilla japonica*). *Gen. Comp. Endocrinol.,* 114, 121-131.

Yu, K. L. and Peter, R. E. (1990). Dopaminergic regulation of brain gonadotropin releasing hormone in male goldfish during spawning behavior. *Neuroendocrinology,* 52, 276-283.

Yu, K. L. and Peter, R. E. (1992). Adrenergic and dopaminergic regulation of gonadotropin-releasing hormone release from goldfish preoptic-anterior hypothalamus and pituitary *in vitro*. *Gen. Comp. Endocrinol.*, 85, 138-146.

Yueh, W. S. and Chang, C. F. (1997). 17*α*,20*β*,21-trihydroxy-4-pregnen-3-one and 17*α*,20*β*-dihydroxy-4-pregnen-3-one stimulated spermiation in protandrous black porgy, *Acanthopagrus schlegeli*. *Fish Physiol. Biochem.,* 17, 187-193.

Yuen, C. W. and Ge, W. (2004). Follistatin suppresses FSHb but increases LHb expression in the goldfish- evidence for an activin-mediated autocrine/paracrine system in fish pituitary. *Gen. Comp. Endocrinol.,* 135, 108-115.

Zhou, L. Y., Wang, D.S., Kobayashi, T., Yano, A., Paul-Prasanth, B., Suzuki, A., Sakai, F. and Nagahama, Y. (2007). A novel type of P450c17 lacking the lyase activity is responsible for C21-steroid biosynthesis in the fish ovary and head kidney. *Endocrinology,* 148, 4282-4291.

Zilli, L., Schiavone, R., Chauvigne, F., Cerda, J., Storelli, C. and Vilella, S. (2009). Evidence for the involvement of aquaporins in sperm motility activation of the teleost gilthead sea bream (*Sparus aurata*). *Biol. Reprod.,* 81, 880-888.

Zohar, Y., Goren, A., Fridkin, M., Elhanati, E. and Koch, Y. (1990). Degradation of gonadotropin-releasing hormones in the gilthead seabream *Sparus aurata* II. Cleavage of native salmon GnRH, mammalian LHRH and their analogs in the pituitary, kidney and liver. *Gen. Comp. Endocrinol.,* 79, 306-319.

Zohar, Y., Goren, A., Tosky, M., Pagelson, G., Leibovitz, D. and Koch, Y. (1989). The bioactivity of gonadotropin-releasing hormones and its regulation in the gilthead seabream, *Sparus aurata*: *in vivo* and *in vitro* studies. *Fish Physiol. Biochem.,* 7, 59-67.

In: Sexual Plasticity and Gametogenesis in Fishes
Editor: Balasubramanian Senthilkumaran
ISBN: 978-1-62618-848-8

Chapter XXIV

Expression Profiling of Various Marker Genes Involved in Gonadal Differentiation of Teleosts: Molecular Understanding of Sexual Plasticity

Chenichery Sudhakumari** and Balasubramanian Senthilkumaran
Department of Animal Sciences, School of Life Sciences-Centre for Advanced Studies, University of Hyderabad, P. O. Central University, Hyderabad, India

Abstract

Undifferentiated gonads transforming into either testis or ovary is an important event in the developmental processes of vertebrate sex determination and differentiation. Molecular mechanisms underlying this phenomenon revealed involvement of several transcription factors, signaling molecules or receptors and steroidogenic enzymes, which are eventually designated as marker genes/factors. Expression profiling of these marker genes using various teleosts models indicated sexually dimorphic or differential expression during gonadal differentiation. In this review, we highlight the functional relevance of certain eventful biomarkers to our molecular understanding of sexual plasticity in teleosts. Among the marker genes, *dmrt1* and *sox9a* showed exclusive expression in differentiating testis while expression of ovarian aromatase, *cyp19a1a* and *foxl2* was prolific in developing ovary. Interestingly, in certain species like tilapia expression of *cyp19a1a* was evident even in undifferentiated gonads of XX females. Similarly, exclusive expression of *sox9b* in ovary was evident in few teleosts including catfish. Pioneering studies in several teleosts demonstrated the expression of *vasa* prior to sexual development which eventually regarded vasa as potential germ cell marker. Transcription factor such as Wt-1, signaling molecule such as Wnt-4 and orphan nuclear receptors such as *Ad4BP/SF-1*, *Dax-1* and *SHP* showed certain degree of differential expression, if not absolute sexual dimorphism during early development. Reports on the fate of these genes after sexual differentiation revealed significant association of some of these correlates during recrudescence and reproductive cycle. One of the necessities in

* Corresponding author: sudhacc@yahoo.com.

the present arena is just not depending on the correlative response of marker genes instead focusing on designing novel experiments to signify specific roles like those determined for *DMY/dmrt1* or *cyp19a1a*. To enable this, innovative studies in several teleosts using morpholino or siRNA or knock-out/exploitation of normal mutants will further contribute on our molecular understanding of sexual development cascade. In this context, natural or artificial sex reversal experiments might also provide new research prospects. Nonetheless, in the present scenario, identification of marker genes provided tremendous application potential to examine impacts of endocrine disruptors at molecular level.

Introduction

Teleosts are classical models to study sex determination and differentiation due to their efficiency to exhibit sexual plasticity during various stages of development or at times even during adulthood (Nakamura et al., 2003; Nagahama et al., 2004; Nagahama, 2005; Angelopoulou et al., 2012; Kobayashi et al., 2013). Earlier times application of wide variety of techniques to analyze sex determination/differentiation, which includes gynogenesis, sex reversal and progeny testing were successful up to a point without using any genetic markers for sex (Penman and Piferrer, 2008). Through these studies, complemented with karyotype analysis, morphological and biochemical analysis of markers for sex have been found in certain species (Penman and Piferrer, 2008). Later the advent of techniques for genome analysis paved way to discover master switch genes of sex determination first in mammals.

From the discovery of *Sry* as male sex determining gene in mammals (Sinclair, 1990), it took a decade to identify the sex determining gene, *DMY* in a lower vertebrate, which happened to be a fish species, the Japanese medaka, *Oryzias latipes* (Matsuda et al., 2002; Nanda et al., 2002). From then on various marker genes were identified to signify their specific role to gonadal differentiation in teleosts. More recently, AMH was also localized on Y chromosome of Patagonian pejerrey, *Odontesthes hatchery* (Hattori et al., 2012). Most of the approach exploited gene expression pattern analysis coupled with few other approaches using morpholino or knock-out strategy or normal sex reversed mutants to demonstrate the roles of several genes with reference to sexual development.

The contribution from lower vertebrates such as fish to understand vertebrate sex determination and differentiation was almost on par with mammalian animal models. In spite of the knowledge gained on this line over the years, it is still debatable that the appearance of certain genes might have been an effect rather than a cause. However, sex change/reversal experiments certainly provided valuable information to propose significant role for several genes which are termed as sexual or gamete development markers. In this review, we highlighted the importance of various marker genes using the reports that demonstrate their localization and expression pattern during normal and altered physiology. We had divided various sections based on the marker genes with an attempt to explain their changes based on gonadal development (ontogeny) and recrudescence. In addition, we had provided a brief account of *vasa* gene, a marker for germ cell differentiation. In teleosts, gonadal maturation is cyclic in annual breeders after initial gonadal development leading to puberty in both sexes while in fortnight breeders female fish but not males undergo cyclicity after initial maturation. In daily breeders cyclicity is not evident to a major extent in both sexes. Thus annual breeders provide us a chance to analyze expression level changes in certain genes related to sex differentiation during gonadal recrudescence. Our review emphasized these aspects explicitly in order to provide more significant information to validate marker genes. In addition, future research directions stress the need for novel experiments to relate the correlative response of marker genes to specific function.

Vasa

Primordial germ cells (PGC) are precursors of gonadal primordia (presumptive gonad), which migrate from the mesonephros to eventually produce gametes (spermatozoa and egg) in sexually reproducing organisms (Saffman and Lasko, 1999; Wylie, 1999). The *vasa* gene was first identified in *Drosophila sp.* as one of the components of germ plasm vital for PGCs commitment (Schüpbach and Wieschaus, 1986; Hay et al., 1988). The vasa gene encodes DEAD box family protein of ATP-dependent RNA helicase (Hay et al., 1988; Liang et al., 1994). Expression of different homologs of *vasa* or its isoforms (Hay et al., 1988; Liang et al., 1994) had been identified in the germ cell lineage of several metazoans starting from *Hydra* to mammals (Mochizuki et al., 2000; Raz, 2000). In teleosts, the migration of progenital or PGC from mesonephors to genetial ridge leads further differentiation into germ cells, which can be characterized by the localization of *vasa* (Knaut et al., 2000). During embryonic development stable expression of *vasa* was evident in zebrafish, catfish and other teleosts till 24 hours post hatch (Braat et al., 2000; Knaut et al., 2000; Raghuveer and Senthilkumaran, 2010a). However, at later time steady increase in *vasa* gene expression was evident which can be equated to prominent increase in the number of germ cells (Raghuveer and Senthilkumaran, 2010a). Interestingly in catfish, differential expression of *vasa* gene was evident between male and female gonads, wherein the levels of *vasa* transcripts were marginally high in females (Raghuveer and Senthilkumaran, 2010a). Considering early meiotic arrest in females, increased expression of *vasa* might have also been contributed by elevated number of primary oocytes in ovary as the differentiation of female sex is slightly ahead of male sex in catfish (Raghuveer et al., 2011). Differential expression was further evident in adult male and female gonads during seasonal reproductive cycle of catfish (Raghuveer and Senthilkumaran, 2010a). In males, *vasa* expression was significantly high during spermatogenesis but it subdued considerably around spermiation (Raghuveer and Senthilkumaran, 2010a). This kind of expression pattern in testis during recrudescence can be best explained based on the localization studies done earlier in tilapia testis wherein *vasa* signal was found to be high in spermatogonia (germ cells), moderate in primary spermatocytes while no signal could be detected in secondary spermatocytes, spermatid and spermatozoa (Kobayashi et al., 2000). In females, expression of *vasa* was predominant during early preparatory phase of recrudescence due to the outnumbering of primary oocytes at that stage in catfish (Raghuveer and Senthilkumaran, 2010a). Barring differences in these periods, *vasa* expression also increased substantially in catfish gonads after inducing with human chorionic gonadotropin (hCG), *in vitro* and *in vivo* (Raghuveer and Senthilkumaran, 2010a). Similar expression level changes were also recorded in other teleosts such as tilapia and gibel carp during reproductive cycle (Kobayashi et al., 2000; Xu et al., 2005). These results warrant *vasa* as a marker gene not only for germ cell lineage but also for early gametogenesis and/or recrudescence, which may be partially dependent on gonadotropin.

DMY/dmrt1

Molecular resemblance in sexual development across different phyla observed hitherto is among *Drosophila doublesex*, *Caenorhabditis mab-3* and vertebrate *dmrt1* (Raymond et al., 1998). Dsx and mab-3 related transcription factor 1 (dmrt1) is one of the gene family of putative transcription factors that has a highly conserved novel zinc finger DNA binding domain (DM domain) in different vertebrates (Raymond et al., 1998; 2000). Cloning and expression analysis of *dmrt1* in several teleosts (Nanda et al., 1999, Guan et al., 2000, Nagahama, 2005; Raghuveer and Senthilkumaran, 2009) unveiled pivotal role in testicular differentiation. Interestingly, DMY, a Y-linked male sex-determination gene, discovered in few species of medaka was shown to be a duplicate of autosomal gene *dmrt1* (Matsuda et al., 2002; 2003; Nanda et al., 2002; Zhang, 2004). After this discovery few reports demonstrated the presence of multiple forms of *dmrt1* in zebrafish and catfish (Guo et al., 2005; Raghuveer and Senthilkumaran, 2009). Nevertheless, none of these forms was localized in Y chromosome like that of medaka. Though autosomal, expression of *dmrt1* is restricted to male gonads during development in several fish species (Fernandino et al., 2006; Xia et al., 2007; Ijiri et al., 2008; Kobayashi et al., 2008; Raghuveer and Senthilkumaran, 2009). In our animal model, catfish, *dmrt1* expression was first evident in undifferentiated gonads and continued throughout the development of testis till adulthood, suggesting an essential role for this correlate in testicular development and recrudescence (Raghuveer and Senthilkumaran, 2009; Raghuveer et al., 2011). Such male-specific expression of *dmrt1* during sex determination/differentiation was also observed in tilapia, orange-spotted grouper, rainbow trout and minnow (Nanda et al., 2002; Fernandino et al., 2006; Xia et al., 2007; Kobayashi et al., 2008; Cao et al., 2012). On the other hand, there are few reports where *dmrt1* expression was also shown in female gonads (Guo et al., 2005; Huang et al., 2005). Interestingly, these reports also confirmed the presence of alternatively spliced forms of dmrt1. Evidences are also available from several other teleost species where *dmrt1* exist in multiple forms and shows profound changes during sex reversal and after treatment with sex steroid mimics or analogues (Nagahama, 2005; Raghuveer and Senthilkumaran, 2009). It is imperative to examine functionality of such alternative spliced forms or isoforms as mere expression did not implicate functional significance. In catfish, methyl testosterone-induced appearance and reproductive cycle related changes in the expression of multiple forms of *dmrt1* authenticate their testis specificity (Raghuveer et al., 2005; Raghuveer and Senthilkumaran, 2009). Hence, such an experimental approach clarifies the importance of *dmrt1* gene in testicular development. Earlier, Matsuda et al. (2007) demonstrated male development in genetic female fish after inducing with *DMY* in medaka. On this line, in another study, *dmrt1* mutation causes a male-to-female sex reversal after the sex determination by *DMY* (Masuyama et al., 2012), which signifies the impact of *dmrt1/DMY* in testicular development. Thus, *dmrt1* is an ancient sex determining gene, found first in invertebrates, and is one of several genes expressed in higher levels in the teleostean embryonic testis than in the embryonic ovary (Ferguson-Smith, 2007; Cao et al., 2012). Thus this gene stands out in this review due to its likely key role in male sex development. Further it is one of the few sex genes that have been mapped in representative species of other vertebrates for a similar function.

SOX9

SOX (SRY-related high mobility group [HMG] box) gene family encodes more than a few transcription factors that have a DNA-binding motif recognized as the HMG domain (Sinclair et al., 1990). The HMG domain of all SOX proteins includes 79 amino acid residues, which is highly conserved across vertebrates. Among different forms of SOX proteins, potential role of SOX9 in testicular development is well established in vertebrates including teleosts. Duplicates of *sox9* genes were identified in zebrafish, rice field eel, fugu, medaka and catfish (Chiang et al. 2001, Zhou et al., 2003, Koopman et al., 2004, Klüver et al., 2005; Raghuveer and Senthilkumaran, 2010b). These two forms of *sox9* gene were considered as orthologs of the tetrapod *sox9* gene that arose during whole-genome duplication (Amores et al., 1998, Postlethwait et al., 2004). In rainbow trout, *sox9* was detected in testis (Takamatsu et al., 1997), whereas in medaka, *sox9* was expressed in ovary (Yokoi et al., 2002). Further reports using medaka model revealed the presence of *sox9a2* at the highest level in adult testis (Nakamoto et al., 2005). In zebrafish and catfish, *sox9a* was found to be expressed in testis while *sox9b* expression was evident in ovary (Chiang et al., 2001; Raghuveer and Senthilkumaran, 2010b). However, in common carp, *sox9b* was expressed in testis (Du et al., 2007). In rice field eel, *sox9a1* and *sox9a2* were expressed in testis, ovary and ova-testis (Zhou et al., 2003). Differential expression of isoforms of *sox9* might have been primarily contributed by species variations or developmental difference (Cresko et al., 2003). In catfish, dimorphic expression of *sox9* orthologs was fairly evident (Raghuveer and Senthilkumaran, 2010b). SOX9 protein was detectable in germ and supporting cells of undifferentiated gonad at 40 days post hatch (dph) in catfish (Raghuveer and Senthilkumaran, 2010b). This kind of expression profiling illustrates flexible gene regulation in siluriforms and few other species. Differential expression was also found in zebrafish and tilapia, which authenticates categorical evolution of isoforms of *sox9* genes to impart testicular and ovarian differentiation through gene duplication (Chiang et al., 2001; Kobayashi et al., 2008). In contrast, *sox9* duplicates failed to show dimorphic expression in the gonads of medaka and eel (Zhou et al., 2003; Kluver et al., 2005). Though *sox9* have been well studied during sexual development, its significance is least determined during reproductive cycle. To understand the fate of *sox9* after sexual differentiation, studies from our laboratory revealed high levels of *sox9a* transcripts during preparatory and prespawning phases of male catfish testis which decreased gradually during spermiation and post-spawning phases (Raghuveer and Senthilkumaran, 2010b). As per the authors, high expression is likely to be contributed by higher number of germ cells expressing *sox9a* (Raghuveer and Senthilkumaran, 2010b). In contrast, during the ovarian cycle, *sox9b* transcripts were copious in the preparatory phase when compared with the prespawning, spawning and post-spawning phases in catfish ovary (Raghuveer and Senthilkumaran, 2010b). Studies from our laboratory further demonstrated up-regulation of *sox9* in response to the treatments of hCG and 11-ketotestosterone, *in vitro* suggesting that gonadotropins might stimulate *sox9* expression either directly or indirectly through androgens (Raghuveer and Senthilkumaran, 2010b). The action of androgens to promote *sox9* expression might have been signaled through androgen receptor. This contention requires additional experimental validation using androgen receptor blockers. Usually *sox9* expression was seen in supporting rather than germ cells (Kobayashi et al., 2008). In catfish, SOX9 was localized in Sertoli cells and different stages of germ cells, but not in spermatid/spermatozoa

(Raghuveer and Senthilkumaran, 2010b). Localization of SOX9 was not uncommon as Dmrt1 which is usually seen in supporting cells was also localized both in supporting and germ cells in catfish (Raghuveer and Senthilkumaran, 2009; 2010b). Similarly, *sox9* was evident in oogonia, indicating the nature of this correlate to express in different cell types (Raghuveer and Senthilkumaran, 2010b). Taken together, it seems *sox9* duplication leading to two different forms seems to categorize gonadal differentiation with reference to sex by specific interplay with other marker genes that happens to co-localize or express during sexual development. Further studies aiming at the specific regulatory mechanism of *sox9* expression might provide important lead to understand its interplaying partners. Despite this fact, potential role of *sox9* orthologs in gonadal differentiation is well documented to consider those as marker genes for testicular and ovarian differentiation (Raghuveer and Senthilkumaran, 2010b).

Wt 1

The Wilms' tumor suppressor gene, *wt1* encodes a transcription factor decisive for development of the urogenital system of zebrafish and medaka likewise in mammals (Bollig et al., 2006; Perner et al., 2007; Klüver et al., 2009). It is one of the genes known to be necessary for the formation of the bipotential, undifferentiated gonad and further testicular differentiation along with *Ad4BP/SF-1*, *GATA4* and *FOG2* (Barrionuevo et al., 2012). Expression of *wt1* in variety of cells and tissues during development ultimately leading to an epithelial to mesenchymal transition or the opposite, mesenchymal to epithelial transition (Essafi et al., 2011). However, in teleosts, two *wt1* genes were identified (Klüver et al., 2009). Studies in medaka by Klüver et al. (2009) found that *wt1a* was expressed in the lateral plate mesoderm during early embryogenesis and in the somatic cells of the gonadal primordium later in the development, while *wt1b* expressed soon after embryogenesis but not in the gonadal primordium. The same report in medaka also revealed that through classical morpholino knockdown experiments, neither *wt1a* nor *wt1b* affected gonad development, whereas the double knockdown of *wt1a* and *wt1b* decreased number of PGC during gonad development. These interesting observations led them to indicate that medaka *wt1* co-orthologs show genetic redundancy in PGC maintenance or survival via responsive backup circuits. Another report using microarray revealed specific up-regulation of *wt1* in male rainbow trout (Hale et al., 2011). It is also known that wt1 regulates wnt-4 activation and repression (Essafi et al., 2011). Further studies using different teleost models might provide more inputs to understand the importance of this gene in sexual plasticity as it shows dimorphic expression before the completion of gonadogenesis due to its key function in the organization of urogenital system.

Orphan Nuclear Receptors

Orphan receptors comprise a diverse component of the nuclear receptor superfamily and possess all the functional domains that characterize classic nuclear receptors (Giguere, 1999). There are 19 families of orphan receptors with unknown physiological ligands or activators than receptors with known ligands (10 families) though they are structurally related to them (Blumberg and Evans, 1998). Some orphan receptors are divergent having only one of the two characteristic domains of the NR superfamily like DAX-1 and short heterodimer partner (SHP), which contain only an ligand-binding domain and lack a classic DNA-binding domain with conserved cysteines as do the other receptors (Benoit et al., 2006).

Dax-1 and SHP

Dax-1 is expressed in the steroidogenic tissues such as adrenal cortex, gonads and its expression overlaps with that of *Ad4BP/SF-1* (Kawabe et al., 1999). Dax1 (NRoB1) is known for its importance in female sex determination in mammals and birds while it plays an important role in testicular development in amphibians (Swain et al., 1996; Smith et al., 2000; Sugita et al., 2001). SHP, also known as NRoB2 is an unusual orphan nuclear receptor that possesses a ligand binding domain but not DNA binding domain (Seol et al., 1996; 1998; Johansson et al., 1999). Interestingly SHP has been cloned in humans and rodents (Seol et al., 1996; Matsuda et al., 1997). For the first time, Wang et al. (2002) cloned *Dax1* and *SHP* cDNAs from the Nile tilapia and found that the expressions of both *Dax1* and *SHP* were fairly detectable in both sexes from 5-90 dph. The expression of *Dax1* was significantly up regulated between 10 and 15 dph while the expression of *SHP* was moderate and consistent throughout the sexual development. DAX1 is known to be up regulated by Wnt4 via the WNT/β-catenin pathway (Mizusaki et al., 2003). Later, no studies were done in other teleost species to categorize the role of Dax-1 and SHP in sexual development.

Ad4BP/SF-1

The steroidogenic factor 1, also known as Ad4BP/NR5A1 (Ad4BP/SF-1), encodes an orphan nuclear receptor that plays an important role in early gonadal development and is expressed in the developing urogenital ridge, hypothalamus, anterior pituitary gland and adrenal glands (Combes et al., 2010). It regulates the expression of a number of genes that are involved in sexual development, including cytochrome P-450 steroid hydroxylase enzymes, cyp19a1a and the 3β-hydroxysteroid dehydrogenase (Yoshiura et al., 2003; Kousta et al., 2010). In tilapia, temporal expression pattern of *Ad4BP/SF-1* by semi-quantitative PCR revealed its presence from 5 dph in female gonads but not in male gonads suggesting the concept that developing females has the potency to produce estrogen to induce ovarian differentiation (Sudhakumari et al., 2005). Later study using same species with real time PCR revealed no sexual dimorphism in between XX and XY gonads with transcript presence from 5-7 dph onwards (Ijiri et al., 2008). The different pattern observed in both the experiments as explained by Sudhakumari et al. (2005) indicated usage of high stringency PCR with whole body template or technical differences. In spite of the differences in the expression pattern of orphan nuclear receptors mentioned above, no categorical dimorphism could be observed. Hence, it is conceivable that orphan nuclear receptors might play an important role in gonadal differentiation rather than imparting sexual differences. Considering the importance of *Ad4BP/SF-1* in the transcriptional regulation of *cyp19a1a* (Wang et al., 2010; Yoshiura et al.,

2003), it is surprising that the former do not exhibit sexually dimorphic expression while the later does. Since *cyp19a1a* is co-regulated by FOXL2 (Wang et al., 2007), over expression of the latter in female gonads might have contributed for sexually dimorphic expression of the former.

Ovarian Aromatase, cyp19a1a

In teleosts, sex steroids influence the development of germ and other cell types during the process of gonadal sex determination/differentiation (Yamamoto, 1969; Devlin and Nagahama, 2002). Estradiol-17β plays a significant role in inducing and maintaining ovarian development, and its levels are found to be higher in females than in males. Ovarian aromatase encoded by cyp19a1a gene, is a key enzyme responsible for the aromatization of androgens to estrogens and it is predominantly expressed in the ovary. It has been suggested that the up-regulation of cyp19a1a triggers and maintains ovarian differentiation while its down-regulation induces a testicular differentiation pathway (Wang et al., 2007; Guiguen et al., 2010; Wang et al., 2010).

The regulation of fish cyp19a1a during ovarian differentiation led to the characterization of the transcription factor FOXL2 as its ovarian specific upstream regulator that would co-activate *cyp19a1a* expression, along with *Ad4BP/SF-1* (Yoshiura et al., 2003; Nagahama, 2005; Ijiri et al., 2008: Wang et al., 2010; Raghuveer et al., 2011). Teleosts exhibit two separate forms of aromatase, i.e. the ovarian specific (cyp19a1a) and brain specific (cyp19a1b) cyp19 genes, which mainly catalyze the aromatization of androgens into estrogens (Nagahama, 2005; Kwon et al., 2001; Sudhakumari et al., 2005; Rasheeda et al., 2010). In teleosts former is found to be expressed very early in primordial/indifferent gonads prior to morphological differentiation of the gonads (Nagahama, 2005; Raghuveer et al., 2011). Later on, the levels were found to be high in ovary throughout development (Rasheeda et al., 2010; Raghuveer et al., 2011). In contrast, its expression was very low in male gonads throughout development (Sudhakumari et al., 2005; Raghuveer et al., 2011). Interestingly, in tilapia both forms of aromatases could be detectable in ovary and even both showed changes during sex reversal (Sudhakumari et al., 2003; 2005; Chang et al., 2005). Based on several reports in teleosts, it is possible to attribute that cyp19a1a play an important role in sex determination and differentiation (Nagahama, 2005; Sudhakumari et al., 2005; Ijiri et al., 2008: Wang et al., 2010; Raghuveer et al., 2011; Cao et al., 2012). Our immunocytochemical localization analysis revealed that Cyp19a1a protein was detectable in the ooplasm of primary oocytes in addition to the follicular layer during ovarian development in catfish (Raghuveer et al., 2011). Similar observations were also evident in killifish using *in situ* hybridization (Dong and Willett, 2008).

However, there are reports showing that Cyp19a1a protein was localized only in the follicular layer of vitellogenic oocytes in the Nile tilapia (Wang et al., 2007). Interestingly, other steroidogenic enzyme genes such as 3β-hsd, cyp17 and 17β-hsd1 and the sterol transfer coding gene star, necessary for the synthesis of estrogens, were expressed at barely detectable levels in undifferentiated gonads and thereafter elevated gradually in both sexes of differentiated gonads as development progressed (Raghuveer et al., 2011). Similarly in the Nile tilapia, these steroidogenic enzyme genes were also found to be expressed in undifferentiated XX gonads (around 5-10 dph) and later gradually augmented in both sexes (Ijiri et al., 2008). In fact, earlier reports on the manipulation/skewing of sex in catfish larva during early development (before sex differentiation) and complete sex reversal of XY tilapia to phenotypic females by following ethynylestradiol treatment established that estrogen is produced in female gonads during the critical period of ovarian differentiation, which decides ovarian development (Nagahama, 2005; Raghuveer et al., 2005).

In the European sea bass, *cyp19a1a* was significantly higher in future females and hence it is revealed as a robust molecular marker to predict future ovarian differentiation (Blázquez et al., 2009).Taken together, cypa19a1a plays a very prominent role in female sex development in teleosts, though it was not localized in the sex chromosome (Harvey et al., 2003; Nagahama, 2005). This characteristic feature specially categorizes teleosts as excellent animal models to study female sex development.

FOXL2

FOXL2 belongs to the forkhead/HNF-3-related family of transcription factors that is involved in several developmental processes including ovarian differentiation. FOXL2 interacts with the ligand-binding domain of Ad4BP/SF-1 through its fork head domain by forming a hetero-dimer that eventually enhances *cyp19a1a* transcription vis-à-vis elevating estradiol production leading to ovarian development and growth in teleosts. Sexually dimorphic expression of *foxl2* was demonstrated in the Nile tilapia (Wang et al., 2004), rainbow trout (Baron et al., 2004) and medaka (Nakamoto et al., 2006). Further, Ijiri et al. (2008) suggested a close relationship between gonadal FOXL2 and *cyp19a1a* during early sex differentiation of XX tilapia.

The existing data denoted ovary specific expression of *foxl2* not only in fish species but also in other vertebrates around the time of female sex differentiation, indicating evolutionarily conserved role of *FOXL2* in vertebrate ovarian development. High levels of *foxl2* mRNA and/or protein were found in ovarian tissues of catfish (Sridevi and Senthilkumaran, 2011), trout (Baron et al., 2004), protogynous wrasse (Kobayashi et al., 2010) and southern catfish (Liu et al., 2007) although the expression was negligible in testis (Baron et al., 2004; Wang et al., 2004; Liu et al., 2007) in these species except wrasse (Kobayashi et al., 2010) wherein the expression was abundant comparable to ovary. This dissimilarity may be attributed to the sex changing nature of wrasse. FOXL2 could be localized in stroma and interstitial cells of the gonads of normal XX and sex-reversed XY Nile tilapia wherein localization of both cyp19a1a and Ad4BP/SF-1 were observed before the occurrence of morphological sex differentiation (Wang et al., 2007). Interestingly, *foxl2* showed high expression in adult catfish signifying its requirement in ovarian recrudescence like that of *cyp19a1a* (Sridevi and Senthilkumaran, 2011). FOXL2 being a co-regulator for *cyp19a1a* along with the principal regulator Ad4BP/SF-1, its importance in implementing ovarian growth and vitellogenesis is not questionable (Wang et al., 2004; 2007). In catfish, the expression of *foxl2* in testis was barely detectable in real time PCR (Raghuveer et al., 2011; Sridevi and Senthilkumaran, 2011). Based on this result and those of others, it is less likely that FOXL2 plays any role in testicular development of teleosts. Further study from our laboratory also identified no role for FOXL2 even in testicular recrudescence (Sridevi and Senthilkumaran, 2011).

Anti-Müllerian Hormone

During embryonic development in mammals, anti-müllerian hormone (amh), a glycoprotein, plays a major role in inducing the caudal to cranial regression of the müllerian ducts during testicular development which necessitates considering amh as an important marker for male sex differentiation (Beau et al., 2001). Though most teleosts lack ductal system including wolffian or müllerian ducts (Skaar et al., 2011), *amh* homologue was found to be expressed during gonadal sex differentiation and/or in adult stage (Hattori et al., 2012). The *amh* ortholog was first identified as a spermatogenesis preventing substance, eSRS21 and it is expressed in sertoli cells of immature testis earlier than the initiation of spermatogenesis in the Japanese eel (Miura et al., 2002). Testis-specific expression was also evident in tilapia (Ijiri et al., 2008).

In teleosts, *amh* down-regulates aromatase gene expression in developing gonads and also modulates the leydig cells differentiation by down-regulating the expression of certain genes involved in steroidogenesis (Wang and Orban, 2007). Intriguingly, amh is found to be co-expressed with *Ad4BP/SF-1* and *sox9a* in the follicular layer of ovary in zebrafish that implies to suggest important role for *amh* during "juvenile ovary-to-testis" transformation (Wang and Orban, 2007). Recently, Hattori et al. (2012) showed *amhy*, a duplicated copy of the *amh* in the single metacentric/submetacentric chromosome of XY individuals of pejerrey and implicated in testicular development. Results on *amhy* knockdown in XY embryos revealed up-regulation of *foxl2* and *cyp19a1a* mRNAs to initiate ovarian development in this species.

These results suggest that there is functional *amh* duplication in vertebrates and hence, *amhy* may consider being sex-determining hormone in teleosts and other vertebrates (Hattori et al., 2012). Further studies on *amh* or its isoforms using other teleosts with an aim to characterize its receptors might provide valuable inputs to link this hormone to teleostean sexual development.

Wnt-4

It is a member of the structurally related WNT (Wingless type MMTV integration site) family of genes that encode the secreted signaling proteins which play an important role in several developmental processes, including regulation of cell fate and patterning during embryogenesis (Schubert et al., 2000; Yu et al., 2009). Wnt4 is known to be the first signaling molecule influencing the sex-determination cascade where it antagonizes the testis-determining factor resulting in female development (Jordan et al., 2001). It was revealed that the male fate is adopted by the active repression of *wnt4* along with the high expression of the male determining genes like *sox9* (Jameson et al., 2012). It is also known that the nephrogenesis and associated tubulogenesis fail in its absence as its signaling induces mesenchyme to epithelium transition revealing its critical role for kidney development

(Murugan et al., 2012). In mammals, *wnt4* is predominantly expressed in the ovary but has no sexually dimorphic expression during female differentiation of non-mammalian vertebrates, including chick, turtles and frogs (Wu and Chang, 2009).

The profile and possible roles of *wnt4* in gonadal differentiation are not well characterized relatively in teleosts as that of mammals except in black porgy, where it is implicated in ovarian growth and sex change (Wu and Chang, 2009). It was demonstrated through phylogenetic and synteny analyses that the teleost fish possess two *wnt4* genes, *wnt4a* and *wnt4b*, resulted as a consequence of the teleost-specific whole-genome duplication (Nicol et al., 2012). The study conducted in rainbow trout by Nicol et al. (2012) identified an additional *wnt4* gene (a wnt4a paralog) that is likely to be resulted from the salmonid-specific whole-genome duplication and these two Wnt4a proteins (Wnt4a1 and Wnt4a2) seem to share high identity (>80%) with other vertebrates where Wnt4a proteins differ from Wnt4b which is more divergent (60% identity) with its restricted expression in the nervous system, suggesting its sub- or a neo-functionalization.

The same report also demonstrated that the *wnt4* expression profile was not ovary-predominant during early gonadal differentiation in rainbow trout, unlike mammals. Further analysis using various teleost models may clarify the importance of *Wnt* family of genes in sex differentiation.

11β-Hydroxylase (cyp11b)

In certain species like European sea bass, tilapia *cyp11b* expression was evident during testicular differentiation and at times even earlier than histological differentiation (Blázquez et al., 2009; Pfennig et al., 2012). However, in catfish steroiodogenic enzymes related to androgen biosynthesis appear late and coincide with histological differentiation (Raghuveer et al., 2011).

Hence, further analysis using several other teleost models may provide more input to denote cyp11b as a potential marker for testicular differentiation, though considered as a robust marker in teleosts that lacks sex chromosomes and for which temperature influences sex ratios (Blázquez et al., 2009).

Testicular or Ovarian Differentiation and Sexual plasticity: Validating Marker Genes

Unlike mammals, molecular mechanisms of sex determination are not understood in teleosts yet role of several genes/factors often pivotal for sex differentiation has been documented. Interestingly in certain teleosts ovarian differentiation precede testicular differentiation due to late onset of androgen biosynthesis in males.

Further, early occurrence and elevation of cyp19a1a vis-à-vis estradiol are prerequisite to ovarian development (Wu et al., 2008; Tang et al., 2010). Though testicular differentiation is marked by several genes, initiation of androgen biosynthesis seems to be slower in several teleosts. It is important to analyze sexual development cascade to reason out slow onset of steroidogenic enzyme genes in males.

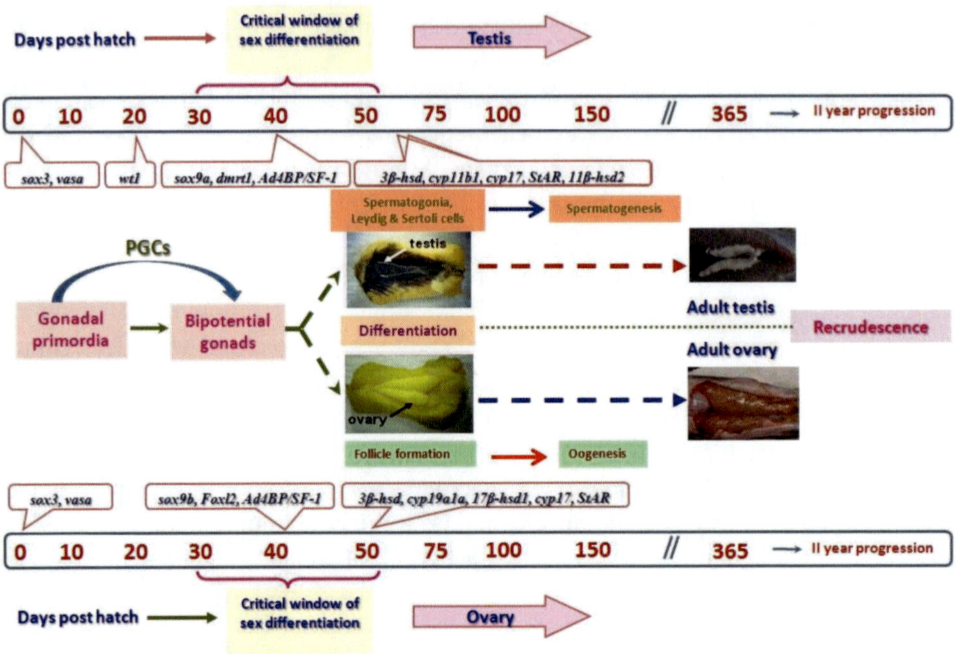

Figure 1. Schematic diagram explaining the ontogeny of marker genes of catfish: A model for teleosts.

Based on the available literature we had provided a schematic diagram (Figure 1) using our model organism, catfish, to depict sexual development with a chronology of marker genes ontogeny.

Further studies on the gene regulation of steroidogenic enzymes and transcription factors regulating urogenital system may provide some interesting observations to explain the core mechanism of sexual development in teleosts. In the present perspective, identification of marker genes definitely yielded reliable tools to understand molecular level changes during natural or artificial sex reversal.

In fact in teleosts, differences in the progression of gonadal differentiation between sexes impart sexual plasticity, which provide significant characteristic feature to adapt to natural or artificial physiological demands. Even so, at times prolong or untimely exposure of endocrine disruptors may cease sexual plasticity leading to intersexuality or developmental or reproduction failure. For example, early exposure of sex steroid analogues at moderate doses skew sexual population, yet high doses leads to intersex (Nagahama, 2005; Raghuveer and Senthilkumaran, 2009). On the other hand exposure of pollutants or pesticides or other compounds exert differential response and most of the time sex based (Chakrabarty et al., 2012; Rajakumar et al., 2012).

Future Research Directions

Most of the studies using various teleost models identified marker genes depending on the correlative response to a physiological event in sexual development. This poses an important question whether these genes have any specified functional role or expressing consequent response. Hence, one of the immediate requirements in this area of research is to design novel experiments like those used in the identification of DMY/Dmrt1 or amhy to signify specific functional roles for marker genes (Matsuda et al., 2002; Nanda et al., 2002; Hattori et al., 2012) instead of merely describing correlative response. To enable this, innovative strategies using genomics techniques, morpholino or siRNA or knock out/exploitation of normal mutants will lead to greater molecular understanding of sexual development. In this context, designing natural or artificial sex reversal experiments might provide new research perspectives to determine the plasticity of gonadal differentiation and implication of marker genes. Identification of marker genes yields immense application in endocrine disruptor research. It will also facilitate application in aquaculture for brood stock selection or for early sex selection of a species where a particular sex has a specific commercial interest.

Conclusion

Expression profiling of molecular marker genes using various teleost models revealed sexually dimorphic or differential expression pattern during gonadal differentiation. In this review, we had provided schematic representation (Figure 1) of ontogeny of marker genes in catfish model. Extensive studies implicate application of these genes in endocrine disruptor research to understand the severity of the exposure. Furthermore, identification of marker genes through ontogeny or sex reversal experiments facilitates application in aquaculture for brood stock selection to select specific gender for profit-based marketability by fish farmers.

Acknowledgments

The research work mentioned in this review was supported by Grant-in-Aids from Department of Biotechnology, India (DBT/TR4567/AAQ/03/219/2003) and Women Scientist Scheme (WOS-A) of the Department of Science and Techonolgy, India (SR/WOS-A/LS-31/2005) for which BS and CCS are thankful. CCS also is grateful to WOS-A (SR/WOS-A/LS-31/2005) for the fellowship.

References

Amores, A., Force, A., Yan, Y. L., Joly, L., Amemiya, C., Fritz, A., Ho, R. K., Langeland, J., Prince, V., Wang, Y. L., Westerfield, M., Ekker, M. and Postlethwait, J. H. (1998). Zebrafish hox clusters and vertebrate genome evolution. *Science*, 282, 1711-1714.

Angelopoulou, R., Lavranos, G. and Manolakou, P. (2012). Sex determination strategies in 2012: towards a common regulatory model? *Reprod Biol. Endocrinol.*, 10, 13.

Baron, D., Cocquet, J., Xia, X., Fellous, M., Guiguen, Y. and Veitia, R. A. (2004). An evolutionary and functional analysis of FoxL2 in rainbow trout gonad differentiation. *J. Mol. Endocrinol.*, 33, 705-715.

Barrionuevo, F. J., Burgos, M., Scherer, G. and Jiménez, R. (2012). Genes promoting and disturbing testis development. *Histol. Histopathol.*, 27, 1361-1383.

Beau, C., Vivian, N., Münsterberg, A., Dresser, D. W., Lovell-Badge, R. and Guerrier, D. (2001). *In vivo* analysis of the regulation of the anti-Müllerian hormone, as a marker of sertoli cell differentiation during testicular development, reveals a multi-step process. *Mol. Reprod .Dev.*, 59, 256-264.

Benoit, G., Cooney, A., Giguere, V., Ingraham, H., Lazar, M., Muscat, G., Perlmann, T., Renaud, J. P., Schwabe, J., Sladek, F., Tsai, M. J. and Laudet, V. (2006). International Union of Pharmacology. LXVI. Orphan nuclear receptors. *Pharmacol. Rev.*, 58, 798-836.

Blázquez, M., Navarro-Martín, L. and Piferrer, F. (2009). Expression profiles of sex differentiation-related genes during ontogenesis in the European sea bass acclimated to two different temperatures. *J. Exp. Zool. B Mol. Dev. Evol.*, 312, 686-700.

Blumberg, B. and Evans, R. M. (1998). Orphan nuclear receptors–new ligands and new possibilities. *Genes Dev.*, 12, 3149-3155.

Bollig, F., Mehringer, R., Perner, B., Hartung, C., Schäfer, M., Schartl, M., Volff, J. N., Winkler, C. and Englert, C. (2006). Identification and comparative expression analysis of a second wt1 gene in zebrafish. *Dev. Dyn.*, 23, 554-561.

Braat, A. K., van de Water, S., Goos, H., Bogerd, J. and Zivkovic, D. (2000). Vasa protein expression and localization in the zebrafish. *Mech. Dev.*, 95, 271-274.

Cao, M., Duan, J., Cheng, N., Zhong, X., Wang, Z., Hu, W. and Zhao, H. (2012). Sexually dimorphic and ontogenetic expression of dmrt1, cyp19a1a and cyp19a1b in Gobiocypris rarus. *Comp. Biochem. Physiol. A Mol. Integr. Physiol.*, 162, 303-309.

Chakrabarty, S., Rajakumar, A., Raghuveer, K., Sridevi, P., Mohanachary, A., Prathibha, Y., Bashyam, L., Dutta-Gupta, A. and Senthilkumaran, B. (2012). Endosulfan and flutamide, alone and in combination, target ovarian growth in juvenile catfish, *Clarias batrachus*. *Comp. Biochem. Physiol. C Toxicol. Pharmacol.*, 155, 491-497.

Chang, X., Kobayashi, T., Senthilkumaran, B., Kobayashi-Kajura, H., Sudhakumari, C. C. and Nagahama, Y. (2005). Two types of aromatase with different encoding genes, tissue distribution and developmental expression in Nile tilapia (*Oreochromis niloticus*). *Gen. Comp. Endocrinol.*, 141, 101-115.

Chiang, E. F., Pai, C. I., Wyatt, M., Yan, Y. L., Postlethwait, J. and Chung, B. (2001). Two sox9 genes on duplicated zebrafish chromosomes, expression of similar transcription activators in distinct sites. *Dev. Biol.*, 231, 149-163.

Combes, A. N., Spiller, C. M., Harley, V. R., Sinclair, A. H., Dunwoodie, S. L., Wilhelm, D. and Koopman, P. (2010). Gonadal defects in Cited2-mutant mice indicate a role for SF1 in both testis and ovary differentiation. *Int. J. Dev. Biol.*, 54, 683-689.

Cresko, W. A., Yan, Y. L., Baltrus, D. A., Amores, A., Singer, A., Rodríguez-Marí, A. and Postlethwait, J. H. (2003). Genome duplication, subfunction partitioning, and lineage divergence: Sox9 in stickleback and zebrafish. *Dev. Dyn.*, 228, 480-489.

Devlin, R. H. and Nagahama, Y. (2002). Sex determination and sex differentiation in fish: an overview of genetic, physiological, and environmental influences. *Aquaculture*, 208, 191-364.

Dong, W. and Willett, K. L. (2008). Local expression of CYP19A1 and CYP19A2 in developing and adult killifish (*Fundulus heteroclitus*). *Gen. Comp. Endocrinol.*, 155, 307-317.

Du, Q. Y., Wang, F. Y., Hua, H. Y. and Chang, Z. J. (2007). Cloning and study of adult tissue- specific expression of Sox9 in *Cyprinus carpio*. *J. Genet.*, 86, 85-91.

Essafi, A., Webb, A., Berry, R. L., Slight, J., Burn, S. F., Spraggon, L., Velecela, V., Martinez-Estrada, O. M., Wiltshire, J. H., Roberts, S. G., Brownstein, D., Davies, J. A., Hastie, N. D. and Hohenstein, P. (2011). A wt1-controlled chromatin switching mechanism underpins tissue-specific wnt4 activation and repression. *Dev. Cell*, 21, 559-574.

Ferguson-Smith, M. (2007). The evolution of sex chromosomes and sex determination in vertebrates and the key role of DMRT1. *Sex Dev.*, 1, 2-11.

Fernandino, J. I., Guilgur, L. G. and Somoza, G. M. (2006). Dmrt1 expression analysis during spermatogenesis in pejerrey, *Odontesthes bonariensis*. *Fish Physiol. Biochem.*, 32, 231-240.

Giguère, V. (1999). Orphan nuclear receptors: from gene to function. *Endocr. Rev.*, 20, 689-725.

Guan, G., Kobayashi, T. and Nagahama, Y. (2000). Sexually dimorphic expression of two types of DM (Doublesex/Mab-3)-domain genes in a teleost fish, the tilapia (*Oreochromis niloticus*). *Biochem. Biophys. Res. Commun.*, 272, 662-666.

Guiguen, Y., Fostier, A., Piferrer, F. and Chang, C. F. (2010). Ovarian aromatase and estrogens:a pivotal role for gonadal sex differentiation and sex change in fish. *Gen. Comp. Endocrinol.*, 165, 352-366.

Guo, Y., Cheng, H., Huang, X., Gao, S., Yu, H. and Zhou, R. (2005). Gene structure, multiple alternative splicing, and expression in gonads of zebrafish Dmrt1. *Biochem. Biophys. Res. Commun.*, 330, 950-957.

Hale, M. C., Xu, P., Scardina, J., Wheeler, P. A., Thorgaard, G. H. and Nichols, K. M. (2011). Differential gene expression in male and female rainbow trout embryos prior to the onset of gross morphological differentiation of the gonads. *BMC Genomics*, 12, 404.

Harvey, S. C., Kwon, J. Y. and Penman, D. J. (2003). Physical mapping of the brain and ovarian aromatase genes in the Nile Tilapia, Oreochromis niloticus, by fluorescence *in situ* hybridization. *Anim. Genet.*, 34, 62-64.

Hattori, R. S., Murai, Y., Oura, M., Masuda, S., Majhi, S. K., Sakamoto, T., Fernandino, J. I., Somoza, G. M., Yokota, M. and Strüssmann, C. A. (2012). A Y-linked anti-Müllerian hormone duplication takes over a critical role in sex determination. *Proc. Natl. Acad. Sci. USA*, 109, 2955-2959.

Hay, B., Ackermann, L., Barbel, S., Jan, L. Y. and Jan, Y. N. (1988). Identification of a component of Drosophila polar granules. *Development*, 103, 625-640.

Huang, X., Guo, Y., Shui, Y., Gao, S., Yu, H., Cheng, H. and Zhou, R. (2005). Multiple alternative splicing and differential expression of dmrt1 during gonad transformation of the rice field eel. *Biol. Reprod.*, 73, 1017-1024.

Ijiri, S., Kaneko, H., Kobayashi, T., Wang, D. S., Sakai, F., Paul-Prasanth, B., Nakamura, M. and Nagahama, Y. (2008). Sexual dimorphic expression of genes in gonads during early differentiation of a teleost fish, the Nile tilapia Oreochromis niloticus. *Biol. Reprod.*, 78, 333-341.

Jameson, S. A., Lin, Y. T. and Capel, B. (2012). Testis development requires the repression of Wnt4 by Fgf signaling. *Dev. Biol.*, 370, 24-32.

Johansson, L., Thomsen, J. S., Damdimopoulos, A. E., Spyrou, G., Gustafsson, J. A. and Treuter, E. (1999). The orphan nuclear receptor SHP inhibits agonist-dependent transcriptional activity of estrogen receptors ERα and ERβ. *J. Biol. Chem.*, 274, 345-353.

Jordan, B. K., Mohammed, M., Ching, S. T., Délot, E., Chen, X. N., Dewing, P., Swain, A., Rao, P. N., Elejalde, B. R. and Vilain, E. (2001). Up-regulation of WNT-4 signaling and dosage-sensitive sex reversal in humans. *Am. J. Hum. Genet.*, 68, 1102-1109.

Kawabe, K., Shikayama, T., Tsuboi, H., Oka, S., Oba, K., Yanase, T., Nawata, H. and Morohashi, K. (1999). Dax-1 as one of the target genes of Ad4BP/SF-1. *Mol. Endocrinol.*, 13, 1267-1284.

Klüver, N., Herpin, A., Braasch, I., Driessle, J. and Schartl, M. (2009). Regulatory back-up circuit of medaka Wt1 co-orthologs ensures PGC maintenance. *Dev. Biol.*, 325, 179-188.

Klüver, N., Kondo, M., Herpin, A., Mitani, H. and Schartl, M. (2005). Divergent expression patterns of Sox9 duplicates in teleosts indicate a lineage specific subfunctionalization. *ev. Genes. Evol., 215,* 297-305.

Knaut, H., Pelegri, F., Bohmann, K., Schwarz, H. and Nüsslein-Volhard, C. (2000). Zebrafish vasa RNA but not its protein is a component of the germ plasm and segregates asymmetrically before germline specification. *J. Cell Biol.*, 149, 875-88.

Kobayashi, T., Kajiura-Kobayashi, H. and Nagahama, Y. (2000). Differential expression of vasa homologue gene in the germ cells during oogenesis and spermatogenesis in a teleost fish, tilapia, *Oreochromis niloticus*. *Mech. Dev.*, 99, 139-142.

Kobayashi, T., Kajiura-Kobayashi, H., Guan, G. and Nagahama, Y. (2008). Sexual dimorphic expression of DMRT1 and Sox9a during gonadal differentiation and hormone-induced sex reversal in the teleost fish Nile tilapia (*Oreochromis niloticus*). *Dev. Dyn.*, 237, 297-306.

Kobayashi, Y., Horiguchi, R., Nozu, R. and Nakamura, M. (2010). Expression and localization of forkhead transcriptional factor 2 (Foxl2) in the gonads of protogynous wrasse, *Halichoeres trimaculatus*. *Biol. Sex Differ.*, 1, 3.

Kobayashi, Y., Nagahama, Y. and Nakamura, M. (2013). Diversity and plasticity of sex determination and differentiation in fishes. *Sex. Dev.*, 7, 115-125.

Koopman, P., Schepers, G., Brenner, S. and Venkatesh, B. (2004). Origin and diversity of the Sox transcription factor gene family genome-wide analysis in *Fugu rubripes*. *Gene*, 328, 177-186.

Kousta, E., Papathanasiou, A. and Skordis, N. (2010). Sex determination and disorders of sex development according to the revised nomenclature and classification in 46, XX individuals. *Hormones (Athens)*, 9, 218-131.

Kwon, J. Y., McAndrew, B. J. and Penman, D. J. (2001). Cloning of brain aromatase gene and expression of brain and ovarian aromatase genes during sexual differentiation in genetic male and female Nile tilapia *Oreochromis niloticus*. *Mol. Reprod. Dev.*, 59, 359-370.

Liang, L., Diehl-Jones, W. and Lasko, P. (1994). Localization of vasa protein to the Drosophila pole plasm is independent of its RNA-binding and helicase activities. *Development*, 120, 1201-1211.

Liu, Z., Wu, F., Jiao, B., Zhang, X., Hu, C., Huang, B., Zhou, L., Huang, X., Wang, Z., Zhang, Y., Nagahama, Y., Cheng, C. H. and Wang, D. (2007). Molecular cloning of doublesex and mab-3-related transcription factor 1, forkhead transcription factor gene 2, and two types of cytochrome P450 aromatase in Southern catfish and their possible roles in sex differentiation. *J. Endocrinol.*, 194, 223-241.

Masuda, N., Yasumo, H., Tamura, T., Hashiguchi, N., Furusawa, T., Tsukamoto, T., Sadano, H. and Osumi, T. (1997). An orphan nuclear receptor lacking a zinc-finger DNA-binding domain: interaction with several nuclear receptors. *Biochim. Biophys. Acta.*, 1350, 27-32.

Masuyama, H., Yamada, M., Kamei, Y., Fujiwara-Ishikawa, T., Todo, T., Nagahama, Y. and Matsuda, M. (2012). Dmrt1 mutation causes a male-to-female sex reversal after the sex determination by Dmy in the medaka. *Chromosome Res.*, 20, 163-176.

Matsuda, M., Nagahama, Y., Shinomiya, A., Sato, T., Matsuda, C., Kobayashi, T., Morrey, C. E., Shibata, N., Asakawa, S., Shimizu, N., Hori, H., Hamaguchi, S. and Sakaizumi, M. (2002). DMY is a Y-specific DM-domain gene required for male development in the medaka fish. *Nature*, 417, 559-563.

Matsuda, M., Sato, T., Toyazaki, Y., Nagahama, Y., Hamaguchi, S. and Sakaizumi, M. (2003). Oryzias curvinotus has DMY, a gene that is required for male development in the medaka, *O. latipes. Zoolog. Sci.*, 20, 159-161.

Matsuda, M., Shinomiya, A., Kinoshita, M., Suzuki, A., Kobayashi, T., Paul-Prasanth, B., Lau, E. L., Hamaguchi, S., Sakaizumi, M. and Nagahama, Y. (2007). DMY gene induces male development in genetically female (XX) medaka fish. *Proc. Natl. Acad. Sci. USA*, 104, 3865-3870.

Miura, T., Miura, C., Konda, Y. and Yamauchi, K. (2002). Spermatogenesis-preventing substance in Japanese eel. *Development*, 129, 2689-2697.

Mizusaki, H., Kawabe, K., Mukai, T., Ariyoshi, E., Kasahara, M., Yoshioka, H., Swain, A. and Morohashi, K. (2003). Dax-1 (dosage-sensitive sex reversal-adrenal hypoplasia congenita critical region on the X chromosome, gene 1) gene transcription is regulated by wnt4 in the female developing gonad. *Mol. Endocrinol.*, 17, 507-519.

Mochizuki, K., Nishimiya-Fujisawa, C. and Fujisawa, T. (2000). Universal occurrence of the vasa-related genes among metazoans and their germline expression in *Hydra. Dev. Genes Evol.*, 211, 299-308.

Murugan, S., Shan, J., Kühl, S. J., Tata, A., Pietilä, I., Kühl, M. and Vainio, S. J. (2012). WT1 and Sox11 regulate synergistically the promoter of the Wnt4 gene that encodes a critical signal for nephrogenesis. *Exp. Cell Res.*, 318, 1134-1145.

Nagahama, Y. (2005). Molecular mechanisms of sex determination and gonadal sex differentiation in fish. *Fish Physiol. Biochem.*, 31, 105-109.

Nagahama, Y., Nakamura, M., Kitano, T. and Tokumoto, T. (2004). Sexual plasticity in fish: a possible target of endocrine disruptor action. *Environ Sci.*, 11, 73-82.

Nakamoto, M., Matsuda, M., Wang, D. S., Nagahama, Y. and Shibata, N. (2006). Molecular cloning and analysis of gonadal expression of Foxl2 in the medaka, *Oryzias latipes. Biochem. Biophys. Res. Commun.*, 344, 353-361.

Nakamoto, M., Suzuki, A., Matsuda, M., Nagahama, Y. and Shibata, N. (2005). Testicular type Sox9 is not involved in sex determination but might be in the development of testicular structures in the medaka, *Oryzias latipes. Biochem. Biophys. Res. Commun.*, 333, 729-736.

Nakamura, M., Bhandari, R. K. and Higa, M. (2003). The role estrogens play in sex differentiation and sex changes of fish. *Fish Physiol. Biochem.*, 28, 113-117.

Nanda, I., Kondo, M., Hornung, U., Asakawa, S., Winkler, C., Shimizu, A., Shan, Z., Haaf, T., Shimizu, N., Shima, A., Schmid, M. and Schartl, M. (2002). A duplicated copy of

DMRT1 in the sex-determining region of the Y chromosome of the medaka, *Oryzias latipes*. *Proc. Natl. Acad. Sci. USA*, 99, 11778-11783.

Nanda, I., Shan, Z., Schartl, M., Burt, D. W., Koehler, M., Nothwang, H., Grützner, F., Paton, I. R., Windsor, D., Dunn, I., Engel, W., Staeheli, P., Mizuno, S., Haaf, T. and Schmid, M. (1999). 300 million years of conserved synteny between chicken Z and human chromosome 9. *Nat. Genet.*, 21, 258-259.

Nicol, B., Guerin, A., Fostier, A. and Guiguen, Y. (2012). Ovary-predominant wnt4 expression during gonadal differentiation is not conserved in the rainbow trout (*Oncorhynchus mykiss*). *Mol. Reprod. Dev.*, 79, 51-63.

Penman, D. J. and Piferrer, F. (2008). Fish gonadogenesis. Part I: Genetic and environmental mechanisms of sex determination. *Rev. Fish. Sci.*, 16, 14-32.

Perner, B., Englert, C. and Bollig, F. (2007). The Wilms tumor genes wt1a and wt1b control different steps during formation of the zebrafish pronephros. *Dev. Biol.*, 309, 87-96.

Pfennig, F., Kurth, T., Meissner, S., Standke, A., Hoppe, M., Zieschang, F., Reitmayer, C., Göbel, A., Kretzschmar, G. and Gutzeit, H. O. (2012). The social status of the male Nile tilapia (Oreochromis niloticus) influences testis structure and gene expression. *Reproduction*, 143, 71-84.

Postlethwait, J., Amores, A., Cresko, W., Singer, A. and Yan, Y. L. (2004). Subfunction partitioning, the teleost radiation and the annotation of the human genome. *Trends Genet.*, 20, 481-490.

Raghuveer, K. and Senthilkumaran, B. (2009). Identification of multiple dmrt1s in catfish: localization, dimorphic expression pattern, changes during testicular cycle and after methyltestosterone treatment. *J. Mol. Endocrinol.*, 42, 437-448.

Raghuveer, K. and Senthilkumaran, B. (2010a). Cloning and differential expression pattern of vasa in the developing and recrudescing gonads of catfish, *Clarias gariepinus*. *Comp. Biochem. Physiol. A Mol. Integr. Physiol.*, 157, 79-85.

Raghuveer, K. and Senthilkumaran, B. (2010b). Isolation of sox9 duplicates in catfish: localization, differential expression pattern during gonadal development and recrudescence, and hCG-induced up-regulation of sox9 in testicular slices. *Reproduction*, 140, 477-487.

Raghuveer, K., Garhwal, R., Wang, D. S., Bogerd, J., Kirubagaran, R., Rasheeda, M. K., Sreenivasulu, G., Bhattachrya, N., Tarangini, S., Nagahama, Y. and Senthilkumaran, B. (2005). Effect of methyl testosterone- and ethynyl estradiol-induced sex differentiation on catfish, Clarias gariepinus: expression profiles of DMRT1, Cytochrome P450aromatases and 3β-hydroxysteroid dehydrogenase. *Fish. Physiol. Biochem.*, 31, 143-147.

Raghuveer, K., Senthilkumaran, B., Sudhakumari, C. C., Sridevi, P., Rajakumar, A., Singh, R., Murugananthkumar, R. and Majumdar, K. C. (2011). Dimorphic expression of various transcription factor and steroidogenic enzyme genes during gonadal ontogeny in the air-breathing catfish, *Clarias gariepinus*. *Sex Dev.*, 5, 213-223.

Rajakumar, A., Singh, R., Chakrabarty, S., Murugananthkumar, R., Laldinsangi, C., Prathibha, Y., Sudhakumari, C. C., Dutta-Gupta, A. and Senthilkumaran, B. (2012). Endosulfan and flutamide impair testicular development in the juvenile Asian catfish, *Clarias batrachus*. *Aquat. Toxicol.*, 110-111, 123-132.

Rasheeda, M. K., Sridevi, P. and Senthilkumaran, B. (2010). Cytochrome P450 aromatases: Impact on gonadal development, recrudescence and effect of hCG in the catfish, *Clarias gariepinus*. *Gen. Comp. Endocrinol.*, 167, 234-245.

Raymond, C. S., Murphy, M. W., O'Sullivan, M. G., Bardwell, V. J. and Zarkower, D. (2000). Dmrt1, a gene related to warm and fly sexual regulators, is required for mammalian testis differentiation. *Genes Dev.*, 14, 2587-2595.

Raymond, C. S., Shamu, C. E., Shen, M. M., Seifert, K. J., Hirsch, B., Hodgkin, J. and Zarkower, D. (1998). Evidence for evolutionary conservation of sex determining genes. *Nature*, 391, 691-695.

Raz, E. (2000). The function and regulation of vasa-like genes in germ-cell development. *Genome. Biol.*, 1, 1017.1-1017.6.

Saffman, E. E. and Lasko, P. (1999). Germline development in vertebrates and invertebrates. *Cell Mol. Life Sci.*, 55, 1141-1163.

Schubert, M., Holland, L. Z. and Holland, N. D. (2000). Characterization of two amphioxus Wnt genes (AmphiWnt4 and AmphiWnt7b) with early expression in the developing central nervous system. *Dev. Dyn.*, 217, 205-215.

Schüpbach, T. and Wieschaus, E. (1986). Maternal-effect mutations altering the anterior–posterior pattern of the *Drosophila* embryo. *Roux's Arch. Dev. Biol.*, 195, 302-317.

Seol, W., Choi, H. S. and Moore, D. D. (1996). An orphan nuclear hormone receptor that lacks a DNA binding domain and heterodimerizes with other receptors. *Science*, 272, 1336-1339.

Seol, W., Hanstein, B., Brown, M. and Moore, D. D. (1998). Inhibition of estrogen receptor action by the orphan receptor SHP (short heterodimer partner). *Mol. Endocrinol.*, 12, 1551-1557.

Sinclair, A. H., Berta, P., Palmer, M. S., Hawkins, J. R., Griffiths, B. L., Smith, M. J., Foster, J. W., Frischauf, A. M., Lovell-Badge, R. and Goodfellow, P. N. (1990). A gene from the human sex-determining region encodes a protein with homology to a conserved DNA-binding motif. *Nature*, 346, 240-244.

Skaar, K. S., Nóbrega, R. H., Magaraki, A., Olsen, L. C., Schulz, R. W. and Male, R. (2011). Proteolytically activated, recombinant Anti-Müllerian hormone inhibits androgen secretion, proliferation, and differentiation of spermatogonia in adult zebrafish testis organ cultures. *Endocrinology*, 152, 3527-3540.

Smith, C. A., Clifford, V., Western, P. S., Wilcox, S. A., Bell, K. S. and Sinclair, A. H. (2000). Cloning and expression of a DAX1 homologue in the chicken embryo. *J. Mol. Endocrinol.*, 24, 23-32.

Sridevi, P. and Senthilkumaran, B. (2011). Cloning and differential expression of FOXL2 during ovarian development and recrudescence of the catfish, *Clarias gariepinus*. *Gen. Comp. Endocrinol.*, 174, 259-268.

Sudhakumari, C. C., Senthilkumaran, B., Kobayashi, T., Kajiura-Kobayashi, H., Wang, D. S., Yoshikuni, M. and Nagahama, Y. (2005). Ontogenic expression patterns of several nuclear receptors and cytochrome P450 aromatases in brain and gonads of the Nile tilapia Oreochromis niloticus suggests their involvement in sex differentiation. *Fish. Physiol. Biochem.*, 31, 129-135.

Sudhakumari, C. C., Senthilkumaran, B., Kobayashi, T., Wang, D. S., Chang, X. T. and Nagahama, Y. (2003). Expression of Cytochrome P-450 Aromatases in the Sex-Reversed Nile Tilapia. *Fish. Physiol. Biochem.*, 28, 177-178.

Sugita, J., Takase, M. and Nakamura, M. (2001). Expression of Dax-1 during gonadal development of the frog. *Gene*, 280, 67-74.

Swain, A., Zanaria, E., Hacker, A., Lovell-Badge, R. and Camerino, G. (1996). Mouse Dax1 expression is consistent with a role in sex determination as well as in adrenal and hypothalamus function. *Nat. Genet.*, 12, 404-409.

Takamatsu, N., Kanda, H., Ito, M., Yamashita, A., Yamashita, S. and Shiba, T. (1997). Rainbow trout SOX9, cDNA cloning, gene structure and expression. *Gene*, 202, 167-170.

Tang, B., Hu, W., Hao, J. and Zhu, Z. (2010). Developmental expression of steroidogenic factor-1, cyp19a1a and cyp19a1b from common carp (*Cyprinus carpio*). *Gen. Comp. Endocrinol.*, 167, 408-416.

Wang, D. S., Kobayashi, T., Senthilkumaran, B., Sakai, F., Sudhakumari, C. C., Suzuki, T., Yoshikuni, M., Matsuda, M., Morohashi, K. and Nagahama, Y. (2002). Molecular cloning of DAX1 and SHP cDNAs and their expression patterns in the Nile tilapia, *Oreochromis niloticus*. *Biochem. Biophys. Res. Commun.*, 297, 632-640.

Wang, D. S., Kobayashi, T., Zhou, L. Y., Paul-Prasanth, B., Ijiri, S., Sakai, F., Okubo, K., Morohashi, K. and Nagahama, Y. (2007). Foxl2 up-regulates aromatase gene transcription in a female-specific manner by binding to the promoter as well as interacting with ad4 binding protein/steroidogenic factor 1. *Mol. Endocrinol.*, 21, 712-725.

Wang, D. S., Zhou, L. Y., Kobayashi, T., Matsuda, M., Shibata, Y., Sakai, F. and Nagahama, Y. (2010). Doublesex- and Mab-3-related transcription factor-1 repression of aromatase transcription, a possible mechanism favoring the male pathway in tilapia. *Endocrinology*, 151, 1331-1340.

Wang, D., Kobayashi, T., Zhou, L. and Nagahama, Y. (2004). Molecular cloning and gene expression of Foxl2 in the Nile tilapia, *Oreochromis niloticus*. *Biochem. Biophys. Res. Commun.*, 320, 83-89.

Wang, X. G. and Orban, L. (2007). Anti-Müllerian hormone and 11β-hydroxylase show reciprocal expression to that of aromatase in the transforming gonad of zebrafish males. *Dev. Dyn.*, 236, 1329-1338.

Wu, G. C. and Chang, C. F. (2009). wnt4 Is associated with the development of ovarian tissue in the protandrous black Porgy, *Acanthopagrus schlegeli*. *Biol. Reprod.*, 81, 1073-1082.

Wu, G. C., Tomy, S., Nakamura, M. and Chang, C.F. (2008). Dual roles of cyp19a1a in gonadal sex differentiation and development in the protandrous black porgy, *Acanthopagrus schlegeli*. *Biol .Reprod.*, 79, 1111-1120.

Wylie, C. (1999). Germ cells. *Cell*, 96, 165-174.

Xia, W., Zhou, L., Yao, B., Li, C. J. and Gui, J. F. (2007). Differential and spermatogenic cell-specific expression of DMRT1 during sex reversal in protogynous hermaphroditic groupers. *Mol. Cell Endocrinol.*, 263, 156-172.

Xu, H., Gui, J. and Hong, Y. (2005). Differential expression of vasa RNA and protein during spermatogenesis and oogenesis in the gibel carp (Carassius auratus gibelio), a bisexually and gynogenetically reproducing vertebrate. *Dev. Dyn.*, 233, 872-882.

Yamamoto, T. (1969). Sex differentiation. In W. Hoar, and D. Randall (Eds.). *Fish. Physiology*, Academic Press, pp. 117-175.

Yokoi, H., Kobayashi, T., Tanaka, M., Nagahama, Y., Wakamatsu, Y., Takeda, H., Araki, K., Morohashi, K. and Ozato, K. (2002). Sox9 in a Teleost Fish, Medaka (Oryzias latipes):

evidence for diversified function of Sox9 in gonad differentiation. *Mol. Reprod. Dev.*, 63, 5-16.

Yoshiura, Y., Senthilkumaran, B., Watanabe, M., Oba, Y., Kobayashi, T. and Nagahama, Y. (2003). Synergistic expression of Ad4BP/SF-1 and cytochrome P-450 aromatase (ovarian type) in the ovary of Nile tilapia, Oreochromis niloticus, during vitellogenesis suggests transcriptional interaction. *Biol. Reprod.*, 68, 1545-1553.

Yu, H., Seah, A., Herman, M. A., Ferguson, E. L., Horvitz, H. R. and Sternberg, P. W. (2009). Wnt and EGF pathways act together to induce *C. elegans* male hook development. *Dev. Biol.*, **327,** 419-432.

Zhang, J. (2004). Evolution of DMY, a newly emergent male sex determination gene of medaka fish. *Genetics*, 166, 1887-1895.

Zhou, R., Liu, L., Guo, Y., Yu, H., Cheng, H., Huang, X., Tiersch, T. R. and Berta, P. (2003). Similar gene structure of two Sox9a genes and their expression patterns during gonadal differentiation in a teleost fish, rice field eel (*Monopterus albus*). *Mol. Reprod. Dev.*, 66, 211-217.

In: Sexual Plasticity and Gametogenesis in Fishes
Editor: Balasubramanian Senthilkumaran
ISBN: 978-1-62618-848-8

Chapter XXV

Membrane Progestin Receptor, a Key Mediator of Final Oocyte Maturation in Fish

Toshinobu Tokumoto [*]
Department of Biology, Faculty of Science,
Shizuoka University, Shizuoka, Japan

Abstract

In fish, fully grown oocytes are arrested at prophase of the first meiotic division prior to oocyte maturation. The oocytes resume meiosis in response to maturation-inducing hormone (MIH) secreted from ovarian follicle cells by stimulating the *de novo* synthesis of cyclin B, a regulatory subunit of maturation promoting factor (MPF), the final mediator of oocyte maturation. MIHs have been identified as a class of steroids called progestins. Steroid hormones can act as chemical messengers mediating not only slow genomic actions but also rapid non-genomic actions in a wide range of target tissues. Fish oocyte maturation is the most well-characterized system mediated through non-genomic actions of steroids. A strong candidate for an MIH receptor, membrane progestin receptor (mPR), has been identified for the first time in fish. Furthermore, three subtypes of mPRs have been identified in other vertebrates and its wide tissue distribution was revealed. Thus, the molecule found as a mediator for fish oocyte maturation has become a target for studying the general mechanism of steroidal non-genomic actions in various tissues of vertebrates.

Almost ten years have passed since the groundbreaking identification of mPR. Although the identification of genes in various organisms and expression patterns have been described since then, the precise physiological roles of mPRs are still unclear. This review summarizes the story of the identification and recent advances in the study of mPR as a receptor mediating non-genomic steroid actions.

Keywords: mPR, non-genomic steroid action, Oocyte Maturation, DES

[*] Email: sbttoku@ipc.shizuoka.ac.jp.

Introduction

In teleost fishes, like in most other vertebrates, full-grown post vitellogenic oocytes are physiologically arrested at the G2/M boundary in the first meiotic prophase and cannot be fertilized. To become fertilizable oocytes, they must complete the first meiotic division and ovulation. The oocytes resume meiosis in response to maturation-inducing hormone (MIH), which is produced by follicle cells under the control of gonadotropins released from the pituitary gland. MIH acts on receptors on the oocyte plasma membrane and induces the *de novo* synthesis of cyclin B, a regulatory subunit of maturation promoting factor (MPF), the final mediator of oocyte maturation (Nagahama, 1987; Yamashita et al., 1995). Although it has not yet been proved, it is thought that MIH also induces ovulation through genomic actions by binding to nuclear progesterone receptor in oocytes and follicle cells. Two steroids have been identified as MIHs, the progestins 17α,20β-dihydroxy-4-pregnen-3-one (17,20β-DHP) and 17α,20β,21-trihydroxy-4-pregnen-3-one (20β-S), in several fish species (Nagahama and Adachi, 1985; Patino and Thomas, 1990; Trant et al., 1986). It was predicted and demonstrated in a series of studies that these progestins do not act via classical nuclear steroid receptors, but instead initiate their actions by binding to specific membrane receptors on the external surface of oocytes (Nagahama et al., 1995). The first evidence that showed receptors for MIH exists on the cell surface had been described in frog oocytes (Ishikawa et al., 1977; Godeau et al., 1978). A membrane impermeable steroid was effective at inducing oocyte maturation when suspended in the medium. Microinjection of progesterone was also ineffective at inducing oocyte maturation (Masui and Clarke, 1979). In addition, a few minutes exposure to MIH was sufficient to initiate oocyte maturation (Thomas and Das, 1997). From these data, it was predicted that the MIH receptor is on the cell membrane. Direct evidence of MIH binding activity on the cell membrane was first characterized in starfish (Yoshikuni et al., 1988). Subsequently, MIH receptors have been reported on the oocyte membranes of several teleost species (Patino and Thomas, 1990; Yoshikuni et al., 1993; King et al., 1997; Rahman et al., 2002;). However, the precise molecular structure and mechanisms of action of these membrane progestin receptors remained unclear until 2003.

Several decades after the initial suggestion of a membrane receptor for steroids, a novel cDNA was cloned and sequenced from a spotted seatrout ovarian cDNA library that encoded a protein (mPRα) with the characteristics of the membrane progestin receptor mediating the MIH-induced initiation of oocyte maturation (Zhu et al., 2003b). The finding that microinjection of zebrafish (*Danio rerio*) oocytes with antisense oligonucleotides to the homologous zebrafish mPRα cDNA blocked the development of maturation competence in response to gonadotropin provided preliminary evidence that mPRα is also involved in the induction of oocyte maturation in this species (Thomas et al., 2004). The physiological functions of mPR in oocyte maturation have been shown in goldfish (*Carassius auratus*), a well-established model of the hormonal control of oocyte maturation, and representative of a major family of freshwater teleosts, the Cyprinidae (Tokumoto et al., 2006; Tokumoto et al., 2012). The mPRα protein was detected on the plasma membranes of oocytes of goldfish. Upregulation of mPRα protein levels during gonadotropin-induction of oocyte maturation was investigated as well as the effects of oocyte microinjection with mPRα antisense oligonucleotides.

The results suggested the mPRα protein is also an intermediary in the MIH induction of oocyte maturation in goldfish. Recently mPR has been recognized as a new target for endocrine disrupting chemicals. Data from binding assays has demonstrated that some chemicals show binding affinity for mPR. Treatment of oocytes with an endocrine disrupting chemical (EDC), diethylstilbestrol (DES), induced maturation in goldfish and zebrafish (Tokumoto et al., 2004). This effect demonstrated that maturation is mediated by mPR (Tokumoto et al., 2007).

Thus, it should be emphasized that the screening of chemical substances for their reactivity with mPRs is very valuable because it is possible that such chemicals are candidates EDCs. Off course, mPRs are attractive pharmaceutical targets, and very recently high expression levels of mRNA for mPRs have been reported in ovarian cancer cells and breast cancer cells (Charles et al., 2010; Dressing et al., 2012). Homologues in yeast have been identified as an antifungal protein, osmotin, receptor (Narasimhan et al., 2005). Thus mPR homologues in fungus are possible targets for anti-fungal drugs (Villa et al., 2011). Therefore, screening systems for mPR-interacting chemicals should be established to aid in the search novel pharmaceuticals.

This review summarizes the story of the identification of mPR, the molecular features of genes and proteins for mPRs, and the possibilities for these proteins as an attractive new target for pharmaceuticals.

Steroid Non-Genomic and Genomic Actions in the Induction of Oocyte Maturation

Steroid hormones can act as chemical messengers mediating not only slow genomic actions but also rapid non-genomic actions in a wide range of target tissues (Figure 1). Numerous actions of steroid hormones cannot be explained by classical genomic actions of steroids through nuclear receptor activated by binding with steroids that penetrate into the cell. Genomic actions of steroids involving new mRNA and protein synthesis are relatively slow, typically occurring over a time scale of hours to days (Losel et al., 2003). Conversely, non-genomic actions through membrane receptors are relatively rapid and cause changes in intracellular protein interactions and post-translational modifications. Oocyte maturation is a well-characterized phenomenon that is induced by steroidal non-genomic actions through a decrease of intracellular cyclic AMP (cAMP) levels within a few minutes (Figure 2). A decrease in cAMP levels has been reported in response to MIH in frog oocytes (Cicirelli and Smith, 1985), and the same results reported in teleost oocytes (Finet et al., 1988). It has been demonstrated in spotted seatrout that MIH activates a pertussis toxin (PTX)-sensitive inhibitory G protein, and that activation of this pathway is necessary for the completion of oocyte maturation (Pace and Thomas, 2005a). Similarly, a decrease in [^{3}H]-17,20β-DHP binding to rainbow trout oocyte membranes was observed after treatment with PTX (Yoshikuni and Nagahama, 1994). These studies suggested that the membrane progestin receptor in seatrout and also probably in rainbow trout is directly coupled to inhibitory G-proteins.

The first evidence that a receptor for MIH is on the surface of oocytes was shown in frog oocytes, in which a non-cell permeable steroid induced oocyte maturation.

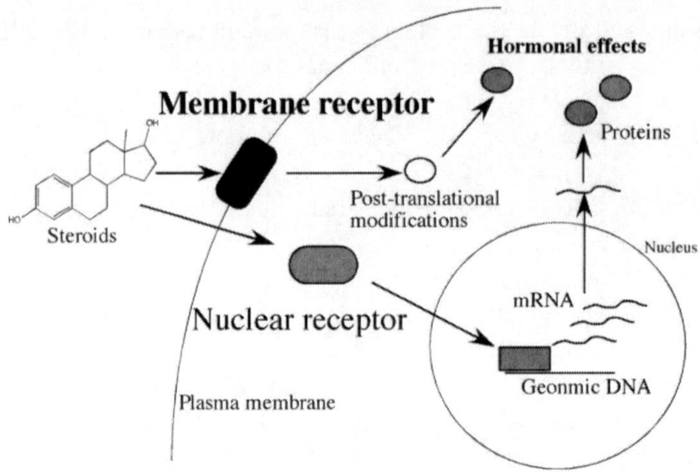

Figure 1. Schematic diagram of non-genomic and genomic actions induced by steroids.

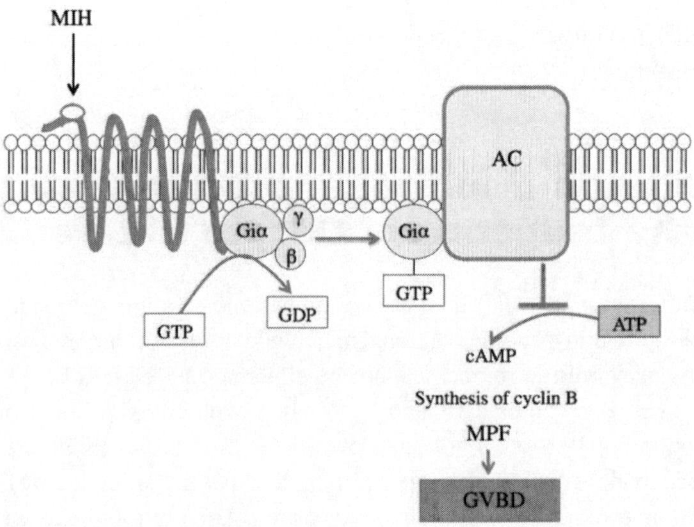

Abbreviations: MIH, maturation-inducing hormone; Giα, inhibitory GTP-binding protein α; AC, adenylate cyclase; cAMP, cyclic AMP; GVBD, germinal vesicle breakdown.

Figure 2. Induction of decreases in intracellular cAMP concentration through mPR.

Although the chemical structure of progesterone itself does not permit conjugation, it was shown that deoxycorticosterone, which has a structure similar to progesterone, exhibited maturation-inducing activity and could be conjugated to agarose beads. When deoxycorticosterone-conjugated agarose (DOC-agarose) was suspended in the medium with immature *Xenopus* oocytes, maturation was induced (Ishikawa et al., 1977; Godeau et al., 1978). Microinjection of MIH was not effective at inducing oocyte maturation (Masui and Clarke, 1979). These results strongly suggested that the receptor for MIH is on the cell surface. Characterization of this membrane receptor was subsequently reported in starfish (Yoshikuni et al., 1988), and evidence has accumulated for the presence of membrane receptors for MIH in other organisms. However, these studies had been conducted using membrane preparations because attempts to obtain highly purified preparations of steroid membrane receptors for amino acid sequencing using classical protein purification approaches had been unsuccessful.

On the contrary, many of the intracellular signaling molecules, such as maturation-promoting factor or metaphase-promoting factor (MPF), involved in the later stages of oocyte maturation in teleosts have been well characterized (Nagahama et al., 1995). MPF activity in fish oocytes was first demonstrated by the induction of germinal-vesicle-breakdown (GVBD) in starfish oocytes by the microinjection of cytoplasm from mature unfertilized goldfish oocytes (Kishimoto et al., 1982). MPF activity was extracted from goldfish oocytes matured by human chorionic gonadotropin (hCG) treatment *in vivo*, and was injected into immature *Xenopus* oocytes (Yamashita et al., 1992a). MPF activity extracted from goldfish oocytes was also effective when injected into immature goldfish oocytes under conditions where protein synthesis was inhibited. Oocytes injected with MPF undergo meiotic maturation much more rapidly than *in vitro* 17, 20β-DHP-induced maturation. MPF activity increases before GVBD and peaks at metaphase I. The activity then decreases during anaphase and telophase I, but was found to be at maximal levels again in metaphase II oocytes (Yamashita et al., 1992b). MPF activity decreases immediately after egg activation (fertilization). In immature goldfish oocytes, there is no detectable cyclin B and cdc2 kinase is monomeric. This is different from *Xenopus* and starfish oocytes. Cyclin B appears during oocyte maturation by *de novo* synthesis (Hirai et al., 1992; Katsu et al., 1993). Cyclin B mRNA is already present in immature goldfish oocytes, and its level is not markedly changed during oocyte maturation. Therefore, regulation of the synthesis of cyclin B protein is under translational control. A cDNA clone of germ cell-specific Y box protein, which is a potential masking protein of cyclin B mRNA in goldfish oocytes, has been cloned (Katsu et al., 1997). It has been suggested that in goldfish oocytes, the synthesis of cyclin B protein is also under translational control and that cytoplasmic 3' poly(A) elongation is involved in 17, 20β-DHP-induced translation of cyclin B mRNA (Katsu et al., 1999). Recent developments in technology have revealed that the translation of cyclin B mRNA takes place in a restricted area at the animal pole where GVBD occurs (Yasuda et al., 2010). In addition, it has been demonstrated that the 5'UTR and coding region of the cyclin B mRNA are required for the temporal initiation of translation.

As described above there had been considerable progress in understanding of the intracellular cascade triggered by MIH.

Despite considerable effort in many laboratories over the past 25 years, however, the identities of steroid membrane receptors for MIH remain unresolved. Recent research in fish ovaries has led to the discovery of novel progestin membrane receptors that are suspected to be of widespread physiological importance in vertebrate reproduction.

Identification of Membrane Progestin Receptor

The purification of membrane progestin receptors for amino acid sequencing using classical protein purification approaches had been unsuccessful. Therefore, a combination of partial receptor purification, antibody generation and expression library screening was used to identify likely cDNA candidates for the membrane progestin receptor (Thomas et al., 2002).

Monoclonal antibodies raised against a partially purified receptor fraction from spotted seatrout ovaries, which were selected by a steroid binding assay, were used for expression library screening (Thomas et al., 2002). By screening cDNA candidates, the membrane progestin receptor was finally isolated. The protein encoded by the cDNA has the major characteristics of a membrane progestin receptor mediating meiotic maturation (Zhu et al., 2003b). The protein had an apparent molecular weight of 40 kDa, and the structure of the protein was predicted to include seven trans-membrane domains by computational analysis, which is characteristic of G protein-coupled receptors. The expression level of the mRNA for the isolated cDNA was relatively high in reproductive tissues, which was consistent with the known regulatory roles of progestins in reproductive and endocrine functions. By immunohistochemical analysis, the protein was detected in the oocyte plasma membrane, and the recombinant protein could bind progestins with the characteristics of a hormone receptor. High affinity, limited capacity, displaceable, and specific binding for progestins typical of steroid membrane receptors was demonstrated in membranes prepared from human breast cancer cells transfected with the cDNA. Rapid activation of the MAP kinase cascade in transfected human cells was observed after treatment with physiological concentrations of progestins. Direct evidence showing that the protein was a receptor for MIH was obtained by knock-down experiments using morpholino-antisense oligonucleotides. Injection of antisense oligonucleotides, but not sense or mis-sense ones, during gonadotropin induction blocked the subsequent response of zebrafish oocytes to MIH (Zhu et al., 2003b). From a series of studies it was concluded that a cDNA cloned from seatrout encoded the MIH receptor and it was named membrane progestin receptor (mPR). Studies *in silico* identified 11 closely related genes in other vertebrates that could be separated into three subtypes on the basis of sequence identity and phylogenetic analysis, and these were designated as α, β and γsubtypes (Zhu et al., 2003a). Genome-wide phylogenetic analyses have shown that mPRs are a new protein family, the progestin and adipoQ receptor (PAQR) family (Tang et al., 2005). In this family, mPRα, β and γ subtypes correspond to PAQR7, 8, and 5, respectively (Thomas et al., 2007). It is generally thought that these progestins do not act via the classical intracellular mechanism of steroid action involving the activation of nuclear steroid receptors, but instead initiate their actions at the external surface of oocytes by binding to specific membrane receptors (Nagahama et al., 1995; Thomas et al., 2002).

The mPRα subtype has also been cloned from a goldfish, *Carassius auratus*, ovarian cDNA library (Tokumoto et al., 2006). Western blotting using a specific goldfish mPRα antibody demonstrated the presence of an immunoreactive band of the appropriate predicted molecular weight (~40 kDa) in oocyte plasma membranes. Similar to the previous results in spotted seatrout, mPRα protein expression in fully grown goldfish oocytes was upregulated by treatment with gonadotropin and was associated with the development of oocyte maturational competence.

Moreover, microinjection of mPRα antisense morpholino-oligonucleotides into goldfish oocytes blocked both the induction of oocyte maturation in response to MIH and the upregulation of mPRα protein levels (Tokumoto et al., 2006). Thus the results obtained with representatives of two distantly related teleosts families, the Scaenidae and Cypriniformes, support the suggestion that mPRα acts as an intermediary in MIH induction of oocyte maturation in teleosts (Figure 2). The finding that microinjection with antisense oligonucleotides to zebrafish mPRβ also partially blocked oocyte maturation in this species suggested that this mPR subtype is also involved in MIH signaling (Thomas et al., 2004). Involvement of mPRβ in oocyte maturation was also demonstrated in goldfish (Tokumoto et al., 2012). The three mPR subtypes have also been cloned from channel catfish (*Ictalurus punctatus*) and goldfish ovarian cDNA libraries (Kazeto et al., 2005; Tokumoto et al., 2012). The pattern of changes observed in gene expression of the mPR subtypes (α, βand γ) in fish ovaries during the final stages of gametogenesis and in response to hormonal treatment suggests that gonadotropin induction of oocyte maturation does not involve upregulation of the mPR message in oocytes (Kazeto et al., 2005; Tokumoto et al., 2006; Tokumoto et al., 2012). Differential upregulation by gonadotropin of mPRα protein levels in the absence of changes in mRNA expression has already been described during the maturation of both goldfish and croaker oocytes (Thomas et al., 2004; Pace and Thomas, 2005b; Tokumoto et al., 2006).

Genomic Structure of the mPR Gene

The gene structures of mPRs were characterized in channel catfish (Kazeto et al., 2005).

The mPRα and β genes have similar structures with intronless coding regions, whereas the mPRγ gene is composed of 8 exons and 7 introns. These features are conserved among fish genomes. In the genome of medaka fish, single coding regions for mPRα (PAQR7) and mPRβ (PAQR8) were annotated. Numerous candidate enhancer sequences have been identified in proximity to mPR genes.

Although the mRNAs for goldfish mPRs are constantly expressed through oogenesis over several months, their tissue specific expression patterns differs among the subtypes; therefore, their enhancer sequences should differ as well.

The Structures of mPR Proteins

As described above, a typical seven transmembrane structure was predicted for mPR proteins by computer analysis of the mPR genes from spotted seatrout (Zhu et al., 2003a).

In this model, the N-terminus of the mPR protein extended into the extracellular environment and the C-terminus into the intracellular compartment, which was consistent with a typical G-protein coupled receptor.

However, the opposite model was predicted for one of the same PAQR family members, with a reversed structural model predicted for adipoQ receptors (PAQR1 and PAQR2) based on the results of immunocytochemical observations using a specific antibody against the N-terminal region of the receptor (Yamauchi et al., 2003). Thus, the N-terminal loop was inside the cell and the C-terminal region was outside of the cell. Using sequence similarity analysis, it was suggested that the structures of mPRs are same as adipoQ receptors (Smith et al., 2008; Villa et al., 2011). As a result, two conflicting models have been proposed for the structure of mPR proteins.

However, the same immunocytochemical analysis was conducted for human mPRα expressed in cultured cells and the results supported the original model (Thomas et al., 2007).

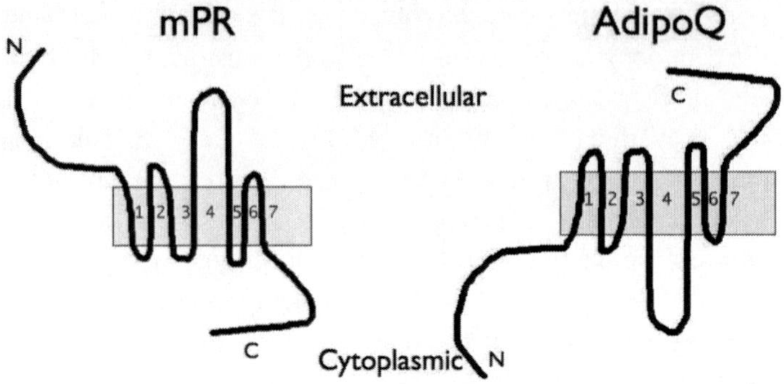

Figure 3. Predicted topology of membrane progestin receptor (mPR) and adipoQ receptor.

Furthermore, analysis using cell sorting suggested that the N-terminal extension was exposed to the outside of the cell (Hanna et al., 2006). A typical seven transmembrane structure was consistently predicted for mPRγ subtypes from goldfish by several computer programs (SOSUI, TMpred, TMHMM). Very recently, we established cell lines that can be used for monitoring intracellular cAMP levels, and a study using these cell lines indicated that cAMP concentration was decreased by mPR ligands (Tokumoto et al., unpublished). It was demonstrated that mPR is coupled to Gi protein, as suggested by a range of evidence.

Thus, the typical seven transmembrane structure depicted in Figure 3 is regarded as the likely structure for mPRs. It is also noteworthy that there are differences in structure among the mPR subtypes. The overall structures of goldfish mPRγ-1 and γ-2 are predicted as notably different from those of goldfish mPRα and β. A larger extracellular loop is composed of the third loop in mPRα and β but the fourth loop in the case of γ subtypes (Tokumoto, 2012). The difference in structure might account for the difference in characteristics between mPRα, βand mPRγ-1, γ-2.

More Subtypes of mPRs

Recent analysis by genome-wide screening for homologous genes to mPR revealed that mPRs are classified in a novel family of G protein-coupled receptor.

The PAQR family is composed of 11 genes, including the originally identified three subtypes of mPR α, β and γ. From experiments involving heterologous expression of human mPRs in yeast, it was suggested that PAQR6 and PAQR9 exhibited progestin receptor activity (Smith et al., 2008). Thus, it was proposed to classify these genes as new subtypes of mPRs as mPRδ and mPRε respectively. Detailed characterization of these genes has not been performed in fish as yet. Nucleotide sequence analysis was also conducted for ancestor genes of mPRs in fungi.

Very interestingly, disruption of the genes of mPR homologues resulted in resistance to antibiotics (Villa et al., 2011). These results suggested that mPR-related genes responsible as a mediator for chemicals might be through the cell surface.

Steroidal Genomic Actions and Non-Genomic Actions in the Induction of Ovulation

As described in above, fish oocyte maturation can be induced by non-genomic actions of steroids. In other words, the essential pathway to induce the activation of MPF could be induced by non-genomic actions. In general, oocyte maturation is assessed by examination of GVBD.

Thus, only the reactions necessary to induce GVBD were assessed by morphological methods. However, the process of oocyte maturation and ovulation in fish to produce fertilizable eggs should be induced coincidently. Immature oocytes in ovary are induced to form mature oocytes by progestins. Matured oocytes ovulate by rupturing from follicle cells along with surrounding oocytes. After ovulation, oocytes are spawned and fertilize for development. There are several reports on the pathway leading to the induction of oocyte maturation and ovulation in fish. MIH secreted from follicular cells acts on mPR on the plasma membrane of oocytes and induces oocyte maturation through non-genomic actions. Moreover, the induction of oocyte maturation by MIH requires protein synthesis but is not blocked by transcription inhibitors, indicating a non-genomic mechanism of action (Jalabert et al., 1976; Goetz and Theofan, 1979; Patino and Thomas, 1990). The main pathway for the induction of oocyte maturation, meiotic cell division, can be induced by non-genomic actions. Conversely, the ovulation-inducing pathway is thought to be activated by genomic actions because actynomycin D, which inhibit the production of mRNA, prevented ovulation in fish (Pinter and Thomas, 1999).

The possible functions of mPRs in the follicle cells surrounding oocytes and their potential interactions with the nuclear progestin receptor are interesting research questions to address in the future.

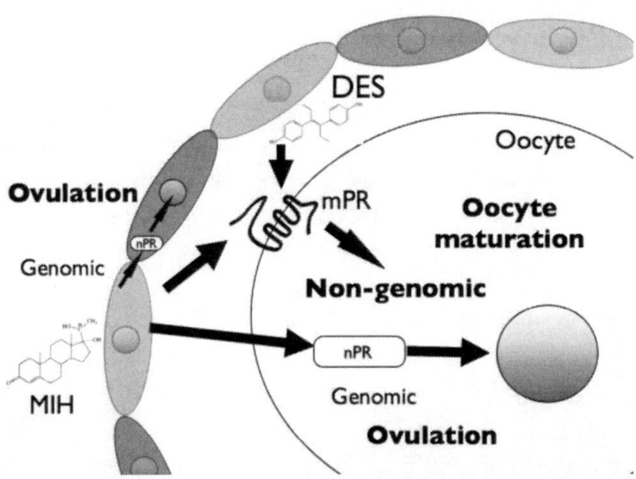

Figure 4. Induction of oocyte maturation by non-genomic actions and ovulation by genomic actions. MIH secreted from follicular cells acts on mPR in the plasma membrane of oocytes and induces oocyte maturation through non-genomic actions. At least the main pathway to induce oocyte maturation, meiotic cell division, can be induced by non-genomic actions. Conversely, the ovulation-inducing pathway is thought to be activated by genomic actions, which induce the production of mRNAs in oocytes and follicular cells.

Recently it was demonstrated *in vitro* study using oocytes from Medaka that MIH can induce ovulation, and very recently a novel method to induce oocyte maturation and ovulation *in vivo* was established in zebrafish (Tokumoto et al., 2011).

In this method, the addition of MIH only into the water induced oocyte maturation and ovulation. This result suggested that MIH induce genomic actions leading to ovulation. At this time, there is a possibility that MIH can activate genomic actions both in oocytes and follicle cells, which is necessary for ovulation. In this *in vivo* system, diethylstilbestrol (DES) induced oocyte maturation only by binding to mPRα.

Thus, non-genomic actions to induce oocyte maturation could be separated from MIH-dependent maturation- and ovulation-inducing actions (Figure 4). Using the *in vivo* system, both samples could be prepared, one of which activated the non-genomic pathway for oocyte maturation and the genomic pathway for ovulation by MIH and the other only activated the non-genomic pathway by DES. By comparing the gene expression profiles between these samples, genes up- or down-regulated to induce ovulation can be found. Gene expression analysis using microarrays is underway and is expected to identify genes for the key regulators in the induction of ovulation.

Induction and Inhibition of Fish Oocyte Maturation by Endocrine-Disrupting Chemicals Through mPR

Several EDCs, such as Kepon and *o,p*-DDD, have been reported to antagonize MIH-induced meiotic maturation of fish oocytes *in vitro* (Das and Thomas, 1999). EDCs such as methoxychlor and ethynyl estradiol also antagonize frog oocyte maturation. DES is an EDC with a non-steroidal structure, which was prescribed during the late 1940s to early 1970s to pregnant women to prevent abortion, preeclampsia, and other complications of pregnancy. Male and female offspring exposed *in utero* to DES developed multiple dysplastic and neoplastic lesions of the reproductive tract, along with other changes, during development (Bern, 1992).

Previously, it was reported that treatment of oocytes with DES induced maturation in goldfish and zebrafish (Tokumoto et al., 2004). Subsequently, the same agonistic activity of tamoxifen and its metabolite 4-hydroxytamoxifen (both chemicals are EDCs) was also demonstrated in zebrafish.

Furthermore, a potent inhibitory effect of pentachlorophenol (PCP) was demonstrated on oocyte maturation induced by 17,20β-DHP, DES and tamoxifen (Tokumoto et al., 2005). PCP is a widely used biocide that has been employed as a wood preservative, herbicide, and defoliant. Its extensive use and persistence have resulted in significant environmental contamination and potential exposure of the general population. Owing to its highly persistent nature, PCP is still one of the dominant phenolic compounds in blood (Sandau et al., 2002).

These results suggested that EDCs might interact with the MIH receptor to induce maturation. The identification of the structure of the putative MIH receptor as mPR permits investigations of the molecular mechanisms by which EDCs induce or inhibit oocyte maturation. Binding of these EDCs to mPRα was examined using membrane fractions prepared from cultured cells transfected with the cDNA for goldfish mPRα. Specific progestin binding was measured in plasma membranes prepared from goldfish mPRα-transfected cells (Tokumoto et al., 2007). Steroid competition studies showed that binding is highly specific for natural MIH, 17, 20β-DHP, in membranes prepared from both ovaries and mPRα-transfected cells. DES was a relatively effective competitor of isotope-labeled 17, 20β-DHP binding to mPRα.

There was a correlation between the relative binding affinities of DES and the DES analogues for mPRα and their potencies in mimicking the actions of 17,20β-DHP on meiotic maturation in the *in vitro* bioassay. These results demonstrate that DES can mimic the non-genomic actions of progestin by binding to mPRα. Binding affinity of DES to three subtypes of mPRs in goldfish has been assessed.

The results demonstrated that of the mPR subtypes, mPRαand β possess relatively higher affinities for DES than the mPRγ types. Combined with the results from knockdown studies using morpholino-antisense oligonucleotides, it was suggested that the mPRα and βsubtypes are the target of DES. By steroid competition studies using cell membranes that were stably transfected with cDNAs of mPR subtypes, the binding affinity of about 60 candidate EDCs listed by Japanese Environmental Agency to mPR was examined. Among these candidate EDCs, 13 compounds were found to react with mPR. This finding emphasizes the need for studies to examine more widely the various non-genomic effects of EDCs in the future.

Other Functions of mPRs

mPRs are likely to be involved in other functions induced by progestins because they are expressed in a wide variety of tissues (Zhu et al., 2003a ; Kazeto et al., 2005; Tokumoto et al., 2006), including sperm and the hypothalamus, where progestins have been shown to exert rapid, non-genomic actions (Thomas et al., 2004; Thomas et al., 2005). Recently, a novel function of mPR was suggested in quail (Ito et al., 2011). Because of the presence of specialized simple tubular invaginations in the oviduct, the ejaculated sperm after entering into the female reproductive tract can survive for a prolonged time in domestic birds (Bakst, 1994). This specialized structure is generally referred to as sperm storage tubules (SSTs).

The intravenous injection of progesterone into the birds stimulated sperm release from the SSTs. This observation clearly demonstrated that progesterone is the factor that triggers the release of the resident sperm from SSTs.

Moreover, as a result of the progesterone injection, a change in SST morphology that closely resembled contraction had been observed. These results demonstrated that progesterone stimulates the release of the resident sperm from the SSTs in bird with a contraction-like morphological change of the SSTs, which is probably initiated via mPRα-mediated signal transduction.

Conclusion

This chapter introduces the story of the identification and characterization of mPRs. Although the molecular basis, gene structure and evolutionary process have now been established, proof of the physiological roles and protein structure of mPRs are still lacking.

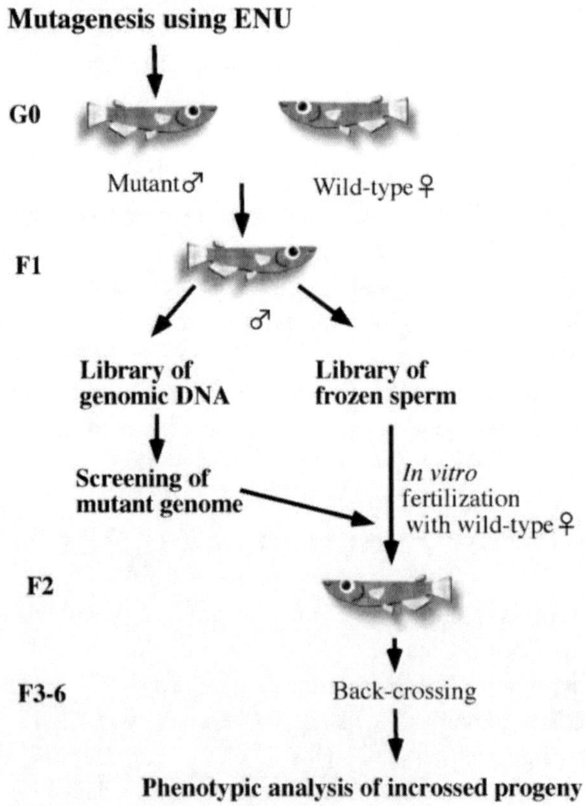

Figure 5. Strategy for establishing mutant strains in medaka by tilling. Initially, random mutations are induced in male fish by *N*-ethyl-*N*-nitrosourea (ENU). An F1 generation of mutant fish is produced by mating with wild-type females. A library of frozen stocks of sperm and genome DNA is prepared from F1 males. Mutations in genes of interest are screened for by DNA sequencing and HRM analysis. Sperm from the F1 fish identified in the screening was used for *in vitro* fertilization with wild-type females. Progenies are developed to adulthood and are genotyped. Then heterozygous fish carrying the mutation of interest are sequentially outcrossed with wild-type females. Finally heterozygous fish are incrossed to obtain homozygous fish for phenotypic analysis.

Thus, the biological significance of non-genomic actions induced by steroids is also still unclear. In medaka fish, a systematic strategy for screening for deficiencies in genes of interest has been established (Figure 5; Taniguchi et al., 2006). From this screening system, we have selected three and five strains in which a point mutation was induced into the coding sequence of mPRα and mPRβ, respectively. We are currently continuing with back-crossing with wild-type fish to establish strains to analyze the phenotype of these mutants. We also plan to establish mutant strains for mPR subtypes. These reverse genetic analyses should elucidate the central roles of mPRs in the near future.

Acknowledgments

I wish to thank Dr. Y. Nagahama, Dr P. Thomas and Dr. M. Tokumoto for their support and encouragement throughout the study. Screening of mutant medaka is being carried out in collaboration with Dr. Y. Taniguchi, Y. Kamei, Miss I. Hara and Dr. K. Naruse supported by a NIBB Individual Collaborative Research Project (11-342 and 12-352 to T.T.) and by National BioResource Project Medaka. This work was supported by Grants-in-Aid for Scientific Research on Priority Areas from the Ministry of Education, Culture, Sports, Science and Technology of Japan and grants from The Naito Foundation (to T.T.), Japanese Environmental Agency (ExTEND2005 FS to T.T.) and a SUNBOR grant from Suntory Institute For Bioorganic Research (to T.T.).

References

Bakst, M. R. (1994). Fate of fluorescent stained sperm following insemination: new light on oviducal sperm transport and storage in the turkey. *Biol. Reprod.*, 50, 987-92.

Bern, H. A. (1992). "Chemically induced alterations in sexual and functional development: The Wildlife/Human Connection." Princeton Sci. Pub., Prinston, NJ.

Charles, N. J., Thomas, P. and Lange, C. A. (2010). Expression of membrane progesterone receptors (mPR/PAQR) in ovarian cancer cells: implications for progesterone-induced signaling events. *Horm Cancer,* 1, 167-76.

Cicirelli, M. F. and Smith, L. D. (1985). Cyclic AMP levels during the maturation of Xenopus oocytes. *Dev. Biol.,* 108, 254-8.

Das, S. and Thomas, P. (1999). Pesticides interfere with the nongenomic action of a progestogen on meiotic maturation by binding to its plasma membrane receptor on fish oocytes. *Endocrinology,* 140, 1953-6.

Dressing, G. E., Alyea, R., Pang, Y. and Thomas, P. (2012). Membrane Progesterone Receptors (mPRs) Mediate Progestin Induced Antimorbidity in Breast Cancer Cells and Are Expressed in Human Breast Tumors. *Horm Cancer.*

Finet, B., Jalabert, B. and Garg, S. K. (1988). Effect of defolliculation and 17 a-hydroxy, 20 b -dihydroprogesterone on cyclic AMP level in full-grown oocytes of the rainbow trout, Salmo gairdneri. *Gamete Res.,* 19, 241-52.

Godeau, J. F., Schorderet-Slatkine, S., Hubert, P. and Baulieu, E. E. (1978). Induction of maturation in Xenopus laevis oocytes by a steroid linked to a polymer. *Proc. Natl. Acad. Sci. U S A,* 75, 2353-7.

Goetz, F. W. and Theofan, G. (1979). In vitro stimulation of germinal vesicle breakdown and ovulation of yellow perch (Perca flavescens) oocytes. Effects of 17 a -hydroxy-20 b -dihydroprogesterone and prostaglandins. *Gen. Comp. Endocrinol.,* 37, 273-85.

Hanna, R., Pang, Y., Thomas, P. and Zhu, Y. (2006). Cell-surface expression, progestin binding, and rapid nongenomic signaling of zebrafish membrane progestin receptors a and b in transfected cells. *J. Endocrinol.,* 190, 247-60.

Hirai, T., Yamashita, M., Yoshikuni, M., Lou, Y. H. and Nagahama, Y. (1992). Cyclin B in fish oocytes: its cDNA and amino acid sequences, appearance during maturation, and induction of p34cdc2 activation. *Mol. Reprod. Dev.,* 33, 131-40.

Ishikawa, K., Hanaoka, Y., Kondo, Y. and Imai, K. (1977). Primary action of steroid hormone at the surface of amphibian oocyte in the induction of germinal vesicle breakdown. *Mol. Cell Endocrinol.,* 9, 91-100.

Ito, T., Yoshizaki, N., Tokumoto, T., Ono, H., Yoshimura, T., Tsukada, A., Kansaku, N. and Sasanami, T. (2011). Progesterone is a sperm-releasing factor from the sperm-storage tubules in birds. *Endocrinology,* 152, 3952-62.

Jalabert, B., Bry, C., Breton, B. and Campbell, C. (1976). Effect of 17 a-hydroxy-20 b dihydroprogesterone and progesterone on maturation and ovulation in vivo and on the plasma level of gonadotropic hormone t-GtH in rainbow trout Salmo gairdneri. *C R Acad. Sci. Hebd Seances Acad. Sci. D,* 283, 1205-8.

Katsu, Y., Yamashita, M., Kajiura, H. and Nagahama, Y. (1993). Behavior of the components of maturation-promoting factor, cdc2 kinase and cyclin B, during oocyte maturation of goldfish. *Dev. Biol.,* 160, 99-107.

Katsu, Y., Yamashita, M. and Nagahama, Y. (1997). Isolation and characterization of goldfish Y box protein, a germ-cell-specific RNA-binding protein. *Eur. J. Biochem.,* 249, 854-61.

Katsu, Y., Yamashita, M. and Nagahama, Y. (1999). Translational regulation of cyclin B mRNA by 17-a, 20 b -dihydroxy-4-pregnen-3-one (maturation-inducing hormone) during oocyte maturation in a teleost fish, the goldfish (Carassius auratus). *Mol. Cell Endocrinol.,* 158, 79-85.

Kazeto, Y., Goto-Kazeto, R., Thomas, P. and Trant, J. M. (2005). Molecular characterization of three forms of putative membrane-bound progestin receptors and their tissue-distribution in channel catfish, Ictalurus punctatus. *J. Mol. Endocrinol.,* 34, 781-91.

King, W. T., Ghosh, S., Thomas, P. and Sullivan, C. V. (1997). A receptor for the oocyte maturation-inducing hormone 17a,20b,21-trihydroxy-4-pregnen-3-one on ovarian membranes of striped bass. *Biol. Reprod.,* 56, 266-71.

Kishimoto, T., Kuriyama, R., Kondo, H. and Kanatani, H. (1982). Generality of the action of various maturation-promoting factors. *Exp. Cell Res.,* 137, 121-6.

Losel, R. M., Falkenstein, E., Feuring, M., Schultz, A., Tillmann, H. C., Rossol-Haseroth, K. and Wehling, M. (2003). Nongenomic steroid action: controversies, questions, and answers. *Physiol. Rev.,* 83, 965-1016.

Masui, Y. and Clarke, H. J. (1979). Oocyte maturation. *Int. Rev. Cytol.,* 57, 185-282.

Nagahama, Y. (1987). Gonadotropin action on gametogenesis in teleost gonads. *Zoological Science,* 4, 209-222.

Nagahama, Y. and Adachi, S. (1985). Identification of maturation-inducing steroid in a teleost, the amago salmon (Oncorhynchus rhodurus). *Dev. Biol.,* 109, 428-35.

Nagahama, Y., Yoshikuni, M., Yamashita, M., Tokumoto, T. and Katsu, Y. (1995). Regulation of oocyte growth and maturation in fish. *Curr. Top. Dev. Biol.,* 30, 103-45.

Narasimhan, M. L., Coca, M. A., Jin, J., Yamauchi, T., Ito, Y., Kadowaki, T., Kim, K. K., Pardo, J. M., Damsz, B., Hasegawa, P. M., Yun, D. J. and Bressan, R. A. (2005). Osmotin is a homolog of mammalian adiponectin and controls apoptosis in yeast through a homolog of mammalian adiponectin receptor. *Mol. Cell,* 17, 171-80.

Pace, M. C. and Thomas, P. (2005a). Activation of a pertussis toxin-sensitive, inhibitory G-protein is necessary for steroid-mediated oocyte maturation in spotted seatrout. *Dev. Biol.,* 285, 70-9.

Pace, M. C. and Thomas, P. (2005b). Steroid-induced oocyte maturation in Atlantic croaker (Micropogonias undulatus) is dependent on activation of the phosphatidylinositol 3-kinase/Akt signal transduction pathway. *Biol. Reprod.,* 73, 988-96.

Patino, R. and Thomas, P. (1990). Gonadotropin stimulates 17a,20 b,21-trihydroxy-4-pregnen-3-one production from endogenous substrates in Atlantic croaker ovarian follicles undergoing final maturation in vitro. *Gen. Comp. Endocrinol.,* 78, 474-8.

Pinter, J. and Thomas, P. (1999). Induction of ovulation of mature oocytes by the maturation-inducing steroid 17, 20- b -21-trihydroxy-4-pregnen-3-one in the spotted seatrout. *Gen. Comp. Endocrinol.,* 115, 200-9.

Rahman, M., Ohta, K., Yoshikuni, M., Nagahama, Y., Chuda, H. and Matsuyama, M. (2002). Characterization of ovarian membrane receptor for 17,20 b -dihydroxy-4-pregnen-3-one, a maturation-inducing hormone in yellowtail, Seriola quinqueradiata. *Gen. Comp. Endocrinol.,* 127, 71-9.

Sandau, C. D., Ayotte, P., Dewailly, E., Duffe, J. and Norstrom, R. J. (2002). Pentachlorophenol and hydroxylated polychlorinated biphenyl metabolites in umbilical cord plasma of neonates from coastal populations in Quebec. *Environ Health Perspect.,* 110, 411-7.

Smith, J. L., Kupchak, B. R., Garitaonandia, I., Hoang, L. K., Maina, A. S., Regalla, L. M. and Lyons, T. J. (2008). Heterologous expression of human mPR a, mPR b and mPRg in yeast confirms their ability to function as membrane progesterone receptors. *Steroids,* 73, 1160-73.

Tang, Y. T., Hu, T., Arterburn, M., Boyle, B., Bright, J. M., Emtage, P. C. and Funk, W. D. (2005). PAQR proteins: a novel membrane receptor family defined by an ancient 7-transmembrane pass motif. *J. Mol. Evol.,* 61, 372-80.

Taniguchi, Y., Takeda, S., Furutani-Seiki, M., Kamei, Y., Todo, T., Sasado, T., Deguchi, T., Kondoh, H., Mudde, J., Yamazoe, M., Hidaka, M., Mitani, H., Toyoda, A., Sakaki, Y., Plasterk, R. H. and Cuppen, E. (2006). Generation of medaka gene knockout models by target-selected mutagenesis. *Genome Biol.,* 7, R116.

Thomas, P. and Das, S. (1997). Correlation between binding affinities of C21 steroids for the maturation-inducing steroid membrane receptor in spotted seatrout ovaries and their agonist and antagonist activities in an oocyte maturation bioassay. *Biol. Reprod.,* 57, 999-1007.

Thomas, P., Pang, Y., Dong, J., Groenen, P., Kelder, J., de Vlieg, J., Zhu, Y. and Tubbs, C. (2007). Steroid and G protein binding characteristics of the seatrout and human progestin membrane receptor a subtypes and their evolutionary origins. *Endocrinology,* 148, 705-18.

Thomas, P., Pang, Y., Zhu, Y., Detweiler, C. and Doughty, K. (2004). Multiple rapid progestin actions and progestin membrane receptor subtypes in fish. *Steroids,* 69, 567-73.

Thomas, P., Tubbs, C., Detweiler, C., Das, S., Ford, L. and Breckenridge-Miller, D. (2005). Binding characteristics, hormonal regulation and identity of the sperm membrane progestin receptor in Atlantic croaker. *Steroids,* 70, 427-33.

Thomas, P., Zhu, Y. and Pace, M. (2002). Progestin membrane receptors involved in the meiotic maturation of teleost oocytes: a review with some new findings. *Steroids,* 67, 511-517.

Tokumoto, M., Nagahama, Y., Thomas, P. and Tokumoto, T. (2006). Cloning and identification of a membrane progestin receptor in goldfish ovaries and evidence it is an intermediary in oocyte meiotic maturation. *Gen Comp Endocrinol,* 145, 101-108.

Tokumoto, T. (2012). Identification of membrane progestin receptors (mPR) in goldfish oocytes as a key mediator of steroid non-genomic action. *Steroids,* 77, 1013-1016.

Tokumoto, T., Tokumoto, M., Horiguchi, R., Ishikawa, K. and Nagahama, Y. (2004). Diethylstilbestrol induces fish oocyte maturation. *Proc. Natl. Acad. Sci. USA,* 101, 3686-3690.

Tokumoto, T., Tokumoto, M. and Nagahama, Y. (2005). Induction and inhibition of oocyte maturation by EDCs in zebrafish. *Reprod. Biol. Endocrinol.,* 3, 69.

Tokumoto, T., Tokumoto, M., Oshima, T., Shimizuguchi, K., Fukuda, T., Sugita, E., Suzuki, M., Sakae, Y. T., Akiyama, Y. I., Nakayama, R., Roy, S. R., Saydur Rahman, M., Pang, Y., Dong, J. and Thomas, P. (2012). Characterization of multiple membrane progestin receptor (mPR) subtypes from the goldfish ovary and their roles in the induction of oocyte maturation. *Gen. Comp. Endocrinol.,* 177, 168-176.

Tokumoto, T., Tokumoto, M. and Thomas, P. (2007). Interactions of diethylstilbestrol (DES) and DES analogs with membrane progestin receptor- a and the correlation with their nongenomic progestin activities. *Endocrinology,* 148, 3459-3467.

Tokumoto, T., Yamaguchi, T., Ii, S. and Tokumoto, M. (2011). In vivo induction of oocyte maturation and ovulation in zebrafish. *PLoS One,* 6, e25206.

Trant, J. M., Thomas, P. and Shackleton, C. H. (1986). Identification of 17 a,20 b,21-trihydroxy-4-pregnen-3-one as the major ovarian steroid produced by the teleost Micropogonias undulatus during final oocyte maturation. *Steroids,* 47, 89-99.

Villa, N. Y., Moussatche, P., Chamberlin, S. G., Kumar, A. and Lyons, T. J. (2011). Phylogenetic and preliminary phenotypic analysis of yeast PAQR receptors: potential antifungal targets. *J. Mol. Evol.,* 73, 134-152.

Yamashita, M., Fukada, S., Yoshikuni, M., Bulet, P., Hirai, T., Yamaguchi, A., Lou, Y. H., Zhao, Z. and Nagahama, Y. (1992a). Purification and characterization of maturation-promoting factor in fish. *Dev. Biol.,* 149, 8-15.

Yamashita, M., Fukada, S., Yoshikuni, M., Bulet, P., Hirai, T., Yamaguchi, A., Yasuda, H., Ohba, Y. and Nagahama, Y. (1992b). M-phase-specific histone H1 kinase in fish oocytes. Purification, components and biochemical properties. *Eur. J. Biochem.,* 205, 537-543.

Yamashita, M., Kajiura, H., Tanaka, T., Onoe, S. and Nagahama, Y. (1995). Molecular mechanisms of the activation of maturation-promoting factor during goldfish oocyte maturation. *Dev. Biol.,* 168, 62-75.

Yamauchi, T., Kamon, J., Ito, Y., Tsuchida, A., Yokomizo, T., Kita, S., Sugiyama, T., Miyagishi, M., Hara, K., Tsunoda, M., Murakami, K., Ohteki, T., Uchida, S., Takekawa, S., Waki, H., Tsuno, N. H., Shibata, Y., Terauchi, Y., Froguel, P., Tobe, K., Koyasu, S., Taira, K., Kitamura, T., Shimizu, T., Nagai, R. and Kadowaki, T. (2003). Cloning of adiponectin receptors that mediate antidiabetic metabolic effects. *Nature,* 423, 762-769.

Yasuda, K., Kotani, T., Ota, R. and Yamashita, M. (2010). Transgenic zebrafish reveals novel mechanisms of translational control of cyclin B1 mRNA in oocytes. *Dev. Biol.,* 348, 76-86.

Yoshikuni, M., Ishikawa, K., Isobe, M., Goto, T. and Nagahama, Y. (1988). Characterization of 1-methyladenine binding in starfish oocyte cortices. *Proc. Natl. Acad. Sci. USA,* 85, 1874-1877.

Yoshikuni, M. and Nagahama, Y. (1994). Involvement of an inhibitory G-protein in the signal transduction pathway of maturation-inducing hormone (17a,20 b-dihydroxy-4-pregnen-3-one) action in rainbow trout (Oncorhynchus mykiss) oocytes. *Dev. Biol.,* 166, 615-622.

Yoshikuni, M., Shibata, N. and Nagahama, Y. (1993). Specific binding of [^{3}H]17a,20b-dihydroxy-4-progen-3-one to oocyte cortices of rainbow trout (*Oncorhynchus mykiss*). *Fish Physiol. Biochem.,* 11, 15-24.

Zhu, Y., Bond, J. and Thomas, P. (2003a). Identification, classification, and partial characterization of genes in humans and other vertebrates homologous to a fish membrane progestin receptor. *Proc. Natl. Acad. Sci. USA,* 100, 2237-2242.

Zhu, Y., Rice, C. D., Pang, Y., Pace, M. and Thomas, P. (2003b). Cloning, expression, and characterization of a membrane progestin receptor and evidence it is an intermediary in meiotic maturation of fish oocytes. *Proc. Natl. Acad. Sci. USA,* 100, 2231-2236.

Index

#

A

B

C

D

E

F

G

H

I

J

K

L

M

N

O

P

Q

R

S

T

U

V

W

X

Y

Z